HANDBOOK OF DIGITAL POLITICS

Handbook of Digital Politics

Edited by

Stephen Coleman

Professor of Political Communication, School of Media and Communication, University of Leeds, UK

Deen Freelon

Assistant Professor, School of Communication, American University, Washington, DC, USA

 Edward Elgar
PUBLISHING

Cheltenham, UK • Northampton, MA, USA

Published by
Edward Elgar Publishing Limited
The Lypiatts
15 Lansdown Road
Cheltenham
Glos GL50 2JA
UK

Edward Elgar Publishing, Inc.
William Pratt House
9 Dewey Court
Northampton
Massachusetts 01060
USA

Paperback edition 2016

A catalogue record for this book
is available from the British Library

Library of Congress Control Number: 2014959459

This book is available electronically in the **Elgar**online
Social and Political Science subject collection
DOI 10.4337/9781782548768

ISBN 978 1 78254 875 1 (cased)
ISBN 978 1 78254 876 8 (eBook)
ISBN 978 1 78643 563 7 (paperback)

Typeset by Servis Filmsetting Ltd, Stockport, Cheshire

Contents

List of figures   viii
List of tables   ix
List of contributors   x

1   Introduction: conceptualizing digital politics   1
    Stephen Coleman and Deen Freelon

PART I   THEORIZING DIGITAL POLITICS

2   The internet as a civic space   17
    Peter Dahlgren

3   The social foundations of future digital politics   35
    Nick Couldry

4   The Fifth Estate: a rising force of pluralistic accountability   51
    William H. Dutton and Elizabeth Dubois

5   Silicon Valley Ideology and class inequality: a virtual poll tax
    on digital politics   67
    Jen Schradie

PART II   GOVERNMENT AND POLICY

6   Online voting advice applications: foci, findings and future of
    an emerging research field   87
    Fadi Hirzalla and Liesbet van Zoonen

7   Internet voting: the state of the debate   103
    Thad Hall

8   Digital campaigning   118
    Daniel Kreiss

9   E-petitions   136
    Scott Wright

10 Argumentation tools for digital politics: addressing the
    challenge of deliberation in democracies   151
    Neil Benn

PART III   COLLECTIVE ACTION AND CIVIC
            ENGAGEMENT

11   The logic of connective action: digital media and the
     personalization of contentious politics   169
     W. Lance Bennett and Alexandra Segerberg

12   Youth civic engagement   199
     *Chris Wells, Emily Vraga, Kjerstin Thorson, Stephanie Edgerly
     and Leticia Bode*

13   Internet use and political engagement in youth   221
     Yunhwan Kim and Erik Amnå

PART IV   POLITICAL TALK

14   Everyday political talk in the Internet-based public sphere   247
     Todd Graham

15   Creating spaces for online deliberation   264
     Christopher Birchall and Stephen Coleman

16   Computational approaches to online political expression:
     rediscovering a 'science of the social'   281
     *Dhavan V. Shah, Kathleen Bartzen Culver, Alexander Hanna,
     Timothy Macafee and JungHwan Yang*

17   Two-screen politics: evidence, theory and challenges   306
     Nick Anstead and Ben O'Loughlin

PART V   JOURNALISM

18   From news blogs to news on Twitter: gatewatching and
     collaborative news curation   325
     Axel Bruns and Tim Highfield

19   Research on the political implications of political
     entertainment   340
     Michael A. Xenos

20   Journalism, gatekeeping and interactivity   357
     Neil Thurman

PART VI INTERNET GOVERNANCE

21 Internet governance, rights and democratic legitimacy 377
Giles Moss

22 Social media surveillance 395
Christian Fuchs

PART VII EXPANDING THE FRONTIERS OF DIGITAL
POLITICS RESEARCH

23 Visibility and visualities: 'ways of seeing' politics in the digital
media environment 417
Katy Parry

24 Automated content analysis of online political communication 433
Ross Petchler and Sandra González-Bailón

25 On the cutting edge of Big Data: digital politics research in
the social computing literature 451
Deen Freelon

Index 473

Figures

6.1	Advice of the Dutch Stemwijzer, the oldest and relatively most popular VAA	91
10.1	A series of screenshots from the PolicyCommons Argument Visualization Tool showing Issue Map and Argument Network styles of visualization	155
11.1	Elements of collective and connective action networks	188
12.1	Voter turnout in American presidential elections, 1972–2012 (by age)	207
16.1	Proportional volume of keyword use for Sandra Fluke and Trayvon Martin cases	291
16.2	Proportional volume of hashtag clusters for Sandra Fluke case	295
16.3	Proportional volume of hashtag clusters for Trayvon Martin case	296
16.4	Retweet network for Sandra Fluke case with major hashtags	298
16.5	Retweet network for Trayvon Martin case with major hashtags	299
17.1	Percentage of people who use their mobile phones while watching television in the United States	317
18.1	Australian Twitter News Index (ATNIX), showing tweets per week linking to key Australian news sites, June 2012 to December 2014	334
20.1	Source, media and audience channels in the gatekeeping process	368

Tables

4.1	Examples of potentially effective forms of Fifth Estate networking	55
4.2	A categorization of networked institutions and individuals	58
13.1	Results of a 4 (citizenship orientation) x 2 (gender) MANOVA examining differences on measures used in the cluster analysis (z-scores)	233
13.2	Results of a 4 (citizenship orientation) x 2 (gender) MANOVA examining citizenship orientation group differences on measures of citizenship competences (z-scores)	234
13.3	Zero-order correlations between political engagement and Internet use	235
13.4	Results of ANCOVAs examining citizenship orientation group differences on measures of Internet use	237
16.1	Follower and following counts at level 1 and 2 of collection	288
16.2	Principal component analysis of hashtag use in Sandra Fluke case	293
16.3	Principal component analysis of hashtag use in Trayvon Martin case	293
22.1	Qualities of Internet surveillance	399
25.1	The methods of 40 highly cited social computing research papers	455

Contributors

Erik Amnå, Örebro University, Sweden

Nick Anstead, London School of Economics and Political Science, UK

Neil Benn, former postdoctoral researcher, University of Leeds, UK

W. Lance Bennett, University of Washington, USA

Christopher Birchall, University of Leeds, UK

Leticia Bode, Georgetown University, USA

Axel Bruns, Queensland University of Technology, Australia

Stephen Coleman, University of Leeds, UK

Nick Couldry, London School of Economics and Political Science, UK

Kathleen Bartzen Culver, University of Wisconsin–Madison, USA

Peter Dahlgren, Lund University, Sweden

Elizabeth Dubois, Oxford Internet Institute, UK

William H. Dutton, Michigan State University, USA

Stephanie Edgerly, Northwestern University, USA

Deen Freelon, American University, USA

Christian Fuchs, University of Westminster, UK

Sandra González-Bailón, Annenberg School for Communication, University of Pennsylvania, USA

Todd Graham, University of Groningen, the Netherlands

Thad Hall, University of Utah, USA

Alexander Hanna, University of Wisconsin–Madison, USA

Tim Highfield, Queensland University of Technology, Australia

Fadi Hirzalla, University of Amsterdam, the Netherlands

Yunhwan Kim, Örebro University, Sweden

Daniel Kreiss, University of North Carolina, Chapel Hill, USA

Timothy Macafee, University of Wisconsin–Madison, USA

Giles Moss, University of Leeds, UK

Ben O'Loughlin, Royal Holloway, University of London, UK

Katy Parry, University of Leeds, UK

Ross Petchler, Oxford Internet Institute, UK

Jen Schradie, University of California, Berkeley, USA

Alexandra Segerberg, Stockholm University, Sweden

Dhavan V. Shah, University of Wisconsin–Madison, USA

Kjerstin Thorson, University of Southern California, USA

Neil Thurman, City University London, UK

Emily Vraga, George Mason University, USA

Chris Wells, University of Wisconsin–Madison, USA

Scott Wright, University of Melbourne, Australia

Michael A. Xenos, University of Wisconsin–Madison, USA

JungHwan Yang, University of Wisconsin–Madison, USA

Liesbet van Zoonen, Loughborough University, UK

1. Introduction: conceptualizing digital politics

Stephen Coleman and Deen Freelon

Power is ubiquitous. People have power over us. We resist power. We exercise power. We complain about the relations of power in which we find ourselves caught up. We collude with power, as if it were a natural force. We stand up to power. We surrender to power. We feel powerful. We feel impotent. Even those who claim to be non-political are engaging in these relationships and experiencing these feelings on a regular basis.

Politics is the organizing dimension of power. It is the language we use for naming and talking about power; the rules we observe in exercising and submitting to power; the sighs, gestures and half-formed sentences we give off in our daily encounters with the structures and routines of power. Politics is not just about what governments do or politicians say. Sometimes the political is deeply embedded in constitutional and institutional protocols; at other times it simmers under the surface, shaping and affecting relations, while co-existing with other cultural dynamics.

It would be difficult to imagine how a development as world-changing as the emergence of the Internet could have taken place without having some impact upon the ways in which politics is expressed, conducted, depicted and reflected upon. It is one thing to say that politics is affected by digital communication. It is quite another thing to say that digital communication fundamentally reshapes politics. The truth lies somewhere between those two statements and the way of getting at that truth is through rigorous, empirical research rather than starry-eyed speculation.

To speak of digital politics is not simply to tell a story about how political routines are replicated online. One feature of all technologies is that they are constitutive: they do not simply support predetermined courses of action, but open up new spaces of action, often contrary to the original intentions of inventors and sponsors. Not only hard technologies, but modes of technical thought, have had profound effects upon governmental strategies. For example, the emergence of the printing press in Europe generated a space in which publics could come together as cohabitants of imagined communities; centralized states could disseminate their propaganda to mass populations; and vernacular idioms and dialects could be systematized into official languages. As Benedict Anderson argued in his study of the rise of nationhood, 'the convergence of capitalism and print technology on the fatal diversity of human language created the possibility of a new form of imagined community, which in its basic morphology set the stage for the modern nation' (Anderson, 1983: 48). Because technologies are not simply employed to replicate power relationships, but play a constitutive role in establishing them, it makes sense to think of 'digital politics' less as an account of how technology serves predetermined political ends than as a complex, ongoing tension between replication and transformation in the social organization of power. Rather than thinking of technology doing something to politics (which, despite non-technological-determinist protestations, still pervades much of the literature), we might think about politics itself as a technology. At the end of the seventeenth century, John Trenchard (1731: 2) argued that 'a government is a mere piece of clockwork, and having such springs and wheels, must act after such a manner: and there the art is to constitute it so that it must move to the public advantage'. To speak metaphorically of the machinery of government, political leaks and re-engineering government is to tacitly acknowledge that Trenchard was right and governance can profitably be understood as a technology. In that sense, this volume is concerned with how one technology (the political process) is affected by another technology (digital forms of producing and circulating information and communication). Each of these technologies is entangled in the history of the other in a constantly reciprocal interplay between structural logic and human reflexivity. Orlikowski (1992: 406) summarizes this complexity well when she states that:

> technology is physically constructed by actors working within a given social context, and technology is socially constructed by actors through the different meanings they attach to it and the various features they emphasize and use. However, it is also the case that once developed and deployed, technology tends to become reified and institutionalized, losing its connection with the human agents that constructed it or gave it meaning, and it appears to be part of the objective, structural properties of the organization.

This volume is largely about how digital politics has been constructed by actors who are motivated by competing meanings and practices and how it has become, on the one hand, reified and institutionalized, and on the other, subversive and transformative. These tensions between maintenance and disruption play out differently for parties to political communication depending upon where they are situated. Political elites might regard digital politics as an opportunity to cut out the critical scrutiny of journalists and appeal directly to the public, but also as a risk because the spread of news sources, discursive spaces and surveillance technologies makes them more visible and vulnerable. Journalists might see the digital mediation of politics as unwelcome competition to their role as authoritative storytellers and gatekeepers, but are increasingly turning to online sources and platforms in order to find ways of sustaining their social role and economic future. Citizens might see digital politics as the same old messages as before, now invading their inboxes and phones; or as entry points to collective – or connective – action that reduces communicative inequalities between the poorest, most dispersed and traditional power elites. In short, generalizations about digital politics as either threat to democracy or political panacea have long outlived whatever usefulness they may once have possessed. As with most historical developments, the significance of these relatively recent innovations in political communication depends upon where one happens to be standing and how one is looking.

In trying to make sense of digital politics and its implications for the broader communication environment, this volume is less interested in making sweeping claims or generalizations than in offering carefully researched observations about what has been going on; what these changes and continuities might mean for political communication in general; and, most importantly, what we do not yet know and need to find new ways of researching. In short, we are less interested here in offering the final, definitive word than in reporting new and original insights and contributing to an agenda for future research and debate.

A CHANGING POLITICAL COMMUNICATION ENVIRONMENT

For most of the past century, the dominant model of political communication focused upon a relationship between politicians, as message-generators and agenda-setters, and journalists, as gatekeepers and message disseminators. As Blumler and Gurevitch (1995: 12) put it:

> if we look at a political communication system, what we see is two sets of institutions – political and media organizations – which are involved in the course of message preparation in much 'horizontal' interaction with each other, while, on a 'vertical' axis, they are separately and jointly engaged in disseminating and processing information and ideas to and from the mass citizenry.

This model remains relevant, but is no longer as stable or insulated as it once was. While political and media organizations remain key players in the circulation of information and ideas, both the independence and interdependence of their roles are changing. In terms of their independent functions as shapers of dominant political narratives, political institutions (such as governments, legislatures and parties) are facing difficulties in commanding public attention, generating authoritative and trusted messages, and competing with other sources of civically rooted agenda-setting. At the same time, the mass media are competing to address a diminishing and ever-fragmenting audience comprising people who can access political information from a variety of sources and no longer regard conventional political narratives as being worthy of dutiful attention. The cosy interdependence between politicians and the media in regard to the preparation of messages of which Blumler and Gurevitch wrote – perceived by many to be a kind of 'establishment' collusion – has been radically interrupted by grassroots articulations of public interests and values that might not in the past have penetrated the mass media agenda, but now simmer and circulate within and across digital networks. Proponents of the so-called 'normalization' thesis argue that none of this destabilization fundamentally changes the systemic logic of elite-driven political communication, which is largely replicated online. A contrasting perspective, sometimes referred to as the 'mobilization' thesis, is based on evidence that the politician–media relationship is weakened by the emergence of grassroots networks that reconfigure the terms of publicness. Normalization theorists are in danger of holding on to an established model of political communication without regard to its inadequacy for explaining manifest empirical changes in the production and circulation of political ideas. Mobilization theorists sometimes exaggerate the force of these changes by failing to distinguish between contextual and general trends.

But there are some general trends that can be identified as characteristic features of an emerging digital political communication environment. Let us attempt to outline five of them.

Firstly, a range of political sources, platforms, channels and formats now make it almost impossible for any one political voice to claim overarching authority or to hope to reach most of the public. Compared with times not so long ago when families gathered around the radio and, later, television to receive 'the news', it is clear that the consumption of political information is now coming from a multitude of directions and at all hours of the day and night. Whereas in the past, audiences for political information tended to deliberately seek it out at set times of the day, it is now much more likely for such messages to be encountered inadvertently. While it is still possible for presidents and prime ministers to issue grand statements at times of national and global crisis or celebration, in the knowledge that they will be delivered by the mainstream media to a mass audience, for most of the time political elites have to compete in a highly dispersed and noisy space to win public attention. Consider for example recent changes to the rhythm and routine of election campaigns. Reaching the electorate involves a combination of communication techniques, from personalized appeals to voters' phones and inboxes to mass media packages, often disguised in the form of non-political genres (Kreiss, Chapter 8 in this volume). Resource-poor citizens and communities seeking to put pressure on governments or corporations were in the past confined to a limited, localized repertoire of actions, whereas now they can utilize digital networks to amplify their own voices and link to others with similar values (Graham, Chapter 14; Kim and Amnå, Chapter 13; Shah et al., Chapter 16; Wells et al., Chapter 12; Wright, Chapter 9 in this volume). So, a first general point to be made about the new political communication environment is that it is much harder to capture or manage; it is more porous and fragmented; it is antithetical to the final word on any subject.

Secondly, the predominantly vertical and linear pathways that characterized political communication in the mass media era are under increasing challenge from the emergence of horizontal networks that allow citizens to evade institutional structures and processes. In some cases (but by no means all), such networks undermine the authority that political elites once enjoyed and diminish the role of mediating interpretation by journalists (Anstead and O'Loughlin, Chapter 17; Bruns and Highfield, Chapter 18; Couldry, Chapter 3 in this volume). In digital environments people come to know about one another in real time and asynchronously. Politically, this makes coordination for collective action much less cumbersome than it was in the past. Political mobilization becomes a matter of mutual visibility (Bennett and Segerberg, Chapter 11 in this volume). The agenda-setting and gatekeeping functions that characterized the traditional political communication system are inadequate for the role of managing the circulation of messages, memes and sentiments in the much more fluid digital environment. Established political institutions are under pressure, therefore, to relate to and even replicate this communicative fluidity. Often, their response is to adopt the cosmetic features of horizontal circulation while maintaining the structural logic of vertical message-management (Kreiss, Chapter 8 in this volume). This is a tension that is played out with increasingly problematic consequences and has led to much soul-searching on the part of institutional communication strategists.

Thirdly, a key imaginary of the traditional political communication system – the audience – is no longer what it once seemed to be. Increasing numbers of people find themselves engaged simultaneously as mass-media consumers and as peer-to-peer message producers – sometimes literally at the same time (Anstead and O'Loughlin, Chapter 17 in this volume).

As horizontal media interactivity practices compete with traditional practices of viewing and listening in, distinctions between audiences and publics are no longer sustainable. People who have their own stake in the production and flow of media content cannot be depended upon to be guided or managed in their perceptions of or relationships to social power. While mass media reception never did entail an inert and passive relationship, it was clearly based upon two assumptions: that a privileged group of well-resourced media-makers would generate and transmit most of the content; and that audiences, however active in their interpretations or responses to such content, would not play a significant feedback role in relation to content originators. In short, mass media was mainly monological. In contrast, digital communication resists the logic of what Postman has called 'the one-way conversation' (Postman, 1986). The default setting for digital media is that messages sent or received are only one part of an ongoing process of sense-making. Consider traditional televised appeals by politicians to voters. The audience or electorate was conceived as a sitting target. It could accept or reject the messages addressed to it, but had nothing like the options for challenging, reformulating or destabilizing the message that is afforded by digital media. So, just as multi-platform media make it hard for political elites (or anyone else) to have the final word, it is also the case that the logic of media interactivity has potential to weaken strategies of argumentative closure and evens out the right to make public claims (Bennett and Segerberg, Chapter 11; Bruns and Highfield, Chapter 18 in this volume). On the other hand, this potential is not evenly distributed: empirical research on the matter has repeatedly shown that online participation (especially in political matters) tends to be dominated by the better-off (Schradie, Chapter 5 in this volume).

Fourthly, what we might call the expressive tone of political communication has become in recent years less constrained by the generic features of official politics. The formal conventions of political interaction that dominated ritual events such as broadcast interviews, expert commentary panels and televised election debates have come to be regarded by many people as stage-managed, self-referential and frequently incomprehensible. A common response to this tendency was noted by Blumler and Kavanagh (1999: 220) in their analysis of 'the third age of political communication': 'Politicians are impelled to speak in a more popular idiom and to court popular support more assiduously. Media organizations are driven to seek ways of making politics more palatable and acceptable to audience members'.

Blumler and Kavanagh's astute observation referred mainly to a broadcast culture in which political and journalistic elites felt under intensifying pressure to pander to audiences that were disinclined to respond deferentially to voices of authority. With the emergence of the Internet as a public space for reflecting and talking about politics (alongside and entangled within a range of other subjects), political discourse has become more vernacular and inclusive and discourse norms that once made it apparently easy to distinguish between political and everyday discussion have become blurred (Dutton and Dubois, Chapter 4; Shah et al., Chapter 16 in this volume). The viral circulation of jokes, images, short comments, ironic pastiches and shared sentiments has become a significant component of contemporary political communication (Xenos, Chapter 19 in this volume). In the digital environment, it is not simply a matter of politicians needing to acknowledge the popular framing of politics, but of such reframing opening up space for hitherto underarticulated policy debates – especially those relating to the intersections of private and public morality – and closing down opportunities for other well-established strategies, such as private negotiation; the suppression of bad news that is likely to affect social behaviour and exacerbate unwanted outcomes; or the maintenance of clear lines between personal and public personas.

Fifthly, while the traditional political communication pyramid was open to forms of regulation, the digital environment is not only largely lawless, but seemingly invulnerable to strong legal regulation. Laws responding to offensive, defamatory or abusive content are extremely difficult to conceive or enforce in an environment where communication flows often bear no relation to the boundaries of individual nation states. One of the effects of globalization is that the free movement of symbolic resources can be as unaccountably and recklessly unbounded as the unchecked movement of capital. This, of course, is one of the great democratic strengths of digital politics, especially in the contexts of political regimes that hitherto found it relatively easy to maintain an authoritarian grip on the flow of public information and discussion. The creation of transnational alliances of jointly affected citizens has often served an important agenda-setting and mobilizing function. At the same time, digital connectivity has strengthened a number of highly undemocratic and destructive movements, often depending upon the unaccountability of their messages to perpetuate vile forms of hate speech. As well as such stark abuses of digital political communication, the absence of regulation has opened spaces for corporate interventions in the public sphere, leaving vast numbers of people who engage in political discussion via social media sites vulnerable to commercial surveillance and even attempts to manipulate their personal information environments (Moss, Chapter 21; Fuchs, Chapter 22 in this volume).

HOW THIS BOOK EXPLORES CHANGES AND CONTINUITIES IN DIGITAL POLITICS

This volume is divided into seven multi-chapter sections, each of which represents a highly active research agenda in digital politics. These sections are not exhaustive – there are surely a few areas we have missed – but we believe they collectively offer a wide-ranging impression of the field's most pressing concerns. The remainder of this chapter describes each section, briefly summarizing each of its constituent chapters.

Part I: Theorizing Digital Politics

The chapters in this section grapple with very broad theoretical issues held in common across many strands of digital politics research. Rather than reviewing empirical findings in detail, as most of the other sections do, these chapters review and compare competing theoretical perspectives and present their own arguments concerning the best paths forward for research. Peter Dahlgren (Chapter 2) discusses the role of the Internet in altering an existing field of 'civic spaces' available to citizens to discuss and engage with politics. His overarching finding is that the Internet's relationship to politics resists succinct summary, and that his notion of 'civic cultures' helps disaggregate the different levels at which the Internet influences politics. Nick Couldry (Chapter 3) focuses on how digital media react with 'the social foundations of political engagement and political action' to alter political opportunity structures. In doing so, he stresses the importance of specific questions concerning how digital media may change who can participate in politics, what they can do, and why. William Dutton and Elizabeth Dubois (Chapter 4) develop a concept they call the Fifth Estate, which refers to uses of the Internet to hold powerful institutions and individuals publicly accountable. In contrast to the Fourth Estate of traditional journalists, the Fifth Estate maintains no professional barriers to entry: anyone can join quickly and easily on an ad hoc basis using widely available tools. And in a provocative departure from the other chapters in this part, Jen Schradie (Chapter 5) argues that 'structural inequalities in the United States create virtual poll taxes for those who do not control the means of digital production in online political spaces'. Thus online politics, like offline politics, comes to be dominated by elite voices and priorities.

Part II: Government and Policy

This section is devoted to research on the roles played by digital media in various aspects of state governance, with each chapter focusing on a distinct governmental function. Fadi Hirzalla and Liesbet van Zoonen (Chapter 6) review research on voting advice applications (VAAs), which help voters decide which political parties to support based on their policy preferences. They focus on three key questions: VAAs' effects on the electoral system, their users' demographics, and how exactly VAAs translate voter input into actionable suggestions. Along very similar lines, Thad Hall (Chapter 7) examines the literature on Internet voting, which promises to lower barriers to democratic participation even as its security risks threaten to limit widespread adoption. The chapter focuses on these dual issues of access and security, ultimately concluding that the current dearth of robust infrastructure will sharply limit the technology's near-term applications. Shifting from voting to soliciting votes, Daniel Kreiss (Chapter 8) reviews recent work on US presidential campaign uses of the Internet and digital media. He finds that many digital tools have already become integrated into long-standing campaign practices, in particular having dramatically amplified and accelerated campaigns' interactions with voters. But citizen–government communication can be bottom-up as well as top-down, and Scott Wright's Chapter 9 focuses on e-petitions, an increasingly popular means of aggregating and communicating support for grassroots policy changes. Taking an international approach, Wright considers who controls e-petitions' agendas, their actual impact on government policies, and the representativeness of their participant base. Neil Benn (Chapter 10) concludes this section with a look at the cutting edge of research on computer-supported argument visualization (CSAV) technology in the specific context of policy deliberation. This chapter helpfully explains the priorities that have historically driven CSAV development and also profiles some of the most prominent CSAV platforms and their outcomes.

Part III: Collective Action and Civic Engagement

Departing from the realm of officially sanctioned citizen–government interaction, the chapters in this section examine the potential for digital tools to help (and at times, hinder) independent collective action and engagement. Lance Bennett and Alexandra Segerberg (Chapter 11) begin with a convenient chapter-length summary of their 'logic of connective action' thesis, which updates the well-known logic of collective action for the digital age. In it they outline their three ideal types of contemporary connective action, which are differentiated by the centrality of digital media, and demonstrate how recent protest cases exemplify each. A more fully realized version of this argument can be found in their recent book *The Logic of Connective Action* (Bennett and Segerberg, 2013). The following two chapters focus on youth civic engagement and politics, albeit in different ways. Chris Wells, Emily Vraga, Kjerstin Thorson, Stephanie Edgerly and Leticia Bode (Chapter 12) review six distinct focus areas of the literature on youth civic engagement. In doing so, they provide a comprehensive and high-level overview of the topics of greatest concern to the field. Yunhwan Kim and Erik Amnå's Chapter 13 overlaps slightly with Wells et al.'s, but whereas Wells et al. focus primarily on US-focused research, Kim and Amnå's literature review is much more international in scope. Moreover, Kim and Amnå present a unique empirical illustration of some of the key points drawn from their own research of Swedish adolescents.

Part IV: Political Talk

Political conversation between citizens has been a fixture of political communication research for decades, but its transition to the digital realm presents a host of theoretical and methodological challenges. The chapters in this part address these challenges from a variety of perspectives. Todd Graham's contribution (Chapter 14) starts with a much-needed defence of the study of online political talk, which can seem to the lay observer as a distraction from 'real' political communication. From there, he proceeds to outline the main distinguishing characteristics of online political talk, including its responsiveness, discursive quality, and opinion diversity. He concludes by briefly sketching an agenda for future research that focuses on understudied areas. Christopher Birchall and Stephen Coleman's Chapter 15 begins by considering the rationale for online deliberation and goes on to set out some key factors likely to determine deliberative quality. They discuss a number of tools that have been developed to facilitate online deliberation and conclude by asking what further research is needed to advance the deliberative agenda. Dhavan Shah, Kathleen Culver, Alexander Hanna, Timothy Macafee and JungHwan Yang (Chapter 16) profile a very exciting and relatively new frontier of online political talk research: the use of computational methods to collect and analyze digital data. After tracing the theoretical warrant for this research back to pioneering social interactionist Gabriel Tarde, they present their own original computational analysis of Twitter conversations about the high-profile Sandra Fluke and Trayvon Martin news stories. This chapter is especially noteworthy for the way it combines a cogent theoretical framework and advanced computational methods; such unions are unfortunately fairly rare as of this writing. Nick Anstead and Ben O'Loughlin's Chapter 17 finishes the part with a similarly cutting-edge review of research on 'two-screen politics', wherein viewers simultaneously watch a television programme while posting content about it to the Internet. The authors devote most of the chapter to reconciling pre-existing political communication theory with two-screen politics, the popularity of which has grown steadily over the past few years. In particular, they dwell at some length on the implications of two-screening for deliberative democracy, hybrid media and networked power.

Part V: Journalism

While journalism is well known as a research domain distinct from digital politics, the two are more closely related than the empirical record would suggest. News is, after all, the medium through which most of us learn about political matters. Thus, in the spirit of encouraging more cross-pollination of theory and methods between the two subdisciplines, we present three chapters focusing on topics of particular interest to their shared intellectual borders. In Chapter 18, Axel Bruns and Tim Highfield apply the theories of gatewatching and collaborative news production to the phenomena of news blogging and curation. Their review of related research reveals that the online news production–consumption cycle is not a hermetic echo chamber as many contend, but rather a complex series of interconnections between parties of varying political preferences and levels of professionalism. Next, Michael Xenos (Chapter 19) shifts the focus from participatory online media to a newer genre of traditional media: political comedy and entertainment. Xenos advances a compelling argument for the relevance of this genre – sometimes derided as politically inconsequential by hard-news producers – and demonstrates its strong connection with digital politics. Much of the chapter is devoted to reviewing the various kinds of effects political entertainment has been observed to exercise on consumers. Finally, Neil Thurman (Chapter 20) presents an overview of journalistic gatekeeping in the digital age with a strong emphasis on his own work on the topic.

Part VI: Internet Governance

In contrast to most of this volume's parts, the topic of Internet governance pertains not to matters of politics and government as typically understood, but to decisions about how the Internet itself will operate and who will own its various components. Our two contributors examine distinct aspects of Internet governance, the implications of which are international in scope. Giles Moss (Chapter 21) offers an overview of the concept, distinguishing principally among the distinct functions of policy-making, regulation and governance. Moss gives a succinct and accessible summary of some of the major areas of controversy in Internet governance, including libertarian optimism, the power of regulation through code, and the deontology of Internet use. At a more concrete level of analysis, Christian Fuchs (Chapter 22) assesses the state of research on online surveillance and privacy, with a particular focus on social media. The power to peer directly into citizens' private lives is an aspect of Internet governance exploited by both democratic and despotic governments, and its relevance to digital politics is self-evident. In his chapter, Fuchs explores both the technical details of social media surveillance as well as its economic, political and cultural implications.

Part VII: Expanding the Frontiers of Digital Politics Research

This volume's final section presents three chapters that look forward to areas of digital politics research that are currently understudied but have substantial growth potential. It is a heterogeneous group, but all make similar attempts to come to grips with types of data that defy traditional research methods. Katy Parry's Chapter 23 takes on the task of reviewing research that examines visual political content online. Of course the history of visual political communication research stretches back decades, but as with most of this book's topics, the emergence of digital communication has occasioned new theoretical and methodological challenges. Dividing her analysis between top-down and bottom-up politics, Parry advocates for a distinct 'visual culture studies' approach that recognizes the unique political power of visual media. Taking a more method-driven approach, Ross Petchler and Sandra González-Bailón (Chapter 24) explore popular techniques of automated text analysis such as lexicon-based analysis, unsupervised learning, sentiment analysis and network analysis. Their chapter is written specifically for a digital politics research audience, and most of the research they cite was chosen to demonstrate the relevance of these methods for that audience. Along similar lines, Deen Freelon (Chapter 25) continues the focus on computational research methods by reviewing digital politics studies from an unlikely source: a field called 'social computing' populated primarily by computer scientists and information scientists. Although they operate under a very different set of research quality criteria than social scientists, their programming-based methods open a plethora of possibilities for adventurous digital politics researchers.

REFERENCES

Anderson, Benedict (1983). *Imagined Communities: Reflections on the Origin and Spread of Nationalism*. London: Verso Books.

Bennett, W. Lance and Alexandra Segerberg (2013). *The Logic of Connective Action: Digital Media and the Personalization of Contentious Politics*. New York: Cambridge University Press.

Blumler, Jay G. and Michael Gurevitch (1995). *The Crisis of Public Communication*. London: Psychology Press.

Blumler, Jay G. and Dennis Kavanagh (1999). 'The third age of political communication: influences and features'. *Political Communication* 16(3): 209–230.

Orlikowski, Wanda J. (1992). 'The duality of technology: rethinking the concept of technology in organizations'. *Organization Science* 3(3): 398–427.

Postman, Neil (1986). *Amusing Ourselves to Death: Public Discourse in the Age of Show Business*. New York: Penguin.

Trenchard, John (1731). *A Short History of Standing Armies in England*.

PART I

THEORIZING DIGITAL POLITICS

2. The internet as a civic space
Peter Dahlgren

Democracy is dependent upon the participation of its citizens, and such participation requires a variety of sites, places, and spaces. When the internet emerged in the mid-1990s as a mass phenomenon, some observers dismissed it as insignificant for politics. However, it soon became apparent that this communication technology was to play an increasingly important role in the life of democracy. Yet today, as we shall see, there remains considerable contention as to just exactly what this role is, and whether or not the internet ultimately is beneficial for democracy. (I signal here at the outset that for ease of exposition I use the term 'internet' in a very broad way, to refer to both the hardware and software of this technical infrastructure, and to include such ancillary technologies as mobile telephony and the various platforms of social media.) As politics in society generally takes on a larger presence online, the prevailing structures of established power in society are increasingly mediated, solidified, negotiated and challenged to a great extent via the internet. From the horizons of democracy, how should we view these developments? This chapter probes answers to that question.

THE SIGNIFICANCE OF THE INTERNET: CIVIC SPACES AND EVOLVING DEMOCRACY

A Conceptual Continuum

At an obvious level, the internet, given its societal ubiquity, has become an understandably significant communication technology for civic space and for the functioning of democracy. However, to grasp this in a more analytic way, and to understand the issues that nonetheless arise in the process, let me begin by very briefly sketching some important conceptual background. First of all, it is important to remember that 'democracy' is both a complex and a contested term. There are not only a range of differing political systems in the world that claim to be democratic, but also, and more pertinent to my purposes here, there are different ideal models (see Held, 2006, for an overview). Without going through an entire inventory, I here simply note a decided polarity between two basic ways of looking at democracy, each with its own view of civic engagement – though it is probably more useful to think of the distinction as a continuum, rather than a simple either–or choice. On the one hand we have what is sometimes called elite democracy; its proponents take the view that the system works best via a rotation of various elite groups who come to power through elections, and where most citizens, aside from voting, do not engage themselves much politically. Here civic participation is seen largely in terms of a formalized system based on elections.

Alternatively there are various versions of republican models (see, for example, Dewey, 1923; Barber, 1984; Mouffe, 2005) that emphasize the ideal that citizens should engage themselves politically as much as possible, not just at election time. It is argued that such engagement is good not only for the vitality of democracy, but also for the individual citizen, since it offers potential for personal growth and development. In this perspective democratic involvement is understood as comprising not just an electoral system, but much larger societal domains. The adjective 'democratic' is as something that should describe a society more generally, not just its voting mechanisms; democratic processes are seen as a part of an ongoing daily reality. Thus, while engagement in elections certainly requires civic spaces of various kinds, elite models put less emphasis on the need for such spaces beyond the context of electoral politics. Republican versions of democracy, on the other hand, underscore the significance of a broad and dynamic array of civic spaces.

The distinction between elitist and republican models manifests itself also in the actual character of participation, that is, what actually goes on in civic spaces. Elite models highlight citizens' needs for information, news, commentary and debate, in order to make (rational) voting decisions. Republican models concur but also demand a more participatory character of civic spaces, seeing them not just as sites where information can be obtained, but also as opportunities where citizens can interact, develop a sense of common interests, sharpen their opinions, and even engage in forms of decision-making. We see, in other words, a distinction in the ideal of the citizenship itself: reactive and restricted, versus proactive and robust. These are of course generalized and abstract conceptions, yet they inform, on a subtle and often unconscious way, the manner in which different kinds of power holders as well as citizens act.

As a further context for the discussion at hand, in the past 25 years or so there has been growing international concern about democracy's difficulties; indeed, the situation is often referred to as a crisis. This crisis is as complex as democracy itself, but for my purposes one basic feature is the decline in civic engagement in both the politics arena and the larger domain of what is often called civil society (which I will come to shortly).

Not least, under the contemporary policy logic of neoliberalism, where representative democratic power is eroded and accumulates increasingly in unaccountable ways in the private corporate sector (Harvey, 2006), the grounds for trust and participation are eroded (Hay, 2007), as are societal norms central for democracy (Sandel, 2012).

Parallel with these challenges, however, we have witnessed a growth in what can loosely be called 'alternative politics' that in various ways bypasses the electoral system (Rosanvallon, 2008, uses the term 'counter-democracy'). Here political engagement lies outside of party structures, and both the issues that become politicized and the modes of engagement are evolving: the political becomes more closely linked to personal meaning, identity processes, and issues that often have to do with cultural matters (see, for example, Bennett and Segerberg, 2012). These transformations have served not least to focus attention specifically on the nature of civic engagement and its circumstances (a concern which is still very much with us; see Schachter and Yang, 2012). In these discussions, the media loom large, even if they are only part of the story.

I can now go further and begin to make the notion of 'civic spaces' itself a bit more concrete by moblizing two key terms that derive from several different trajectories in political theory. They can provide some helpful roadmaps, and they function well together, pointing to two kinds of civic spaces: the first is civil society, the second is the public sphere.

Civil Society and the Public Sphere

With 'civil society' (see Edwards, 2009, for an introductory overview) I refer to an eclectic tradition in democracy theory that accentuates citizens' free association for common purpose outside the private sphere of the home, and independent of the market and the state. There are undeniably some unresolved issues with the concept, but the idea of civil society emphasizes that in a democracy people can exercise the freedom to interact in pursuit of their shared interests, in settings that are protected by the rights of expression and assembly. For example, dealing with friends, colleagues, communities, associations and social networks for non-commercial purposes are all a part of civil society. On the internet, and especially in the context of social media, there is an almost infinite realm of shared engagement in meaningful and pleasurable activities around sports, hobbies, music (for example, amateur contributions on YouTube), fandom, wikis and so forth – though it is often not possible to keep market logics completely out.

Thus, on one border, civil society has a porous demarcation between itself and what we can broadly call consumption, that is, commercial logics. Its other border is with politics: the political may arise in civil society settings, transforming them into what we call the public sphere, that is, the communicative space of politics. At what point the political actually emerges can be difficult to specify; most fundamentally, it materializes through talk: as people speak, topics may turn to – and become – political. At that point, conceptually, one could say that the discussion has entered the public sphere. Indeed civil society is important not just for the interaction and association it facilitates, but also precisely because it is in a sense a precondition for a functioning public sphere: without that free association, the public sphere could not survive (Cohen and Arato, 1992, underscore the links between the two domains in their classic treatment). Civil society comprises the sites where people can enact their roles as citizens, talk and work together; for this to happen there must exist a minimal foundation of trust and shared democratic values. Without such a sense of civic community and solidarity, civil society evaporates, undercutting democracy's communicative dynamics (Alexander, 2006). If civil society atrophies, so does the vitality of democracy, as Putnam (2000) has famously argued.

The concept of the public sphere, while having a somewhat mixed lineage, is more cohesive than the notion of civil society. The key text in English is Habermas (1989), although since its first publication in the early 1960s there has been much debate on the theme, and Habermas himself has modified his views somewhat over the years. However, in simplified terms we can say that today the notion of the public sphere has become a key conceptual pillar in linking the media to democracy in a normative and critical way.

As a normative ideal the public sphere is seen as the institutionalized communicative spaces that are accessible to all citizens and that help to promote the development of public opinion and political will formation. These public processes are to take place through the unhindered access to pertinent information, ideas and debate. In the modern world, much of the public sphere is comprised of media, especially in the form of journalism, yet face-to-face contexts remain essential, since this is where discussion and debate between citizens take place (and we can readily understand that the internet has been offering mediated extensions to such civic deliberation). Habermas in his book proceeds to examine how various historical factors have served to constrict this ideal, not least the commercial logic of the media. Analysts have continued to use the concept as both a normative horizon and an empirical referent to be critically evaluated, especially with a strong emphasis on the affordances, limitations and actual modes of use of the internet.

Habermas and others make clear that the public sphere is far from unitary; empirically, it is comprised of many sprawling communicative spaces of considerable variety (see Habermas's update from 2006). At the same time, these heterogeneous spheres are by no means equal in terms of access or political impact. Some are socially and politically more main-stream, and situated closer to decision-making power. Others are more geared toward the interests and needs of specific groups; emphasizing, for example, the need for collective group identity formation and/or the ambi-tion to offer alternative political orientations, that is, subaltern, counter-public spheres (see Fraser, 1992). If one of the key normative elements of the public sphere is the ideal of universal access which permits citizens to participate in democracy, it is precisely on this point where much difficulty is encountered: ostensibly democratic societies have a variety of formal – but often informal – mechanisms that hinder democratic participation in civic spaces.

With their emphasis on participation in broader societal contexts, republican versions of democracy push for a broader understanding of the political, one that readily extends beyond electoral politics and can potentially insert itself in just about any societal context where contention can arise. This ties in with another question that I will look at: if the public sphere is the communicative space of politics, what should civic engage-ment look like – how should political talk ideally proceed? As we shall see, the internet becomes very salient in these discussions.

To pull together the discussion thus far, we have the ideal of democracy, which can be understood as leaning towards more elite or more republican versions; the latter tendency underscores the importance of civic participa-tion not only in elections but in the larger societal terrain as well. The role of citizens is today cogently actualized by the current crisis in democracy, where civic participation has become a central theme. Electoral politics is going through a difficult time; alternative politics, although seemingly more robust, has a tenuous track record of success. Moreover, the char-acter of participation, and of politics itself, is transforming, as social and cultural foundations of democracy become refigured; not least, there exist a variety of exclusionary mechanisms that obstruct universal participation in civic spaces. At some point the issue of the power relations that shape civic spaces becomes pertinent. Clearly the internet figures prominently here, residing in a force-field of different premises and views about the political world generally and civic spaces in particular. The manner in which we might perceive the internet's normative and actual role is inevi-tably to some extent linked to how we view the contested and moving ana-lytic target of democracy. I now turn to the key currents of research on the internet as a civic space.

RESEARCH FINDINGS – AND CONTESTATIONS

No Techno-Fix

Since the mass-circulation printed press became an essential feature of democratic life in the nineteenth century, the media have been entwined with power structures, serving both to promote and encourage civic participation as well as to limit and deflect it. The specifics of course have varied greatly between different contexts and with changing circumstances. Almost every major revolution in media technology – radio, television, CB radios, desktop publishing, computers, internet, Web 2.0 – has been accompanied by a rhetorical promotion of the respective technology's democratic benefits. While such claims are not necessarily wrong, there is often a basic fallacy involved, namely technological determinism; that is, a view that analytically puts technology in the driver's seat and discounts the modifying impact of socio-cultural settings. This, in short, is the vision of the quick techno-fix, which implicitly suggests that democracy's ills at bottom have to do with an insufficiency of apparatus.

This was certainly noticeable in the first few years of the internet, when the technology was so new, so startling in its affordances. 'Armchair theorists' could uninhibitedly proclaim all sorts of wondrous developments for society, and democracy in particular, that would derive from the net – or, alternatively, predict the end of both democracy and civilization as we know it. Gradually, however, the research findings began coming in towards the end of the 1990s, and the discussions began to take on sharper contours.

Beyond Business as Usual

Most researchers have from the start explicitly or implicitly suggested that the internet is a boon for civil society: it permits and indeed promotes horizontal communication in society. Individuals, groups and organizations can get in touch with each other, even on a global level, and exchange ideas, experiences and support. While abuse of such communicative freedom can never be fully eliminated, various efforts (with varying degrees of success) have attempted to regulate or discourage such behaviour (for example, harassment, privacy violations, child pornography). The debates became more pointed when the issue had to do with the internet's impact on the public sphere.

One major trajectory here was 'business as usual', that the internet's role in in the public sphere, and democracy more broadly, was and would remain quite modest (for example, Margolis and Resnick, 2000). This view from the late 1990s acknowledged that the major political actors may engage in online campaigning, lobbying, policy advocacy, organizing, and so forth, but did not see the net as a significant space for civic activity: overall, the political landscape would remain basically the same. A few years later it was also noted that various experiments, usually on a rather small scale, to incorporate the internet as 'e-democracy' or 'e-participation' into local governments had not been hugely successful (see, for example, Malina, 2003; Gibson et al., 2004; Chadwick, 2006).

What should be emphasized is that this overall perspective was anchored in the formal political system, and coloured by the traditional role of the mass media in that system. Indeed, much of the evidence is based on American electoral politics (for example, Hill and Hughes, 1998). This view, however, began to change, as it became more and more apparent that citizens were using the internet for political engagement in various discussion forums and so-called news groups, and that this could have consequences for how they vote. Certainly by the time of Barack Obama's election victory in 2008, where it was clear that the strategic use of social media had played a major role, the internet was firmly in place as a terrain of relevance for established political parties.

In regard to alternative politics, one point emerged quite early: without the internet, the sprawling landscape of activist groups, advocacy organizations, social movements and political networks would have a very difficult time of it. How effective their impact was, has been, and is, remains contested; sceptics point to the (probable) low numbers of people who are actually involved in these activities, and the (generally) low impact they have. There are exceptions, of course: the Occupy movement in the autumn of 2011 spread from New York City across the USA and went global. However, by the spring of 2012 there was not much left – it had dissipated. On the other hand, the crisis within the European Union (EU) has mobilized many people to alternative politics in recent years, especially in Southern Europe, and these manifestations, where the net has an important role to play, have thus far had more longevity.

Online Civic Spaces and Practices

In the evolution of the internet itself, three areas or domains of convergence can be specified (Meikle and Young, 2012). There is the fundamental – and incessant – technological convergence of computers and digital media, where older media are constantly being reformatted and upgraded to be compatible with the ever-evolving new possibilities. This leads to the second area of convergence: organizational ones, fusing convergences, the older institutions of the mass media and the newer online actors, with constantly new mergers, new trade-offs and bankruptcies taking place, with a very few giants emerging to dominate the web landscape. Finally, there are convergences of form and content: multimedia (where words, images and sounds can be integrated on the same device by virtue of the shared digital language); transmedia (where the same content is dispersed across a variety of platforms); and mash-ups (which involve sampling, remixing and reconstituting texts). Thus, we need to think of the internet as a dynamic, highly protean milieu; which in some ways becomes problematic for democracy, since it requires a degree of institutional stability.

For researchers it was becoming clear not only that the internet had become a prime site of civic spaces, but also that citizens' practices were becoming very diversified; the affordances that allowed for easily achieved user-generated content (UGC) were promoting more active modes of participation. Many citizens active in online civic spaces were moving from mere interactivity to full-fledged 'produsers', where UGC was becoming all the more relevant for politics, in both its electoral and, especially, its alternative variants. Moreover, the internet has become inseparable from the daily life and social worlds of citizens; it is hyper-ubiquitous: it is everywhere, used by (almost) everyone in democratic societies, for a seemingly endless array of purposes. For many people it is no longer something they merely visit or occasionally check: we see especially the younger age cohorts spending significant amounts of time on the net, socializing, pursuing all manner of information, engaged in consumption, entertainment, and so on. Everyday life is increasingly embedded to a great extent on the internet; it is where much of it takes place.

These developments are predicated on the interplay between the transformation of the internet and the uses to which it is put. Lievrouw (2011) underscores the continuing interplay between the affordances of communication technologies and the practices by which people utilize them for their own purposes, resulting in a sort of dialectic between technological innovation and creative adaption. Strict adherence to the formal criteria of deliberative democracy, while laudable and relevant for specialized contexts, seemed far removed from the realities of today's political communication on the net.

Further, the mobile character of the net has important consequences for how we live: while the importance of place does not simply vanish, its relevance is in many circumstances diminished by mediated connectivity. We are more accessible than before, and we become more portable and flexible. A good deal of our social coordination and organization can be carried out from a distance. Surveillance can also be enhanced by mobile technologies, by authorities for a variety of purposes (crime-fighting, political suppression, routine monitoring), by peers and by parents (who often want their children to carry a mobile phone).

These developments have significance for civic space: civil society can take on a more ambulatory character, obviously enough, but for the public sphere the changes become more profound. At bottom, the boundaries between public and private space have become negotiable (Meikle and Young, 2012). Public space can now be 'reformatted' in a variety of ways and for different purposes, including the interjection of the private (for example, a personal conversation). Such modulations begin to alter the basic coordinates of our social geography. While mobile devices are often used for personal purposes, crossing the thin line to public and political contexts is easily and often done. The public sphere becomes less demarcated from other domains, a development many of the republican persuasion support, since it allows politics to more easily seep into other, less traditional areas; and indeed, with social media, to go viral.

Enthusiasts and Sceptics

In the large and diverse literature from recent years are found enthusiasts such as Benkler (2006), Castells (2010) and Shirky (2008). More sceptical and critical voices, who argue that the democratic possibilities of the web have been seriously oversold, are found in Fuchs (2011), Hindman (2009) and Morozov (2011). The enthusiasts are no doubt easier to understand (and to like). They pick up on the horizontal, civic society character of the internet, with its open quality and participatory affordances. They note how this in turn meshes with the ideas of social networks as the new organizing logic of society (Castells, 2010), and how the sharing and collective wisdom typified by wiki-logics can empower citizens and strengthen democracy (Shirky, 2008). Such authors point to social media in particular as spaces where interaction can readily shift from personal encounters to commerce, to civil society activity, and not least to public sphere communication. They highlight the almost infinite amount of information and views available online, and how this empowers citizens and broadens the spectrum of the public sphere.

Sceptics, for their part, contend that using the internet for political activities (at least defined in traditional terms) is certainly one of the less frequent usages; politics generally comes far behind consumption, entertainment, social connections, pornography, and so on. Today the opportunities for such kinds of involvement are overwhelmingly more numerous, more accessible and more enticing for most people, compared to civic or political pursuits. Moreover, it has been shown that access to the internet in itself does not turn people towards political issues; in fact, younger cohorts, who are the most net savvy, are less likely to do politics on the net than older age cohorts. It is also argued that the very density of the symbolic environments on the online public sphere becomes a distraction, and they lead to massive competitions for attention. For political actors using the internet, getting and holding an audience is a constant challenge (the case of political bloggers is often mentioned: most seem to fizzle out after a short time, while the big heavy ones, tied to major media organizations online, have more staying power).

Hindman (2009) and Morozov (2011) are among the voices who are adamant that the benefits of the internet for democracy have been much exaggerated; the latter author in particular makes a strong case for seeing the net as a tool for authoritarian control, as witnessed in places such as China and Belarus, and he asserts that similar patterns in web use by the authorities are also emerging in the Western democracies. The revelations that the National Security Agency in the USA, and similar organizations in other countries, engage in massive surveillance on citizens in democracies suggests that we have entered the post-privacy era (Greenwald, 2014). Other critics point to structural issues about the internet. They argue that the net, the regulation around it, the major operators that define how it functions, and the various platforms available on it – not least social media – are shaped by the commercial imperatives of political economy and the power relations that derive from them, to the detriment of the character of these civic spaces.

On an even more fundamental level, other authors such as Carr (2010) argue that the architectural logic of the net and its impact on our modes of cognitive functioning have a deleterious impact on our capacity to think, read and remember. If many observers laud how the participatory 'wisdom of the many' (as manifested, for example, in Wikipedia and the blogosphere) is producing new and better forms of knowledge, others such as Keen (2008) warn of the dangers, asserting that it erodes our values, standards and creativity, as well as undermining cultural institutions. In a related vein, the argument is often made that the strongly affective character of the multimedia internet, particularly social media, also contributes to the decline of rationality in the public sphere. While emotionality is of course essential for political engagement, many observers note that it often tips over in a manner that is counterproductive for sound democratic politics (for example, populist discourses). Moreover, affect can become an easy way to bypass what seems like infinite amounts of information yielding ambivalence from sources one may not fully trust (Andrejevic, 2013).

Problematic Political Economy

The internet is not just a technological device, it is also a socially organized institution, enmeshed in power relations; these features of its political economy, as mentioned above, impact greatly on its character. For instance, the internet is profoundly affected by Google, which greatly shapes how the net operates and what we can do with it (Vaidhyanathan, 2011; Fuchs, 2011; Cleland and Brodky, 2011). Moreover it has become the largest holder of information in world history, shaping not only how we search for information, but also what information is available, and how we organize, store and use it. In many ways it is an utterly astounding development, yet it has also grown into an enormous concentration of power that is largely unaccountable, hidden behind the cheery corporate motto 'Don't be evil'. We all strew daily personal electronic traces; these are gathered up, stored, sold and used for commercial purposes by Google (and other actors). This selling of personal information is done with our formal consent, but if we refuse we effectively cut ourselves off from the major utilities of the internet. Increasingly very serious questions are being raised, and those struggling to defend the interests of the public in regard to privacy have begun, at lest indirectly, to confront Google's agenda to organize knowledge on a global scale.

All this is not to detract from Google's truly impressive accomplishments; rather, the issue is that the position it has attained, and the activities it pursues (which are quite logical given its position), raise questions about information, democracy, accountability and power in regard to the internet. Just to take one example: given the logic of personal profiling – the filtering of results to 'fit your known locality, interests, obsessions, fetishes, and points of view' (Vaidhyanathan, 2011: 183) – the answers that two people will receive based on the same search words may well differ significantly. This can wreak havoc with the whole idea of shared public knowledge (Pariser, 2011), which in the long run can potentially undermine the democratic culture of debate between differing points of view.

Facebook, now with about 1 billion users, also compiles massive amounts of data on individuals, largely freely given. As with Google, the data gathered is for commercial purposes, but again, changing social contexts can generate new uses and meanings of personal information. With Facebook, the spillover from private to public is much easier, resulting in embarrassment, entanglements, defamation or even death. Data theft is also easier; digital storage systems are simply not fail-safe, as witnessed when hackers have even entered high-security military databases. Thus, to participate in Facebook and similar social media is to expose oneself to surveillance and to have one's privacy put at risk. Moreover, such digital information is not erased; it is archived, and can be retrieved and inserted into new – and troubling – contexts of a person's life.

As noted above, social media sites such as YouTube, Facebook and Twitter have became incorporated into political communication. They have become important outlets and sources for journalism, and are increasingly a part of the public sphere of both electoral and alternative politics. Not least, they have become the sites for massive marketing efforts, as Dwyer (2010) underscores. In Facebook's role as a site for political discussion, one can reflect on the familiar mechanism of 'like': one clicks to befriend people who are 'like' oneself, generating and cementing networks of like-mindedness. As time passes, people increasingly habituate themselves to encountering mostly people who think like they do, and getting their biases reinforced. The danger arises that citizens lose the capacity to discursively encounter different views; the art of argument erodes, and differences to one's own views can become incomprehensible.

What is ultimately required, as MacKinnon (2012) argues, is a global policy that can push regulation of the net such that it will be treated like a democratic, digital commons. We have a long way to go.

A New Kind of Civic Space?

Other implications of using screen-based social media for political life have been explored by a number of authors. Dean (2010) and Papacharissi (2010), for example, argue that it is not just a question of people choosing politics or consumption or popular culture, but that the internet environment in its present form promotes a transformation of political practices and social relations whereby the political becomes altered and embodied precisely in the practices and discourses of privatized consumption. Political practices become entangled with the drive for personalized visibility, self-promotion and self-revelation. When (especially) younger people do turn to politics, it seems that the patterns of digital social interaction increasingly carry over into the digital. Papacharissi (2010) argues that this is engendering a new form of civic space. I call it the 'solo sphere', and it can be seen as a historically new habitus for internet-based political participation, a new social milieu for political agency. A networked, often mobile, yet oddly privatized sociality emerges, a personalized space from which the individual engages with the complex political outside world. Operating in this comfort zone often results in what is disparagingly called 'slacktivism' or 'clicktivism'. It is easy to understand this stance as a safe retreat into an environment that many feel they have more control over. To the extent that this is true, however, it introduces a historically new – and problematic – set of circumstances for civic agency.

SYNTHESIZING AND SPLITTING THE DIFFERENCES

The internet has been contributing to the massive transformations of contemporary society at all levels for about two decades now, and it would be odd if it did not also alter the premises and infrastructure of political life. In making available vast amounts of information, fostering decentralization and diversity, facilitating interactivity and individual communication, while providing seemingly limitless communicative space for whoever wants it, at speeds that are instantaneous, it has redefined the practices and character of political engagement. Also, while politics remains a minor net usage, the vast universe of the internet and its various (and ever evolving) technologies make it easier for the political to emerge in online communication, especially within the new kind of alternative politics that is on the rise.

Contingencies as Dynamic Configurations

There are thus grounds for optimistic views about the internet's significance for civic spaces, and there is a good deal of research which supports this view, some of which I have mentioned above. At the same time, as noted above, other voices are cautionary (and a very few are outright dismissive): once we leave the mythical realm of technological determinism and enter complex socio-cultural realities, the role of the internet becomes more equivocal. Clearly it is not a question of coming to some simple resolution, a neat, all-purpose truth about the internet as a civic space. The diverse approaches, assumptions and horizons in the extensive literature signal the complexity of the issues involved.

A key theme that unites many of the diverse sceptical views is precisely their insistence on socio-cultural contingencies: that is, seeing the internet (and all social phenomena) as products of circumstances that both engender and delimit them. There are only possibilities, nothing is necessary; any concrete phenomenon is shaped by a series of other factors, in processes of dynamic configurations. We should keep in mind that the sceptics for the most part are not categorically rejecting any possible positive dimension in regard to the internet as a civic space; rather, they are often reacting against the excessively enthusiastic and/or naïve view that has been circulated by some commentators, and not least by internet industries themselves. Thus, the sceptical position challenges us to look critically at the contingencies of whatever social phenomenon we are addressing.

A first step in such a direction is conceptual clarity in regard to the phenomenon and its dynamic configurations. In regard to the internet, we should specify which aspects, services or platforms are relevant. Thus, for example, in regard to social media, different platforms can offer different forms of civic participation. For example, an activist group may need to: (1) internally discuss ideas and debate; (2) develop collective identities; (3) mobilize members; (4) strive to reach out to new members; (5) try to get mass media coverage; and (6) coordinate on-site during a demonstration. Facebook could well serve (1) and (2), Twitter may be very serviceable for (3) and (5), YouTube might be useful for (4), and mobile phone calls and SMS texts be especially useful for (6). There is nothing hard and fast here, yet one should be aware of how different platforms offer divergent affordances, and how this may shape the patterns of use in specific settings. Moreover, the various platforms can be and are used in convergent ways, with relays, feeds and sharing across the platforms (see, for example, Thorson et. al., 2013).

Among the contingencies to clarify are the zones of interface between on- and offline settings; to illuminate the contexts of use, the modes of usage, the social actors involved, their circumstantial settings, the overarching power relations, the links to their media and communicative spaces, and so forth. This kind of mapping of dynamic configurations, and the elucidation of the consequent contingencies at work, will provide a more rigorous and useful portrait of the specific civic spaces in question than sweeping generalizations. The actual technology itself is of course highly relevant, but it must be understood as being adapted for particular uses by certain actors; it does not operate as an independent, ahistorical force. An extended example of this kind of approach is found in Mattoni's (2012) study of the media practices of activist workers in today's crisis-ridden Italy; this movement in fact used such an analysis to devise its own media strategies.

In my own work (Dahlgren, 2009) I have followed a version of this logic in looking at how the media may contribute to, or hinder, civic practices. My basic supposition is that for people to participate politically, to engage in civic spaces, they must be able to see engagement as both possible and meaningful. In other words, people need some kind of an empowering civic identity. Yet such identities cannot flourish in a vacuum; they need to be nourished by what I call 'civic cultures'. Civic cultures is a way of answering, analytically and empirically, the question of what facilitates or hinders people from acting as political agents, from engaging in civic spaces. If we insert the internet into this framework, we would want to highlight how various aspects interface with everyday life, how the particular citizens in question use it for political purposes, what the political means for them, the power relations in which they find themselves, and so forth. Civic cultures serve as taken-for-granted resources that people can draw upon, while citizens in turn also contribute to the civic cultures development via their practices; that is, their political uses of the internet.

Further, civic cultures are comprised of a number of distinct dimensions that interact with each other. Participatory practices themselves constitute one key dimension of civic cultures; others include suitable knowledge about the political world and one's place in it, democratic values to guide one's actions, and appropriate levels of trust. A minimal level of 'horizontal' trust – that is, between citizens – is necessary for the emergence of the social bonds of cooperation between those who collectively engage in politics; there is an irreducible social dimension to doing politics. These dimensions could be elucidated in regard to specific affordances and usages of the net, in concrete situations. Moreover, civic cultures require communicative spaces where such agency can take place; the internet as a civic space would be critically evaluated in relation to, for example, its political economy and technical architecture to clarify its democratic assets and drawbacks. Finally, forms of identity as political agents – my starting point above – are also a major dimension of civic cultures: people must be able to take on a civic self, to see themselves as actors who can make meaningful interventions in relevant political issues. Clarifying how these dimensions operate configurationally with each other (or not) in specific contexts would enhance our understanding of the internet as a civic space.

Proposals for Future Research

A great deal of research on the internet as a civic space has been done over the years, from varying perspectives. Yet there is still so much we need to know; indeed, from the broader perspective of mediatization and political participation, there is a need for developing a further research agenda, as suggested recently in Dahlgren and Alvares (2013). Based on that collective effort, I would propose that for the theme of the internet as a civic space, researchers would do well to explore questions that continue some key trajectories in current research, such as the following:

- How does the use of the internet contribute – in concrete situations – to the development of civic agency, knowledge, practices and identities? This would include a particular focus on alternative politics in the face of the continuing crises and the inadequacies of mainstream politics in dealing with them.
- How do these use strategies tend promote or hinder actual political engagement, and shape its subjective perceptions and its concrete manifestations and expressions?
- How might existing engagement in popular culture, consumption

and sociality be linked to the political, as civic spaces on the net (for example, social media) intersect all the more with societal domains beyond both civil society and the public sphere?

FURTHER READING

Bennett, W. Lance and Alexandra Segerberg (2013) *The Logic of Connective Action: Digital Media and the Personalization of Contentious Politics*. Cambridge: Cambridge University Press.
Castells, Manuel (2010) *Communication Power*. Oxford: Oxford University Press.
Coleman, Stephen and Jay Blumler (2009) *The Internet and Democratic Citizenship*. New York: Cambridge University Press.
Dahlgren, Peter (2013) *The Political Web*. London: Palgrave.
Dewey, John (1923) *The Public and its Problems*. Chicago, IL: Swallow Press.
van Dijck, José (2013) *The Culture of Connectivity: A Critical History of Social Media*. Oxford: Oxford University Press.
Gerbaudo, Paolo (2012) *Tweets and the Streets: Social Media and Contemporary Activism*. London: Verso.
Hindman, Mathew (2009) *The Myth of Digital Democracy*. Princeton, NJ, USA and Oxford, UK: Oxford University Press.
Loader, Brian and Dan Mercea (eds) (2012) *Social Media and Democracy*. Abingdon: Routledge.
McChesney, Robert W. (2013) *Digital Disconnect*. New York: New Press.
Morozov, Evgeny (2011) *The Net Delusion: How Not to Liberate the World*. London: Allen Lane.
Papacharissi, Zizi (2010) *A Private Sphere: Democracy in a Digital Age*. Cambridge: Polity Press.

REFERENCES

Alexander, Jeffrey C. (2006) *The Civil Sphere*. New York: Oxford University Press.
Andrejevic, Mark (2013) *Infoglut: How Too Much Information Is Changing the Way We Think and Know*. Abingdon: Routledge.
Barber, Benjamin (1984) *Strong Democracy: Participatory Politics for a New Age*. Berkeley, CA: University of California Press.
Benkler, Y. (2006) *The Wealth of Networks: How Social Production Transforms Markets and Freedom*. New Haven, CT: Yale University Press.
Bennett, W. Lance and Alexandra Segerberg (2012) 'The logic of connective action: digital media and the personalization of contentious politics'. *Information, Communication and Society* 15(5): 739–768.
Carr, Nicholas (2010) *The Shallows: How the Internet is Changing the Way We Think, Read and Remember*. London: Atlantic Books.
Castells, Manuel (2010) *Communication Power*. Oxford: Oxford University Press.
Chadwick, Andrew (2006) *Internet Politics: States, Citizens and New Communication Technologies*. New York: Oxford University Press.
Cleland, Scott and Ira Brodky (2011) *Search and Destroy: Why You Can't Trust Google*. St Louis, MO: Telescope Books.
Cohen, Jean and Andrew Arato (1992) *Civil Society and Political Theory*. Cambridge, MA: MIT Press.

Dahlgren, Peter (2009) *Media and Political Engagement*. New York, USA and Cambridge, UK: Cambridge University Press.

Dahlgren, Peter and Claudia Alvares (2013) 'Political participation in an age of mediatisation: toward a new research agenda'. *Javnost/The Public* 20(2): 47–66.

Dean, Jodi (2010) *Blog Theory*. Cambridge: Polity Press.

Dewey, John (1923) *The Public and its Problems*. Chicago, IL: Swallow Press.

Dwyer, Tim (2010) 'Net worth: popular social networks as colossal marketing machines'. In Gerald Sussman (ed.) *Propaganda Society: Promotional Culture and Politics in Global Context*. New York: Peter Lang, pp. 77–92.

Edwards, Michael (2009) *Civil Society*. 2nd edn. Cambridge: Polity Press.

Fraser, Nancy (1992) 'Rethinking the public sphere: a contribution to the crtitique of actually existing democracy'. In Craig Calhoun (ed.) *Habermas and the Public Sphere*. Boston, MA: MIT Press, pp. 109–142.

Fuchs, Christian (2011) *Foundation of Critical Media and Information Studies*. London: Routledge.

Gibson, Rachel K., Andrea Römmele and Stephen J. Ward (eds) (2004) *Electronic Democracy: Mobilisation, Organisation and Participation via New ICT's*. London: Routledge.

Greenwald, Glenn (2014) *No Place to Hide: Edward Snowden, NSA, and the Surveillance State*. New York: Macmillan.

Habermas, Jurgen (1989) *Structural Transformation of the Public Sphere*. Cambridge: Polity.

Habermas, Jürgen (2006) 'Political communication in mediated society'. *Communication Research* 16(4): 411–426.

Harvey, David (2006) *A Brief History of Neoliberalism*. Oxford: Oxford University Press.

Hay, Colin (2007) *Why We Hate Politics*. Cambridge: Polity Press.

Held, David (2006) *Models of Democracy*. 3rd edn. Cambridge: Polity Press.

Hill, Kevin A. and John E. Hughes (1998) *Cyberpolitics: Citizen Activism in the Age of the Internet*. Lanham, MD: Rowman & Littlefield.

Hindman, Mathew (2009) *The Myth of Digital Democracy*. Princeton, NJ, USA and Oxford, UK: Oxford University Press.

Keen, Andrew (2008) *The Cult of the Amateur*. New York: Doubleday.

Lievrouw, Leah A. (2011) *Alternative and Activist New Media*. Cambridge: Polity Press.

MacKinnon, Rebecca (2012) *Consent of the Networked: The Worldwide Struggle for Internet Freedom*. New York: Basic Books.

Malina, Anna (2003) 'e-Transforming democracy in the UK: considerations of developments and suggestions for empirical research'. *Communications: The European Journal of Communication Research* 28(2): 135–155.

Margolis, Michael and Resnick, David (2000) *Politics as Usual: The Cyberspace 'Revolution'*. London: Sage.

Mattoni, Alice (2012) *Media Practices and Protest Politics: How Precarious Workers Mobilise*. Farnham: Ashgate Publishing.

Meikle, Graham and Sherman Young (2012) *Media Convergence: Networked Digital Media in Everyday Life*. Basingstoke: Palgrave Macmillan.

Morozov, Evgeny (2011) *The Net Delusion: How Not to Liberate the World*. London: Allen Lane.

Mouffe, Chantal (2005) *On the Political*. London: Routledge.

Papacharissi, Zizi (2010) *A Private Sphere: Democracy in a Digital Age*. Cambridge: Polity Press.

Pariser, Eli (2011) *The Filter Bubble: What the Internet is Hiding from You*. London: Penguin.

Putnam, Robert (2000) *Bowling Alone: The Collapse and Revival of American Community*. New York: Simon & Schuster.

Rosanvallon, Pierre (2008) *Counter-Democracy: Politics in an Age of Distrust*. Cambridge: Cambridge University Press.

Sandel, Michael (2012) *What Money Can't Buy: The Moral Limits of Markets*. London: Allen Lane.

Schachter, Hindy Lauer and Kaifeng Yang (eds) (2012) *The State of Citizen Participation in America*. Charlotte, NC: Information Age Publishing.

Shirky, Clay (2008) *Here Comes Everybody: The Power of Organizing Without Organizations*. London: Allen Lane.

Thorson, K., K. Driscol, B. Ekdale, S. Edgerly, L.G. Thompson, A. Schrock, L. Swartz, E.K. Vraga and C. Wells (2013) 'YouTube, Twitter, and the Occupy movement: connecting content and circulation practices'. *Information, Communication and Society* 16(2): 1–31. Available at http://www.tandfonline.com/doi/abs/10.1080/1369118X.2012.756051 (accessed 6 March 2013).

Vaidhyanathan, Siva (2011) *The Googlization of Everything: And Why We Should Worry*. Berkeley, CA: University of California Press.

3. The social foundations of future digital politics

Nick Couldry

Some argue that we are witnessing, through digital media (and especially social media), an entirely new type of politics, embodied in the 2011 Arab Spring (McDonald, 2011). Others more cautiously understand those events as exceptional moments of mobilization that leave few traces on long-term political structures (Gladwell, 2010). Still others acknowledge as unresolved the question of whether 'digital civics is all that different from older models' (Zuckerman, 2013). Underlying this disagreement is the question of why citizens engage and act politically. To make progress on these difficult questions requires giving more attention to the social foundations of political engagement and political action; optimistic accounts of how digital media have suddenly revolutionized political processes tend to neglect that social dimension.

Predictions of new forms of political and social connection, even radical politics, have accompanied many previous waves of technological change. In the past half-century, accounts of the internet have been distorted by what Vincent Mosco (2004) sardonically calls the 'digital sublime'. Yet, hype or no hype, we must acknowledge that the internet is potentially a major source of institutional innovation, because digital communication practices, just like the newspaper two centuries ago, constitute resources with the force of institutions (Chadwick, 2006: 3).

Let us remember at the outset two things. Firstly, governments, media, corporations and many elements in civil society may have a vested interest in avoiding such fundamental reorganization of the political process. Secondly, predictions of political and cultural change based on media change tend to rely on a rather thin account of how social processes operate, change and are organized. There are factors which suggest a less optimistic assessment of digital media's implications for democratic politics. While there are many competing theories (liberal, republican, deliberative or elitist) of how democracy should function, my argument here does not depend on taking a particular position on the ideal form of democracy, only on asking what stable social conditions and resources are supportive of political engagement and action.

THE MISSING SOCIAL

The internet, because of its basic networked features, has without question generated new possibilities for political association, mobilization and action. Sara Bentivegna (2002: 54–56) sums up the democratic potentials of the internet as interactivity, co-presence, disintermediation, reduced costs, speed and the lack of boundaries. We can now meet and organize politically with people we do not know and cannot see at great speed across local, regional and even national boundaries. Some see this as the beginning of a new more conversational, less formal mode of politics; others are more sceptical. On any view, there are new mechanisms of political socialization to investigate.

These basic facts about the digital conditions for politics have encouraged some scholars to build bigger theories about digital politics. They have argued that Web 2.0 has transformed the larger configuration of individual, group and institutional strivings that makes up public culture, including political culture. Important here have been cultural studies scholar Henry Jenkins's account of 'convergence culture', legal scholar Yochai Benkler's book *The Wealth of Networks*, and sociologist Manuel Castells's recent book *Communication Power*, which builds on his earlier three-volume work on The Information Age.[1] I am going to focus here on Castells's work, because of these three he offers the most developed and comprehensive theory of digital politics; its limitations are therefore of wider significance, but similar problems, it can be argued, run through those other texts and indeed many more.

An interesting point however is worth making in passing about Yochai Benkler's book. Benkler argues that lower entry costs to cultural production and lower transaction costs for communications generate faster, more targeted possibilities for political mobilization without any need for physical coordination: 'emerging models of information and cultural production, radically decentralized and based on emergent patterns of cooperation and sharing, but also of simple coordinate existence, are beginning to take on an ever-larger role in how we produce meaning' (Benkler, 2006: 32–33).

True, but what follows from this? Benkler has a vision of a completely new model of social storytelling, and he is right to believe that storytelling (the accounts we generate of the world we live in) is at the basis of any democratic notion of politics. 'We have an opportunity', he argues, 'to change the way we create and exchange information, knowledge and culture' (2006: 473). Is this, Andrew Chadwick (2012) argues, the start of a change in people's opportunities to contribute to larger political processes? If so, it must be at the heart of any account of the social foundations of digital politics.

Benkler offers an eloquent account of how the internet's distributive capacity enables production and distribution tasks to be broken down into modules through what he calls the 'granularity' of cultural action (Benkler, 2006: 105–106). Potentially this makes new types of political action possible through the enabling of certain types of coordination at a speed and scale that previously were impossible (compare Kavada, 2010 on the European Social Forum). But Benkler's argument highlights only a very small part of the terrain of information production necessary for politics. As Clay Shirky (2010: 107–109) notes, political or social change require much more than technological opportunity: there must be the motive for new types of media production to support such change, and our motives only change when dictated by our needs, which may not change as much as some think. Whose needs are we discussing? Are the full range of citizens (not just the already empowered and motivated elite) inclined to use new opportunities to build new forms of positive politics? This remains unresolved.

While Benkler starts out from the consequences of online networks, it is Manuel Castells who has most fully theorized the role of those networks in politics and society generally. Castells's argument in *Communication Power* (Castells, 2009) is, put simply, that in recent decades the organization of society and politics has changed radically: firstly, the emergence of networks has provided the fabric that connects people, across old nation-state boundaries; secondly, there have been transformations derived from the construction of meaning within those networks, because power always needs to be legitimated and culturally translated. The first factor is no doubt important (no one would deny that the internet makes it easier to sustain networks), but the key question is: what meaning is constructed within those networks, and how is this achieved?

Castells approaches the social foundations of digital politics via his well-known concept of the 'network society'. By this he means 'a social structure constructed around (but not determined by) digital networks of communication' across national borders: Castells believes global networks 'structure all societies' (Castells, 2009: 24, 53). He rightly dismisses the notion that societies are currently, if they ever have been, 'based on shared values'. Instead it is 'relational power' that counts, and 'relational power' is based not only on force but also on 'communicative resources' captured and amplified in networks. So Castells helpfully introduces a cultural element into his analysis of social structure. The question remains whether he tells us enough about how to understand the culture of networks, and its social basis, and exactly how that socially grounded culture might alter the dynamics of political engagement and political action.

It is worth looking more closely at Castells's argument. Castells rightly highlights the growth of 'mass self-communication' (online social networks, blogs and the like). No one would doubt that this is a striking new feature of public culture, political process and even everyday life. But the question is: who is involved in mass self-communication and for what purposes? Mass self-communication gives the appearance of lots of new engaged activity, but what makes it political? What if our relentless communications on Facebook and Twitter occur in an overall context that tends to orient people overwhelmingly towards non-political contexts? Jodi Dean (2010) argues that such communication is co-opted into corporate frames which divert communicative energies away from politics. That is surely too sweeping a conclusion in light of the explicit role that social media played in the political events of 2011, but we certainly cannot ignore the long literature on the multiple factors which block people from interest in politics, or from the belief that politics is 'for them', particularly hierarchies of class and gender.[2] What grounds are there for believing such factors are no longer shaping peoples' judgements about whether their potential political actions would matter?

Castells's examples of networked politics (the 2008 Obama presidential campaign, the decades-long growth of global environmental politics) steer clear of everyday power struggles in the economy (rights in and to work). This neglect of labour activism is symptomatic of a wider problem in Castells's account: his insufficient attention to how non-political forms of power (economic, legal, social) shape individuals' framings of their opportunities for political action, even within networks. We might look to Castells's concept of 'network society' to help here, pointing to new horizons of political action that were simply not available to individuals and isolated groups before, but its underlying notion of 'society' is surprisingly underdeveloped. Castells tries to invoke a new 'social': 'could it be that the technological and organizational transformation of the network society provides the material and cultural basis for the anarchist utopia of networked self-management to become a social practice' (2009: 346).

But what does this mean? Is the horizontal network *ipso facto* equivalent to the social? If so, what exactly qualifies it as a social practice rather than a purely practical arrangement? How does it relate to other types of thick social context that we still encounter (the family, the club, the local crowd)? It is worth remembering that in his earlier work Castells insisted that 'people, locales and activities lose their structural meaning' in a network society (1996: 477). But if so, what is the basis now for being optimistic about the social potential of networks? Networks are, by definition, structures that most people can leave at no cost, and with little consequences for others; only if people are key network nodes is the position reversed, but such nodes are necessarily few in number (Barabasi, 2003). So, if we are to think about the sustainable consequences of networks for politics, we must know more about the long-term contexts that drive and sustain individual action for all members of networks. I return to this issue in relation to social media's contribution to political movements later.

My argument so far has focused on a foreshortening of 'the social' that is, I would argue, common to many writers about digital politics. It is present even in those who proclaim the social as a site of new politics, based in our experience of: 'a kind of social flesh, a flesh that is not a body, a flesh that is common, living substance . . . The flesh of the multitude is pure potential, an unformed life force, and *in this sense* an element of social being, aimed constantly at the fullness of life' (Hardt and Negri, 2005: 191).

What is lacking here, much more drastically than in Castells, is any sense of the specific resources, contexts of action, and historical opportunity structures required for sustained political mobilization: in short, the 'social institutions' (Turner, 2005: 136) that underlie politics. It is those same social institutions which, in spite of all the transformative hopes surrounding past communication technologies, have ensured that in the long run they reinforced existing prior structures and social networks (see, for example, Fischer, 1992 on the telephone). Is there any serious reason for thinking that the outcome from today's technological innovations will be different?

In thinking about this, we should avoid the blanket pessimism of some writers such as Evgeny Morozov (2011) on 'the net delusion'. Rather we should recall the work of a different theorist, the late Charles Tilly. Tilly defined democracy as 'the extent to which the state behaves in conformity to the expressed demands of its citizens', and was interested in clarifying what factors, across many societies and periods, tend to encourage the emergence of democracy in practice. He identified three macro-conditions (Tilly, 2007: 13, 23):

1. the integration of trust networks into public politics;
2. the insulation of public politics from categorical inequality; and
3. the reduction of major non-state power centres' autonomy from public politics.

Note condition (1) in particular (I come back to (2) and (3) later). Tilly's point is not that we need more trust, but more subtly that our trust networks (for example, with friends, family, work colleagues) must be integrated into the bargaining processes of public politics. Such integration does not require people to trust rulers more (often the opposite), but rather to commit the resources, assets and projects they most prize to the risks inherent in the sorts of 'mutually binding consultation' on which democratic process depends (Tilly, 2007: 74). Such trust is not easy. Indeed, Tilly argues, democratization itself is not easy, and there is always the risk of it slipping back into reverse: 'Democratization is a dynamic process that always remains incomplete and perpetually runs the risk of reversal – de-democratization' (Tilly, 2007: xi).

If so, let us be cautious in assessing the sorts of factors that might, over the longer term, shape the social foundations of political engagement and action in the digital age. The buzz created around technological innovation may be deeply misleading here: as Clay Shirky points out, 'Communications tools don't get socially interesting until they get technologically boring' (Shirky, 2008).

SOCIAL CONDITIONS FOR NEW POLITICAL ENGAGEMENT

How can we rethink the relations between digital media and politics in a way that, rather than starting out from the abstract potential of networks, looks more seriously at the consequences digital media environments have for the concrete contexts and resources likely to support engagement in politics?

To make progress on these difficult questions, I will start by looking at the who of politics (what kinds of people or things now count as political agents?), moving quickly onto the what of politics (what kinds of things count as political in various modes: deliberation, action, decision?) and finally the why of politics (what larger contexts make certain kinds of political agent/action possible or likely?).

Who is Organizing Digitally, and for What?

The set of political actors has always been much larger than the official list of mainstream political institutions and their representatives. Does the digital media environment substantially change who can be political actors?

On the face of it, yes. The internet has made possible new kinds of legitimate political actors. First there are network actors: distributed agents of political coordination that link multiple persons, groups and positions across space, without the need for a physical headquarters or a bounded social membership. These became increasingly prominent in the 1990s with the rise of international online networks in the non-governmental organization (NGO) sector and insurgent actors such as the Zapatistas.[3] The internet also brings new possibilities for informal political actors to form and build communities of practice online including across national borders. Well-designed campaign websites can, simply by existing, name a terrain of contest, provide networking resources for local campaigns, and help orient future mobilization and long-term action; in so doing, they extend the scale and repertoire of political action, and in the process enable an expansion of the range of political actors. National political actors such as the UK mothers' lobby group Mumsnet quickly grew online from a small local group based on an immediate community of interest into something much broader (Coleman and Blumler, 2009: 127–134). Similar transformations have occurred across national borders in global civil society (Kavada, 2014), with movement forms replicating internationally at great speed as with the Occupy movement in late 2011 (Juris, 2012).

What explains this? The possibility of anonymous action-at-a-distance reduces some of the barriers to action such as people's fear of reprisal or embarrassment (Chadwick, 2006: 202). More generally, the internet, through the links it enables between networks, enables actors to link easily and quickly with each other to form larger networks. The ease with which informal connections can be created online is also important (Lim and Kann, 2008).

In addition, digital politics involves new kinds of individual political actors: no longer just the party leader, or the journalist commentator, or the demonstrator in a crowd, but the individual – without any initial political authority – who can suddenly acquire status as a political actor online. Blogging is one way of doing this, although its long-term implications remain uncertain. Are individual bloggers genuinely new voices in politics? Some no doubt are. In China the number of active blog users has grown very fast: the Chinese scholar Yin Haiqing sees a playful relationship to mainstream political ideology in the Chinese blogosphere, but even playful commentary is politically challenging to an authoritarian government. In Iran, explicit contention against mainstream politics generates a very lively blogosphere involving both individual and collective authors. Is it coincidence that one person killed by Egyptian police in the January–February 2011 uprising was a blogger, Khaled Said? But bloggers in many other contexts are members of old-style political elites whose voices are simply now archived in print. In the UK's highly centred politics, the leading political bloggers (Guido Fawkes, Ian Dale) are arguably just digital versions of the commentators who have always hovered in political backrooms, although if their stories get picked up regularly by mainstream media, they acquire authority. But their influence still looks more like a reinforcement of, rather than a counter-logic to, mainstream political journalism.[4]

There are also broader questions about who the new digital political actors actually are. Web visibility is hard to achieve, and certainly not open to everyone. It depends (Hindman, 2009) on being picked up by search engines (which prioritize on the basis of numbers of incoming links), or on personal recommendations by web users. If so, then success depends on factors not so different from older ways of entrenching political power. The much-hyped democratic potential of Twitter is also significantly concentrated: most tweeters around the failed Iranian revolution of 2009 tweeted just once, and the top 10 per cent of users generated two-thirds of tweets.[5]

On the other hand, the set of latent political actors (those who might, given certain circumstances, enter the field of political action) has expanded. In Britain, for example, many people now blog or tweet about aspects of institutional life who have never had a public voice before (Couldry, 2009). They are forced to remain anonymous, but blogging doctors, teachers, policemen and women, army officers, magistrates and employees of big pharmaceutical companies occupy an important edge of the political terrain. When such disclosure activities are coordinated across international borders through a wider infrastructure (as with Wikileaks), this has major political consequences (Brevini et al., 2013).

So in spite of some doubts, the spectrum of political actors has expanded in an age of what Bruce Bimber (2003) calls 'information abundance'; but with what sustained effect? Sustained effect would mean converting a temporary online presence into legitimacy and visibility beyond a small circle, which is hard. For Hindman (2009: 102), blogs remain 'the new elite media', not a radical expansion in who can act effectively in politics. Western political agendas have, arguably, changed very little in the past two decades in spite of an undoubted extension in the range of political actors: a neoliberal policy agenda has remained largely constant during this period (Leys, 2001; Crouch, 2011). The explanation may lie in three underlying dimensions: authority, evaluation and framing. Regardless of the expansion of political actors, political change requires changes in the distribution of political authority, which in turn must be grounded in transformations of how people evaluate political action, which in turn depends on how society's concerns and needs get framed. Such framing of 'the political' requires alliances across the divides of gender, age, ethnicity and class. Changing how 'the political' is framed is difficult, and this reduces the chances for even the most resourceful new networked actors to wield wider political influence. Jeffrey Juris's excellent study (Juris, 2008) of the anti-globalization movement brings out the resulting frustrations for such actors. When we consider the best candidates for large political actors which have emerged online, the US Tea Party movement had only 67000 members at its peak (in 2010) and even so two crucial advantages: its campaigning worked with, not against, the grain of dominant regimes of evaluation in US society (pro-market, anti-state, pro-local action), and it benefited from rich corporate supporters.[6] By contrast, the Occupy movement in 2011–2012 (which lacked the Tea Party's major advantages) did benefit from online resources (social media, particularly Twitter), but was just as crucially based on the physical gathering of people in prominent meeting places.

How do we understand this hybridity (Chadwick, 2007)? Juris (2012) argues that Occupy was characterized by two parallel logics: a decentred 'logic of aggregation' which draws people to congregate in physical sites, running in parallel with the long-familiar and more centrally driven 'logic of networking'. For Juris, Twitter is distinctive for its encouragement of the former, but why is not exactly clear. After all, the fast international growth in February 2003 of the protests against the Iraq War was based on websites and mass media, before social media even existed. Bennett and Segerberg go further, arguing that it is the personalized sharing of content via social media that is crucial, changing the 'core dynamics of [political] action' (2012: 739). The personalized feel of 'mass self-broadcasting' (Castells, 2009), when carried out intensively, can mobilize people on the basis of 'weak ties' (Granovetter, 1973), without any context of institutional membership, so bypassing the normal obstacles to institutional recruitment and accelerating the scaling-up of collective action-frames (Bennett and Segerberg, 2012: 746). This is an interesting proposal for a sociological grounding to new forms of 'connective action' (as Bennett and Segerberg, 2012 call them), but it will need to be tested out in more empirical research.

Meanwhile, it is easy to forget two key points. Firstly, that the same forces that have potentially extended the cast-list of politics benefit all political institutions, including established political actors: they too can use social media, and organize short-term disruptive campaigns. Secondly, other aspects of digital media make the survival of political institutions harder. Leaks of information are easier: digital media make political institutions porous in new ways (Bimber, 2003). This poses challenges for existing political institutions (parties, campaigns, NGOs, governments), but it also makes it harder for new political actors to build institutions that are stable. And in the long run it is new institutional actors that are needed if new directions of political action are to be sustained. Interruptive action is not enough.

To sum up, there are clear new opportunities for collaborative, indeed distributed, social and political production online, as part of an online activist public sphere. Under favourable conditions, today's public sphere can look less like a slow process of institutional referral and more like a rapid feedback loop, moving with alarming speed, like fire (*huo*) as the Chinese put it (Qiang, 2011). But the possibilities for transformative political action may, even so, be weighted towards short-term disruptive interventions and away from long-term projects of political construction.

Motives for Digital Politics

What about the 'why?' of politics, the reasons for political engagement and action? Is the digital environment changing why people get involved in politics? As Lance Bennett (2008: 3, 5) notes, introducing a recent study of young people's 'civic life online', we need to acknowledge a broad range of 'pathways to political engagement'. Political engagement is not an isolated phenomenon but has deep roots in underlying social contexts (of empowerment or disempowerment). 'Most people simply do not believe', Bennett (2008: 19) notes, 'that following and learning about various issues will translate into power to help decide them'. In this context, it remains striking that less than 50 per cent of those aged under 50 in Britain regard themselves as certain to vote in the next general election, although that figure has recently risen for the overall UK population (Hansard, 2011: 66). Apparently interactive sites may contribute little, for example, to young people's sense of engagement if those same young people do not feel that their contributions 'are being listened to' (Livingstone, 2007: 180). This gap between the promise of voice and the reality of political (non)-participation matters a great deal for whether people are motivated in the long term to act politically in the digital media environment (Couldry, 2010).

The interactivity of today's information interfaces is not itself the same as political engagement: what matters are genuine possibilities of sustained collaborative action. The French political theorist Pierre Rosanvallon (2008) makes the important point that today's politics involves more and more incitements to participate in forms of 'counter-politics' – the people as watchdogs, veto-wielders, judges (of politicians, of bureaucracy, of celebrities) – but less and less opportunities to participate in 'ordinary democracy', that is, sustained political action to achieve positive political goals. Rosanvallon's point is not just the obvious one that negative coalitions are always easier to organize than positive coalitions. He is concerned with a shift in the balance of incentives and disincentives that structure from below the landscape of political possibilities. Although Rosanvallon says little in detail about media, they are surely key enablers and amplifiers of this shift.

How can we think about this broad shaping role of media on political opportunity? The hugely increased incitements to discourse, by anyone online, create a supersaturated environment of media consumption from which individuals are even less likely to select a particular theme for attention and engagement. As Danilo Zolo (1992) pointed out for the age when multichannel TV was new, the premium on all forms of attention increases, which means that the likelihood of specifically political attention is even more heavily reduced. Yes, if we all stopped watching TV, we would have more time to follow the welter of political communication available online, but will that 'cognitive surplus' (Shirky, 2010) be enough in an age when the flow of information and opinion is effectively infinite? Inherited loyalties and 'interest-based political affiliation' may now count for less than short-term 'event-based' loyalties, with media institutions having a vested interest in pushing the latter (Bimber, 2003: 103). Meanwhile, Lance Bennett (2008: 11–14) argues, older patterns of citizen duty have been destabilized by institutional change. So clear signals about why we should engage politically become scarcer overall and on average.

The evidence about how people are actually using digital media about politics is disturbing. In a rich study of Argentinian news production and audiences, Pablo Boczkowski (2010) shows that a vast majority of the news audience read no blogs or online commentary, and barely do more than glance at the headlines on their homepage, leaving news links unclicked. Indeed as the internet unties the bundle of information that was the traditional newspaper, it has never been easier to move on (click) past political knowledge on the way to online entertainment (Prior, 2002: 145; Starr, 2009: 8).

If so, specific possibilities for political action are now greater and better resourced than in the pre-digital age, a shift that extends beyond rich countries; yet, for reasons that are connected, attention to and awareness of the political becomes on average harder to sustain, undermining long-term strategies of positive politics and new political institution-building. A huge amount of 'noise' in today's augmented media environment fills up the news cycle, just as it did in old-style media (Sambrook, 2010: 93), but overall, even if in absolute terms the incitements towards political engagement have increased, as a proportion of the wider media environment to which most people are exposed they have reduced, not increased. Put another way, the practical bases of 'public connection' (Couldry et al., 2010), while diversified, have overall been weakened.

What are the implications of this analysis for democratization? Let us return to Tilly's three conditions of democratization (Tilly, 2007). In the digital media age, Tilly's third condition (the reduction of the autonomy of non-state power centres from public politics) is partly enhanced. All institutions become more porous and open to media scrutiny and scandal, and so it becomes increasingly difficult for any institutional power centre to insulate itself from public politics. The status of Tilly's second condition (the openness of public politics to participants from any social category) is unclear: there is certainly evidence that participation in online politics and contention creation is skewed towards the wealthy and well educated (Schlozman et al., 2010; Schradie, 2011). In some locations (where gender inequality drastically restricts entry into politics), media spectacle may provide openings for challenges to old regimes of evaluation,[7] and the expanded range of political actors is potentially important, though its long-term significance is still not fully clear. Tilly's first condition (the integration of trust networks into public politics) is however unlikely to be enhanced in the digital media age, because the increasing information saturation of politics means there are fewer reasons to trust the institutions which offer consultation. Meanwhile more and more political decision-making is pulled beyond the spaces of democratic decision-making (Fraser, 2007).

The potential of digital media for democratization is at best ambiguous and partial, because the implications of a digital media environment for political engagement are unclear. The possibility of faster and more effective networks of the engaged will not itself increase average levels of political engagement.

CONCLUSION AND PRIORITIES FOR FUTURE RESEARCH

None of what I have argued so far denies the significance of the movements of political protest since 2010, the age of social media. These 'networks of outrage and hope' (Castells, 2012) have expressed the voices of many, interrupted politics-as-usual, and given hope, for a while, to countless others. In some Arab countries, they have led to changes in government, at the same time transforming the iconography of popular politics. But the evidence that, as a result, politics itself has fundamentally changed, or that the balance and nature of political power has changed decisively and permanently in any particular country (let alone those with historically more stable political institutions), is not yet available. As in other areas where everyday cultural and discursive practice appears to be being transformed by 'the digital', it is now, as Henry Jenkins notes, 'more and more urgent to develop a more refined vocabulary that allows us to better distinguish between different models of participation and to evaluate where and how power shifts may be taking place' (Jenkins, 2014). This cannot be disconnected from understanding 'how voters feel' (Coleman, 2013), that is, the social bases of individual political engagement.

Under these circumstances, there are three priorities for new research: (1) sustainability, that is, the resources and contexts through which individuals tend to become and stay active in networks; (2) demographics, that is, whether the type of people participating in digital networks is significantly different from those who were politically active in movements and campaigns of the pre-digital era; and (3) institutionalization, that is, the conditions under which online networks have transformed into stable political forces that directly compete for political power. It is interesting when leading researchers of social movements see recent social-media-based political actions as potentially antithetical to the emergence of common political positions (Juris, 2012: 272).

In addressing those questions against the often dramatic background of current events, we should bear in mind that, in the end, most dictators fall for one reason or another. Certainly social media were an accompanying factor on some recent occasions (Tunisia, Egypt), but that does not mean they were decisive. It is certainly not clear whether the conditions under which dictators remain in power have fundamentally changed as a result of digital media (Syria). It is unwise to draw long-term conclusions from short-term, if vivid, movements of despair. A sociology of media and politics must recognize the multiple forces of inertia that ordinarily (outside exceptional conditions of political anger) make decisive change difficult. The more important question long term is whether those inertia conditions have themselves been strengthened by digital media environments. If so, it is much too soon to tell whether digital media have overall changed the terms of political action.

ACKNOWLEDGEMENT

This chapter draws on some material from Chapter 5 of the author's *Media Society World* (Cambridge: Polity, 2012). Polity's permission to reproduce this material is gratefully acknowledged here.

NOTES

1. Jenkins (2006), Benkler (2006) and Castells (1996, 1997, 1998, 2009).
2. Dean (2010); and on broader constraints, see Pateman (1970), Croteau (1995) and LeBlanc (1999).
3. Keck and Sikkink (1998), Castells (1996), Lievrouw (2011: Ch. 6).
4. Haiqing (2007) and Qiang (2011) on Chinese blogs; Khiabany and Sreberny (2009) on

blogging in Iran; *Guardian*, 'Mohamed ElBaradei joins Egyptian sit-in over police death case', 25 June 2010, on Khaled Said; for sceptical views on blogging and elites, see Meikle (2009: Ch. 4); on elite political bloggers in the UK, see Davis (2009).
5.   http://webecologyproject.org, quoted in Sambrook (2010: 92).
6.   Chinni (2010), Monbiot (2010) and Skocpol and Williamson (2012) on the Tea Party.
7.   See Kraidy (2010) on reality TV in the Middle East.

FURTHER READING

For an important early essay, see Bentivegna (2002). Fundamental books are Castells (2009), Chadwick (2006), Coleman (2013) and Hindman (2009). Useful recent articles on the latest social-media-influenced movements are Bennett and Segerberg (2012) and Juris (2012).

REFERENCES

Barabasi, A.-L. (2003). *Linked*. Harmondsworth: Penguin.
Benkler, Y. (2006). *The Wealth of Networks*. New Haven, CT: Yale University Press.
Bennett, L. (2008). Changing Citizenship in a Digital Age. In Bennett, L. (ed.) *Civic Life Online* (pp. 1–24). Cambridge, MA: MIT Press.
Bennett, L. and Segerberg, A. (2012). The Logic of Connective Action: Digital Media and the Personalization of Contentious Politics. *Information, Communication and Society*, 15(5), 739–768.
Bentivegna, S. (2002). Politics and New Media. In Lievrouw, L. and Livingstone, S. (eds) *Handbook of New Media*, 1st edn (pp. 50–61). London: Sage.
Bimber, B. (2003). *Information and American Democracy*. Cambridge: Cambridge University Press.
Boczkowski, P. (2010). *News at Work*. Chicago, IL: Chicago University Press.
Brevini, B., Hintz, A. and McCurdy, P. (eds) (2013). *Beyond Wikileaks*. Basingstoke: Palgrave Macmillan.
Castells, M. (1996). *The Rise of the Network Society*. The Information Age: Economy, Society and Culture, Vol. 1. Oxford: Blackwell.
Castells, M. (1997). *The Power of Identity*. The Information Age: Economy, Society and Culture, Vol. 2. Oxford: Blackwell.
Castells, M. (1998). *End of Millennium*. The Information Age: Economy, Society and Culture, Vol. 3. Oxford: Blackwell.
Castells, M. (2009). *Communication Power*. Oxford: Oxford University Press.
Castells, M. (2012). *Networks of Outrage and Hope*. Cambridge: Polity.
Chadwick, A. (2006). *Internet Politics*. Oxford: Oxford University Press.
Chadwick, A. (2007). Digital Network Repertoires and Organizational Hybridity. *Political Communication*, 24, 283–301.
Chadwick, A. (2012). Recent Shifts in the Relationship Between Internet and Democratic Engagement in Britain and the US: Granularity, Informational Exuberance and Political Learning. In Arviza, E., Jensen, M. and Jorba, L. (eds) *Digital Media and Political Engagement Worldwide*. Cambridge: Cambridge University Press.
Chinni, D. (2010). Tea Party Mapped: How Big is it and Where is it Based?. 21 April. http://www.pbs.org/newshour/rundown/2010/04.
Coleman, S. (2013). *How Voters Feel*. Cambridge: Cambridge University Press.
Coleman, S. and Blumler, J. (2009). *The Internet and Democratic Citizenship*. Cambridge: Cambridge University Press.

Couldry, N. (2009). New Online News Sources and Writer-gatherers. In Fenton, N. (ed.) *New Media, Old News* (pp. 138–152). London: Sage.

Couldry, N. (2010). *Why Voice Matters*. London: Sage.

Couldry, N., Livingstone, S. and Markham, T. (2010). *Media Consumption and Public Engagement*. Rev. edn. Basingstoke: Palgrave Macmillan.

Croteau, D. (1995). *Politics and the Class Divide*. Philadelphia, PA: Temple University Press.

Crouch, C. (2011). *The Strange Non-death of Neoliberalism*. Cambridge: Polity.

Davis, A. (2009). Elite News Sources, New Media and Political Journalism. In Fenton, N. (ed.) *New Media, Old News* (pp. 121–137). London: Sage.

Dean, J. (2010). *Democracy and Other Neoliberal Fantasies*. Durham, NC: Duke University Press.

Fischer, C. (1992). *Calling America*. Berkeley, CA: University of California Press.

Fraser, N. (2007). Transnationalizing the Public Sphere. *Theory, Culture and Society*, 24(4), 7–30.

Gladwell, M. (2010). Small change: why the revolution will not be tweeted. 4 October. *New Yorker*.

Granovetter, M. (1973). The Strength of Weak Ties. *American Journal of Sociology*, 78, 1360–1380.

Haiqing, Y. (2007). Blogging Everyday Life in Chinese Internet Culture. *Asian Studies Review*, 31(4), 423–433.

Hansard Society (2011) *8th Audit of Political Engagement*. http://www.hansardsociety.org.uk/blogs/publications/archive/2011/04/08/audit-of-political-engagement-8.aspx.

Hardt, M. and Negri, T. (2005). *Multitude*. Harmondsworth: Penguin.

Hindman, M. (2009). *The Myth of Digital Democracy*. Princeton, NJ: Princeton University Press.

Jenkins, H. (2006). *Convergence Culture*. New York: New York University Press.

Jenkins, H. (2014). Rethinking 'Rethinking Convergence/Culture'. *Cultural Studies*, 28(2), 267–297.

Juris, J. (2008). *Networking Futures*. Durham, NC: Duke University Press.

Juris, J. (2012). Reflections on #Occupy Everywhere: Social Media, Public Space and Emerging Logics of Aggregation. *American Ethnologist*, 39(2), 259–279.

Kavada, A. (2010). Email Lists and Participatory Democracy in the European Social Forum. *Media Culture and Society*, 32(3), 355–372.

Kavada, A. (2014). Transnational Civil Society and Social Movements. In Wilkins, K., Tufte, T. and Obregon, R. (eds) *Handbook of Development Communication and Social Change* (pp. 351–369). Malden: Blackwell-Wiley.

Keck, M. and Sikkink, K (1998). *Activists Beyond Borders*. Ithaca, NY: Cornell University Press.

Khiabany, G. and Sreberny, A. (2009). The Internet in Iran: The Battle over an Emerging Virtual Public Sphere. In Goggin, G. and McLelland, M. (eds), *Internationalizing Internet Studies* (pp. 196–213). London: Routledge.

Kraidy, M. (2010). *Reality Television and Arab Politics*. Cambridge: Cambridge University Press.

LeBlanc, R. (1999). *Bicycle Citizens*. Berkeley, CA: University of California Press.

Leys, C. (2001). *Market-driven Politics*. London: Verso.

Lievrouw, L. (2011). *Alternative and Activist New Media*. Cambridge: Polity.

Lim, M. and Kann, M. (2008). Politics: Deliberation, Mobilization and Networked Practices of Agitation. In Varnelis, K. (ed.) *Networked Publics* (pp. 77–107). Cambridge, MA: MIT Press.

Livingstone, S. (2007). The Challenge of Engaging Youth Online: Contrasting Producers' and Teenagers' Interpretations of Websites. *European Journal of Communication*, 22(2), 165–194.

McDonald, K. (2011). The Old Culture of Rigid Ideologies is Giving Way to Individual Activism. 18 February. www.smh.com.au.

Meikle, G. (2009). *Interpreting News*. Basingstoke: Palgrave.
Monbiot, G. (2010). The Tea Party is Deluded and Funded by Billionaires. *Guardian*, 25 October.
Morozov, E. (2011). *The Net Delusion*. London: Allen Lane.
Mosco, V. (2004). *The Digital Sublime*. Cambridge, MA: MIT Press.
Pateman, C. (1970). *Participation and Democratic Theory*. Cambridge, Cambridge University Press.
Prior, M. (2002). Efficient Choice, Inefficient Democracy. In Cranor, L. and Greensetin, S. (eds) *Communications Policy and Information Technology* (pp. 143–179). Cambridge, MA: MIT Press.
Qiang, X. (2011). The Rise of Online Public Opinion and its Political Impact. In Shirk, S. (ed.) *Changing Media, Changing China* (pp. 202–224). Oxford: Oxford University Press.
Rosanvallon, P. (2008). *Counter-Democracy*. Cambridge: Cambridge University Press.
Sambrook, R. (2010). *Are Foreign Correspondents Redundant? The Changing Face of International News*. Oxford: Reuters Institute.
Schlozman, K., Verba, S. and Brady, H. (2010). Weapon of the Strong? Participatory Inequality and the Internet. *Perspectives on Politics*, 8, 487–510.
Schradie, J. (2011). The Digital Production Gap: The Digital Divide and Web 2.0 collide. *Poetics*, 39, 145–168.
Skocpol, T. and Williamson, V. (2012). *The Tea Party and the Remaking of American Conservatism*. Oxford: Oxford University Press.
Shirky, C. (2008). *Here Comes Everybody*. London: Allen Lane.
Shirky, C. (2010). *Cognitive Surplus*. London: Penguin.
Starr, P. (2009). Goodbye to the Age of Newspapers (Hello to a New Era of Corruption). *New Republic*, 4 March. http://www.tnr.com/print/article/goodbye-the-age-nespapers-hello-new-era-corruption.
Tilly, C. (2007). *Democracy*. Cambridge: Cambridge University Press.
Turner, B. (2005). Classical Sociology and Cosmopolitanism: A Critical Defence of the Social. *British Journal of Sociology*, 57(1), 133–151.
Zolo, D. (1992). *Democracy and Complexity*. Cambridge: Polity.
Zuckerman, E. (2013). Beyond 'The Crisis in Civics'. http://dmlcentral.net/blog/ethan-zuckerman/beyond-%E2%80%9C-crisis-civics%E2%80%9D.

4. The Fifth Estate: a rising force of pluralistic accountability
William H. Dutton and Elizabeth Dubois

INTRODUCTION

Digital technologies offer new options for the practice of politics. The concept of a Fifth Estate is based on findings that show how citizens are enabled by the Internet to hold institutions like the government and the press more accountable. Members of the Fifth Estate independently source information and connect with each other in ways that enhance their communicative power vis-à-vis institutions, such as government. Though potentially empowering, a focus on the role of networked citizens of the Fifth Estate proposes has rarely been the approach of scholars in studies of digital politics or in the practice of introducing digital tools to politics. Some scholars have identified the potential of the Internet to enable the many to monitor the few, what Bauman (1999) has called a 'synopticon', but this is only part of a broader range of strategies that can enhance the communicative power of the Fifth Estate.

The debate over direct versus representative democracy has flourished as scholars investigate the political ramification of increasing Internet use by citizens. For example, the impetus behind proposals for electronic initiatives for more direct forms of democracy is most often tied to dissatisfaction with the limitations of voting every four or five years for a candidate or a political party. How can such a vote provide any meaningful direction to an elected official on the myriad public policy choices presented to them in the course of holding office? This has always been the case, but it has struck many as increasingly unnecessary given that digital infrastructures are becoming more available and permit individuals to register their opinions on specific issues at almost any time. This is approximated already by the activities of online polling organizations and by initiatives such as Her Majesty's (HM) Government's e-petitions, which allows the public to create and sign petitions online.[1] Capabilities for registering opinions on policy issues are also being developed outside of government, by networked individuals, as with a petition site called '38 Degrees'.[2]

Nevertheless, opposition to calls for more direct online democracy is overwhelming, most often tied to two general concerns. One is a fear of direct forms of participation undermining the dialogue and debate that occurs with more representative forms of democracy. The second is a concern over equity and the disenfranchisement of citizens who are not online, which is about one-fifth of citizens in Britain, and nearly a quarter of US citizens, two countries with relatively high levels of Internet access.

Such reservations over direct forms of digital governance have shifted efforts to design technical innovations towards those that will reinforce and enhance representative democracy, such as consultation between citizens and their elected officials (Coleman and Shane, 2011). Open government initiatives have been another target of developments that could support democratic institutions, such as enabling governments to publish more information online in forms that the public can more easily access, such as through innovations in open linked data.

As enduring as this debate over direct versus representative democracy may be, it has ignored the rise of a new organizational form, which we have called the 'Fifth Estate' (Dutton, 2007, 2009; Dutton and Dubois, 2013). This new form poses the most realistic potential for enhancing democratic governance in the digital age. The following sections explain the concept of a Fifth Estate, providing an illustration of how the Internet does indeed create the potential for enhancing democratic accountability, but in ways that the advocates of direct and representative democracy have largely ignored. It does not replace representative government, but promises to hold representatives more accountable. It does not require universal access to the Internet, only a critical mass of citizens online.

However, this democratic potential is not an inevitable outcome of technical change, and could well be at risk from a number of developments that could undermine the Fifth Estate. Therefore, after describing the Fifth Estate and the empirical research that supports the emergence of this new organizational form, we briefly outline some of the most serious risks posed by contemporary developments. Knowing the potential of a Fifth Estate, and the developments that threaten its vitality, it might be possible to realize the Internet's 'gift to democracy' (Dutton, 2011).

THE CONCEPT OF A FIFTH ESTATE

The concept of a Fifth Estate is defined here to refer to the ways in which the Internet is being used by increasing numbers of people to network with other individuals and with information, services and technical resources in ways that support social accountability across many sectors, including business and industry, government, politics and the media. Just as the press created the potential for a Fourth Estate in the eighteenth century, the Internet is enabling a Fifth Estate in the twenty-first.

The concept of 'estates of the realm' originally related to divisions in feudal society between the clergy, nobility and the commons. Much licence has been taken with their characterizations over the centuries. For example, American social scientists linked the three estates with Montesquieu's tripartite conception to be identified with the legislative, executive and judicial separation of powers. However, the eighteenth-century philosopher Edmund Burke first identified the press as a Fourth Estate, arguing (according to Carlyle, 1905): 'there were Three Estates in Parliament; but, in the Reporters' Gallery yonder, there sat a Fourth Estate more important far than they all. It is not a figure of speech, or witty saying; it is a literal fact – very momentous to us in these times'.

The rise in the twentieth century of press, radio, television and other mass media consolidated this reality as a central feature of pluralist democratic processes in countries where the press gained an independence from government and commercial control. Growing use of the Internet, Web 2.0, mobile information and communication technologies (ICTs) and other digital online capabilities is creating another new 'literal fact': networked individuals reach out across traditional institutional and physical boundaries into what Castells (2001) has called a 'space of flows', rather than a 'space of places', reflecting contemporary perspectives on governance processes as 'hybrid and multijurisdictional with plural stakeholders who come together in networks' (Bevir, 2011: 2).

Within this hybrid space of flows, people increasingly go to find information and services. These could be located anywhere in the world and relate to local issues (for example, taxes, political representatives, schools) and regional, national or international activities. This is significant because it signals that more and more people are likely to go first to a search engine or to a site recommended through their favourite social networking site, rather than to a specific organization's site or to a place, such as to a local government office, library, newspaper, university or other institution (Dutton, 2009).

When access to information is no longer under the control of institutions the door is open for others to exert their own influence and control in new ways. In other words, there are potential shifts in communicative power that the Internet is enabling citizens to advance. In short, members of the Fifth Estate are Internet-enabled networked individuals who, independent of institutions, access resources and people in ways that enhance their communicative power. This power allows them to hold other institutions more accountable, such as government, the press and businesses.

Challenging Traditional Approaches

The concept of a Fifth Estate challenges major perspectives on the political role of the Internet in society. Some view the Internet as an unimportant political resource; a technical novelty or passing fad. A number of other perspectives assert some form of determinism, such as seeing computer-mediated communication as inherently a technology of freedom (de Sola Pool, 1983) or, quite the opposite, as inherently centralizing (Hindman, 2008). A third set of perspectives essentially assert some level of 'reinforcement politics' (Danziger et al., 1982). This perspective is that the Internet can be designed and used to support and reinforce many different political structures and processes, whether more or less democratic or autocratic. For example, Evgeny Morozov (2011) falls into this perspective, arguing that authoritarian states have used the Internet to reinforce their control of citizens.

Fifth Estate theory differs from each of these more traditional perspectives. It takes the Internet as a significant political resource that is changing patterns of governance across multiple sectors, but it does not view this impact as inevitable or an inherent feature of the technology, but as a pattern of use observed over time that can be undermined by other estates. It differs from reinforcement politics by not seeing any single actor to be in control of the Internet and its political use and implications. Fifth Estate theory is anchored in a realization that control over the Internet as a complex, large-scale, global ensemble of technologies, is shaped by an ecology of actors. Pluralistic interplay enables the Fifth Estate to claim a new role in governance, but like any other actor, it is limited by this larger ecology of strategies and choices by being made by multiple actors.

The concept of a Fifth Estate suggests that the choices and uses of the Internet by multiple actors will lead to differing political outcomes depending on the situation. Through a growing range of digital and Internet-enabled technologies, including search, social networking sites, e-mail, texting and tweeting, individuals are reshaping not only how they connect with information, people and services, but also what they know, who they communicate with, and what services they access (Dutton, 1999). In doing so, individuals seamlessly move across the boundaries of existing institutions, thereby sourcing their own information and networks in ways that open new opportunities for calling to account politicians, journalists, experts and other loci of power and influence.

Communicative Power Strategies

Members of the Fifth Estate are able to hold others to account, within sectors ranging from government to media to business, by strategically using the Internet to support their communicative power. These members of the Fifth Estate are able to source information in new ways, access different people and groups, and connect with others in order to advance goals, independent of institutions.

Table 4.1 provides an overview of the five main ways in which members of the Fifth Estate can enhance their communicative power in relation to institutional actors. A growing variety of cases demonstrate the potential for these strategies to enable new forms of accountability.

Searching and sourcing information are perhaps the most obvious forms of Fifth Estate action. The Internet enables individuals to source information and groups by reconfiguring access: changing the ways people seek out information, services, and people as well as the outcome of these activities (Dutton, 1999). The ability to search and access desired

Table 4.1   Examples of potentially effective forms of Fifth Estate networking

Fifth Estate forms	Description	Examples
Searching	Individual finds information through search or their social network	Voter locates information about an elected official and policy issue
Sourcing	Individual creates information	Bloggers post original information, images, observations, such as during the Arab Spring
Networking	Individuals join self-selected network(s)	Networks form around local or global political issues, such as Move On, or health and medical care, such as through social networks of patients
Leaking	Individuals leak information in ways that can be publicly accessible online	Whistleblowers reach sites that distribute information, such as WikiLeaks, and OpenLeaks
Collective intelligence	Platform aggregates information from individuals, or organizations	Environmental sensing in Beijing; churnalism; 38 Degrees
Collective observation	Platform aggregates observations, experiences	Bribery websites in India

information from a variety of sources, paired with the ability to produce original information, highlight the fact that members of the Fifth Estate can act independently of existing institutions.

A local-level example comes from a primary school. In 2012, a nine-year-old school girl in Scotland created a blog for her school writing project, called 'NeverSeconds'. She took a photo of her school lunches on her mobile phone, and posted them on her blog with commentary. NeverSeconds became popular, leading her school to request that she not bring her camera to lunch. This fuelled greater interest in her site and by 2015 garnered nearly 9 million page views, and generated funding for charities and a book with her father about her project (Payne and Payne, 2012). A single primary school student was able to produce her own information, distribute it globally and foster a debate that has led to change in the practices of her school and well beyond.

This example illustrates that the Fifth Estate is not equivalent to a social or political movement. One networked individual can play a critical role in holding an institution to higher standards of accountability. Even social movements, which often have strong hierarchical structures, must cope with the potential of networked individuals to source their own information and counter the narrative of a movement.

Networking as an effective form of Fifth Estate action is also important, and becoming more so. Networked individuals seek like-minded people online through such means as social networking sites, blogs and e-mail. With the Internet they bypass traditional barriers, such as space and time, in order to access people who lie beyond common geographical and institutional boundaries. By reconfiguring how people source information and networks of networked individuals, the Internet enhances their communicative power vis-à-vis institutions in the other estates of the Internet realm and sets the stage for increased accountability.

For example, a patient within a health service can use the Internet to source their own information, or join networks of other individuals with similar health or medical issues, such as a network of more than 500 families linked to the UK Children With Diabetes Advocacy Group. Likewise, physicians can rely on online institutional resources, but also explore new sources of information online and network with other physicians who are outside their own institutional setting, such as Sermo, one of the largest online communities for physicians in the USA.

The Fifth Estate is not equivalent to the mass media, but the strategies of bloggers and other Fifth Estate activities often entails gaining media coverage as one means to enhance their communicative power. There is a clear complementarity between the traditional media and the Fifth Estate. While the Internet's broad social roles in government, politics and other sectors may parallel those of traditional media at times (for example by whistleblowing) the Fifth Estate differs from traditional media in how it helps networked individuals drive opportunities for greater social accountability across institutions, occasionally setting the mass media agenda.

Likewise, the distributed and collective nature of the Fifth Estate means networked individuals are able to exert communicative power without relinquishing power to, or being dependent on, an established institutional body. This collective nature is illustrated most vividly by the ability of individuals to contribute to collective intelligence and collective or distributed observation. For example, collective intelligence such as through smartphone apps have enabled the creation of independent sources of information about pollution levels in cities. Individuals can pool their observations, such as used in the creation of bribery websites to monitor and map incidences of corruption. Such forms of distributed collaboration are not directly dependent on any one institutional source or any single estate. This is highlighted in Table 4.1 by references to collaborative intelligence and observation.

The Fifth Estate might be described in terms of networks of accountability rather than a formally organized institution. Internet use enables the creation of alternative sources of information and collaboration that are not directly dependent on any one institutional source or any single estate. Internet-enabled individuals, even those whose primary networking activities are social, can often create networks that span across standard geographical, organizational and institutional networks, to link with others online.

Networked individuals are enabled to build and exercise their 'communicative power' (Garnham, 1999) by using ICTs to reconfigure networks in ways that can lead to real-world power shifts. For instance, the relationship between media producers, gatekeepers and consumers are changed profoundly when previously passive audiences generate and distribute their own content and when search engines point to numerous sources reflecting different views on a topic.

Access to online resources that incorporate and go beyond more traditional institutions is supported by the increasing ubiquity of the Internet and related technology. Individuals can then network with information and people to change their relationships with more institutionalized centres of authority in the other estates, thereby holding them more socially accountable through the interplay between ever-changing networks of networks.

NETWORKED INSTITUTIONS AND INDIVIDUALS

Individuals are not the only ones making use of the Internet and related technologies. Indeed, institutions are also becoming increasingly

Table 4.2   A categorization of networked institutions and individuals

Arena	Networked individuals of the Fifth Estate	Networked institutions of the other estates
Governance and democracy	Social networking, net-enabled political and social movements, and protests	E-government, e-democracy
Press and media	Bloggers, online news aggregators, Wikipedia contributors	Online journalism, radio and TV, podcasting
Business and commerce	Peer-to-peer file sharing (for example, music downloads), collaborative network organizations	Online business-to-business, business-to-consumer (for example, e-shopping, e-banking)
Work and the organization	Self-selected work collaborations, systems for co-creation and distribution (for example, open source software)	Flatter networked structures, networking to create flexible work location and times
Education	Informal learning via the Internet, checking facts, teacher assessment	Virtual universities, multimedia classrooms, online courses
Research	Collaboration across disciplinary, institutional and national boundaries	Institutional ICT services, online grant proposal submissions

Source: Adapted from Dutton (2009: 7).

networked, using digital technology to advance their goals. Table 4.2 outlines examples of the comparative activities of networked individuals and networked institutions across a variety of sectors.

There are a variety of digital tools and many ways in which these technologies are used within social and political systems. This section takes a deeper look into the ways in which the networked individuals of the Fifth Estate and networked institutions interact within governance, the press, business, work and organization, and education and research to illustrate a pattern that extends to many other sectors not covered here.

Government and Democracy on the Line

Digital government initiatives, such as enabling citizens and businesses to obtain public services via the Internet, such as in paying taxes, have been paralleled by innovations in digital democracy and efforts to use the Internet to support democratic institutions and processes. Some critics suggest that digital democracy could erode traditional institutions of representative deliberative democracy by offering oversimplistic 'point and click' participation. These criticisms are based on an institution-centric view of the Internet, focused on voting or parliamentary consultations, for example, rather than considering the role of the individual outside of existing institutional processes, such as in sourcing their own information.

The Fifth Estate is not necessarily reliant on institutions and therefore presents new opportunities and threats. Networked individuals can challenge institutional authority and provide a novel means for holding politicians and mainstream institutions to account through ever-changing networks of individuals, who form and re-form continuously depending on the issue generating the particular network (such as to form ad hoc 'flash mob' meetings at short notice through social networks and mobile communication). An example is the use of texting after the March 2004 Madrid train bombings to organize anti-government rallies challenging the government's claims, contributing to unseating that administration by quickly providing people with important information and instruction which enabled mobilization. Pro-democracy protests across the Middle East and North Africa in early 2011 or Ukraine in 2014 further illustrate the potential for networked individuals to challenge governmental authorities and other institutions.

Governments, such as in the US and the UK, are making information available online in user-friendly forms as a key element of open government initiatives promoting greater transparency and accountability. These initiatives illustrate how other estates can support the role of the Fifth Estate. Examples of governments using technology to limit the power of the Fifth Estate are also extensive, such as by filtering and censoring the press and the Internet. Several cases described by Huan et al. (2013) show how networked individuals in China have been able to create and distribute their information online in ways that held the press and government authorities in China more accountable. Attempts to control access to digital information often fail, but such threats to the Fifth Estate remain real.

The Press

As citizen journalists, bloggers, information seekers and producers, networked individuals are increasingly able to contest claims made by traditional media sources, provide new details and perspectives on specific news items, and thus hold the Fourth Estate to account. Simultaneously, traditional media outlets are beginning to incorporate products of the Fifth Estate into their own reporting. Many news programmes include reviews of social media responses to an issue. Other programmes create and promote specific Twitter hashtags in order to generate and follow conversation related to their programme.

Yet, the relationship between the Fourth and Fifth Estate is murky. The Internet and its users are criticized for eroding the quality of the public's information environment and undermining the integrative role of traditional Fourth Estate media in society. This includes claims that the Internet is marginalizing high-quality journalistic coverage by proliferating misinformation, trivial non-information and propaganda created by amateurs (Keen, 2007), and creating 'echo chambers' where personal prejudices are reinforced as Internet users choose to access only a narrow spectrum from the vast array of content at their fingertips (Sunstein, 2007).

The traditional mass media embodies equivalent weaknesses (for example, a focus on sensational negative news stories, poor-quality reporting and celebrity trivia). There is an unjustified assumption that the Internet substitutes for, rather than complements, traditional media. Many Internet users read online newspapers or news services, although not always the same newspaper as they read offline (Dutton et al., 2009). Thus, the Internet is indeed a source of news that in part complements, or even helps to sustain, the Fourth Estate (Hindman, 2008).

Business Organizations and Work

The Fifth Estate has transformative potential at all levels in businesses and other private sector organizations. Geographically distributed individuals cooperating together to form collaborative network organizations (CNOs) to co-create or co-produce information products and services (Dutton, 2008) are one example. The online encyclopaedia Wikipedia and open source software products such as the Firefox web browser are examples of this phenomenon, becoming widely used and trusted despite initial doubts about the merits of their methods of creative co-production.

There are concerns that CNOs may blur the boundaries and operations of the firm, or undermine the firm's productivity. Instead, evidence suggests that individuals generally choose to join CNOs primarily to enhance their own productivity, performance or esteem (Benkler, 2006; Dutton, 2008; Surowiecki, 2004). As consumers become increasingly empowered to hold businesses accountable, such as through Internet-orchestrated boycotts, or better-informed consumer groups, the role of the Fifth Estate in business and industry will increase. Already, networked individuals are challenging the information practices of major Internet companies, such as Facebook and Google, and leading them to alter their approaches to protecting the privacy of their users and many other information policies, driven by what can easily be identified as Fifth Estate accountability.

Education and Research

E-learning networks often follow and reinforce prevailing institutional structures (for example, the teacher as the primary gatekeeper in a multimedia classroom), but they can move beyond the boundaries of the classroom and university (Dutton and Eynon, 2009). Students challenge teachers by introducing alternative authority positions and views through their networking with one another and with a variety of sources of knowledge. This can be a positive force, better engaging students in the learning process, or a disruption in teaching, depending on how well preparations have been made to harness online learning networks.

Universities are building campus grids, digital library collections, institutional repositories and online courses, such as massive open online courses (MOOCs), to maintain and enhance their productivity and competitiveness. Researchers are also increasingly extending their collaborations through Internet-enabled networking, often across institutional and national boundaries (Dutton and Jeffreys, 2010). These researchers are more likely to go to an Internet search engine before their library; as likely to use the Internet to support network-enabled collaboration as to meet their colleagues in the next office; and tend to post work on websites, such as disciplinary digital repositories, and blogs rather than in institutional repositories. Freely available social networking sites offer tools for collaboration that could be as, or more, useful to researchers than systems for collaboration in which universities and governments have invested much money. Academics are engaged in their own subsection of the Fifth Estate, for instance by online mobilization around both local issues (for example, university governance) and more international topics (for cxample, copyright and open science).

Interaction within the Fifth Estate

The Fifth Estate is not a homogenous group. Multiple actors within the Fifth Estate may interact in varying ways. The citizen response to a piece of Canadian legislation dubbed C30, which aimed to allow police access to subscriber information, is a prime example. When C30 was proposed by a minister, Vic Toews, in February 2012, a non-profit organization called OpenMedia had already established an online 'Stop Spying' petition in order to contest the bill. Framing C30 as 'An Act to Protect Children from Online Predators', the government and Minister Toews hoped to avoid conflict; however, networked individuals of the general public reacted swiftly and strongly in opposition to the bill despite the new framing (see Dubois and Dutton, 2013).

Multiple efforts contributed to the ultimate death of the bill. In addition to the online petition, Vikileaks, a Twitter account, was created with the motto 'Vic Toews wants to know about you, let's get to know him'. A less sinister Twitter hashtag, #TellVicEverything, became very popular as a humorous way for Canadians to make it easier for Toews by providing him with all the mundane details of their lives by tagging him in all of their tweets. And, the hacker group Anonymous produced a threatening YouTube video. Links and references to these various acts were shared across multiple media including social networking sites and within the traditional media. In this case interaction among estates is evident, but so too is interaction among various members of the Fifth Estate. Ultimately it was this interaction that helped demonstrate to Minister Toews and the government the extent to which the Canadian public was against C30, contributing to the death of the bill.

IMPLICATIONS OF A FIFTH ESTATE

The evidence supporting the emergence of a Fifth Estate is growing, and a consideration of its potential implications for both the citizenry and for governance systems is much needed. While the Internet could be an empowering tool for networked individuals, because it enables them to harness communicative power and thus hold other institutions to account, there are threats as well.

For example, the enhanced communicative power of networked individuals has led to efforts to censor and otherwise control the Fifth Estate, including calls for disconnecting the Internet, blocking sites such as Twitter, and arresting bloggers. This mirrors familiar forms of governmental control of traditional media, such as bans, closures and the arrest of journalists. The Internet's opening of doors to an array of user-generated content equally allows in techniques deployed by governments and others to block, monitor, filter and otherwise constrain Internet traffic (Deibert et al., 2008). These include, notably, government efforts to control Internet content, such as the 'Great Firewall of China', the Burmese government's closing down of the country's Internet service during political protests in 2007, and efforts by a number of governments to block Internet access and create a 'kill switch' to block the Internet.

Networked individuals continue to challenge attempts to control Internet access and circumvent censorship. The site www.herdict.org accepts and publishes reports from Internet users of inaccessible websites around the world, and the OpenNet Initiative and Reporters Sans Frontiers support worldwide efforts to sustain and reinforce the Internet's openness. At times networked individuals use the Internet in order to hold governments to account when the traditional media is not in a position to do so. In the case of a July 2010 gas explosion in Nanjing, the capital of Jiangsu Province in China, individuals contested the traditional media's reporting of the incident as well as the government's explanation of how the accident happened (Huan et al., 2013). Although the blast was the biggest in the area since 1949, little to no coverage of the event was found within the traditional media. Reports that did appear across newspapers were inconsistent, and bloggers and other netizens challenged many claims, rapidly disseminating pictures of the explosion, challenging information and proactively encouraging people to donate blood as the local hospital quickly ran through its supply. Members of the online public used the Internet and other ICTs – specifically blogs, Twitter and text messages – to criticize the media, provide alternative information and support to each other, and call the government to account.

This example from China points to another important consideration. Because Internet use has become an increasingly central aspect of everyday life and work in networked societies, disparities in access to the Internet are of substantive social, economic and political significance. However, despite digital divides, the Internet has achieved a critical mass in many nations, enabling networked individuals to become a significant force for accountability. The existence of a Fifth Estate does not depend on universal access, but on reaching a critical mass of users. This enables the Fifth Estate to play an important political role even in nations such as India with relatively low proportions of Internet users, but many users.

Similarly, when considering the implications of a Fifth Estate on governance systems broadly, we see that not all citizens need harness the communicative power of the Fifth Estate for it to be a meaningful political force. Individuals make use of the Internet and related tools in varying ways, some of which will contribute to demands for accountability. It is these members of the Fifth Estate who are challenging governance structures, at times indirectly, when holding others within the governance system to account. Members of the Fifth Estate are inserting themselves into political conversations that were once reserved for institutional bodies and the political elite, by using the Internet and related tools to harness their communicative power.

The implications of the Fifth Estate range from empowering individuals, to impacting upon policy, to potentially changing the balance of power among political players in governance systems. The following section briefly summarizes our thesis and discusses important next steps for Fifth Estate research.

SUMMARY AND DIRECTIONS FOR FUTURE RESEARCH

This role of the Internet in 'reconfiguring access' illustrates the centrality of the information politics of the digital age (Dutton, 1999). How the Internet potentially reconfigures access is shaped by patterns of 'digital choices' (Dutton et al., 2007), which can impact upon the communicative reach of individuals, groups and nations.

The impact of networked individuals seeking to shape access to and from the outside world, in local and global contexts, has supported the rise of the Fifth Estate, but also the role of the Internet in all the other estates. In this sense, the Fifth Estate is the unintended outcome of an ecology of choices by multiple actors, rather than a specific organizational form that people seek to join. This is demonstrated, for instance, in the strategies of government agencies, politicians, lobbying groups, news media, bloggers and others trying to gain access to citizens over the Internet, countered by networked individuals seeking to source their own information and networks. What individuals know is one outcome of this ecology of choices and strategies that creates a Fifth Estate role. This chapter has argued that distinctive features of the Fifth Estate make it worthy of being considered a new estate of at least equal importance to the Fourth, in addition to being the first estate not to be essentially institution-centric.

There are three main directions for research on the Fifth Estate. Firstly, an increased understanding of how members of the Fifth Estate interact and collaborate is important, such as might be gained through case studies of its role in particular political events. While many researchers focus on the ways in which social movements and campaigns make use of the Internet, these are only one strategic form of Fifth Estate activity. Few researchers have examined the many other ways in which individuals use the Internet to hold accountable institutions from a variety of sectors. The focus on the individual in their personal network is an important perspective and one which potentially provides insight into political systems.

Second, research into how governance systems and specific institutions do and do not respond to and integrate the Fifth Estate is needed. While research looking at how various institutions use the Internet and how individuals use the Internet exists, there is a lack of understanding about how each might integrate the other into their strategic planning, daily activities, and/or organizational structure.

Finally, systematic research on the emergence of the Fifth Estate in varying contexts is paramount. The potential implications of economic, social, cultural and political factors are not well known in the context of Fifth Estate research and must be developed further. Policy changes, such as in Internet regulation, could dramatically reshape the vitality of the Fifth Estate and need to be tracked over time.

ACKNOWLEDGEMENT

This chapter progresses work on the Fifth Estate by the authors (Dutton, 2009 and 2011; Dubois and Dutton, 2013).

NOTES

1. http://epetitions.direct.gov.uk/.
2. http://www.38degrees.org.uk/.

FURTHER READING

Dubois, E. and Dutton, W.H. (2013), 'The Fifth Estate in Internet Governance: Collective Accountability of a Canadian Policy Initiative', *Revue française d'Etudes Américaines RFEA*, 4, 81–97.
Dutton, W.H. (2009), 'The Fifth Estate Emerging through the Network of Networks', *Prometheus*, 27(1), 1–15.
Dutton, W.H. (2011), 'A Networked World Needs a Fifth Estate', *Wired Magazine*, 22 October, http://www.wired.co.uk/magazine/archive/2011/11/ideas-bank/william-dutton.
Dutton, W.H. and Eynon, R. (2009), 'Networked Individuals and Institutions: A Cross-Sector Comparative Perspective on Patterns and Strategies in Government and Research', *Information Society*, 25(3), 1–11.
Newman, N., Dutton, W.H. and Blank, G. (2012), 'Social Media in the Changing Ecology of News: The Fourth and Fifth Estates in Britain', *International Journal of Internet Science*, 7(1), 6–22.

REFERENCES

Bauman, Z. (1999), *In Search of Politics*, London: Polity.
Benkler, Y. (2006), *The Wealth of Networks: How Social Production Transforms Markets and Freedom*, New Haven, CT: Yale University Press.
Bevir, M. (2011), 'Governance as Theory, Practice, and Dilemma', pp. 1–16 in Bevir, M. (ed.), *The SAGE Handbook of Governance*, London: Sage.
Carlyle, T. (1905), *On Heroes: Hero Worship and the Heroic in History*. Repr. of the *Sterling Edition of Carlyle's Complete Works*, Teddington, UK: Echo Library.
Castells, M. (2001), *The Internet Galaxy*, Oxford: Oxford University Press.
Coleman, S. and Shane, P.M. (eds) (2011), *Connecting Democracy: Online Consultation and the Flow of Political Communication*, Cambridge, MA: MIT Press.
Danziger, J.N., Dutton, W.H., Kling, R. and Kraemer, K.L. (1982), *Computers and Politics*, New York: Columbia University Press.
de Sola Pool, I. (1983), *Technologies of Freedom*, Cambridge, MA: Harvard Press.

Deibert, R., Palfrey, J., Rohozinski, R. and Zittrain, J. (eds) (2008), *Access Controlled*, Cambridge, MA: MIT Press.

Dubois, E. and Dutton, W.H. (2013), 'The Fifth Estate in Internet Governance: Collective Accountability of a Canadian Policy Initiative', *Revue française d'Etudes Américaines RFEA*, 4, 81–97.

Dutton, W.H. (1999), *Society on the Line: Information Politics in the Digital Age*, Oxford, UK and New York, USA: Oxford University Press.

Dutton, W.H. (2007), 'Through the Network (of Networks) – the Fifth Estate', Inaugural lecture, Examination Schools, University of Oxford, 15 October. Available at http://webcast.oii.ox. ac.uk/?view=Webcast&ID=20071015_208.

Dutton, W.H. (2008), 'The Wisdom of Collaborative Network Organizations: Capturing the Value of Networked Individuals', *Prometheus*, 26(3), 211–230.

Dutton, W.H. (2009), 'The Fifth Estate Emerging through the Network of Networks', *Prometheus*, 27(1), 1–15.

Dutton, W.H. (2011), 'A Networked World Needs a Fifth Estate', *Wired Magazine*, 22 October, http://www.wired.co.uk/magazine/archive/2011/11/ideas-bank/william-dutton.

Dutton, W.H. (forthcoming), 'The Internet's Gift to Democratic Governance: The Fifth Estate', forthcoming in Coleman, S., Moss, G. and Parry, K. (eds), *Can the Media Save Democracy? Essays in Honour of Jay G. Blumler*, London and Abington: Palgrave.

Dutton, W.H. and Dubois, E. (2013), 'The Fifth Estate of the Digital World', pp.131–143 in Youngs, G. (ed.), *Digital World: Connectivity, Creativity and Rights*, London: Routledge.

Dutton, W.H. and Eynon, R. (2009), 'Networked Individuals and Institutions: A Cross-Sector Comparative Perspective on Patterns and Strategies in Government and Research', *Information Society*, 25(3), 1–11.

Dutton, W.H., Helsper, E.J and Gerber, M.M. (2009), 'The Internet in Britain: The Oxford Internet Survey 2009', Oxford: Oxford Internet Institute.

Dutton, W.H., Shepherd, A. and di Gennaro, C. (2007), 'Digital Divides and Choices Reconfiguring Access', pp.31–45 in Anderson, B., Brynin, M., Gershuny, J. and Raban, Y. (eds), *Information and Communication Technologies in Society*, London: Routledge.

Garnham, N. (1999), 'Information Politics: The Study of Communicative Power', pp.77–78 in Dutton, W.H. (ed.), *Society on the Line*, Oxford, UK and New York, USA: Oxford University Press.

Hindman, M. (2008), *The Myth of Digital Democracy*, Princeton, NJ: Princeton University Press.

Huan, S., Dutton, W.H. and Shen, W. (2013), 'The Semi-Sovereign Netizen: The Fifth Estate in China', pp.43–58 in Nixon, P.G., Rawal, R. and Mercea, D. (eds), *Politics and the Internet in Comparative Context: Views From the Cloud*, London: Routledge.

Keen, A. (2007), *The Cult of the Amateur: How Today's Internet is Killing Our Culture*, New York: Doubleday.

Morozov, E. (2011), *The Net Delusion: How Not to Liberate the World*, London: Penguin Books.

Payne, M. and Payne, D. (2012), *NeverSeconds: The Incredible Story of Martha Payne*, Glasgow: Cargo Publishing.

Sunstein, C.R. (2007), *Republic.com 2.0*, Princeton, NJ: Princeton University Press.

Surowiecki, J. (2004), *The Wisdom of Crowds: Why the Many Are Smarter than the Few and How Collective Wisdom Shapes Business, Economies, Societies and Nations*, New York: Doubleday.

5. Silicon Valley Ideology and class inequality: a virtual poll tax on digital politics[1]
Jen Schradie

A growing and widely held assumption is that digital communication technologies enable average citizens to participate in politics more easily, actively and directly than in traditional ways. The Internet in this framework, from blogs and video posts to social media and mobile apps, breaks down barriers for political participation. The idea is that the Internet is a non-hierarchical democratic space where people can access, create and act on political information in a broader range of activities, whether as part of election campaigns, online petitions, digital activism or even revolutionary change. In this thinking, it is increasingly the individual who participates in digital politics without the involvement of a civic group or political party. Put together, this creates a more democratic scenario in which individuals exercise freedom of expression in a networked, horizontal and participatory digital network without bureaucratic, organizational or state intervention.

I call this philosophy around the Internet 'Silicon Valley Ideology', which has proliferated along with the mass diffusion of social media technologies. The contradiction in this ideology, however, is that these institutionalized beliefs in egalitarian political participation mask the realities of structural inequalities. Silicon Valley Ideology builds on theories that challenge the folly of free markets (Somers, 2008), Internet utopianism (Barbrook and Cameron, 1995) and egalitarian citizenship (Marshall, 1950).

Silicon Valley Ideology is part of a broader 'free market fundamentalism', which Somers (2008) described as a belief that free markets fix everything. In this case, though, it is a free market fundamentalism that the Internet inevitably produces more democratic forms of participation. It is a free market fundamentalism embedded in society, corporations and other institutions. And it is a free market fundamentalism tethered to neoliberalism, as both the Internet and Silicon Valley Ideology grew symbiotically with the rise of neoliberalism over the last few decades.

Certainly, not everyone who celebrates the non-hierarchical architecture of the Internet aligns with the neoliberal agenda of capitalism. Still, this egalitarian philosophy of the Internet clashes with social class inequalities in a free market economy. People without resources fall through the digital cracks. The question, then, is whose voices are part of this online political platform and who is left out?

DIGITAL CITIZENSHIP AS POLITICAL, CIVIL AND SOCIAL

Up until the 1960s many southern US counties forced voters to pay poll taxes and take literacy tests to discourage voting. While some working class whites were excluded, the intention was to prevent African-Americans from going to the polls. As a result, their issues and concerns were not represented in city councils, county commissions, and state legislatures, nor in congressional and presidential elections. But the civil rights movement was about more than just the right to vote, or political citizenship (Marshall, 1950), and it was eventually about more than civil citizenship, which Marshall described as equal and individual freedoms under the law. When Martin Luther King, Jr was assassinated in Memphis, Tennessee in 1968, he was supporting striking sanitation workers for economic justice, an example of what Marshall called social citizenship. The political struggle for all of these rights goes hand in hand with social class divisions in capitalist societies: contradictions between social inequality and citizenship rights are inevitable with market-driven economies (Marshall, 1950; Somers, 2008).

While the ballot box is still a critical space for all levels of citizenship, many forms of political activity have moved and merged online into a form of digital politics via the Internet and other forms of new media technologies. Political news, information, debates, as well as ways to communicate with politicians and engage in political and social movement activism, have both sped up and spread across digital networks. Digital politics go beyond the ballot box into Marshall's more wide-ranging notions of citizenship. Yet social class inequalities prevent many from exercising their citizenship rights in this digital system that is open only to those with the access, knowledge, labor and power to use them.

Digital inequality, often called the digital divide, is a way to talk about how some groups of people do not have the means to use the Internet, or other digital technologies, at the same rate as other groups. While race, ethnicity, gender and age are strong factors in predicting Internet use (Hargittai, 2008; Jones et al., 2009), social class differences are the most consistent over time (Martin and Robinson, 2007; Schradie, 2012). These class inequalities are persistent in Internet access and consumption of online content, as well as in digital content production (Correa, 2010; Hargittai and Walejko, 2008; Schradie, 2011). Inequality in online politics and activism is no different. These class inequalities, then, drive a digital politics gap. This chapter traces these differences.

Structural inequalities in the United States create virtual poll taxes for those who do not control the means of digital production in online political spaces. Digital political inequalities are not deliberate like historical and contemporary voter suppression tactics are, such as poll taxes. These class inequalities are rooted in and inextricably linked to neoliberal capitalism (Couldry, 2010; Harvey, 2005; Somers, 2008). This economic and political system of personal liberty and initiative fosters the Silicon Valley Ideology of the Internet's entrepreneurial spirit of individuation and atomization. If you can pull yourself up by your digital bootstraps, you can engage in digital politics. Often absent from the idea of digital citizenship (for example, Mossberger et al., 2008a) is any situated and structural understanding of digital inequality. People are left to their own digital devices, if they have them, to participate in online politics. The Internet as a level playing field for exercising citizenship rights is part of the neoliberal framework, but the reality is the exacerbation, not the amelioration, of political inequalities.

Rather than directly causing digital inequality, then, Silicon Valley Ideology is part of an evolving neoliberal system (Tugal, 2012), much like Gramsci's (2005) fluid conceptualization of how the economy, state and civil society are tied together and foster hegemonic ideologies. In this case, then, the broader neoliberal system creates and is sustained by inequality, and Silicon Valley Ideology serves as justification for the exclusionary segregation of the digital politics gap.

ORIGINS OF SILICON VALLEY IDEOLOGY

Silicon Valley, a former fruit farming area in Northern California, is likewise fertile ground for the corporations that control the most popular Internet platforms, such as Google and Facebook. It is also home to Silicon Valley Ideology. In 1995, Barbrook and Cameron critiqued *Wired* magazine for promoting what they called a California Ideology, which is a 'profoundly anti-statist dogma'. They argued that California embodies this Internet utopian philosophy of individualism and the free market, at the expense of those from more marginalized classes.

I build on their conceptualization of this California Internet dream but call it 'Silicon Valley Ideology', not just to specify the corporate headquarters of digital neoliberalism but also to expand the definition to include the development of an institutionalized attachment to non-hierarchical and horizontal online participation. With the proliferation of social media platforms over the past two decades, Silicon Valley Ideology also differs from California Ideology with the critical addition of the massification, institutionalization and corporatization of networked technology platforms and devices. In 1995, it was mostly digital elites who used the Internet, and *Wired* was a prominent yet niche magazine of California Ideology. But 2006 marked the dawn of the mass diffusion of social media: Twitter launched; Google bought YouTube; Facebook became available to the general public; and *Time Magazine* named 'you' as person of the year for participating in this online content production system. This represented a transition to widespread, yet still classed, public practices. Critically, though, these public digital practices have also become connected to institutions in civil and corporate society, enabled by the hands-off neoliberal state. For instance, since then, many political movements embraced and celebrated digital technology, from anti-globalization protests to the Arab Spring; many technology companies became the most highly valued in the world, such as Apple; and governments, especially in the United States, kept the digital economy largely free from taxes and regulations. In other words, an idea of egalitarian participation became an ideology, as it became embedded in pervasive social media practices and institutions. This is not to say that all of these digital practices have been non-hierarchical, just that social media use has been broadly diffused and that social media use is tethered to the *belief* in horizontal participation.

The digital age has often been called a networked society of individuals (Benkler, 2006; Castells, 2010; Raine and Wellman, 2012). A critical piece of this argument is that the architecture of the Internet is horizontal and non-hierarchical. The implication, therefore, is that the artifact itself has the politics of democratic participation (Winner, 1980). Others contend that technology more directly shapes political freedoms and democracy (Pool, 1983), which is a form of technological determinism. On the other hand, some scholars and journalists have pointed out how the Internet's corporate ownership and capitalist features (Fuchs, 2013; McChesney, 2013; Morozov, 2011) counter democratic claims. What is missing in all of these arguments, however, is that it ultimately depends on whether people can and do go online and participate. While most digital media theorists acknowledge the existence of inequalities, the emphasis is on the general affordances of digital technology. The Internet, as networked and non-hierarchical by design, presupposes that people are already wired, engaged and have the skills, practices and social, political and economic support to be active digital citizens. However, at the core of any non-hierarchical digital network are digital elites, who are early adopters, have more resources and dominate the digital public sphere. A networked neoliberal society privileges the individual user, leaving it up to someone's personal circumstances as to whether or not that individual has the resources and motivation to engage in digital politics.

Despite Silicon Valley Ideology's namesake, the broader Bay Area played a role in this ideology. Turner (2006), in his historical account of the political and cultural origins of the Internet, chronicled how the New Communalist movement, as part of the San Francisco hippy countercul- ture movement of the 1960s in northern California, spawned participa- tory Internet communities, including the WELL, an early and influential online discussion forum. Turner took issue with Barbrook and Cameron's claim and the popular understanding that the New Left and other radical movements inspired techno-utopianism. Turner explained that while both movements rejected hierarchy, it was the New Communalists who also challenged the bureaucratic order with the use of collaborative technol- ogy, rather than completely rejecting the technical and military–industrial complex. According to Turner, the Cold War between the US and the former Soviet Union inspired the 1960s youth who grew into Internet pioneers, especially Stuart Brand, co-founder of the WELL, an early and influential online discussion forum: 'The liberation of the individual was simultaneously an American ideal, an evolutionary imperative, and, for Brand and millions of other adolescents, a pressing personal goal' (Turner, 2006: 45).

And, in fact, the hacking culture of the Defense Advanced Research Projects Agency (DARPA), the military origin of the Internet, influenced this belief that individuals can creatively build a more accessible intercon- nected technological system, such as the University of California–Berkeley UNIX hacker Bill Joy, who co-founded Sun Microsystems (McKusick, 1998).

This post-Fordist ideology of distributed, non-bureaucratic systems has had a strong libertarian bent. The evangelists, corporate or otherwise, of the Internet promoted how we can all be ourselves, communicate, network, share information and even engage in online political action without state interference. Still, proponents of the networked non-hierarchical aspects of the Internet have various political orientations and are not all adherents to each facet of neoliberalism. In fact, much of the scholar- ship on the Internet's non-hierarchical characteristics has not come from conservative thinkers. Therein lies the contradiction inherent in Silicon Valley Ideology; or perhaps, then, it is Silicon Valley Ideologies. On the one hand, there is the democratic belief in diverse equality and broad participation on the political left. On the other hand, the political right's democracy emphasizes free markets and individual liberties. The Internet crosses this political divide, as does Silicon Valley Ideology, and is where the more technological deterministic theories of the affordances-spectrum fall flat. Affordances are what an object allows people or systems to do. The concept of affordances can take into account societal differences, yet how can the architecture of the Internet beg for more egalitarianism yet also rely on an individualism that creates inequalities? Silicon Valley Ideology is not a simple political left or political right orientation, then, nor is it simply a belief in these ideas. It is part of a broader articulation, or connection, of a broader neoliberal state, economy and society. Rather than floating in its own democratic space, Silicon Valley is embedded in everyday practices that are rife with political contradictions.

The Internet is often misconstrued as a blanket statement of being more democratic. The question remains: more democratic than what? The 'other' in this oft-repeated analysis is usually the mainstream one-to-many media. Certainly, the Internet has social media platforms for more participatory and, therefore, democratic communication than traditional media outlets. With other political formats, such as interactions with the state or social movements, the claim of the Internet as being more democratic becomes much fuzzier because of the digital politics gap.

THE MYTHS OF DIGITAL INEQUALITY

What does it take for someone to participate in any sort of online political activity? I will explain the body of research on how mainstream politics, as well as digital activism, or online collective action, intersects with inequality. First, though, it is essential to tackle some oft-repeated myths of the digital divide in order to understand better the mechanisms of the digital politics gap.

Myth #1: The Digital Divide is Over

Three arguments are often presented to bolster the myth that digital inequalities, or in this case online politics gaps, are or will soon be over in the United States: firstly, digital differences are simply a lagged effect as a process of diffusion; secondly, digital natives will completely replace older non-users; and thirdly, digital inequality is just a Third World problem. Data, however, show otherwise.

With a lens of market-based microeconomics, a common story of Internet adoption is based on an individual making a rational choice whether or not to go online. Eventually, the story of diffusion goes, everyone will be using the Internet. Indeed, Internet adoption has generally followed the traditional path of technology diffusion (Rogers, 1962). First, a small group of early elite adopters began using the Internet in the early 1990s. Then, more and more people went online through market saturation as costs went down and the utility went up. Still, in 2011, 94 percent of college-educated Americans used the Internet but only 43 percent of people without a high school education were online, and 62 percent of people who make less than $30 000 per year were online, while 97 percent of those making over $75 000 used the Internet. Simply, one in five American adults did not use the Internet (Zickuhr and Smith, 2012). For many of these 'late adopters' it is not simply an individual rational choice as to whether or not to go online. Constraints of connectivity and hardware costs, as well as limited digital skills and literacy, prevent people from being online consistently. For instance, Robinson (2009) tracked low-income students and found that one-quarter of the high school students in her study were online less than an hour per week and struggled to find and get to publicly available computers, whether at school or in libraries.

Yet some argue that it is just a matter of time before digital inequality disappears along the adoption curve because of market forces (for example, Compaine, 2001), and others contend that these inequalities are not even consequential (Block, 2004). Differential Internet connectivity rates in this model are expected initially, as any technological innovation has a small group of early adopters (Fischer, 1992). Yet gaps are supposed to taper off over time in a typical S-curve, which would make digital inequality simply a lagged effect. Early users tend to hail from privileged backgrounds, but according to diffusion models, eventually everyone will catch up. Others, though, challenge this predictive model, given that Internet connectivity diffusion rates are dramatically slower for the lowest income levels in the USA (Martin and Robinson, 2007). Also, unlike the eventual widespread adoption of televisions and telephones, Internet platforms are not just about access and ownership of a technology but also about the participation in and production of online content. For the Internet, it is not only a question of the persistence of the Internet access gaps, as newer gadgets or social media platforms, especially for content production, continue to emerge. This leaves the poorest Americans on a treadmill, never able to catch up.

Another contention is that once older, less wired people die, then the digital divide will be over, as younger digital natives take over. Well, not quite. Certainly, young people are online more than those from older generations (for example, Lenhart et al., 2007), and youth are more likely to participate in online politics than older generations (Mossberger et al., 2008a). In fact, one study focused on how the digital divide has disappeared, yet it only studied youth and sampled primarily from people already online (Cohen and Kahne, 2012). But class divides persist across age brackets and types of online activities. For instance, even among people who are online, blogging gaps do not close over time between high school and college-educated Americans (Schradie, 2012). And even at the start of their online lives, youth from more marginalized backgrounds have different, and less beneficial, digital practices than their wealthier counterparts (Robinson, 2009; Sims, 2014).

Another claim that is made to show that digital inequality is not consequential in the United States or other more developed countries is to compare Internet use rates with less developed countries. Yes, the divide is stark between the global North and South. For instance, in 2013, 77 percent of Europeans had Internet access while just 7 percent of Africans did, although the United States, for instance, overall ranked fourteenth for broadband per capita (Telecommunication Development Bureau, 2013). Certainly, gaps are larger in the Global South, but that does not mean that everyone in the Global North has access. The gaps between elites and non-elites are consistent over time in the United States (Witte and Mannon, 2010). While digital inequality is more pronounced in developing countries, class differences persist in countries like the United States because of the structural inequalities inherent in neoliberal economic systems, especially those with less social support for the poor.

Myth #2: The Digital Divide is a Divide

Online political inequality is not a simple divide between those who have any level of Internet access and those who lack it altogether, or between those with broadband access and those without it. These are questions of the consumption of digital content. Inequality is also based on the production of online content (Correa, 2010; Hargittai and Walejko, 2008; Schradie, 2011). Production is about the creation of blogs, YouTube videos and Tweets, for instance. In fact, it is this more active and agentic online participation that is a hallmark of one's political voice within the neoliberal Silicon Valley Ideology. Posting political content is critical to expressing one's views and opinions. While some see the blurring between production and consumption (Ritzer and Jurgenson, 2010), this division is not so blurry for the poor and working class, who are much less likely to ever have participated in content creation.

Scholars who have examined the general adult population have found digital production inequality between people with more advanced educational degrees and those with less education. Even among people online with Internet access, production gaps persist. Hargittai and Walejko

(2008) find that social class backgrounds affect content creation both online and offline.

A variety of factors lead to a digital production gap between elites and non-elites. One main reason for this inequality is not being able to control the means of digital production. More important for digital content production than broadband access is the number of gadgets one owns (Schradie, 2011). It is about whether you have one desktop computer shared among multiple households, or if people in a household have individual gadgets to use at their own disposal. Having online access at a variety of locations (that is, home, work, mobile) and owning many gadgets allow people high levels of autonomy to control the means of digital political production and content creation. Owning one cell phone is not enough. People who have autonomous access in multiple places are more likely to create content (Hargittai, 2007; Hassani, 2006; Schradie, 2011), and it is people with higher education levels who have control over their digital labor practices. Conceptualizing the Internet as a space for labor production (Fuchs, 2013; Terranova, 2000) sheds light on the often-ignored yet intuitive fact that people who do not have a surplus of time are not producing content at the same rate as others.

Other factors for digital content production inequality are variations in skills or confidence levels in using digital technologies (Correa, 2010; Hargittai, 2009; Van Deursen and Van Dijk, 2011). But as Sims (2014) pointed out, having access and skills does not always lead to the same levels of digital practices. Simply put, there are many people whose voices are consistently absent from political discussions in the Twitter-sphere.

While the term 'inequalities' is an apt description for a range of skills, practices, connectivity or gadgets, what does remain is a singular digital divide over time: an online politics gap based on social class with almost every measure of online engagement.

Myth #3: Race Gaps have Disappeared

On the one hand, African-Americans' digital citizenship practices can be viewed as a success story. African-Americans who are online are often more likely to use social media than whites. For instance, a higher proportion of blacks who are online tweet than whites (Hargittai and Litt, 2011). However, what is often missing in this discourse is that blacks are still less likely to be online in the first place. Blacks are more likely than whites to produce content, such as blogging at a rate twice as much as whites, if they are already online, but overall they are less likely to consume content, or simply have Internet access (Schradie, 2012). Poor African-Americans are rarely able to view what elite African-Americans post online, nor produce content themselves.

In fact, a weakness in some of the digital inequality literature, let alone broader new media scholarship, is the failure to adequately study those who are not online. While it is impossible to report on Internet use among people who do not have access, making claims of racial or class equality based solely on what people online are doing is inaccurate at best, as it further marginalizes and biases people who are not included in research findings.

A related myth about the African-American community is that there is no digital divide because they are 'leapfrogging'. This is a concept that is often used to describe people, usually in developing countries, who never had landlines or desktop computers because of the lack of infrastructure but are able to 'leapfrog' over these technologies with mobile devices, for instance. But the global leapfrog effect is more myth than reality (Howard, 2007), so even though African-Americans are more likely than whites to have a smart phone, as sociologist Shelia Cotton asked, 'Could you type a 10 page paper on your phone?'[2] However smart it might be, newer, smaller, sleeker gadgets, such as tablets, are designed more for consumption rather than for producing and engaging and participating with online political content (Jarvis, 2010). Certainly, many people tweet and post status updates with their phones, but mobile devices are not always 'smart', nor do they substitute for the range of gadgets that the political elite use.

In other words, while American society is quite racialized, there are also strong class divisions, or the lack of equality in social citizenship. Even if everyone had access to a laptop, tablet or a smart phone, other levels of class inequalities would persist. Gaps vary by digital medium, but also by what people do with that medium. Are they simply consuming political content, or are they participating in politics and social movements online, or are they perhaps heavy producers of multimedia political content? The next section traces the research on class inequality with both electoral politics and online collective action. For the individual citizen in a market economy, who may not have the means to go online, or participate or produce online, engaging in digital politics may not always be viable. The result is a virtual poll tax and digital literacy test.

THE DIGITAL POLITICS GAP

Some of the ways in which poor and working-class people are denied full civic and political citizenship using the Internet is rooted in what theorists celebrate as the Internet's agentic participatory potential (Jenkins, 2006; Negroponte, 1995; Rheingold, 1993). Despite the Internet being hailed as a non-hierarchical space for political participation opened up to the masses, differences persist between those who participate in digital politics and those who do not (Neuman et al., 2010), creating a digital politics gap.

Research on how inequality maps onto digital politics has generally focused on how individuals participate in mainstream electoral politics, from learning about candidates online to e-mailing political information to friends, or contacting officials online to making campaign contributions (Mossberger et al., 2008a; Mossberger et al., 2003; Norris, 2001; Schlozman et al., 2012; Smith, 2013). Smith (2013) finds social class divisions with political activity on social media. Given class-based online gaps with both consumption and production, these inequalities are part of the digital political space. Hindman (2009) found that online political content is mostly produced by elites, and as Mossberger et al. (2008a: 50) noted, 'Existing disparities are simply replicated in cyberspace.' Political elites can better harness digital tools (Jennings and Zeitner, 2003; Margolis and Resnick, 2000). Other scholars, though, argue that there is an expansion of inequalities in the digital age as more political activity goes online (van Dijk, 2005; di Gennaro and Dutton, 2006). There is general agreement, though, that the poor and working class are less likely to participate in online electoral politics than those with higher education and income levels.

The virtual poll tax extends beyond electoral politics. Most of the inequality scholarship on digital political participation focuses on the individual as the unit of analysis. Inequality researchers often ask some variant of 'What is the likelihood that individuals from a certain class (or race or gender) engage in digital politics?' But civic and political citizenship also encompasses protest and civil society organizing. Another framework to examine structural inequalities and digital politics, therefore, is from the standpoint of collective action, or digital activism. Earl and Schussman (2003) argue that digital activism is about individual users protesting, rather than members of organizations.[3] Many studies have examined the individual in relationship to digital activism. People with more income and education are more likely to participate in online civic engagement activities (Brodock et al., 2009; Smith, 2013). 'The Internet is principally used by the higher educated, and those with a full-time job, with a lot of interest in politics and with more experience in previous demonstrations' (Van Laer 2010: 356). In a case study, Le Grignou and Patou (2004) found that online organizing simply reinforces societal hierarchies of experts. In fact, they contend that electronic tools even help to increase the gaps between experts and non-experts.

Studies of social movement organizations, however, have also more directly challenged the neoliberal atomization of individualistic proclivities.

But the results are the same. Activist groups with fewer resources tend to use Internet tools, including social media platforms, less than organizations with more resources (Eimhjellen et al., 2013; Merry, 2011; Schradie, 2013). Some organizations attempt to include people without access by having offline materials available, or even sometimes providing digital technology for activists without access (Pickerill, 2003).

In 2001, Norris argued that political movements of the have-nots could potentially harness the power of the Internet for social change. Others contend that social inequalities would prevent more marginalized communities from using the Internet to challenge those in political power (Tilly, 2004; Warf and Grimes, 1997), as elites also have access to the Internet (Donk et al., 2004; Dordoy and Mellor, 2001). In other words, if the Internet is a weapon of the elite (Schlozman et al., 2010), how could the weak win if they had access to fewer weapons?

The data so far do not show that the Internet can overcome offline political inequalities and increase participation. Much like the knowledge gap thesis (Eveland and Scheufele, 2000; Tichenor et al., 1970; Tolbert and McNeal, 2003), the consensus is that the gaps in political participation persist rather than narrow in the digital age. Some suggest that the Internet could deepen inequalities (van Dijk, 2005), not simply reflect and reproduce inequality. Comparative scholarship, however, is scant on the topic. Certainly, the Internet is a much more open architecture than the gatekeepers of county boards of elections, which can restrict voting. Still, structural inequalities in online spaces result in broad-based poll taxes beyond the voting booth. Social citizenship, then, is at stake in the digital age.

NEOLIBERALISM, THE VIRTUAL POLL TAX AND CITIZENSHIP

The rise of the Internet has not only coincided with the rise of technological advancements, but it has also run parallel with the rise of neoliberal, market-based politics and economies (Hassan, 2008) in which individual rights are at the core of neoliberal ideology. However, class-based capitalist societies are not based on the individual but on hegemonic forces linking the market, the state and civil society (Gramsci, 2005; Harvey, 2005; Somers, 2008). Yet part of the Silicon Valley Ideology is to keep the state's hands off the Internet and not to intervene, censor or monitor personal Internet activities. Yet this philosophy is in conflict with any state support for full social citizenship, Internet or otherwise. Citizenship in a neoliberal age of digital politics often marginalizes the poor and working class, resulting in a digital politics gap.

The Internet platforms that most Americans use, however open their initial architecture or design, are now mostly owned, controlled and commodified by corporations (Fuchs, 2013; McChesney, 2013; Youmans and York, 2012). This is the irony in Silicon Valley Ideology: the belief in the individual as primary, yet institutions, especially corporate, continue to dominate. Therefore, I extend Somers's argument of how 'market fundamentalism' has created 'increasing numbers of people [who] have lost meaningful membership in civil society and political community – that which confers recognition and rights – through a process of the contractualization and commodification of citizenship' (Somers, 2008: 118). I argue that the Internet has only accelerated this process of losing citizenship rights for people from lower classes due to digital inequality. Virtual poll taxes encompass costs of capital, labor, time, education, information, motivation and other class constraints that block online participation in politics for the poor and working class.

Poll taxes were intentionally meant to keep people from voting. The Internet, despite the commercialization of various online platforms, is the opposite: its design and architecture were not meant to restrict but to expand participation. The connection between the Internet and poll taxes is the Silicon Valley Ideology of free markets, free speech and the 'marketplace of ideas': the Internet itself has no constraints, like historical poll taxes or literacy tests, in restricting political participation. In reality, though, because of its neoliberal architecture that does not dissolve class boundaries, it is not entirely democratic (Harvey, 2005). So in this sense the poll tax is indeed 'virtual'. It excludes invisibly, whereas the previous poll taxes were an active and very visible means of exclusion.

So what of citizenship and digital inequality? The term 'digital citizenship' (Mossberger et al., 2008b) aptly describes the ways in which people use, and cannot use, digital technology for political engagement. Absent from this conceptualization of Marshall's citizenship framework is how the digital is part of a broader economic and political system, rather than solely a communicative practice in mainstream politics. The neoliberal market-based system itself is what prevents equality in digital politics. The connections across citizenship, class, the state and the economy are inextricably linked to neoliberalism. We cannot disentangle the digital from society and society from the digital.

Both Marshall and Somers theorized that citizenship and class have an unresolved tension in market-based economies. I extend this to the digital age: with the Internet, if you are connected, in the broad sense, you can engage with certain levels of political and civic citizenship. Otherwise, social citizenship, based on market 'failures', leaves you offline politically: with the failure to make Internet access available for all, the failure to give people enough leisure time to participate online, and the failure to give people enough education and capital to engage politically online.

While much of the Internet in general, and social media tools in particular, were built for commercial purposes, they secondarily gained political signif-icance as spaces for political discussion and action. Yet, as Schattschneider pointed out in 1960, 'The flaw in the pluralist heaven . . . is that the heavenly chorus sings with a strong upper-class accent' (Schattschneider, 1960: 35). These structural inequalities predate the Internet, as well as predate the rise of neoliberalism in the 1980s. Yet these gaps are almost harder to pin down since they are hidden beneath the veneer of Silicon Valley Ideology. The extent to which the Internet and its inequalities marginalize certain classes of people are less the insidious work of elites focused on political sup-pression, like poll taxes and literacy tests, but rather the continuation and exacerbation of inequality that persists and widens online. Digital politics are not the panacea for the limitations of a (neo)liberal democracy. This ideology, then, does more than mask the digital politics gap. Often couched in emancipatory and egalitarian language, Silicon Valley Ideology actively perpetuates and reproduces inequalities.

FUTURE RESEARCH ON THE DIGITAL POLITICS GAP

Any research on politics and the Internet, especially digital democracy questions and claims, must acknowledge, at minimum, various levels of digital inequalities. In more than a footnote, sociologists, communication scholars and other Internet researchers need to recognize the limitations of studying digital elites or even simply those with consistent connectivity. Questions such as the effects of digital politics on the 'unwired' or how digital activism and inequality intersect are but two of the many ways in which scholarship could expand in the digital political inequality area. This line of research could lead to broader questions of power that chal-lenge Silicon Valley Ideology. We also need more precise comparative research on whether or not the digital era may ameliorate, exacerbate or reproduce specific political inequalities.

Furthermore, understanding class inequality, and any intersections with race, ethnicity or gender, is particularly essential with any analysis of Big Data. The enormous reams of information from digital repositories, such as social media platforms, have become a goldmine for researchers of political communication. But generalizable claims for this deluge of data are limited by those who do not create online content or produce very little. In essence, Big Data is too small.

Finally, analyses of digital politics and inequality need anchoring in theoretical constructs, with an eye toward understanding the connections between the state, economy and civil society. Without insight into broader questions of how society operates vis-à-vis power and inequality, any claims of the Internet's effect on politics is an argument made in a vacuum, rather than in society.

NOTES

1. Support through the ANR – Labex IAST is gratefully acknowledged.
2. From a presentation at the American Sociological Society Annual Meeting, Las Vegas, NV, August 22, 2010.
3. This also further extends Skocpol's argument of the trend from civic groups in the middle of last century to membership-based advocacy organizations.

FURTHER READING

To further understand the theoretical ideas behind Silicon Valley Ideology, neoliberalism and various forms of citizenship, the following books are invaluable: Marshall (1950), Somers (2008) and Turner (2006). For solid foundations in online political inequality, consult the following books: Mossberger et al. (2008a), as well as Schlozman et al. (2010). For understanding the nuances of digital inequality, the following articles are useful: Robinson (2009) and Sims (2014).

REFERENCES

Barbrook, R. and Cameron, A. (1995). The California Ideology. Mute Magazine, August.
Benkler, Y. (2006). The Wealth of Networks. New Haven, CT: Yale University Press.
Block, W. (2004). The 'Digital Divide' is Not a Problem in Need of Rectifying. Journal of Business Ethics, 53(4), 393–406.
Brodock, K., Joyce, M. and Zaeck, T. (2009). Digital Activism Survey Report. Boston, MA: DigiActiv.
Castells, M. (2010). The Rise of the Network Society. Chichester: Blackwell Publishing.
Cohen, C.J. and Kahne, J. (2012). Participatory Politics: New Media and Youth Political Action. Los Angeles, CA: MacArthur Research Network on Youth and Participatory Politics.
Compaine, B. (2001). The Digital Divide: Facing a Crisis or Creating a Myth? Cambridge, MA: MIT Press.
Correa, T. (2010). The Participation Divide Among 'Online Experts': Experience, Skills and Psychological Factors as Predictors of College Students' Web Content Creation. Journal of Computer-Mediated Communication, 16(1), 71–92. doi:10.1111/j.1083-6101.2010.01532.x.
Couldry, N. (2010). Why Voice Matters: Culture and Politics After Neoliberalism. London: Sage Publications.
Di Gennaro, C. and Dutton, W. (2006). The Internet and the Public: Online and Offline Political Participation in the United Kingdom. Parliamentary Affairs, 59(2), 299–313.
Donk, W.B.H.J. van de, Loader, B.D., Nixon, P.G. and Rucht, D. (2004). Introduction:

Social Movements and ICTs. In van de Donk, W.B.H.J., Loader, B.D., Nixon, P.G. and Rucht, D. (eds), Cyberprotest: New Media, Citizens and Social Movements (pp. 1–22). London, UK and New York, USA: Routledge.

Dordoy, A. and Mellor, M. (2001). Grassroots Environmental Movements: Mobilisation in an Information Age. In Webster, F. (ed.), Culture and Politics in the Information Age: A New Politics (pp. 167–182). London: Routledge.

Earl, J. and Schussman, A. (2003). The New Site of Activism: On-Line Organizations, Movement Entrepreneurs, and the Changing Location of Social Movement Decision-Making. Research in Social Movements, Conflict, and Change, 24, 155–187.

Eimhjellen, I., Wollebæk, D. and Strømsnes, K. (2013). Associations Online: Barriers for Using Web-Based Communication in Voluntary Associations. VOLUNTAS: International Journal of Voluntary and Nonprofit Organizations, 123. doi:10.1007/s11266-013-9361-x.

Eveland, W.P. and Scheufele, D. (2000). Connecting News Media Use with Gaps in Knowledge and Participation. Political Communication, 17(3), 215–237. doi:10.1080/105846000414250.

Fischer, C.S. (1992). America Calling: A Social History of the Telephone to 1940. Berkeley, CA: University of California Press.

Fuchs, C. (2013). Class and Exploitation on the Internet. In Scholz, T. (ed.), Digital Labor: The Internet as Playground and Factory (pp. 211–224). New York: Routledge.

Gramsci, A. (2005). Selections from the Prison Notebooks of Antonio Gramsci. Hoare, Q. and Nowell-Smith, G. (eds), New York: International Publishers.

Grignou, B. Le and Patou, C. (2004). ATTAC(k)ing Expertise: Does the Internet really Democratize Knowledge? In van de Donk, W.B.H.J., Loader, B.D., Nixon, P.G. and Rucht, D. (eds), Cyberprotest: New Media, Citizens, and Social Movements (pp. 145–158). London, UK and New York, USA: Routledge.

Hargittai, E. (2007). Whose Space? Journal of Computer-Mediated Communication, 13(1), 14.

Hargittai, E. (2008). The Digital Reproduction of Inequality. In D. Grusky (ed.), Social Stratification (pp. 936–944). Boulder, CO: Westview Press.

Hargittai, E. (2009). Skill Matters: The Role of User Savvy in Different Levels of Online Engagement. July 1, Cambridge, MA: Berkman Center for Internet and Society.

Hargittai, E. and Litt, E. (2011). The Tweet Smell of Celebrity Success: Explaining Variation in Twitter Adoption among a Diverse Group of Young Adults. New Media and Society, 13(5), 824–842.

Hargittai, E. and Walejko, G. (2008). The Participation Divide: Content Creation and Sharing in the Digital Age. Information, Communication and Society, 11(2), 239–256.

Harvey, D. (2005). A Brief History of Neoliberalism. Oxford: Oxford University Press.

Hassan, R. (2008). The Information Society: Cyber Dreams and Digital Nightmares. Cambridge: Polity Press.

Hassani, S.N. (2006). Locating digital divides at home, work, and everywhere else. Poetics, 34(4–5), 250–272.

Hindman, M. (2009). The Myth of Digital Democracy. Princeton, NJ: Princeton University Press.

Howard, P.N. (2007). Testing the Leap-Frog Hypothesis: The Impact of Existing Infrastructure and Telecommunications Policy on the Global Digital Divide. Information, Communication and Society, 10(2), 133–157. doi:10.1080/13691180701307354.

Jarvis, J. (2010). iPad Danger: App v. Web, Consumer v. Creator. Buzzmachine, April 4.

Jenkins, H. (2006). Convergence Culture: Where Old and New Media Collide. New York: New York University Press.

Jennings, M.K. and Zeitner, V. (2003). Internet Use and Civic Engagement. Public Opinion Quarterly, 673, 311–334.

Jones, S., Johnson-Yale, C., Millermaier, S. and Pérez, F.S. (2009). US College Students' Internet Use: Race, Gender and Digital Divides. Journal of Computer-Mediated Communication, 14(2), 244–264. doi:10.1111/j.1083-6101.2009.01439.x.

Lenhart, A., Madden, M., MacGill, A. and Smith, A. (2007). Teens and Social media. Pew Research Internet Project, December 19.

Margolis, M. and Resnick, D. (2000). Politics as Usual: The Cyberspace 'Revolution'. Thousand Oaks, CA: Sage Publications.

Marshall, T.H. (1950). Citizenship and Social Class and Other Essays. Cambridge: Cambridge University Press.

Martin, S.P. and Robinson, J.P. (2007). The Income Digital Divide: Trends and Predictions for Levels of Internet Use. Social Problems, 54(1), 1–22. doi:10.1525/sp.2007.54.1.1.

McChesney, R. (2013). Digital Disconnect: How Capitalism is Turning the Internet Against Democracy. New York: New Press.

McKusick, M.K. (1998). Twenty Years of Berkeley Unix: From AT&T-Owned to Freely Redistributable. In DiBona, C., Ockman, S. and Stone, M. (eds), Open Sources: Voices from the Open Source Revolution (pp. 31–46). Malvern, PA: Free Software Foundation.

Merry, M.K. (2011). Interest Group Activism on the Web: The Case of Environmental Organizations. Journal of Information Technology and Politics, 8(1), 110–128. doi:10.10 80/19331681.2010.508003.

Morozov, E. (2011). Net Delusion. London: Allen Lane, Penguin Books.

Mossberger, K., Tolbert, C.J. and McNeal, R.S. (2008a). Digital Citizenship: The Internet, Society, and Participation. Cambridge, MA: MIT Press.

Mossberger, K., Tolbert, C.J. and McNeal, R.S. (2008b). Digital Citizenship: The Internet, Society and Participation. Journal of the American Society for Information Science and Technology, 59: 221. doi:10.1002/asi.

Mossberger, K., Tolbert, C.J. and Stansbury, M. (2003). Virtual Inequality: Beyond the Digital Divide, American Governance and Public Policy. Washington, DC: Georgetown University Press.

Negroponte, N. (1995). Being Digital. New York: Random House.

Neuman, R.W., Bimber, B. and Hindman, M. (2010). The Internet and Four Dimensions of Citizenship. In Jacobs, L.R. and Shapiro, R.Y. (eds), Oxford Handbook of American Public Opinion and Media (pp. 22–42). Oxford: Oxford University Press.

Norris, P. (2001). Digital Divide: Civic Engagement, Information Poverty, and the Internet Worldwide. Communication, Society and Politics. Cambridge: Cambridge University Press.

Pickerill, J. (2003). Cyberprotest: Environmental Activism Online. Manchester: Manchester University Press.

Pool, I. de S. (1983). Technologies of Freedom. Cambridge, MA: Belknap Press of Harvard University Press.

Raine, L. and Wellman, B. (2012). Networked: The New Social Operating System. Cambridge, MA: MIT Press.

Rheingold, H. (1993). The Virtual Community: Finding Connection in a Computerized World. Boston, MA: Addison-Wesley.

Ritzer, G. and Jurgenson, N. (2010). Production, Consumption, Prosumption: The Nature of Capitalism in the Age of the Digital 'Prosumer'. Journal of Consumer Culture, 10(1), 13–36. doi:10.1177/1469540509354673.

Robinson, L. (2009). A Taste for the Necessary. Information, Communication and Society, 12(4), 488–507. doi:10.1080/13691180902857678.

Rogers, E. (1962). The Diffusion of Innovations. New York: Simon & Schuster.

Schattschneider, E.E. (1960). The Semisovereign People: A Realist's View of Democracy in America. New York: Holt, Rinehart & Winston.

Schlozman, K.L., Verba, S. and Brady, H.E. (2010). Weapon of the Strong? Participatory Inequality and the Internet. Perspectives on Politics, 8(02), 487–509. doi:10.1017/S1537592710001210.

Schlozman, K.L., Verba, S. and Brady, H.E. (2012). The Unheavenly Chorus: Unequal Political Voice and the Broken Promise of American Democracy. Princeton, NJ: Princeton University Press.

Schradie, J. (2011). The Digital Production Gap: The Digital Divide and Web 2.0 Collide. Poetics, 39(2), 145–168. doi:10.1016/j.poetic.2011.02.003.

Schradie, J. (2012). The Trend of Class, Race, and Ethnicity in Social Media Inequality: Who Still Can't Afford to Blog? Information, Communication and Society, 15(4), 555–571.

Schradie, J. (2013). The Digital Activism Divide: Social Media, Social Movements and Social Class. American Sociological Association Annual Meeting, New York.

Sims, C. (2014). From Differentiated Use to Differentiating Practices: Negotiating Legitimate Participation and the Production of Privileged Identities. Information, Communication and Society, 17(6), 670–682. doi:10.1080/1369118X.2013.808363.

Smith, A. (2013, April 25). Civic Engagement in the Digital Age. Pew Research Internet Project. Washington, DC.

Somers, M. (2008). Genealogies of Citizenship: Markets, Statelessness, and the Right to have Rights. Cambridge: Cambridge University Press.

Telecommunication Development Bureau (2013). The World in 2013: ICT Facts and Figures. New York: United Nations.

Terranova, T. (2000). Free Labor: Producing Culture for the Digital Economy. Social Text, 18(2 63), 33–58. doi:10.1215/01642472-18-2_63-33.

Tichenor, P.J., Donohue, G.A. and Olien, C.N. (1970). Mass Media Flow and Differential Growth in Knowledge. Public Opinion Quarterly, 34(2), 159–170.

Tilly, C. (2004). Social Movements, 1768–2004. Boulder, CO: Paradigm Publishers.

Tolbert, C.J. and McNeal, R.S. (2003). Unraveling the Effects of the Internet on Political Participation? Political Research Quarterly, 56(2), 175–185. doi:10.1177/106 591290305600206.

Tugal, C. (2012). 'Serbest Meslek Sahibi': Neoliberal Subjectivity among Istanbul's Popular Sectors. New Perspectives on Turkey, 46, 65–93.

Turner, F. (2006). From Counterculture to Cyberculture: Stewart Brand, the Whole Earth Network, and the Rise of Digital Utopianism. Chicago, IL: University of Chicago Press.

Van Deursen, A.J.A.M. and Van Dijk, J.A.G.M. (2011). Internet Skills and the Digital Divide. New Media and Society, 13(6), 893–911. doi:10.1177/1461444810386774.

Van Dijk, J.A.G.M. (2005). The Deepening Divide: Inequality in the Information Society. Thousand Oaks, CA: Sage Publications.

Van Laer, J. (2010). Activists Online and Offline: The Internet as an Information Channel for Protest Demonstrations. Mobilization: An International Quarterly, 15(3), 347–366.

Warf, B. and Grimes, J. (1997). Counterhegemonic Discourses and the Internet. Geographical Review, 87(April), 259–274.

Winner, L. (1980). Do Artifacts Have Politics? Daedalus, 109(1), 121–136.

Witte, J. and Mannon, S. (2010). The Internet and Social Inequalities. New York: Routledge.

Youmans, W.L. and York, J.C. (2012). Social Media and the Activist Toolkit: User Agreements, Corporate Interests, and the Information Infrastructure of Modern Social Movements. Journal of Communication, 62, 315–329.

Zickuhr, K. and Smith, A. (2012). Digital Differences. Pew Internet and American Life Project, April 13. Washington, DC.

PART II

GOVERNMENT AND POLICY

6. Online voting advice applications: foci, findings and future of an emerging research field
Fadi Hirzalla and Liesbet van Zoonen

INTRODUCTION

In this chapter, we will review the state-of-the-art literature about voting advice applications (VAAs). VAAs are interactive internet platforms that run during elections. They operate short questionnaires to gauge users' policy preferences, which are subsequently compared with the policy preferences of political parties (or, in some instances, candidates). Based on these comparisons, VAAs provide advice to users about the political parties that supposedly come closest to their policy preferences (Triga et al., 2012b).

The political potential of VAAs to inform and mobilize voters during election time is apparent in view of what these internet applications are designed to do, but also in light of their growing popularity. There are increasingly more VAAs being used by more people in more countries and elections. The political potential and popularity of VAAs, however, have also raised serious concerns about whether they can be relied upon as trustworthy indicators of party preference. While these concerns have been debated within popular media ever since VAAs were launched online, academics (political scientists and media scholars in particular) have more recently been inclined to investigate the 'hows' and 'whats' and the pros and cons of VAAs (Wall et al., 2012).

Academic studies about VAAs are embedded (though not always elaborately or explicitly) in a broader theoretical context that takes an interest in the health of democracy and investigates elections and voter turnout rates as indicators of how well a democracy and its citizens are doing. Which factors influence and increase electoral participation? How is it possible to promote the 'informedness' of voting? However 'old' such intricate questions are, they have remained relevant in light of persisting concerns about people's political (and civic) participation generally – and particularly suboptimal or declining voter turnouts in consolidated Western democracies in the last few decades (Gallego, 2009; Wattenberg, 2002). Furthermore, these questions have gained new relevance and meaning with the proliferation of new media, specifically the internet.

Since the advent of the internet, social scientists have been theorizing and researching the impact of internet-based technologies on politics, asking among other things whether and how political participation may be boosted and facilitated by online applications (Chadwick and Howard, 2010; Coleman and Blumler, 2009). VAA studies form a subsection of this research about the role of internet applications in people's political participation.

Our goal in this chapter is to outline the findings of extant research on VAAs. This research has focused on three major issues: the socio-demographic or political background of VAA users; the effects of VAA usage on the electoral process; and the methods that VAAs use. First, however, we shall explain in more detail how VAAs work, how they are developed, which benefits they potentially bring and their increased usage in elections worldwide. Directions for further research will be discussed in the closing section of the chapter.

VAAs

The basic function of VAAs is to compare the policy preferences of voters and political parties. Therefore, these applications rely on three parameters in rendering their output, the voting recommendation: the policy profile of users, the policy profile of political parties, and the issues in the application that are selected to compare the policy profiles of users and parties. Firstly, the issues that determine the policy profiles are generally selected on the basis of two criteria: their saliency and their variability. That is, the questionnaires operated by VAAs focus on issues which their designers consider to be salient at election time, as well as variant among the competing parties. Secondly, the policy profile of political parties on these issues is determined in several different ways. For example, while in some VAAs parties are invited to state their position on the issues that are included in the VAA questionnaire, in other instances the designers of the VAA themselves determine the position of the competing parties, based on analyses of official party documents, such as manifestos. Thirdly, the policy profile of users is based on their answers to a sequence of questions (or, more specifically, their responses to statements) about policy issues, but VAAs may also request that users indicate which issues are more or less important to them. The responses of users to the questionnaire and the weight users give to their responses are generally the two main sources of information that VAAs use to determine the extent to which the policy profiles of voters and political parties differ. The less a party differs from a voter, the higher the party ranks in the voting advice that VAAs provide, and vice versa (Garzia and Marschall, 2012; Triga et al., 2012b).

The automated comparisons on policy issues constitute what is regarded as the practical value of VAAs for voters. This value, more specifically, pertains to the reduction of the 'costs' (time, effort) that voters would normally need to invest to become informed about the views of political parties. By using VAAs, voters are relieved from acquiring information about the policy views of competing parties; from systematically comparing their own policy views with those of the political parties; and from deciding how these comparisons between the policy views of themselves and political parties translate into a rank-order of parties that meet their own policy preferences most and least (Edwards, 1998, in Garzia, 2010: 19; see also Hirzalla et al., 2011; Walgrave et al., 2008). Potentially, the reduction of these costs facilitates informed voting and boosts voter turnouts (Popkin, 1991; see, for a more detailed discussion, Garzia, 2010: 23–27).

VAAs have the potential to reduce electoral information costs due to their 'accessibility'. Access in this regard has three dimensions. Firstly, VAAs are physically accessible to a broad public. In most instances, users do not need to make additional financial or material investments to use VAAs. As physical internet access disparities have diminished to a high extent in developed countries, most people can readily access VAAs online. Secondly, VAAs are accessible in terms of simplicity of usage. That is, they are relatively easy to employ and do not assume a great deal of digital skill. As VAAs have user-friendly interfaces, it is expected that few users will misunderstand what they need to do (filling out the questionnaire) and what the meaning is of the VAA output (the voting recommendation). A third kind of accessibility has to do with political competence. VAAs employ surveys with questions or statements that are limited in number (in most instances, 30 to 40 questions about various policy issues) and that have closed-ended answer categories. Using a VAA, therefore, does not require a great deal of time (on average, perhaps ten minutes or so) and does not call on users to build on elaborate pre-existing political knowledge or insights (Triga et al., 2012b; Walgrave et al., 2008).

Although there are many differences between VAAs – regarding, for instance, their popularity, the phrasing and structure of questions, and the procedural design of the questionnaires (see for a detailed discussion: Garzia and Marschall, 2012: 205–210; Ladner and Fivaz, 2012: 181–183) – on a general level, they all share the common goal of helping voters to cast a better-informed vote. The relevance and importance of such help, however, varies by electoral context. Given the comparisons that VAAs make between the policy issues that are preferred by voters and political parties, the help that VAAs provide is especially appreciated by floating voters who cast their vote on the basis of issue preferences (instead of party allegiances or strategic considerations, for instance) and in elections in which there is multi-party (rather than two-party) competition (Hooghe and Teepe, 2007; Mendez, 2012).

For these voters and in these contexts in particular, VAAs have been said to contribute to informed voting and, more generally, to the health of democracy. Fivaz and Schwarz (2007: 12–15), for instance, argue that VAAs contribute to three different pillars of democracy. Firstly, since VAAs urge parties to 'reveal their issue positions', the authors argue that VAAs promote 'transparency' during the election process. The authors maintain that VAAs may also be helpful instruments after elections are held, since these applications can link 'pre- with post-voting spheres and thus establish an "accountability cycle" in which pre-election pledges are systematically monitored in the legislative field'. The third pillar of democracy to which VAAs contribute, according to the expectations of Fivaz and Schwarz, is political participation, as VAAs might mobilize those people who normally are not interested in politics or elections.

It is the potential of VAAs to promote democracy by reducing voters' costs of information processing during elections that has enthused policy-makers, non-governmental organizations (NGOs) and social scientists across the globe to develop a VAA of their own. Specifically, after the year 2000, many VAAs were launched in the run-up to local, national and transnational elections. In Europe, where VAAs have been far more popular than elsewhere, more than 40 VAAs have been launched (Garzia and Marschall, 2012). Almost each European Union (EU) member state has had one or more VAAs running. Viewing the number of times VAAs provided advice (not necessarily unique) in relation to the size of electorates, VAAs in Belgium, Finland, the Netherlands and Switzerland have been particularly successful. To date, the Dutch 'Stemwijzer' has been the most popular VAA of all. At the 2012 parliamentary elections, it provided 4.9 million recommendations to an electorate of 13 million voters. In absolute terms, the German 'Wahl-O-Mat' has been the most successful. This VAA was consulted 6.7 million times (among 62 million voters) prior to the German 2009 national elections (Garzia and Marschall, 2012; see for more: Ladner and Fivaz, 2012).

The most pertinent determinants of any explanation of the increasing popularity of VAAs have been the extent of 'free publicity' given to VAAs, the media saliency of elections, the importance of elections as perceived by voters, the issue-orientedness and volatility of voters, the party system context in which VAAs operate, and the accessibility and usability of VAAs (Ladner and Fivaz, 2012; Garzia, 2012; Walgrave et al., 2009).

As the pervasiveness of VAAs during elections has grown, however, more debate among politicians, journalists and scientists has emerged about their accurateness, neutrality or autonomy in various countries

U bent het op 7 punten **oneens** met Piratenpartij

Piratenpartij	
SP	
Partij voor de Dieren	
PvdA	
ChristenUnie	
GroenLinks	
PVV	
50Plus	
CDA	
D66	
SGP	
Democratisch Politiek Keerpunt	
VVD	

U bent het op 13 punten **eens** met VVD

Note: Advice of the Dutch Stemwijzer to vote for Piratenpartij, the Dutch Pirate Party. Stemwijzer is relatively the most popular VAA, and it was the first VAA ever launched online (in 1998). The output as shown was generated at http://www.stemwijzer.nl (15 June 2013) on the VAA that was designed for the Dutch 2012 parliamentary elections.

Figure 6.1   Advice of the Dutch Stemwijzer, the oldest and relatively most popular VAA

(for example, Hooghe and Teepe, 2007; Ramonaité, 2010; Wagner and Ruusuvirta, 2011; Wall et al., 2009; Wall et al., 2012). Most controversy has arisen around the (relatively) most popular VAA of all, the Dutch Stemwijzer (Figure 6.1). Critics have argued that this VAA too frequently produced recommendations for extreme right or left wing parties, and there have been worries that the managers of Stemwijzer were too prone to adjust their questionnaires to the wishes and whims of political leaders. A more fundamental critique concerned the rationale by which Stemwijzer translates users' policy preferences into voting recommendations. Such concern prompted the development of a second VAA in the Netherlands, called 'KiesKompas'. Like Stemwijzer, KiesKompas develops voting recommendations based on users' policy preferences, but it also takes into account how people view the incumbent Cabinet and the performance of political leaders in terms of competence and reliability (Hirzalla et al., 2011; Kleinnijenhuis et al., 2007; Van Praag, 2007).

EXTANT RESEARCH

While the first few studies about VAAs were conducted soon after these applications appeared on the political scene in the 2000s, only since 2010 has the literature on VAAs developed into a distinct research area, as studies have grown in number and became more empirically grounded, with the employment of elaborate statistical methods. The number of published studies on VAAs was boosted by a special double issue of the *International Journal of Electronic Governance* in 2012, guest edited by Triga, Serdült and Chadjipadelis, and two volumes edited by Cedroni and Garzia (2010) and Dziewulska and Ostrowska (2012). Garzia and Marschall (2014) have recently produced a third volume. Nonetheless, the literature on VAAs is still young and therefore limited. The scholarly work that has been done to date has focused on three main research themes: the usage of VAAs, the effects of VAA use and the methods of VAAs (Garzia and Marschall, 2012). We will briefly discuss the foci and findings of these three research themes below.

Usage

The usage of VAAs is relevant against the backdrop of disparities in (political) internet use: the 'digital divide' problem (Norris, 2001). One of the different aspects of the internet's political potential depends on the profile of the people who actually use it for political purposes, such as participation in forums, taking part in e-democracy projects, writing blogs or reading news about political issues. Who are these people? The internet would serve a reinforcing purpose if these people are already politically active offline or belong to socio-demographic groups that are considered as traditionally empowered and active in terms of political participation, knowledge and interest (for example, male, older, white, well-educated people). Insofar as this is the case, political participation online entails a continuation of (offline) business as usual. On the other hand, the internet fulfils a mobilizing function insofar as these people belong to disadvantaged groups (Jennings and Zeitner, 2003; Norris, 2001).

Thus, the big question is: do VAAs 'mobilize the mobilized'? First of all, the straightforward statistics on the number of times that VAAs were consulted indicate that they have been used by a minority (in most countries or elections, 20 per cent or less) of the electorate (Garzia and Marschall, 2012). Further research points to the conclusion that, compared to the population at large and/or internet users in general, young, male, well-educated, leftist or urban persons are over-represented in this minority of VAA users in various countries (Wall et al., 2009), such as the Netherlands (Boogers, 2006), Belgium (Hooghe and Teepe, 2007), Germany (Marschall, 2005), Finland (Ruusuvirta, 2010), Switzerland (Fivaz and Schwarz, 2007), Italy (De Rosa, 2010) and Ireland (Wall et al., 2009).

Socio-demographics in and of themselves, however, do not explain why a person does or does not use political internet applications, such as VAAs. The simple fact of being a man or woman, for example, does not 'do' anything in this regard. Instead, socio-demographic disparities in political internet use result from corresponding differences in mental or physical resources, such as political interest or digital skills (Hirzalla et al., 2011).

Studies that have focused on the role of these resources suggest that VAAs are mostly used by people who are politically interested, knowledgeable or active (De Rosa, 2010; Fivaz and Nadig, 2010; Marschall and Schmidt, 2010), although the persistency of such disparities (as well as socio-demographic differences) may vary between contexts and generations (for example, Marzuca et al., 2011). Another possible factor that influences the usage of VAAs may be their (perceived) 'usefulness'. On this matter, based on their study of the 'EU Profiler' VAA during the 2009 European Parliamentary elections, Alvarez et al. (2012) found that 'individuals who believe new technology should be used to facilitate political participation perceive higher utility from using the VAA'; that 'individuals who think politics is "regularly" complicated instead of only "occasionally" complicated are significantly more likely to perceive higher utility from the VAA'; and that 'centrally-located users – who are likely to feel weaker partisan attachments and ideology commitments relative to respondents with clear left–right ideology positions – experience lower utility from using the VAA' (p. 23).

Although the results of existing research about the usage of VAAs are quite uniform, there is some difference between the ways in which scholars interpret their findings. Some analysts remain optimistic about the political potential of VAAs. They expect that disparities in usage will disappear in time as increasingly more people grow familiar with the exist-ence of these applications and the internet more generally (for example, Fivaz and Schwarz, 2007). Nonetheless, until a broader and more varied section of the population decides to consult VAAs, current findings tend to curtail the extent to which VAAs meet the expectations of the mobiliza-tion hypothesis. As Ruusuvirta (2010: 63) maintains unequivocally, 'the old challenge of bringing politics to those who are not interested in it still remains . . . Those who would benefit the most from the information in online voting advice applications are the least likely to seek it'.

Effects

A second focal point in the VAA literature concerns the effects that VAAs may have on their users. The effects potential is realized in the extent to which VAAs yield an electorally relevant impact on the behaviours or mind-sets of VAA users, such as increased voter turnout and more informed voting, which in turn should contribute to better democratic elections. Without such effects, VAAs may fulfil an entertainment function rather than facilitate and promote voting (for example, Walgrave et al., 2008).

For this reason, some studies have asked if VAA usage influences whether people decide to cast their vote. While most studies on voter turnout impact arrive at the conclusion that VAAs do mobilize people to vote, how significant this mobilization is seems to vary with the context or study. For instance, in the Dutch and Swiss contexts, studies found that the use of VAAs raised voter turnouts by 3 per cent and 5 per cent, respec-tively; while other studies found that VAAs increased the probability of voting in Finland and Germany by about 20 per cent and 11 per cent, respectively (Garzia and Marschall, 2012; see Fivaz and Nadig, 2010; Marschall and Schmidt, 2010; Mykkänen and Moring, 2006; Ruusuvirta and Rosema, 2009).

Extant research has also investigated whether consulting a VAA influ-ences how people vote; that is, for which political party or candidate. The results of these studies are also diverse. For example, Fivaz and Schwarz (2007) found on the basis of their research about the role of 'SmartVote' during 2006 regional elections in Switzerland that 74 per cent of users report that this VAA influenced their vote. Less abundant, yet still impres-sive, statistics come from Kleinnijenhuis and Van Hoof (2008). They conclude that 'roughly one out of ten voters changed their mind in accord-ance with the advice' they received from the Dutch VAAs Stemwijzer or 'KiesKompas' during the Dutch 2006 Parliamentary elections (p. 7). Boogers (2006) reports a slightly higher outcome on the same elections, with about 15 per cent of respondents saying that the advice they received from Stemwijzer made them change their mind about their vote. In contrast, Walgrave et al. (2008) found that merely 1 per cent of users said that they were influenced by the VAA called 'DoeDeStemTest' during the Belgian 2004 regional elections. Similarly, De Rosa (2010) found that 3 per cent of users of the Italian 'Cabina-elletorale' during the 2009 EU parliamentary elections declared that they would change their vote in accordance with the advice that was given to them by the VAA.

As the aforementioned statistics vary with the context and the election, it seems hard to arrive at a general conclusion about the extent to which VAAs influence people's voting intentions or decisions. Furthermore, only few studies have asked whether people's voting intentions are varyingly sensitive to VAAs. Political knowledge and interest seem to form two of the indicators that affect how prepared voters are to follow the advice given by VAAs. More politically knowledgeable and interested voters seem to be less inclined to change their vote due to VAA use (Dumont and Kies, 2012; Kleinnijenhuis and Van Hoof, 2008). The extent to which voters 'swing' seems to be another indicator. Volatile voters may be more prone to follow the advice of VAAs (Ladner et al., 2012; Ruusuvirta and Rosema, 2009). More specifically, voters may be less inclined to change a pre-existing voting preference due to VAA use when the advice they receive from a VAA differs strongly from that preference (Wall et al., 2012).

Along with the research on the effects of VAA use on voter turnout and voting intentions or decisions, there are also studies that have paid attention to the impact of VAA use on other behaviours that are also relevant to elections. For instance, various studies reveal that quite large proportions of VAA users, in some instances more than 40 per cent, report that they were inspired to seek further information about the elections or politics generally due to their use of a VAA (for example, Boogers, 2006; Ladner and Pianzola, 2010; Marschall and Schmidt, 2010). It has also been observed that political awareness and 'light' forms of political participation may be positively related to VAA use. Many users of the Italian Cabina-elletorale in 2009, for instance, said that they were inclined to discuss the advice they received with others (reported by 41 per cent of users), that the VAA made them aware of current political issues (reported by 40 per cent), and that it helped them with identifying the differences between the competing political parties (reported by 34 per cent) (De Rosa, 2010).

Methods

A less substantial part of the literature focuses on the quality of VAA methods. Quality in this regard concerns the way in which the input of respondents (that is, their responses to the questionnaire) is translated into a voting recommendation. This process relies generally on the configuration of three matters: the content of the questions, the structure of the answer categories and the calculation of the voting advice.

VAAs' methodological quality is important in that it forms, or ought to form, a point of reference of any normative evaluation of VAA usage and effects; and it is a pertinent matter in view of apparent inconsistencies in the recommendations that VAAs yield (Van Praag, 2007). 'VAAs' outcome is not stable', as Walgrave et al. (2009: 1167) say, based on a review of past research on this issue, 'but seems to change haphazardly from one year to another and from one VAA to the other'. As the quality of the methods of VAAs is part of any appraisal of VAAs' role in elections, it is safe to say that this quality matters principally. Various studies have aimed to demonstrate how it matters. One of the few studies that focused on the questions in VAAs was conducted by Kleinnijenhuis et al. (2007, in Walgrave et al., 2009: 1168). Their investigation in the Dutch context demonstrates that an imbalance in the political denotation of the phrasing of questions, as well as the selection of issues, in VAA questionnaires affects the voting advice that a VAA provides. Another point is made by Gemenis (2013), who demonstrates how items in VAAs are sometimes 'double-barrelled'. A double-barrelled item asks a respondent to answer more than one question at once. Consequently, which of the questions (if any) an answer corresponds to remains unclear.

With regard to the answer categories, Baka et al. (2012) discuss the middle 'neither agree nor disagree' 'neutral' response option that is used in VAAs with five-point Likert scale answer categories, ranging from 'strongly agree' to 'strongly disagree'. Such an answer scale – which has been used frequently in the social sciences generally – has been critiqued before, mainly from within a statistical paradigm (Gemenis, 2013). Baka et al. (2012) take a different approach, as they employ a qualitative analysis to 'explore the variety of meanings attributed to the middle category' in VAA questionnaires (p. 245). The varieties they find indicate that the middle category may be an overly ambiguous answer option. Respondents are seen as using this answer option for different reasons, such as conveying dilemmas or rejecting the assumptions on which an item is based, rather than making clear that they are 'neutral' in regard to a particular issue.

Studies which focus on the questions and answer categories of VAAs generally indicate that the designers of VAAs still have work to do if their product is to be considered as reliable. Škop (2010), for instance, points out that among two VAAs during the 2006 elections for the Czech

Lower House there were problems with the relevance and selection of the questions, which leads the author to conclude that 'the existing and established VAAs may have serious problems with the trustworthiness of their advices' (p. 216). A study by Walgrave et al. (2009) shows that the statements that are selected to gauge a user's policy profile has an impact on the vote advice a VAA renders. More specifically, parties benefit and suffer differently from different configurations. Hence, according to the authors, 'statement selection is the crux of the VAA-building exercise . . . The care with which political scientists design their scientific surveys stands in sharp contrast to the carelessness with which some of them engage in devising VAAs' (Walgrave et al., 2009: 1178).

Next to the research on the contents of the questionnaire, some attention has also been paid to the ways in which users' answers are transformed into a recommendation (for example, Mendez, 2012). This research points out, unsurprisingly yet importantly, that the recommendations offered by VAAs depend intrinsically on how the advice is calculated. One of the implications of this finding is, in the words of Kleinnijenhuis and Krouwel (2008: 8), that 'party advice websites should not pretend to point towards a single party that would match the preferences of the voter perfectly'. A more general implication is that the decision rules based on which VAAs determine their advice should be part of any fundamental appraisal of the meaning of that advice.

NEXT STEPS

The review of the literature in the previous section shows that quite some research effort has been invested already in explaining how online voting advice applications work. Within a short period of time, and since 2010 in particular, scholars have studied a wide range of topics regarding the usage, effects and methods of VAAs. As straightforwardly acknowledged by these VAA scholars themselves, however, extant findings should not be seen as final and clear-cut answers to the complex questions that have been investigated. In this section, we will propose some directions for future studies, to develop a more profound understanding and appraisal of the role of VAAs during elections.

First of all, a marked feature of the existing body of literature is the rather fragmented way in which research questions have been formulated and research findings have been interpreted. That is, studies in this field have often approached issues of VAA usage, effects and methods separately, instead of in relation to each other, at the expense of wider theoretical and normative questions about the role and potential of VAAs. For instance, what light does VAA usage, insofar as it is 'digitally divided', shed on significant effects of VAAs on electoral outcomes? What is the meaning of VAA usage and effects if and when these applications in and of themselves are 'unreliable'? What does VAAs' reliance on official election slogans mean for a fundamental valuation of VAA effects on political 'knowledge'? Contemplation of such questions that go beyond the focus on uses, effects or methods is likely to yield the still-needed bigger story about the role of VAAs during elections. As the research on usage and methods of VAAs has yielded quite critical findings, it is particularly the interpretation of the more sanguine findings of effect studies, indicating that VAA use does significantly affect vote decisions or intentions, that may require further nuance.

Another, more specific way in which the literature is still fragmented concerns the methodological design of studies. Different studies have relied on different measurements, particularly with regard to VAA effects on voting behaviour. These effects have been measured either through self-reports from respondents about the extent to which a VAA stimulated them to vote (for example, Marschall and Schmidt, 2010) or change their vote (for example, Boogers, 2006), or are based on data about actual voting turnout and decisions (for example, Ruusuvirta and Rosema, 2009). Such different measurements may, in part, explain why studies agree that VAA usage has influence on electoral outcomes, but disagree on how strong this influence is.

Disagreements about the impact of VAAs should not, however, necessarily be attributed merely to measurement issues. Another problem is that studies have often focused on single cases (that is, one VAA in one country during one election) rather than researching the role of VAAs during elections in an internationally or historically comparative perspective. Few studies, in other words, have been designed to compare VAAs. Such comparative research, however, is imperative if scholars aim at arriving at more general theorizations that take into account where differences between VAAs' usage and impact come from. Returning to our earlier suggestion, comparative research may be fruitful particularly when it seeks to approach issues on usage, effects and methods holistically. One of the questions that seems to be pertinent against this backdrop is whether and (if so) how intercontextual disparities in VAA usage and effects relate to differences in the design of the applications.

Both comparative and case studies may also be advanced when they dive deeper into possible micro- and macro-level contextual factors that may 'compete' with VAAs in influencing electoral behaviour. People's consumption of other media than VAAs in the run-up to elections may be considered as one of the most notable factors that requires more attention. People's interpersonal communication with family, friends or others may be seen as another potential 'competitor' in this sense. More fundamental factors may relate to prolonged socialization processes that have no direct link with election time per se and are more complex than personal traits such as age or gender. Taking into account such factors in studying VAAs hints at the importance of premising voting behaviour as something that is more plausibly moulded over time than determined at a particular moment of VAA usage. Accepting that premise imposes serious restraint on ascribing to VAAs the muscle to change or determine people's voting behaviour, as long as the role of VAAs is insufficiently studied in relation to other probable determinants of voting outcomes.

Qualitative research, to date a rare phenomenon in this research field, can aid with developing a more detailed and deep understanding of how VAA usage and impact relate to their context. In-depth interviews in particular could be employed to gain a greater understanding of a more specific issue that has thus far remained relatively neglected in the literature about VAAs: namely, why, or with what aims, people use VAAs. While some studies have sought to tap into the aims of VAA users based on predefined and short lists of answers (for example, Boogers, 2006), an inductive approach will yield much richer and detailed insights that can be used to elaborate further on possible causes of VAA usage and effects.

Qualitative research may also yield new findings with regard to VAAs' methods, particularly users' interpretation of VAA questionnaires. An example might illustrate the small and big issues that may arise in this area. Suppose that somebody 'agrees' with a policy issue in one of the items in a VAA on the assumption that it was propagated by his favourite political party; and suppose that the policy in question was in fact opposed, not propagated, by his favourite party. This person would thus favour a policy issue based on a non-issue argument, relying on a wrong assumption about which party favours what policy. What such simple and practical instances of 'incorrect' practices mean for the output and impact of VAAs, and to what extent, among whom, and with regard to what items such practices are prone to occur, are relevant questions that deserve more attention in the literature. It may well turn out that, in part, instrumentally 'increasing the knowledge' of voters in the ways intended by the designers of VAAs constitutes one part of the value of VAAs. For another part, VAAs' contribution to elections may lie in the moments that they merely stimulate voters to spend in thinking about the vote they plan to cast.

ACKNOWLEDGEMENTS

We would like to thank Dr Diego Garzia for his comments on an earlier draft of this chapter.

FURTHER READING

Cedroni, L. and Garzia, D. (eds) (2010). *Voting Advice Applications in Europe: The State of the Art*. Naples, Italy: Civis/Scriptaweb.
Dziewulska, A. and Ostrowska, A. (eds) (2012). *Europeisation of Political Rights: Voter Advice Application and Migrant Mobilisation in 2011 UK Elections*. Warsaw, Poland: Centre for Europe.
Trechsel, A. and Mair, P. (2011). When parties also position themselves: an introduction to the EU Profiler. *Journal of Information Technology and Politics*, 8(1), 1–20.
Triga, V., Serdült, U. and Chadjipadelis, T. (eds) (2012a). *International Journal of Electronic Governance*, 5(3–4), 194–412.
Walgrave, S., Van Aelst, P. and Nuytemans, M. (2008). 'Do the Vote Test'. The electoral effects of a popular vote advice application at the 2004 Belgian elections. *Acta Politica*, 43(1), 50–70.
Walgrave, S., Nuytemans, M. and Pepermans, K. (2009). Voting aid applications and the effect of statement selection. *West European Politics*, 32(6), 1161–1180.
Wall, M., Sudulich, M.L., Costello, R. and Leon, E. (2009). Picking your party online: an investigation of Ireland's first online voting advice application. *Information Polity*, 14(3), 203–218.

REFERENCES

Alvarez, R.M., Levin, I., Trechsel, A.H. and Vassil, K. (2012). Voting advice applications: how useful? For whom? Paper presented at the conference on Internet, Policy, Politics 2012: Big Data, Big Challenges?, 20–21 September, Oxford, UK. Retrieved from: http://microsites.oii.ox.ac.uk/ipp2012/sites/microsites.oii.ox.ac.uk.ipp2012/files/LevinIPPConfPaper_250812.pdf.
Baka, A., Figgou, L. and Triga, V. (2012). "Neither agree, nor disagree": a critical analysis of the middle answer category in voting advice applications. *International Journal of Electronic Governance*, 5(3–4), 244–263.
Boogers, M. (2006). *Enquête bezoekers Stemwijzer*. Retrieved from: http://www.publiek-politiek.nl.
Cedroni, L. and Garzia, D. (eds) (2010). *Voting Advice Applications in Europe: The State of the Art*. Naples, Italy: Civis/Scriptaweb.
Chadwick, A. and Howard, P.N. (eds) (2010). *The Routledge Handbook of Politics*. Abingdon, UK: Routledge.
Coleman, S. and Blumler, J.G. (2009). *The Internet and Democratic Citizenship: Theory, Practice and Policy*. Cambridge, UK: Cambridge University Press.
De Rosa, R. (2010). Cabina-elletorale.it (Provides advice to Italian voters since 2009). In Cedroni, L. and Garzia, D. (eds), *Voting Advice Applications in Europe: The State of the Art* (pp. 187–198). Naples, Italy: Civis/Scriptaweb.
Dumont, P. and Kies, R. (2012). Smartvote.lu: usage and impact of the first VAA in Luxembourg. *International Journal of Electronic Governance*, 5(3–4), 388–410.
Dziewulska, A. and Ostrowska, A. (eds) (2012). *Europeisation of Political Rights: Voter*

Advice Application and Migrant Mobilisation in 2011 UK Elections. Warsaw, Poland: Centre for Europe.

Edwards, A.R. (1998). Towards an informed citizenry? Information and communication technologies and electoral choice. In Snellen, I. and Donk, W. van de (eds), *Public Administration in an Information Age: A Handbook* (pp. 191–206). Amsterdam, The Netherlands: IOS Press.

Fivaz, J. and Nadig, G. (2010). Impact of voting advice applications (VAAs) on voter turnout and their potential use for civic education. *Policy and Internet*, 2(4), 167–200.

Fivaz, J. and Schwarz, D. (2007). Nailing the pudding to the wall: E-democracy as catalyst for transparency and accountability. Paper presented at International Conference on Direct Democracy in Latin America, 14–15 March, Buenos Aires, Argentina. Retrieved from: http://www.ipw.unibe.ch/unibe/wiso/ipw/content/e2425/e2439/e7615/e7616/files7631/naili ng_the_pudding_to_the_wall_ger.pdf.

Gallego, A. (2009). Where else does turnout decline come from? Education, age, generation and period effects in three European countries. *Scandinavian Political Studies*, 32(1), 23–44.

Garzia, D. (2010). The effects of VAAs on users' voting behaviour: an overview. In Cedroni, L. and Garzia, D. (eds), *Voting Advice Applications in Europe: The State of the Art* (pp. 13–33). Naples, Italy: Civis/Scriptaweb.

Garzia, D. (2012). Understanding cross-national patterns of VAA-usage: Integrating macro- and micro-level explanations. In Dziewulska, A. and Ostrowska, A. (eds), *Europeisation of Political Rights: Voter Advice Application and Migrant Mobilisation in 2011 UK Elections* (pp. 25–32). Warsaw, Poland: Centre for Europe.

Garzia, D. and Marschall, S. (2012). Voting advice applications under review: the state of research. *International Journal of Electronic Governance*, 5(3–4), 203–222.

Garzia, D. and Marschall, S. (eds) (2014). *Matching Voters with Parties and Candidates: Voting Advice Applications in a Comparative Perspective*. Colchester, UK: ECPR Press.

Gemenis, K. (2013). Estimating parties' policy positions through voting advice applications: some methodological considerations. *Acta Politica*, 48(3), 268–295.

Hirzalla, F., Van Zoonen, L. and De Ridder, J. (2011). Internet use and political participation: reflections on the mobilization/normalization controversy. *Information Society*, 27(1), 1–15.

Hooghe, M. and Teepe, W. (2007). Party profiles on the web: an analysis of the logfiles of non-partisan interactive political internet sites in the 2003 and 2004 election campaigns in Belgium. *New Media and Society*, 9(6), 965–985.

Jennings, M.K. and Zeitner, V. (2003). Internet use and civic engagement: a longitudinal analysis. *Public Opinion Quarterly*, 67(3), 311–334.

Kleinnijenhuis, J. and Van Hoof, A.M.J. (2008). The influence of internet consultants. Paper presented at conference on Voting Advice Applications: Between Charlatanism and Political Science, 16 May, Antwerp, Belgium. Retrieved from: http://citation.allacademic. com//meta/p_mla_apa_research_citation/2/3/4/5/4/pages234549/p234549-1.php.

Kleinnijenhuis, J. and Krouwel, A.P. (2008). Simulation of decision rules for party advice websites. Retrieved from: http://www.iiis.org/cds2008/cd2008sci/pista2008/paperspdf/ p125lt.pdf.

Kleinnijenhuis, J., Scholten, O., Van Atteveldt, W., Van Hoof, A., Krouwel, A., Oegema, D., De Ridder, J.A., Ruigrok, N. and Takens, J. (2007). *Nederland vijfstromenland: De rol van de media en stemwijzers bij de verkiezingen van 2006*. Amsterdam, The Netherlands: Uitgeverij Bert Bakker.

Ladner, A. and Fivaz, J. (2012). Voting Advice Applications. In Kersting, N. (ed.), *Electronic Democracy: The World of Political Science* (pp. 177–198). Opladen, Germany: Barbara Budrich Publishers.

Ladner, A., Fivaz, J. and Pianzola, J. (2012). Voting advice applications and party choice: evidence from smartvote users in Switzerland. *International Journal of Electronic Governance*, 5(3–4), 367–387.

Ladner, A. and Pianzola, J. (2010). Do voting advice applications have an effect on electoral participation and voter turnout? Evidence from the 2007 Swiss federal elections.

In Tambouris, E., MacIntosh, A. and Glassey, O. (eds), *Electronic Participation* (pp. 211–224). Berlin, Germany: Springer.

Marschall, S. (2005). Idee und Wirkung des Wahl-O-Mat. *Aus Politik und Zeitgeschichte*, 51–52, 41–46.

Marschall, S. and Schmidt, C.K. (2010). The impact of voting indicators: the case of the German Wahl-O-Mat. In Cedroni, L. and Garzia, D. (eds), *Voting Advice Applications in Europe: The State of the Art* (pp. 65–90). Naples, Italy: Civis/Scriptaweb.

Marzuca, A., Serdült, U. and Welp, Y. (2011). Questão Pública: first voting advice application in Latin America. In Tambouris, E., MacIntosh, A. and Glassey, O. (eds), *Electronic Participation*. Third IFIP WG 8.5 International Conference (pp. 216–227). Delft, The Netherlands.

Mendez, F. (2012). Matching voters with political parties and candidates: an empirical test of four algorithms. *International Journal of Electronic Governance*, 5(3–4), 264–278.

Mykkänen, J. and Moring, T. (2006). Dealigned politics comes of age? The effects of online candidate selectors on Finnish voters. Paper presented at conference on Politics on the Internet: New Forms and Media for Political Action, 24–25 November, Tampere, Finland. Retrieved from: http://www.edemocracy.uta.fi/eng/haefile.php?f=132.

Norris, P. (2001). *Digital Divide: Civic Engagement, Information Poverty and the Internet Worldwide*. Cambridge, UK: Cambridge University Press.

Popkin, S. (1991). *The Reasoning Voter: Communication and Persuasion in Presidential Campaigns*. Chicago, IL: University of Chicago Press.

Ramonaité, A. (2010). Voting advice applications in Lithuania: promoting programmatic competition or breeding populism? *Policy and Internet*, 2(1), 117–141.

Ruusuvirta, O. (2010). Much ado about nothing? Online voting advice applications in Finland. In Cedroni, L. and Garzia, D. (eds), *Voting Advice Applications in Europe: The State of the Art* (pp. 47–63). Naples, Italy: Civis/Scriptaweb.

Ruusuvirta, O. and Rosema, M. (2009). Do online vote selectors influence electoral participation and the direction of vote? Paper presented at ECPR General Conference, 10–12 September, Potsdam, Germany. Retrieved from: http://www.utwente.nl/mb/pa/staff/rosema/publications/working_papers/paper_ruusuvirta_and_rosema_ec.pdf.

Škop, M. (2010). Are the voting advice applications (VAAs) telling the truth? Measuring VAAs' quality. Case study from the Czech Republic. In Cedroni, L. and Garzia, D. (eds), *Voting Advice Applications in Europe: The State of the Art* (pp. 199–216). Naples, Italy: Civis/Scriptaweb.

Triga, V., Serdült, U. and Chadjipadelis, T. (2012b). Introduction. *International Journal of Electronic Governance*, 5(3–4), 194–202.

Van Praag, P. (2007). De Stemwijzer: Hulpmiddel voor de kiezers of instrument van manipulatie? Paper presented at the Amsterdam Academy Club, 24 May, Amsterdam, The Netherlands.

Wagner, M. and Ruusuvirta, O. (2011). Matching voters to parties: voting advice applications and models of party choice. *Acta Politica*, 47(4), 400–422.

Walgrave, S., Van Aelst, P. and Nuytemans, M. (2008). 'Do the Vote Test'. The electoral effects of a popular vote advice application at the 2004 Belgian elections. *Acta Politica*, 43(1), 50–70.

Walgrave, S., Nuytemans, M. and Pepermans, K. (2009). Voting aid applications and the effect of statement selection. *West European Politics*, 32(6), 1161–1180.

Wall, M., Sudulich, M.L., Costello, R. and Leon, E. (2009). Picking your party online: an investigation of Ireland's first online voting advice application. *Information Polity*, 14(3), 203–218.

Wall, M., Krouwel, A. and Vitiello, T. (2012). Do voters follow the advice of voter advice application websites? A study of the effects of kieskompas.nl on its users' vote choices in the 2010 Dutch legislative elections. *Party Politics*, 1–21. doi: 10.1177/1354068811436054.

Wattenberg, M.P. (2002). *Where Have All the Voters Gone?* Cambridge, MA: Harvard University Press.

7. Internet voting: the state of the debate
Thad Hall

Over the past two decades, the debate over Internet voting has developed from one centered on the theoretical benefits and perils related to its adoption to a debate where actual data can be used to inform this discussion. During this time, Internet voting trials have been conducted in countries across the world.[1] Its use is most prominent in Estonia, where the Internet has become a standard platform for voting, and Switzerland, where Internet voting is used in three cantons for referenda elections (Alvarez et al., 2009). Several countries, including Ireland and the Netherlands, have conducted pilot Internet voting trials but discontinued its use. Canada, Norway, the United Kingdom and the United States have conducted a series of pilot trials as well, with the American trials generally focused on enfranchising military personnel and overseas civilians (Alvarez and Hall, 2004, 2008; EAC 2011).

With each Internet voting trial that is conducted, the same concerns are raised about the use of the Internet for voting and the same arguments are made for its introduction. The primary concerns center on security – security of the ballot and the system – as well as voter privacy. A key secondary concern is the digital divide; Internet voting can enfranchise the 'haves' and make it easier for them to vote but not help the 'have nots', who already typically vote at lower rates (Gray and Caul, 2000; Norris 2001). The primary benefit that accrues from the use of Internet voting is that it improves accessibility to the ballot box, especially for difficult-to-serve populations such as individuals with disabilities, the elderly, expatriate and military personnel, and younger individuals.

In many respects, the claims about Internet voting are similar to those made in regards to other facets of the voting process. For example, Riker and Ordeshook's (1968) work on the cost of voting centers on the trade-off between the ease of voting and the benefits that accrue from casting a ballot. More recently, arguments regarding the need for voters to show a form of photo identification in order to authenticate themselves before casting a ballot are often framed as a trade-off between the accessibility of the voting process to voters and the need for greater security against fraud in the voting process (for example, Atkeson et al., 2010). Voting at polling places comes with many costs and the fact that more voters now vote by-mail or during in-person early voting, especially in the United States, may indicate less willingness to pay the costs associated with turning out to vote on Election Day. Inclement weather, long lines, missed work, costs of transportation; all of these inconveniences are eliminated when Internet voting is implemented. However, Internet voting turns cost-of-voting concerns to concerns of security.

This simple trade-off between accessibility and security defines the debate over Internet voting. The basic question regarding Internet voting is this: do the benefits related to accessibility outweigh the security risks associated with Internet voting? This chapter will look at this trade-off, considering the advantages and concerns regarding Internet voting and its possible effects on turnout. We start by considering how youth participation is affected by Internet voting, then consider issues related to ballot being counted correctly in the by-mail absentee voting process, as compared to over the Internet. We then consider variations in Internet access across different countries and across age, racial and ethnic divides. Finally, we discuss concerns that Internet voting represents a security issue.

TURNOUT AND INTERNET VOTING

The data regarding the impact of Internet voting on voter turnout are somewhat inconclusive. Analyses of the many small pilot studies of Internet voting that have been conducted indicate that Internet voting does not increase turnout. As Pammett and Goodman (2013: 9) note, 'any hoped-for "turnout effect" [from the introduction of Internet voting] has been elusive; in some cases it was minor and in others non-existent'. One reason why this may be the case is that turnout in general has been declining in many Western countries and Internet voting has been introduced in an attempt to stop the decline in turnout.

There is some evidence that Internet voting may benefit certain marginal voters. For example, in Estonia, where Internet voting has been firmly in place for almost a decade, there are indications that some who may not have voted otherwise did so because it was convenient. As Trechsel et al. (2010: 4) note, 'Our simulation showed that turnout in the 2009 local elections might have been up to 2.6 percent lower in the absence of Internet voting. A technologically induced change in turnout by 2.6 percent is far from negligible'. Similarly, a Norwegian study of Internet voting in 2012 found that Internet voting there was very popular, especially with people with disabilities; 72.4 percent of early voters voted on the Internet and people with disabilities expressed excitement that they could vote without assistance for the first time.[2]

There are several problems with trying to evaluate the effects of Internet voting on turnout. First, it may be the case that pilot projects do not have the same effect on turnout as would occur from implementing Internet voting on a permanent basis. For example, Internet voting typically requires a voter to engage in new security measures such as having an online identity (for example, Estonia) or using some similar security code system, as well as to be Internet-savvy. A person might not want to go through the effort of learning this new system for it to be used only once. Second, it may be difficult to actually measure changes in turnout for sub-populations of interest. Epstein (2011: 892–893) points out that Internet voting does not increase turnout overall in most elections studied, but notes that it may boost turnout among special populations, such as individuals in the United States covered by the Uniform and Overseas Civilian Absentee Voting Act (UOCAVA), who might not have voted due to ballot transit time and inconvenience.[3] However, this is a very difficult population to survey. Likewise, individuals with specific disabilities are also often difficult to survey unless a survey is done with a very large number of respondents. In such cases, identifying benefits of Internet voting on turnout among special populations of voters would be difficult.

The closest analogue to Internet voting is by-mail voting and the research here is illuminating. If we consider the traditional research on convenience voting – pre-election in-person voting, by-mail absentee voting, and elections where all votes are by mail – the evidence is mixed but generally suggests that any gains in turnout are quite modest (for example, Gronke et al., 2008). Research on voting by mail indicates that having a ballot delivered to you automatically increases turnout only in low participation elections; in general elections, it does little to increase turnout (Karp and Banducci, 2000; Kousser and Mullin, 2007). There has been some controversy regarding whether elections increase turnout when the only mode of voting is postal voting, as is now done in Washington and Oregon in the United States.[4] In general, voting by mail and other convenience voting tools tend to make voting easier for those individuals who would have voted anyway; they do not increase turnout among those individuals who would not have otherwise been mobilized (Berinsky, 2005).

INTERNET VOTING AND YOUNG VOTERS

One common argument made about Internet voting is that it has the potential to boost turnout among young people. As Alvarez and Hall (2004, 2008; Hall, 2012) note, young people often wonder why they can bank online, shop online and do many other tasks online, but cannot vote online. These comments occur at the same time as the Internet is playing a key role in American politics. There are certain contradictions among young people and the Internet when it comes to politics. As Smith et al. (2009) note, even though most young people have some online presence, they are less likely than are other age cohorts to engage in some basic political actions, such as making an online political donation or sending an email to public officials. However, young people (aged 18–24) post large amounts of information online. Smith et al. (2009: 51) notes:

> those under age 35 represent 28% of the respondents in our survey but make up fully 72% of those who make political use of social networking sites, and 55% of those who post comments or visual material about politics on the Web. The youngest members of this group – those under age 25 – constitute just 10% of our survey respondents but make up 40% of those who make political use of social networking sites and 29% of those who post comments or visual material about politics online.

There is evidence that political activity online can translate into voting. Cohen et al. (2012: 10) found that acting politically, whether through online or other activities, makes it more likely that youth will engage in other political acts. Youth who engaged in at least one act of participatory politics were almost twice as likely to report subsequently voting compared to those who did not engage in such acts. Internet voting would give some young people easier access to the polls, especially those who are away from home for post-secondary education. Likewise, political parties and candidates could use social media to prompt young people to vote, and these prompts could be acted on almost instantly by the voter, who could just click over to the voting application on their computer (or appropriate device, like a smartphone or tablet) and cast their ballot.

Because of the importance of the Internet in politics and society, Internet voting is often viewed as the way forward for the future of elections (Alvarez and Hall, 2004, 2008; EAC 2011). Trials of Internet voting in the United Kingdom, Norway and Estonia have all found that when voters are offered the opportunity to vote online, many voters will make the choice to switch to online voting. However, studies of Internet voting have not found that adding the Internet as a mode of voting causes a significant turnout boost among young people (Pammett and Goodman, 2013). The results noted previously about turnout generally hold true for younger people; in countries which have engaged in Internet voting trials, it has made voting easier for citizens who were already planning to vote.

In survey research, younger people both in the USA and internationally express attitudes about Internet voting that are more favorable than those expressed by older and less educated people. However, in the US context, the higher level of support for Internet voting among young people does not translate into large majorities wanting it. In fact, using data from a survey of 10 000 registered voters conducted after the 2008 presidential election, Alvarez et al. (2009) found that only 38 percent of respondents aged 18–30 support Internet voting, compared to 24 percent of respondents over age 55. The overall support in the United States for Internet voting in 2008 was only 30.1 percent. In 2012, a similar survey found that national support for Internet voting was 36.7 percent and support for Internet voting among younger people had risen to 47.1 percent.[5] Although Internet voting is often touted as a reform that will improve youth participation in elections, young people are not overwhelmingly supportive of such reforms.

HAVING YOUR VOTE COUNT

Ease of voting – often measured as turnout – is just one metric to use when evaluating a voting technology. Yao and Murphy (2007: 107) point out that other factors – such as privacy protection, accuracy of the system, convenience, and the resources and tools needed to use the technology – all play roles in citizens' willingness to use new technologies to vote. Their study indicates that availability, ease of use, accuracy and mobility all affect a voter's intention to use remote voting systems (Yao and Murphy, 2007: 114).

The issue of accuracy in voting technologies is often overlooked when considering which voting system to use. Even if Internet voting does not increase turnout, it could increase the percentage of ballots that are cast and counted. For example, one study of absentee voting in Los Angeles County, California found that military personnel, individuals living overseas, voters who request a ballot in a language other than English, and permanent absentee voters were much less likely to return a requested absentee ballot. Moreover, individuals in these groups were also less likely to have their ballot counted in the final election results compared with the entire population of absentee voters (Alvarez et al., 2008).

For some of these individuals, their paper absentee ballots were not returned because voters realized that the ballot transit time – from when the election official sent the ballot to the voter, to when the voter returned it to the election office – would not allow their ballot to be received in time to be included in the election results. The ballot transit time problem has long existed for military and expatriate voters and the Internet is an obvious way of ameliorating this problem. By sending ballots electronically, the time needed for sending a ballot from overseas can be cut from more than a week to being instantaneous.

For those voters who did return a ballot, many of the ballots were rejected and not included in the final election results because of problems with the ballot. For example, absentee ballots received after the close of the polls on Election Day are rejected, as are ballots that are missing certain information on the outside of the ballot envelope, such as the voter's signature and the voter's address. Even if the ballot envelope is signed, if the signature does not match the signature on file, the ballot is challenged and not included in the final results (Alvarez et al., 2008). In a study of ballots cast in California between 1990 and 2010, Alvarez et al. (2012) found that voters who cast a by-mail ballot are more likely to cast ballots with higher percentages of residual votes (the total number of votes that cannot be counted for a specific contest on a ballot). Voters who use absentee ballots do not get feedback that they have made a mistake voting as they would in a voting precinct, where an electronic ballot scanner or a direct recording electronic (DRE) voting system would tell them so. In California, all of the gains that were made in reducing residual votes by the adoption of modern voting technologies have been wiped out by the increased usage of absentee balloting (Alvarez et al., 2012).

Internet voting creates a system whereby computers can keep voters from making these simple mistakes when casting a ballot. The computer can inform a voter when they have failed to complete a required text field or when they have ' skipped a race' and failed to vote for one of the elected offices on a ballot. Thus, Internet voting provides a platform where the absentee voter can still vote remotely – as they would with a paper ballot – but they can do so in an environment where they can take advantage of technologies that help prevent mistakes. The convenience of by-mail absentee voting is only a convenience when the ballot that the voter casts is actually counted. Internet voting can help to ensure both that ballots are cast accurately and that the vote choices are captured correctly.

DIGITAL DIVIDES

One of the biggest questions related to Internet voting is one of fairness. Is it fair to allow people to vote online, given how easy it can be, when some people do not have equal access to computers or the Internet? The idea of the Internet facilitating participation in elections can only work if the public has broad access to the Internet. This is somewhat less of a concern in the developed world but is a major concern in the developing world. In the USA, there has been a steady increase in Internet usage over the past decade. In 2003, just over half of US households had Internet access but by 2011 that number was up to 71.7 percent (US Census Bureau, 2013: 1).

The US Census continues to find that the digital divide has a racial dimension – minority groups are less likely to be online compared to white Americans – but this gap has narrowed.

However, given the rise of mobile Internet, the Pew Internet and American Life Project finds that only 15 percent of Americans have no Internet access and most of these people just do not think the Internet is necessary or relevant to them.[6] The picture of Internet use is highly skewed toward the young; only 2 percent of respondents aged 18–29 and 8 percent of those aged 30–49 were not online. Cohen et al. (2012) go so far as to say that the 'digital divide', where access to the Internet was the main concern, no longer exists in the USA. They find that individuals 34 and older are more likely to use the Internet for non-entertainment purposes. Internet users in Generation X (those aged 34–45) and older cohorts are more likely than Millennials (those born after 1980) to engage in several online activities, including visiting government websites and getting financial information online. Younger Americans are more frequent Internet users but older individuals are quite adept at using the Internet for political purposes (Cohen et al., 2012: 20–21).

Europe has seen a similar increase in usage. In 2012, 76 percent of the respondents in a Eurostat 28-country European Union (EU) survey had regular access to the Internet and 22 percent of respondents had never been on the Internet.[7] The Eurostat data show that there are differences across European nations related to Internet access. Iceland, the Netherlands, Luxembourg, Norway, Denmark and Sweden each have household Internet access rates that exceed 90 percent. On the other end of the spectrum, Bulgaria, Greece, Romania and the countries of the former Yugoslav Republic have household Internet access rates that are below 60 percent. However, 90 percent of young people across Europe (aged 16–24) access the Internet daily (Seybert, 2011: 3).

In the developing world, access to computers and the Internet is much less common than in developed countries. However, cellular telephone penetration rates in developing countries tend to be quite high. For example, in developing countries approximately 89 percent of people have mobile cellular subscriptions.[8] For comparison, the subscription rate is more than 128 percent in developed countries, since personal cell phones are ubiquitous and many people have a second work cellular device. Mobile broadband is relatively rare in developing countries (~6.2 percent) and even has low penetration rates in developing countries (~27 percent). However, simple texting applications are becoming widely used for banking services in some developing countries, especially in Africa.[9] The World Bank recently reported that, in Kenya, two-thirds of adults have used a mobile telephone to receive money and 60.5 percent have used mobile phones to send money.[10] These data suggest that, with the proper incentives and infrastructure, even simple cellular phones can be used for important transactions such as banking. Making an application for voting via a cellular phone could also become feasible, with time.

SECURITY CONCERNS

The digital divide concerns related to Internet voting are important but also have the potential to be overcome. The biggest concerns about Internet voting relate to security. In 2013, these concerns became most apparent with the revelations related to US surveillance conducted on the Internet by the National Security Agency (NSA). As ProPublica reported:

> The National Security Agency is winning its long-running secret war on encryption, using supercomputers, technical trickery, court orders and behind-the-scenes persuasion to undermine the major tools protecting the privacy of everyday communications in the Internet age, according to newly disclosed documents. The agency has circumvented or cracked much of the encryption, or digital scrambling, that guards global commerce and banking systems, protects sensitive data like trade secrets and medical records, and automatically secures the e-mails, Web searches, Internet chats and phone calls of Americans and others around the world, the documents show.[11]

For critics of Internet voting, the efforts of the NSA to break Internet security illustrate why the idea of Internet voting is so dangerous. These critics note that the problem is not with Internet voting per se but with 'vulnerabilities associated with the Internet itself' (Lauer, 2004: 182), such as the threats of 'worms, viruses, and Trojan horses' (Schyren, 2004: 2). The NSA efforts to crack Internet security protocols suggest that the US government or other governments willing to invest high levels of resources into attacking any Internet platform could aim those resources at Internet voting, with the goal of either actually changing the outcome of an election or undermining public confidence in the outcome of an election.

The threats to the integrity of an election held online have analogous threats in the traditional world of elections, especially absentee voting. Alvarez and Hall (2004, 2008) provide a systematic review of the 'pathologies' related to remote voting – either paper-based or Internet-based – and note that the Internet realm replicates many of the by-mail voting threats. Volkamer (2009) notes that, with any activity online, the question is not whether it can be secure but whether it can be 'secure enough', since all Internet-based activities have fundamental threats associated with them. People often underestimate the risks of engaging in Internet-based activities such as voting, by pointing out the activities in which they already engage that involve transmitting highly sensitive or valuable personal information online. For example, the fact that a person can bank or shop online may give the impression that the Internet is a secure platform and that Internet voting would be one more convenience to be added to daily life. Simons and Jones (2012) argue that the threats associated with online banking are serious, and note that banks replace funds stolen through online fraud schemes. However, there is some evidence that, as both banks and consumers become more aware of the threats associated with online banking, the amount of fraud declines.[12]

The US Election Assistance Commission (EAC, 2011) commissioned a review of voting in the United States, which noted that there are several questions related to Internet voting that are difficult to answer, such as:

- Given that no system can be 100 percent secure, what level of risk can be accepted for such a fundamental democratic process as voting?
- How can a jurisdiction that wants to implement Internet voting and a company that wants to build an Internet voting platform minimize the level of risk associated with various methods and technologies?
- How do we create and implement standards for this technology and reliably test to those standards?

The EAC's report notes that Internet voting is often viewed as a possible replacement for the paper-based system that is currently used by overseas and military voters. They recommend – as Alvarez and Hall (2004, 2008) suggest – that the baseline for evaluating Internet voting is that it should at least be as secure as current absentee voting systems. However, no risk assessment has ever been performed on the by-mail absentee voting process for either domestic or overseas voters, so there is no baseline for making a comparison. Therefore, it is not possible to compare the threat profile for voting by mail to the threat profile for Internet voting. Interestingly, reviews of electronic voting technologies that were conducted by states because of concerns related to electronic voting machines (for example, the California 'Top-to-Bottom Review') never evaluated the threats associated with paper-based voting in precincts or paper-based by-mail absentee voting. Without these comparative data, it has not been possible to determine whether the threat profiles associated with electronic voting or Internet voting vary from those related to paper-based voting.

Although there is not a true baseline for comparing Internet and by-mail voting, scholars have often compared the two modes of voting. For example, Alvarez and Hall (2004) note that the risks associated with

Internet voting are often analogous to the risks associated with absentee mail voting. For example, attacks on ballots cast online are similar to mail tampering in absentee voting. With both types of voting, a person can potentially be coerced into voting in a desired way.

One factor that is critical for having secure Internet voting is for there to be a reliable way for individuals to authenticate themselves online. In the by-mail voting process in the US, voters authenticate themselves by signing the back of their ballot envelope and the signature on the envelope can be compared to the signature on file for that voter, either on a voter registration card or on the voter's driver's license record (which today is typically in an electronic file). In Estonia, where Internet voting is one of three modes of voting in all elections (in addition to in-person early voting and in-person Election Day voting), the country's residents are given digital identities. The digital identity operates much like an automated teller machine (ATM) card for banking. When an Estonian is engaging in an activity that requires online identity, they insert their identity card into a card reader attached to a computer (the card readers are inexpensive and common in Estonia) and then enter a personal identification number (PIN) into the system. The identification card reader takes information from the chip on the card and the system matches that information with the PIN. This combination of a 'hard' token (the identity card) and a 'soft' token (the PIN number) provides security that verifies that the person logging onto the system is, in fact, the correct person. In Switzerland, the country uses a different version of digital identity but this is a critical aspect of having secure Internet voting (Alvarez et al., 2009; Vinkel, 2012: 6–7).

Olsen and Nordhaug (2012: 37) point out that voters are the most vulnerable part of the voting system. Personal computers are not necessarily secure, and voters are vulnerable to misinformation about how to use the voting system correctly and this misinformation could take them to spoof websites that have already been hacked. Voters are all the more vulnerable because, although they may be likely to notice if something goes wrong with their online banking because of money lost, the anonymity of the voting process means that it is almost impossible for them to notice if their vote has been changed (see also Epstein, 2011: 900–902).

Computer scientists have made an effort to define the scope of the problems associated with the security of Internet voting. Hassan and Zhang (2013) have identified six areas that need to be addressed in security. The first is confidentiality, meaning that no one but the voter can be able to identify for whom they voted. The second area, anonymity, addresses concerns of identity theft and possible coercion by focusing on making sure that voters' identities cannot be tracked. Third, authenticity looks at ensuring that voters are eligible to vote, and only eligible voters can vote. The integrity of the system ensures that no one else can change a voter's ballot, and that if such tampering does occur, the system is able to detect the change. The fifth area, auditability, is tricky because it demands that the election must be auditable without violating confidentiality or anonymity. If the system cannot be properly audited then room for improvement is greatly diminished. The final area of Internet security that needs to be taken into consideration is that of verifiability. This area, closely tied to the others, allows the system to count and verify the votes, and to detect any missing or fraudulent votes. Any system that fails to address each of these areas makes itself more prone to attack.

INTERNET VOTING: NEXT STEPS

Internet voting is a concept that is often viewed as very black and white. Some people see it as a panacea for making voting easier and for bringing younger people into the electoral process. Others see Internet voting as impossible because of fundamental flaws with the Internet. As with most things, it is likely that the actual answer is much more nuanced. It is unlikely that, in the United States, Internet voting will be a widespread voting platform in the next several years. The infrastructure needed to make Internet voting work – widespread use of digital identification, public confidence in this voting system, as well as a secure voting platform – are not likely to appear in the next several years because of the costs associated with developing them. However, this does not mean that citizens of the United States will never vote over the Internet. Instead of widespread use of Internet voting, there instead will be targeted uses of Internet voting that address very specific voting problems for very specific voting populations.

The two most prominent examples where Internet voting efforts will continue to develop are with UOCAVA voters and individuals with disabilities. The UOCAVA voter continues to have problems related to ballot transit time. The difficulty remains of getting ballots to people living overseas, to military personnel deployed away from home, and to the dependents of military personnel, and then getting those ballots back to the election official so that they can be counted. The Military and Overseas Voter Empowerment (MOVE) Act required states to send ballots to UOCAVA voters at least 45 days prior to an election so that voters can receive the ballot, vote and return the ballot with more time. However, there will still be benefits associated with the electronic transmission of ballots and the electronic return of ballots. For example, voters who can vote online, print the ballot and mail it back are less likely to make a mistake marking the ballot. Some states and localities may experiment with full Internet voting for UOCAVA voters but others will experiment with the one-way electronic transmission of ballots. These efforts will provide important information that can improve knowledge of how to make Internet voting useful.

For people with disabilities, one of the biggest issues is actually getting to the polling place to vote.[13] For these voters, Internet voting can be very helpful because the ballot box can be brought into their homes. As the populations of the USA and the rest of the countries in the Organisation for Economic Co-operation and Development (OECD) continue to age, making voting easier for this population will continue to be a pressing concern. In Oregon and Denver, Colorado, individuals with disabilities can vote on an iPad and then have their ballot printed out, which provides both an electronic version and a paper version of the ballot. The paper ballot can be placed in a mobile ballot box and the electronic ballot can be audited against the paper ballot.[14]

Voting on iPads can be taken outside a specific voting jurisdiction, even overseas, and made into 'kiosk' Internet voting. With kiosk Internet voting, a voter casts a ballot electronically and it is sent to the voter's local election official to be included in the final tabulation of votes. However, unlike traditional Internet voting, the voter has to physically travel to a kiosk – an electronic voting machine located in a secure location, such as an embassy – to cast their vote. The ballot would then be transmitted electronically, but a paper record of the vote may be produced as well and kept securely for auditing purposes. With kiosk voting, many of the risks associated with Internet voting can be resolved, especially authentication and auditing, but the benefits of speed of ballot delivery and return, and of ballot accuracy, can be addressed.

Internet voting will continue to be an area of experimentation in elections for some time. Its widespread use will not likely occur in the next decade but expanded application of this technology will provide more data and learning so that, in the future, voting remotely will be much more common.

NOTES

1. http://www.ifes.org/Content/Publications/News-in-Brief/2012/June/~/media/Files/Publ icatioms/Reports/2012/EVote_International_Experience_2012.pdf (accessed September 29, 2013).
2. http://www.regjeringen.no/en/dep/krd/press/press-releases/2012/e-vote-project-evaluati on-is-ready.html?id=685023 (accessed September 29, 2013).

3. As defined by the federal government, 'UOCAVA citizens are US citizens who are active members of the Uniformed Services, the Merchant Marine, and the commissioned corps of the Public Health Service and the National Oceanic and Atmospheric Administration, their family members, and US citizens residing outside the United States', http://www.fvap.gov/reference/laws/uocava.html (accessed September 29, 2013).
4. Gronke and Miller (2012) summarize this controversy well.
5. The replication results come from the Survey of the Performance of American Elections (2012), http://thedata.harvard.edu/dvn/dv/measuringelections/faces/study/StudyPage.xhtml?globalId=hdl:1902.1/21624&studyListingIndex=0_8508ab729c048b7e9ba27c369457 (accessed January 20, 2015).
6. http://www.pewinternet.org/Reports/2013/Non-internet-users/Summary-of-Findings.aspx (accessed September 29, 2013).
7. http://epp.eurostat.ec.europa.eu/portal/page/portal/information_society/data/main_tables (accessed September 29, 2013).
8. http://mobithinking.com/mobile-marketing-tools/latest-mobile-stats/a#subscribers (accessed September 29, 2013).
9. http://www.economist.com/node/21553510 (accessed September 29, 2013).
10. http://blogs.worldbank.org//publicsphere/media-revolutions-how-many-kenyans-use-mobile-money (accessed September 29, 2013).
11. http://www.propublica.org/article/the-nsas-secret-campaign-to-crack-undermine-internet-encryption. See also ProPublica's array of other work on this topic, http://www.propublica.org/series/surveillance (accessed September 29, 2013).
12. Data on losses associated with online fraud are not easy to find. However, see http://www.dutchnews.nl/news/archives/2013/09/losses_due_to_internet_banking.php for one example of the decline in rates of such fraud (accessed September 29, 2013).
13. http://elections.itif.org/reports/AVTI-001-Hall-Alvarez-2012.pdf (accessed September 29, 2013).
14. http://investigations.nbcnews.com/_news/2012/08/30/13570208-vote-on-an-ipad-technology-could-supplant-voter-ids-at-polls?lite (accessed September 29, 2013).

FURTHER READING

Alvarez, R. Michael and Thad Edward Hall. 2004. *Point, Click, and Vote: The Future of Internet Voting*. Washington, DC: Brookings Institution Press.
Alvarez, R. Michael, Thad E. Hall and Alexander H. Trechsel. 2009. Internet Voting in Comparative Perspective: the Case of Estonia. *Political Science and Politics*, 42(3): 497–505.
Elections Canada. n.d. Establishing a Legal Framework for E-voting in Canada: Literature and References. http://www.elections.ca/content.aspx?section=res&dir=rec/tech/elfec&document=ab&lang=e.
International Foundation for Electoral Systems. 2012. International Experience with E-voting – Norwegian E-Vote Project. Washington, DC: International Foundation for Electoral Systems. http://www.regjeringen.no/upload/KRD/Prosjekter/e-valg/evaluering/Topic6_Assessment.pdf.
OSCE-ODIHR. 2009. Discussion Paper in Preparation of Guidelines for the Observation of Electronic Voting. http://www.osce.org/odihr/elections/34725?download=true.
Pammett, Jon H. and Nicole Goodman. 2013. Consultation and Evaluation Practices in the Implementation of Internet Voting in Canada and Europe. Quebec: Elections Canada. http://www.elections.ca/res/rec/tech/consult/pdf/consult_e.pdf.
US Election Assistance Commission. 2011. Testing and Certification Technical Paper #2: A Survey of Internet Voting. http://www.eac.gov/assets/1/Documents/SIV-FINAL.pdf.

REFERENCES

Alvarez, R. Michael, Dustin Beckett and Charles Stewart. 2012. Voting Technology, Vote-By-Mail, and Residual Votes in California, 1990–2010. *Political Research Quarterly*, 66(3): 658–670.

Alvarez, R. Michael and Thad Edward Hall. 2004. *Point, Click, and Vote: The Future of Internet Voting*. Washington, DC: Brookings Institution Press.

Alvarez, R. Michael and Thad E. Hall. 2008. *Electronic Elections: The Perils and Promises of Digital Democracy*. Princeton, NJ: Princeton University Press.

Alvarez, R. Michael, Thad E. Hall and Betsy Sinclair. 2008. Whose Absentee Votes are Returned and Counted: The Variety and Use of Absentee Ballots in California. *Electoral Studies*, 27(4), 673–683.

Alvarez, R. Michael, Thad E. Hall and Alexander H. Trechsel. 2009. Internet Voting in Comparative Perspective: the Case of Estonia. *Political Science and Politics*, 42(3): 497–505.

Atkeson, L.R., L.A. Bryant, T.E. Hall, K. Saunders and M. Alvarez. 2010. A New Barrier to Participation: Heterogeneous Application of Voter Identification Policies. *Electoral Studies*, 29(1): 66–73.

Berinsky, A.J. (2005). The Perverse Consequences of Electoral Reform in the United States. *American Politics Research*, 33(4): 471–491.

Cohen, Cathy J., Joseph Kahne, Benjamin Bowyer, Ellen Middaugh and Jon Rogowski. 2012. New Media and Youth Political Action. http://www.civicsurvey.org/YPP_Survey_Report_FULL.pdf (accessed July 2, 2013).

Election Assistance Commission (EAC). 2011. A Survey of Internet Voting. http://www.eac.gov/assets/1/Documents/SIV-FINAL.pdf (accessed July 2, 2013).

Epstein, Jeremy. 2011. Internet Voting, Security, and Privacy. *William and Mary Bill of Rights Journal*, 19: 885–906.

Gray, Mark and Miki Caul. 2000. Declining Voter Turnout in Advanced Industrial Democracies, 1950 to 1997: The Effects of Declining Group Mobilization. *Comparative Political Studies*, 33(9): 1091–1122.

Gronke, Paul, Eva Galanes-Rosenbaum, Peter A. Miller and Daniel Toffey. 2008. Convenience Voting. *Annual Review of Political Science*, 11: 437–455.

Gronke, Paul and Peter Miller. 2012. Voting by Mail and Turnout in Oregon Revisiting Southwell and Burchett. *American Politics Research*, 40(6): 976–997.

Hall, Thad E. 2012. Electronic Voting. In Norbert Kersting (ed.), *Electronic Democracy*. Leverkusen, Germany: Barbara Budrich Publishers.

Hassan, Ahmed and Xiaowen Zhang. 2013. Design and Build a Secure E-voting Infrastructure. Systems, Applications, and Technology Conference, 2013. IEEE.

Karp, Jeffrey A. and Susan A. Banducci. 2000. Going Postal: How All-Mail Elections Influence Turnout. *Political Behavior*, 22(3): 223–239.

Kousser, Thad and Megan Mullin. 2007. Does Voting by Mail Increase Participation? Using Matching to Analyze a Natural Experiment. *Political Analysis*, 15: 428–445.

Norris, Pippa 2001. *Digital Divide: Civic Engagement, Information Poverty, and the Internet Worldwide*. Cambridge University Press.

Olsen, Kai A. and Hans Fredrik Nordhaug. 2012. Internet Elections: Unsafe in Any Home? *Communications of the ACM*, 55(8): 36–38.

Pammett, Jon H. and Nicole Goodman. 2013. Consultation and Evaluation Practices in the Implementation of Internet Voting in Canada and Europe. Quebec: Elections Canada. http://www.elections.ca/res/rec/tech/consult/pdf/consult_e.pdf.

Riker, William H. and Peter C. Ordeshook. 1968. A Theory of the Calculus of Voting. *American Political Science Review*, 62(1): 25–42.

Schryen, Guido. 2004. Security Aspects of Internet Voting. *Proceedings of the 37th Annual Hawaii International Conference on System Sciences*. http://epub.uni-regensburg.de/21303/1/Schryen_-_Security Aspects_of_Internet_Voting_-_HICSS.pdf.

Seybert, Heidi. 2011. Internet Use in Households and by Individuals in 2011. http://epp.

eurostat.ec.europa.eu/cache/ITY_OFFPUB/KS-SF-11-066/EN/KS-SF-11-066-EN.PDF (accessed July 2, 2013).

Simons, Barbara and Douglas W. Jones. 2012. Internet Voting in the US. *Communications of the ACM*, 55(10): 68–77.

Smith, Aaron Whitman, Kay Lehman Schlozman, Sidney Verba and Henry Brady. 2009. *The Internet and Civic Engagement*. Washington, DC: Pew Internet and American Life Project.

Trechsel, Alexander H., Kristjan Vassil, Guido Schwedt, Fabian Breuer, Michael R. Alvarez and Thad E. Hall. 2010. Report: E-Voting in Estonia, 2005–2009. SaverioCor. http://www.scribd.com/doc/246048564/Report-E-Voting-in-Estonia-2005-2009#scribd.

US Census Bureau. 2013. Computer and Internet Use in the United States. Washington, DC: US Census Bureau. http://www.census.gov/prod/2013pubs/p20-569.pdf (accessed July 2, 2013).

Vinkel, P. 2012. Internet Voting in Estonia. *Lecture Notes in Computer Science*, 7161: 4–12.

Volkamer, M. 2009. *Evaluation of Electronic Voting*. Lecture Notes in Business Information Processing. Berlin: Springer.

Yao, Yurong and Lisa Murphy. 2007. Remote Electronic Voting Systems: an Exploration of Voters' Perceptions and Intention to Use. *European Journal of Information Systems*, 16: 106–120.

8. Digital campaigning
Daniel Kreiss

Political campaigning has often been the vessel for democratic hopes amid the vast changes in media that have taken shape during the past two decades. Since the emergence of the World Wide Web in the early 1990s, scholars, practitioners, pundits, and the public have debated, forecast, hyped and celebrated the Internet's potential to revolutionize campaigning and speculated about the implications for civic engagement, citizenship, expression and ultimately power in a democracy. In the USA, John McCain's small-dollar fundraising during the 2000 primaries, Howard Dean's spectacular online grassroots mobilization, George W. Bush's online field organizing during his re-election bid, and the user-generated outpouring around Barack Obama's two campaigns have fueled democratic desires for a more participatory polity.

Despite apparent changes in the ways candidates contest elections, how and even whether networked media are reshaping electoral campaigning is a matter of considerable and long-standing debate among scholars. As Rasmus Nielsen (2013) has noted, many early scholarly perspectives that posited strong distinctions between the real and virtual and often saw revolutionary changes afoot have gradually given way to contemporary accounts that view electoral campaigning (and movement campaigns of all stripes) as being 'internet-assisted' (Nielsen, 2011) or 'digitally-enabled' (Earl and Kimport, 2011). This captures the idea that digital media are not creating a world separate and apart from the social and material contexts within which they are taken up. Citizens, campaigns, parties and other political actors create and use digital tools in the service of their political aims that take shape in institutionalized contexts. Rather than radically transforming them, digital tools often further existing electoral practices, from fundraising and canvassing to phone banking and attack advertising.

There are, however, changes in campaigning that have been enabled by the advent and widespread uptake of digital media. As the scale of contemporary campaign activities clearly demonstrates, the Internet has significantly amplified certain electoral practices. It is much easier and faster for citizens to donate to their favored candidates and causes online. Citizens now have a range of commercial and campaign-provided digital tools that enable them to more easily find those with similar political interests and organize and manage events for their favored candidates.

Volunteers can now access campaign voter lists online, canvass high-priority targets among their neighbors, and record the outcomes of these conversations back into the data-coffers of campaigns. Social media platforms, such as Twitter and Facebook, provide citizens with new capacities to express themselves politically, affording them the ability to create, share and remix political content with their friends and neighbors as well as global publics (Rainie et al., 2012). In other words, even if the formal legal codes of electoral institutions and political representation have not changed (which still require forms of geographic-based mobilization and the fundraising necessary to contest elections) there are still significant changes in the organization, practice, and experience of campaigning for many of the actors involved.

This chapter proceeds as follows, focusing exclusively on the thoroughly researched United States context. First, it details the major findings of the literature to date, traces the broad contours of scholarly disagreement, and discusses the limitations of the literature. Second, it outlines my approach to analyzing campaigning and offers some directions for future research, focusing on: exogenous shifts in digital platforms and applications that take shape outside of, but affect, the political field; the strategic action of political actors within particular technological contexts; and infrastructure building by campaigns, parties and other actors such as consultants that shape the capacities and contexts for digital campaigning. The chapter also offers additional readings for students of digital campaigning.

THE RESEARCH LITERATURE

A number of scholars have examined how the affordances of technologies have affected political campaigning and the electorate more generally.[1] At considerable risk of simplifying, within this literature there are two broad traditions of research into digital campaigning. One has looked at the organizational layer of politics through the lens of the types of collective political action in electoral contexts that new media afford. The other has examined the effects of digital campaign content and new media environments on political knowledge, attitudes and participation.

Collective Action and Political Organization

While digital technologies such as e-mail have been a part of governance processes at the presidential and congressional levels dating from 1992, it was the 1996 presidential election that featured the first candidate websites (Foot and Schneider, 2006). The Clinton and Dole campaigns both provided e-mail updates to supporters and content that scholars have dubbed 'brochureware': generally simple HTML versions of printed literature providing information to voters, although the Dole campaign's site also had some interactive features (ibid.).

During the 2000 presidential primaries the first contemporary uses of the Internet emerged, as campaigns began to focus on volunteer recruitment and mobilization, fundraising, and strategic messaging through social networks. In other words, campaigns began using the Internet to speak primarily to their pre-existing supporters, not undecided voters. Bruce Bimber and Richard Davis (2003) were among the first scholars to clearly identify these practices, arguing that campaigns used the Internet to shore up support, drive fundraising, and register and mobilize their voters. Political staffers themselves drove many of these changes, recognizing for the first time that the primary users of their websites were supporters, not undecided voters seeking detailed policy statements.[2] In response, the presidential campaigns of the cycle began explicitly encouraging supporter participation, in part through the design of dynamic, interactive pages for users.

The losing primary campaigns of Democratic presidential candidate Bill Bradley and Republican presidential candidate John McCain in 2000 provide two examples of the potential of small-dollar online fundraising. McCain raised record amounts of money online after his New Hampshire primary victory over George W. Bush. During the cycle, campaigns also began using the Internet to involve supporters in activities such as promoting the visibility of candidates through providing printable literature and signs for supporters to distribute in their communities, as well as tips for contacting local news outlets to promote the candidate. Candidates also increasingly used the Internet to fashion supporters into the conduits of strategic communications. Al Gore's campaign enabled supporters to create their own customized webpages based on template policy content so that they could e-mail them to their friends and family.

These uses of digital media demonstrate that campaigns have largely taken up the Internet in ways that amplify existing electoral practices, given the stability of political institutions and the legal frameworks for political representation. This was true even during the 2004 campaign cycle, despite some real innovations in digital campaigning. The Howard Dean primary campaign was innovative on a number of levels, largely in response to the mobilization spurred by the then emerging anti-war, intra-party social movement known as the 'netroots' (Kreiss, 2012a). The netroots drove much early attention, money and volunteers to the Dean campaign. It also prompted many of the campaign's technological innovations as staffers struggled to harness and direct this energy around the candidate, such as by creating new social platforms that enabled supporters to plan events, donate money, network among themselves and contact voters (ibid.). And yet, while the Dean campaign was the product of the massive mobilization of supporters online facilitated by the lowered costs of organizing and taking action that digital media affords, this volunteerism was directed towards institutionalized ends: fundraising, voter contact, and messaging. As Matt Hindman (2005) pointed out, even on the Dean campaign the Internet's effects have generally come at the operational 'back end', enabling staffers to more efficiently perform routine electoral functions. While the Internet has certainly enabled more citizens to get involved and express themselves politically in public, digital media have not brought about revolutionary changes in institutionalized electoral practices, such as more radical forms of participatory policy-crafting that many observers had hoped for.

In the end, the Dean effort came up short precisely in those domains where digital media met the demands of electoral institutions. While the national online effort was exceptional at garnering financial resources and journalistic attention for the campaign, it failed to translate into effective on-the-ground voter identification, persuasion, and turnout efforts (Kreiss, 2012a). Although it has received comparatively less attention in the literature on new media and politics, the campaign of the 2004 cycle that most successfully incorporated digital media into larger electoral strategy and successfully harnessed database and networked technologies for the purposes of field campaigning was George W. Bush's re-election effort. Obama's campaign manager David Plouffe (2009) acknowledged as much, citing the Bush team's re-election effort as a model for the historic 2008 run. The Bush campaign created a 'virtual precinct captain' program, where online volunteers stewarded electoral districts in geographic areas that were a priority for the campaign. Bush's re-election effort also revealed that the Republican Party and its consultancies had far more robust online platforms, voter databases, and volunteer and voter mobilization efforts than the Democratic Party (Nielsen, 2012).

As I have shown in previous work (Kreiss, 2012a), the 2008 Obama campaign was an extension of many developments in digital campaigning that took shape across earlier electoral cycles, not a radical or revolutionary break with history. I focus on the 2008 and 2012 Obama campaigns here because practitioners on both sides of the aisle view them as the most technologically sophisticated electoral efforts to date, and the standards to be emulated. This has occurred to such an extent that the national Republican Party and its allied consulting firms have extensively studied the Obama 2012 effort and made considerable investments to match and advance the campaign's digital strategies and tactics (see, for instance, Engage's extensive report on the 2012 Obama campaign, *Going Inside the Cave*).[3]

On the one hand, the 2008 and 2012 Obama campaigns were the product of extensive infrastructure-building efforts launched after John Kerry's defeat in 2004. As head of the Democratic Party, Howard Dean recruited his former staffers to steward the creation of a powerful new voter database and online interface system, called VoteBuilder (or 'the VAN' after Voter Activation Network, the firm that built the interface). At the same time, Democratic consultancies and progressive non-profit and advocacy organizations launched an extraordinary array of infrastructure-building efforts that laid the foundation for Obama's initial run. This included the development of the consultancy Blue State Digital's electoral platform rebuilt from the Dean campaign's toolset, which the firm provided to the Democratic Party ('Partybuilder') and the Obama campaign (My. BarackObama.com), among many other campaigns in 2006 and 2008.

On the other hand, the 2008 and 2012 Obama campaigns were products of the development of knowledge, skills, and practice around these database and electoral technologies. A number of Dean's former staffers including Joe Rospars, Blue State Digital co-founder and director of the Obama campaign's New Media Division in 2008 and chief digital strategist for 2012, carried with them their experiences from the failures of the 2004 bid. This resulted in a 2008 Obama campaign that made the organizational decisions and technological investments that supported an extraordinarily effective electoral effort. For example, one hallmark of Obama's 2008 effort was the integration of digital and on-the-ground field efforts, a strategy that paid dividends during the 23-state contests that took place on 'Super Tuesday' when Obama was able to remain competitive with Hillary Clinton. Again, while the 2008 Obama campaign was the vessel for many democratic aspirations by scholars, pundits, and citizens alike, the ethos of the campaign is best captured in the New Media Division's mantra of 'money, message, and mobilization'. While there was extraordinary popular participation around the campaign, particularly on platforms such as YouTube and Facebook, at the end of the day the campaign sought to leverage this mobilization in the service of its electoral goals rather than providing supporters voice into the policy or strategy of the campaign.

The 2012 Obama effort, in turn, featured the extension of many of the digital practices of the 2008 campaign, especially the 'computational management' (Kreiss, 2012a: 144) style of organizational decision-making within the campaign. This involved the delegation of managerial, allocative, messaging, and design decisions to analysis of the campaign's multiple data streams. As a number of journalistic accounts of the 2012 Obama campaign suggest, data was central to all aspects of the re-election effort, a development facilitated by the affordances of digital media. For example, extending the voter modeling efforts of the 2008 campaign, the re-election bid assigned numerical scores of likely political attitudes and behavior to every member of the electorate. These scores are the outgrowth of an enormous proliferation of data about citizens over the last decade and, as importantly, new analytical techniques that render them meaningful. The re-election campaign used four scores that on a scale of 1 to 100 modeled voters' likelihood of supporting Obama, turning out to vote, being persuaded to turn out, and being persuaded to support Obama on the basis of specific appeals (Beckett, 2012). These modeling scores were the basis for the entire voter contact operation, which ranged from making 'personalized' appeals on the doorsteps (Nielsen, 2012) and through the social media accounts of voters (Judd, 2012), to running advertisements on the cable television screens of swing voters (Rutenberg, 2012).

To date, there has been little systematic research focused on the diffusion, or lack thereof, of the 2008 or 2012 Obama campaigns' data and digital campaign tactics to Democratic campaigns at other levels of office or the Republican Party more generally, save for some limited work on why Republicans failed to adopt Obama's innovations after the 2008 cycle (Kreiss, 2014). That said, journalistic reports suggest both that the data-driven approach of the 2008 campaign was further developed in 2010 for the Democratic Party during the midterm elections (Issenberg, 2012b), and that since 2012 a host of new Democratic firms founded by alumni of the re-election bid carried similar data, analytics, and targeted media message strategies to gubernatorial races (Daileda, 2013) and Organizing for Action (Stirland, 2013). Republicans in turn have made large investments in their own data and digital organizing efforts since the 2012 cycle (Ball, 2014).

Given a two-decades-long trend towards mobilization-based campaigns featuring 'personalized political communication' (Nielsen, 2012) across many different interpersonal contexts and media platforms, it is likely that data and analytics, which increase the efficiency of voter contacts, will play more prominent roles in electoral politics in the years ahead. Even more likely is that the organizational efficiencies and resource gains that result from computational management will lead to these practices being more widely adopted across the two parties. That said, as Nielsen and Vaccari (2013) argue, while 'push' strategies of delivering campaign content to people seem broadly transferrable across campaigns at all levels of office, 'pull' strategies that are premised on voter interest in 'opting in' are contingent upon electoral contexts. As such, while candidates have widely taken them up, there is limited utility of Facebook, Twitter, and YouTube for the vast majority of campaigns, given generally low interest in politics in a high media choice environment (ibid.). This suggests a broader cautionary note, that scholars need to carefully analyze how campaign digital media practices may diffuse in terms of adoption – especially from comparatively well-financed presidential campaigns – but have differential effects in varying electoral contexts given seemingly fixed constraints such as generally low interest in contemporary elections.

In sum, the history of the uptake of digital media in campaigning is a story of institutional amplification, not necessarily democratic revolution. Digital media have dramatically amplified some forms of political collective action in institutionalized contexts. Campaigns use digital media to significantly lower the cost to supporters of making small-dollar contributions online. Supporters have more opportunities to volunteer, and it is far easier to do so than it once was, as phone banking, event planning, and fund-raising have gone online. Meanwhile, from Twitter to Facebook, supporters have new vehicles for political expression and engagement. What new media have not necessarily done, however, is make campaigns more responsive to their mobilized supporters outside of the generally shared ends of getting a candidate elected (Kreiss, 2012a).

Political Knowledge, Attitudes and Participation

As Neuman et al. (2010: 24–26) point out in their review of the literature, students of the effects of digital media on politics need to be attentive to three factors that complicate the research findings: diffusion effects, differential effects, and conditional effects. All three factors have significant implications for digital campaigning. Technologies diffuse at different rates throughout societies, meaning that we cannot presume that what we see now will be the case 20 years from now as more people come online. Diffusion effects suggest that what we see happening in digital campaigning today may be different in the future. Differential rates of adoption based on factors such as culture, class, age, and social position in turn suggest that the technologically rich may often be getting richer. Differential effects suggest that there are likely significant differences among the citizens who engage in digital campaigning based on pre-existing political interest and knowledge. Finally, conditional effects refer to the fact that some groups of people will be advantaged by the affordances of technology more than others, given the contexts within which they adopt it. Conditional effects suggest that digital campaigning may ultimately benefit those already most engaged in political life or organized to adopt new technologies.

With these in mind, we can consider the literature on digital campaigning from the standpoint of the types of political communication digital media afford among citizens and campaigns, the effects of digital media on knowledge and attitudes about politics, and the motivations and contexts for participation in digital campaigns. These are significant areas of focus for scholars and reflect, even if they are not stated explicitly, normative theories of democracy (for a review, see Freelon, 2010). Studies of digital campaigning that look at the effects on the electorate are animated by concerns over citizen deliberation and knowledge, polarization, and participation. Within each of these domains there is significant debate and a continual evolution in findings, in part given the diffusion, differential, and conditional effects detailed above.

From the very beginnings of the uptake of digital media in campaign contexts, scholars have seen the possibility for a more deliberative polity, in terms of both citizen dialogue and, more broadly, access to information for improved political knowledge. Prevalent in much of the early literature was the idea that new media would bring about a new era of deliberative democracy as citizens took to the Internet, accessed information, and debated the merits (and demerits) of parties, candidates, and policies, especially in electoral contexts (for a review of this literature and scholarly debate, see Chadwick, 2006; Neuman et al., 2010).

A decade of empirical scholarship, however, has generally found the notion that a new era of deliberation was reshaping democratic processes to be overstated. On a structural level, political scientist Matthew Hindman (2008) has influentially shown that very few voices are heard online and shape public debate, and they are often elites and those already most well resourced to contribute politically. Farrell and Drezner (2008) show that the professional media still largely set the agenda for blogs and other non-professional outlets. On a social and psychological level, scholars have shown that deliberation has to be 'designed-in' to online platforms (Wright and Street, 2007); and this is exceedingly rare in the context of campaigning. As the body of empirical literature suggests, online discussion in electoral contexts features a range of communicative styles, including emotional, moral, and partisan appeals, not simply the rational, generalized, and respectful discourse that lies at the heart of normative deliberative theory. As scholars such as Michael Schudson (2003) have suggested, this means that political communication online looks much the same as offline, and the normative ideal of rational, critical debate is the outlier, not the norm (even across historical periods).

Meanwhile, studies suggest that while there have been some new formats for campaign communications, they have generally furthered the instrumental ends of electoral politics, and have not necessarily created a more deliberative polity. Foot and Schneider's (2006) extensive content analysis of campaign websites during the 2000, 2002, and 2004 election cycles found that campaigns sought to inform, involve, connect, and mobilize users in the context of electing candidates. This finding was echoed by a unique cross-national comparative study that found markedly similar uses of websites for electoral purposes (Foot et al., 2009). A number of scholars have studied campaign use of the video-sharing site YouTube, finding that the platform supports a range of new campaign strategic content such as inspirational videos and footage of gaffes from the campaign trail. That said, the diffusion of these videos (and citizen-produced content) is premised on how campaigns and other actors interact around these platforms in pursuit of their strategic political ends (Karpf, 2010; Wallsten, 2010).

All of these forms of campaign content are publicly available, but there are considerable methodological difficulties in studying campaign communications that are targeted. The marriage of data, content, and inter-activity online has made it difficult if not impossible to study some new forms of campaign communications through conventional methods such as content analysis and experiments. For one, unlike in the mass communication era, online advertising, e-mails, and even webpages are tailored on the basis of data and analytics through website optimization and targeting. Howard (2006) noted many of these tactics a decade ago, and recent work (Serazio, 2014) suggests that these practices have only grown more sophisticated. In-depth interview studies suggest that e-mails are targeted and tailored based on the geography, behavior, and demographics of citizens, as well as their prior history with campaigns and the sequences of appeals (Kreiss, 2012a). The challenge is getting inside campaigns to discover the processes behind targeting, or figuring out how to draw a sample of such narrowly tailored content. For instance, Barnard and Kreiss (2013) conducted a comprehensive survey of the research literature on online advertising and evaluated it in light of interview data from practitioners working in digital campaign advertising. Their conclusion is that the experimental literature is premised on manipulations that were commonplace in campaigning more than a decade ago, and that conventional experimental methods and content analysis of digital campaign advertising may be ill-equipped to discover the nuances of ads and targeting when 'individualized information flows' between campaigns and citizens are increasingly the norm (ibid.).

Many of these studies suggest that digital campaign communications fail to meet normative deliberative standards, and also that they are often used to appeal to likely supporters of candidates and those who are already committed and engaged. This accords with the findings of a generation of political scientists who have argued that those who are most interested in and knowledgeable about politics, and routinely engaged in political discussion online, are generally those who are the most ideologically committed and partisan (Abramowitz, 2010). Increased media choice reinforces this phenomenon. As Prior (2007) has demonstrated, with the rise of a much more fragmented media environment individuals can turn away from politics entirely if they are uninterested. The loss of 'inadvertent exposure' (ibid.) with the increased capacity for media choice has exacerbated information and knowledge inequalities, with significant consequences for who participates in electoral politics. In other words, given that political interest generally shapes political knowledge and behavior, in the era of cable television and the Internet citizens uninterested in politics can avoid seeking out and even seeing much political information during a campaign. The Internet, on its own, has not necessarily made for better-informed or more knowledgeable citizens on the whole; a finding echoed by Zukin et al. (2006) who find markedly stable average patterns of political knowledge and interest over the past 50 years.

Alongside political interest shaping political knowledge is the phenomenon of polarization. Citizens not only turn away from political information entirely, but can also self-select partisan media (Baum and Groeling, 2008). That said, scholars consistently find that the most knowledgeable and committed partisans not only consume general interest news, but also tend to be the most aware of arguments across the ideological spectrum, which makes them look like model citizens (Prior, 2013). The roots of polarization are many and diverse (see Abramowitz, 2010). In the context of media choice, there is debate over whether consuming the information of certain media outlets or platforms actually leads to more extreme positions, or whether media use comes after pre-existing partisanship but also has its own differential effects in terms of issue publics (Prior, 2013; Stroud, 2011). Either way, campaigns have adopted strategies to appeal narrowly to their base of supporters through new partisan digital outlets. Studies of the netroots show how campaigns such as Obama's treat bloggers as important conduits of information to likely Democratic voters during primaries, and to the legacy press during general election campaigns when the electorate is more ideologically diverse (see Kreiss, 2012b).

Scholars also find that political interest shapes not only media choice and knowledge, but also the decision to participate in digital campaigning, although research that specifically accounts for social media is just now emerging. Research suggests that participation in campaigning online generally reflects the ways that political motivation, interest, and resources shape all forms of political engagement (Xenos and Moy, 2007). Scholars argue that the Internet has not generally brought about massive increases in donations, attendance at political events, or voting when political interest is considered (Boulianne, 2009). Meanwhile, recent studies found that it is the wealthy and well educated who are the most politically active online (Schlozman et al., 2010).

That said, other studies that account for social media use suggest a potential positive causal effect of digital media use on political participation. One contingent of scholars argues that certain kinds of Internet use increase the likelihood of voting, and this has more pronounced effects among youth, given that they are the most likely to be online and the least likely to be politically engaged (Mossberger et al., 2008). De Zúñiga et al. (2010) argue that consuming new forms of political information online predicts political engagement; Warren and Wicks (2011) argue that online media use socializes youth into civic engagement; and Pasek et al. (2009) find that social media use builds social capital that translates into political engagement (see also Shah et al., 2001). The literature has suggested a number of potential causal factors, including the lowered costs of communication, new forms of online mobilization, targeted communications from campaigns and other political actors, inadvertent exposure to political content shared through social media, the blending of new audio and video formats, the accessibility of information, and social affordances of online engagement (for a review, see Mossberger and Tolbert, 2010). In addition, the 2012 Obama campaign's 'targeted sharing' program – where supporters on Facebook were asked to contact friends who the campaign targeted to ask them to do things such as register to vote and volunteer – suggest that campaigns may be able to take up social media as a push medium, creating new 'digital two-step flows' of information to online opinion leaders for strategic purposes.

NEW DIRECTIONS IN RESEARCH

In addition to the diffusion, differential, and conditional effects cited above, it is necessary to add the potential for 'medium effects' to the literature on digital campaigning. Despite the considerable insights of the literature cited above, many of its limitations stem from the fact that studies about new media and politics are often premised on assumptions that the medium itself, or at least its dominant and salient features, remains stable. And yet, as Karpf (2012) argues:

> The Internet is unique among Information and Communications Technologies (ICTs) *specifically because* the Internet of 2002 has important differences from the Internet of 2005, or 2009, or 2012. It is a suite of overlapping, interrelated technologies. The medium is simultaneously undergoing a social diffusion process and an ongoing series of code-based modifications. Social diffusion

brings in new actors with diverse interests. Code-based modifications alter the technological affordances of the media environment itself . . . What was costly and difficult in 2004 is cheap and ubiquitous in 2008. That leads, in turn, to different practices. The Internet's effect on media, social, and political institutions will be different at time X from that at time $X + 1$, because the suite of technologies we think of as the Internet will itself change within that interval.

In other words, given a medium undergoing significant and ongoing changes, it is exceptionally difficult to generalize findings about digital campaigning from one time period to another. Scholars can look for continuities in social and technological practice, such as the development of computational management practices in digital campaigning cited above, but it is difficult to generalize claims about how candidates and citizens use the Internet across election cycles from studies conducted at one moment in time. For example, arguments and empirical findings about Facebook and political knowledge and participation are only valid to the extent that Facebook remains the same; a tenuous claim given that new functionalities are added to the platform almost daily (while others are taken away). Meanwhile, scholars who have located the Internet's effects on the operational 'back end' of campaigns were certainly correct on a number of levels in looking at the 2004 cycle, but this view dramatically understates how the 2012 Obama campaign used digital technologies to engage in 'front end' strategic campaign communications on an individualized level (see, for example, Judd, 2012; Issenberg, 2012b).

To account for medium effects along with the ongoing diffusion of digital media throughout society and their uptake into many domains of social life, my current conceptualization of researching digital campaigning lies in looking at the interplay of changes in the medium and its application layer, the strategic actions of campaigns and parties, and the infrastructure that shapes the background contexts of action for campaigns and citizens. First, the history of digital media and campaigning reveals a gradual and continual evolution in the affordances of the Internet developed outside of the political field. Second, there is the continual development of the strategies and practices of campaigns, parties, and consultants in response to changing socio-technical contexts. And, third, there is the ongoing crafting of infrastructures of technical artifacts, organizations, knowledge, skills, and practices that provide background contexts of action for campaigns endogenous to politics and that afford action across electoral cycles. This enables us to situate the research findings detailed above in terms of these three conceptual areas. It also provides a set of research questions for scholars conducting research on digital campaigning.

First, and most familiar, is the focus on technologies developed outside of politics for commercial or other purposes that have impacted digital campaigning. On a macro level, scholars have suggested that technological changes are shaping broad shifts in social structure towards a 'networked society' (Benkler, 2006; Castells, 2011), providing the contexts for the disembedding of individuals from the group structures of earlier eras (Bennett and Manheim, 2006; Bennett and Iyengar, 2008), and eroding the capacities of long-standing civic organizations and institutions (Karpf, 2012). On the more micro level of digital applications, these dynamics are driven in part by a seemingly endless array of commercial platforms that support much contemporary social and political life. Platforms such as Facebook, Twitter, Tumblr, and Pinterest form part of the ubiquitous and invisible technological context for the conduct of social and political life, and shape expectations for sociality and civic engagement (Bimber et al., 2012). These macro- and micro-level changes have 'spillover' effects in campaign contexts, shaping everything from the diffusion of political messages and inadvertent exposure, to political content to the social contexts within which people communicate about politics.

We must also seek to understand how campaigns and parties perceive and experience their media environment and take strategic action within it to realize their goals. In other words, digital campaigning is not simply driven by underlying changes in technology, social structure, and social practice exogenous to the political field, but also by the strategic actions of campaigns. The core goal of campaigning is to secure a majority of votes, and this end has remained the same despite significant changes in political culture that have shaped how campaigns are waged, from candidates standing for office and the whiskey-fueled spectacles of the strong party era, to the candidate-centric campaigns of our own historical moment (Schudson, 1998). In a broad sense, campaigns appropriate digital tools more or less well to use to their advantage (such as MeetUp, Facebook, Twitter, and YouTube), from creating new efficiencies at their operational back ends, to leveraging new opportunities for strategic communications to supporters and undecided voters.[4] They do so by reading broader shifts in technological, social, and cultural contexts, which may be shaped far afield from politics but impact the potential range of actions campaigns can take in their efforts to elect candidates. This is one reason why 'field-crossers' from the technology industry have been so central to innovations in digital campaigning, from the hackers and 'dot.commers' on the Dean campaign in 2004 to the Facebook (Kreiss, 2012a) and Threadless (Issenberg, 2012a) executives who helped to power Obama's runs in 2008 and 2012, respectively. Particularly important, given rapidly changing technical contexts, consultants and staffers who come to campaigns from contexts outside of politics, particularly the technology industry, are able to bring their knowledge of larger changes in technology and social structure to electoral politics.

At the same time, scholars need to consider the long-term efforts of candidates, parties, and other political actors to build their capacity to act and achieve their goals across election cycles. This is where infrastructure comes in: the background context of action that campaigns and parties shape and that affords organizational capacity for candidates during elections. Campaigns and parties build infrastructure (or fail to) to afford future electoral action. The most taken-for-granted forms of digital campaigning, such as donating money and contacting voters, are premised upon years of technical development and knowledge creation, as well as enormous investments of financial and human resources. Strategic political actors draw on these background social and technical resources to support their digital campaign efforts, from contacting voters online, to e-mailing supporters urging them to give money. For example, as noted above, as Sasha Issenberg (2012b) suggests, the 2012 Obama campaign's success in digital organizing, given advances in voter modeling, was the product of work that took shape within the Democratic Party in-between presidential election cycles around the 2010 midterm elections. In other words, we cannot fully understand digital campaigning by only examining discrete electoral cycles or changes in technologies that are exogenous to politics, such as the emergence of new social media platforms such as Facebook.

In sum, analysis of digital campaigning must move across three levels of conceptualization: changes in underlying social and technological contexts; the strategic actions of campaigns; and the background capacities to act that campaigns have. These, in turn, suggest a number of avenues for future research. As suggested by Bimber et al. (2012), scholars can research how the expectations and practices of citizenship and engagement around campaigning change (or do not) in conjunction with shifts in media. Scholars can, in turn, research how campaigns and parties respond to these shifts organizationally and technologically. More research is needed into how social media affect the agenda-setting role of the press, digital two-step flows of political communication, and the actors involved in political communication more broadly in the context of campaigns. We also need more empirical studies and conceptual work on how campaigns and other institutional actors respond to these changes to realize their strategic goals, from adapting new organizational forms and work practices (Bimber, 2003), to networked ways of strategically distributing political communication (Kreiss, 2012b). We also know very little about the infrastructure of party networks and how they shape the capacity of campaigns to contest elections from state to presidential races; and even less about the role this infrastructure plays at a time of rapid media and social change.

As scholars begin to think more broadly about participation to include citizens' creation, distribution, and interaction with political content online it seems clear that we lack firm categories for many contemporary aspects of political communication. For instance, how should scholars conceptualize retweets or sharing campaign content through Facebook, actions that are low cost and perhaps not even done with much forethought but that may be highly meaningful or consequential forms of political speech in terms of inadvertent exposure (see Freelon, 2014, on this point)? Finally, how do people encounter, create, and express public opinion in everyday life away from institutional political settings (Walsh, 2004), but with implications for campaigns?

NOTES

1. 'Affordance' broadly refers to the capacities for action that technologies make possible.
2. Another significant factor was changes in Federal Election Commission rules that enabled campaigns to collect online donations.
3. Available online at http://enga.ge/fundraising/going-inside-the-cave/.
4. In extraordinary cases, such as Obama's two runs, campaigns are also the incubators of entirely new political tools.

FURTHER READING

Bennett, W.L. and Segerberg, A. (2012). The logic of connective action. *Information, Communication and Society*, 15 (5), 739–768.
Bimber, B. (2003). *Information and American Democracy: Technology in the Evolution of Political Power*. New York: Cambridge University Press.
Chadwick, A. (2013). *The Hybrid Media System: Politics and Power*. New York: Oxford University Press.
Farrell, H. (2012). The consequences of the internet for politics. *Annual Review of Political Science*, 15, 35–52.
Foot, K. and Schneider, S. (2006). *Web Campaigning*. New York: Oxford University Press.
Howard, P. (2006). *New Media Campaigns and the Managed Citizen*. New York: Cambridge University Press.
Kreiss, D. (2012a). *Taking Our Country Back: The Crafting of Networked Politics from Howard Dean to Barack Obama*. New York: Oxford University Press.
Neuman, W.R., Bimber, B., Hindman, M. (2010). The internet and four dimensions of citizenship. In Edwards, G.C., Jacobs, L.R. and Shapiro, R.Y. (eds), *The Oxford Handbook of American Public Opinion and Media* (pp. 22–42). New York: Oxford University Press. Available at: http://www.wrneuman.com/nav_pub_92_275693743.pdf (accessed May 29, 2013).
Nielsen, R. (2012). *Ground Wars: Personalized Communication in Political Campaigns*. Princeton, NJ: Princeton University Press.

REFERENCES

Abramowitz, A. (2010). *The Disappearing Center: Engaged Citizens, Polarization, and American Democracy*. New Haven, CT: Yale University Press.

Ball, M. (2014). Does the Republican Party have to change? *Atlantic*. http://www.theatlantic.com/politics/archive/2014/01/does-the-republican-party-have-to-change/283312/ (accessed February 25, 2014).

Barnard, L. and Kreiss, D. (2013). A research agenda for online advertising: surveying campaign practices, 2000–2012. *International Journal of Communication*, 7, 2046–2066.

Baum, M.A. and Groeling, T. (2008). New media and the polarization of American political discourse. *Political Communication*, 25 (4), 345–365.

Beckett, L. (2012). Everything we know (so far) about Obama's big data tactics. *Propublica.org*. http://www.propublica.org/article/everything-we-know-so-far-about-obamas-big-data-operation (accessed May 28, 2013).

Benkler, Y. (2006). *The Wealth of Networks: How Production Networks Transform Markets and Freedom*. New Haven, CT: Yale University Press.

Bennett, W.L. and Iyengar, S. (2008). A new era of minimal effects? The changing foundations of political communication. *Journal of Communication*, 58, 707–731.

Bennett, W.L. and Manheim, J.B. (2006). The one-step flow of communication. *Annals of the American Academy of Political and Social Science*, 608 (1), 213–232.

Bimber, B. (2003). *Information and American Democracy: Technology in the Evolution of Political Power*. New York: Cambridge University Press.

Bimber, B. and Davis, R. (2003). *Campaigning Online: The Internet in US Elections*. New York: Oxford University Press.

Bimber, B., Flanagin, A.J. and Stohl, C. (2012). *Collective Action in Organizations: Interaction and Engagement in an Era of Technological Change*. New York: Cambridge University Press.

Boulianne, S. (2009). Does Internet use affect engagement? A meta-analysis of research. *Political Communication*, 26 (2), 193–211.

Castells, M. (2011). *The Rise of the Network Society: The Information Age: Economy, Society, and Culture*, Vol. 1. New York: Wiley-Blackwell.

Chadwick, A. (2006). *Internet Politics: States, Citizens, and New Communication Technologies*. New York: Oxford University Press.

Daileda, C. (2013). In Virginia, McAuliffe crawls personal data to win undecided votes. *Mashable*. http://mashable.com/2013/11/05/virginia-mcauliffe-data/ (accessed February 25, 2014).

de Zúñiga, H.G., Veenstra, A., Vraga, E. and Shah, D. (2010). Digital democracy: reimagining pathways to political participation. *Journal of Information Technology and Politics*, 7 (1), 36–51.

Earl, J. and Kimport, K. (2011). *Digitally Enabled Social Change: Activism in the Internet Age*. Cambridge, MA: MIT Press.

Farrell, H. and Drezner, D.W. (2008). The power and politics of blogs. *Public Choice*, 134 (1–2), 15–30.

Foot, K. and Schneider, S. (2006). *Web Campaigning*. New York: Oxford University Press.

Foot, K.A., Xenos, M., Schneider, S.M., Kluver, R. and Jankowski, N.W. (2009). Electoral web production practices in cross-national perspective: The relative influence of national development, political culture, and web genre. In Chadwick, A. and Howard, P.N. (eds), *Routledge Handbook of Internet Politics* (pp. 40–55). New York: Routledge.

Freelon, D. (2010). Analyzing online political discussion using three models of democratic communication. *New Media and Society*, 12 (7), 1172–1190.

Freelon, D. (2014). On the interpretation of digital trace data in communication and social computing research. *Journal of Broadcasting and Electronic Media*, 58 (1), 59–75.

Hindman, M. (2005). The real lessons of Howard Dean: reflections on the first digital campaign. *Perspectives on Politics*, 3 (1), 121–128.

Hindman, M. (2008). *The Myth of Digital Democracy*. Princeton, NJ: Princeton University Press.

Howard, P. (2006). *New Media Campaigns and the Managed Citizen*. New York: Cambridge University Press.

Issenberg, S. (2012a). *The Victory Lab: The Secret Science of Winning Campaigns*. New York: Random House.

Issenberg, S. (2012b). A more perfect union. *TechnologyReview.com*. http://www.techno logyreview.com/featuredstory/508836/how-obama-used-big-data-to-rally-voters-part-1/ (accessed May 29, 2013).

Judd, N. (2012). Obama's Targeted GOTV on Facebook Reached 5 Million Voters, Goff Says. *TechPresident*. http://techpresident.com/news/23202/obamas-targeted-gotv-facebook-reached-5-million-voters-goff-says (accessed May 29, 2013).

Karpf, D. (2010). Macaca moments reconsidered: electoral panopticon or netroots mobilization? *Journal of Information Technology and Politics*, 7 (2–3), 143–162.

Karpf, D. (2012). *The MoveOn Effect: The Unexpected Transformation of American Political Advocacy*. New York: Oxford University Press.

Kreiss, D. (2012a). *Taking Our Country Back: The Crafting of Networked Politics from Howard Dean to Barack Obama*. New York: Oxford University Press.

Kreiss, D. (2012b). Acting in the public sphere: the 2008 Obama campaign's strategic use of new media to shape narratives of the presidential race. *Research in Social Movements, Conflict and Change*, 33 (2012), 195–223.

Kreiss, D. (2014). Explaining technical breakdown: data, analytics, and the Mitt Romney presidential campaign. Presented at the International Communication Association Annual Meeting, Seattle, Washington.

Mossberger, K. and Tolbert, C.J. (2010). Digital democracy: how politics online is changing electoral participation. In Leighley, J.E. (ed.), *The Oxford Handbook of American Elections and Political Behavior* (pp. 200–218). New York: Oxford University Press.

Mossberger, K., Tolbert, C.J. and McNeal, R.S. (2008). *Digital Citizenship: The Internet, Society, and Participation*. Cambridge, MA: MIT Press.

Neuman, W.R., Bimber, B. and Hindman, M. (2010). The Internet and four dimensions of citizenship. In Edwards, G.C., Jacobs, L.R. and Shapiro, R.Y. (eds), *The Oxford Handbook of American Public Opinion and Media* (pp. 22–42). New York: Oxford University Press.

Nielsen, R.K. (2011). Mundane internet tools, mobilizing practices, and the coproduction of citizenship in political campaigns. *New Media and Society*, 13 (5), 755–771.

Nielsen, R.K. (2012). *Ground Wars: Personalized Communication in Political Campaigns*. Princeton, NJ: Princeton University Press.

Nielsen, R.K. (2013). Mundane internet tools, the risk of exclusion, and reflexive movements: Occupy Wall Street and political uses of digital networked technologies. *Sociological Quarterly*, 54 (2), 173–177.

Nielsen, R.K. and Vaccari, C. (2013). Do people 'like' politicians on Facebook? Not really. Large-scale direct candidate-to-voter online communication as an outlier phenomenon. *International Journal of Communication*, 7 (2013), 2333–2356.

Pasek, J., More, E. and Romer, D. (2009). Realizing the social Internet? Online social networking meets offline civic engagement. *Journal of Information Technology and Politics*, 6 (3–4), 197–215.

Plouffe, D. (2009). *The Audacity to Win: The Inside Story and Lessons of Barack Obama's Historic Victory*. New York: Viking Adult.

Prior, M. (2007). *Post-Broadcast Democracy: How Media Choice Increases Inequality in Political Involvement and Polarizes Elections*. New York: Cambridge University Press.

Prior, M. (2013). Media and political polarization. *Annual Review of Political Science*, 16 (1), 101–127.

Rainie, L., Smith, A., Schlozman, K.L., Brady, H. and Verba, S. (2012). Social media and political engagement. *Pew Research Internet Project*. http://pewinternet.org/Reports/2012/Political-engagement.aspx (accessed May 29, 2013).

Rutenberg, J. (2012). Secret of the Obama victory? Rerun watchers, for one thing. *New York Times*. http://www.nytimes.com/2012/11/13/us/politics/obama-data-system-targeted-tv-viewers-for-support.html (accessed May 29, 2013).

Schlozman, K.L., Verba, S. and Brady, H.E. (2010). Weapon of the strong? Participatory inequality and the Internet. *Perspectives in Politics*, 8 (2), 487–509.

Schudson, M. (1998). *The Good Citizen: A History of American Civic Life*. New York: Martin Kessler Books.

Schudson, M. (2003). Click here for democracy: a history and critique of an information based model of citizenship. In Jenkins, H., Thorburn, D. and Seawell, B. (eds), *Democracy and New Media* (pp. 49–60). Cambridge, MA: MIT Press.

Serazio, M. (2014). The new media designs of political consultants: campaign production in a fragmented era. *Journal of Communication*, 64 (4), 743–763.

Shah, D.V., Kwak, N. and Holbert, R.L. (2001). 'Connecting' and 'disconnecting' with civic life: patterns of Internet use and the production of social capital. *Political Communication*, 18 (2), 141–162.

Stirland, S.L. (2013). Organizing for action is ramping up. *TechPresident*. https://techpresident.com/news/23639/organizing-action-hiring (accessed February 25, 2014).

Stroud, N.J. (2011). *Niche News: The Politics of News Choice: The Politics of News Choice*. New York: Oxford University Press.

Wallsten, K. (2010). 'Yes we can': how online viewership, blog discussion, campaign statements, and mainstream media coverage produced a viral video phenomenon. *Journal of Information Technology and Politics*, 7 (2–3), 163–181.

Walsh, K.C. (2004). *Talking About Politics: Informal Groups and Social Identity in American Life*. Chicago, IL: University of Chicago Press.

Warren, R. and Wicks, R.H. (2011). Political socialization: modeling teen political and civic engagement. *Journalism and Mass Communication Quarterly*, 88 (1), 1156–1175.

Wright, S. and Street, J. (2007). Democracy, deliberation and design: the case of online discussion forums. *New Media Society*, 95 (5), 849–869.

Xenos, M. and Moy, P. (2007). Direct and differential effects of the Internet on political and civic engagement. *Journal of Communication*, 57 (4), 704–718.

Zukin, C., Keeter, S., Andolina, M., Jenkins, K. and Delli Carpini, M.X. (2006). *A New Engagement?: Political Participation, Civic Life, and the Changing American Citizen*. New York: Oxford University Press.

9. E-petitions
Scott Wright

INTRODUCTION

In the field of digital politics, electronic petitioning, or 'e-petitioning' as it is more commonly abbreviated, is controversial. Some people consider e-petitions to be one of the most successful e-democracy tools ever – at least quantitatively in terms of citizen uptake (Chadwick, 2012: 61) and its capacity to enhance representative democracy (Bochel, 2013: 798) and empower individuals (Wright, forthcoming a; Cotton, 2011). Tens of thousands of petitions are being created on government-led e-petition platforms every year, far more than were submitted by paper (Bochel, 2013: 803; Wright, 2012a). They are seen as a way to bring the centuries-old and relatively unchanged right to petition Parliament, government or the monarch into the twenty-first century.[1] However, critics lament e-petitions as little more than 'slacktivism' or 'clictivism' which has limited, if any, impact on politics (Shulman, 2009; Morozov, 2009). This analysis is supported by a UK Labour government minister, who argued that the Downing Street e-petitions platform (closed down in 2010) was 'useless, pointless and pernicious' because it claimed to influence policy when it had none (cited in Wright, forthcoming a). Another Labour minister was also negative, but for the exact opposite reason, arguing that the creator of Downing Street e-petitions was a 'prat', allegedly because it did have an influence, leading to the shelving of a key government policy they were responsible for (Wright, 2012a). These starkly different analyses can be linked to different normative conceptions of democracy (Navarria, 2011), but there is also a fundamental disagreement about what impact e-petitions actually have.

E-petitions are controlled, or at least sponsored, by a wide range of bodies, and each platform has its own nuances that need to be considered. Prominent examples include: national government and executive branches (Downing Street e-petitions, We the People); parliaments (German Bundestag, Scottish Parliament); hybrid models that sit between government and parliament (Direct.gov); local governments and parliaments (Queensland Parliament; virtually all local governments in the UK, for example South Derbyshire, Dover, Brighton and Hove); platforms that are independent of formal political systems funded through charity

(38 Degrees, GetUp!, Avaaz); and commercial entities relying on advertising (Change.org). While this often gets lost in debates, there are significant variations between the different e-petition systems. Focusing primarily on government-led e-petition systems, this chapter will make sense of this diversity by identifying key issues and how they cut across the different platforms using case study analysis. The chapter will argue that e-petitions do have the potential to impact upon policy and empower individuals, but how e-petitions are institutionalized is crucial, and there are a number of challenges that can serve to limit their impact.

AGENDA-SETTING AND MODERATION

The ability to set or control the agenda is considered crucial in many theories of democracy. For example, Smith (2009) includes agenda-setting as one aspect of his six-part theoretical framework for analysing democratic innovations. In this context, agenda-setting works in two directions. First, there is a question of who gets to choose the topics of petitions. Second, there is a question of whether e-petitions actually shape the agenda of governments and parliaments if they are successful.

E-petitions are managed in myriad ways by their controllers. Government-led systems initially allow people to create their own petitions, and they are then moderated before being published on the website or blocked. In the most popular government-led systems, such as Downing Street, We the People and Direct.gov, decisions over what petitions to accept are typically made by individual civil servants and this gives them the potential to control the agenda. Government-led moderation can create controversy, as it is sometimes perceived as unjustified censorship (Wright, 2006: 558–560). However, it is considered a necessity even in countries with strong protection of free speech such as the United States, and moderation can be seen as an attempt to encourage free speech and enhance the chances of policy impact if designed and implemented well (Blumler and Coleman, 2001: 17–18; Wilhelm, 2000: 140; Wright, 2006: 563). In the US case, moderators enforce the terms of participation[2] and moderation policy[3] alongside a user-led system for flagging petitions that might breach the rules. With Downing Street e-petitions, a small team of civil servants (normally around three), largely from within the web team in the Prime Minister's Office and thus not policy experts, vetted the petitions as an addition to their existing work. At peak times, casuals were employed to cope with the volume of petitions being received. As there were limited existing examples to draw on, the rejection criteria were 'educated guesses . . . it was very, very difficult to be consistent'.[4] Perhaps unsurprisingly there was a steep learning curve with many problems and they had 'to just fly through it as much as we can, making what we hope are informed and consistent decisions – but we are not experts about every single area of national life'.[5] This is an interesting statement, because it highlights both the speed at which decisions had to be made and the difficult nature of such decisions. When we consider that this was being squeezed in around other work, a few minutes here and there, and they received more than 70 000 petitions in less than four years and rejected over half, it is questionable how much time could be allocated to making decisions – and it is unsurprising that there were many issues such as petitions being published that contained incorrect material (which became 'true' once it had the legitimacy of appearing on the Downing Street website) and repeat petitions that created confusion (see Wright, 2012a).[6]

The transparency of moderation is often important: silent moderation, where government just removes user-generated content, has come in for significant criticism because it creates anger and cynicism (Coleman et al., 2002) and can lead to bad publicity (Wright, 2006). E-petition platforms vary quite significantly with regard to the transparency of their moderation decisions. Downing Street, for example, was unusually transparent: all accepted and rejected petitions were published (in the latter case, offensive petitions were edited), with a reason given for the rejection. Most platforms do not publish rejected petitions (which must be quite resource intensive if thousands of petitions are being submitted), though they do give a list of moderation criteria, which is helpful.

The resources invested into e-petition systems vary hugely, though. In the German system, for example, there are around 80 members of staff (Lindner and Riehm, 2011: 9), what Bochel (2012) describes as a substantive e-petition system (as opposed to a descriptive system) and one that Jungherr and Jurgens (2010: 8) see as having 'strong agenda-setting potential' based on case studies of quantitatively successful petitions. Similarly, in the Scottish Parliament and Welsh Assembly (where far fewer petitions are submitted), there are civil servants 'whose job it is to assist petitioners, give advice about the process and how to word the petition itself' (Scottish Parliament, 2010: 6). A committee of seven members of the Scottish Parliament then adjudicates on each petition. It could be argued that e-petitions are being taken more seriously by these institutions and they help to set agendas and start new debates (Thomson, 2009: 45; see Birrell, 2012 for an excellent comparison).

There are other ways in which governments can shape the agenda of petitions, such as by defining how many signatures must be received to get an official reply and, particularly, how long is given to achieve this.

Downing Street e-petitions empowered individual activists, with 19 of the 20 most signed petitions created by individuals (Wright, forthcoming a). With Downing Street e-petitions, only 500 signatures (increased from 250 when it was realized how many petitions were reaching the threshold) were required to receive an official reply and people could choose to leave their petitions open for a year or more. We The People, on the other hand, does not make a petition searchable unless it receives 150 signatures in a month, and to receive an official reply a petition must receive 100000 signatures in a month. This was changed from 5000, to 25000, before being quadrupled. Even before this change was made, the chance of receiving a reply was less than 0.5 per cent (Snider, 2013). The epetitions.direct.gov.uk petition platform requires 100000 signatures for a petition to be considered for debate in parliament, though people are allowed up to one year to garner the signatures. Nevertheless, a tiny proportion of petitions have achieved the necessary signatures to be debated: 32 of 26672 closed petitions have achieved 100000 signatures, and 211 had received the 10000 signatures required to receive an official reply (although 54 of these, largely from 2012 had not received the reply, though other petitions with less than 10000 had).

Deciding how many signatures will be needed to receive an official reply requires a delicate balance to be struck. Set the barrier too low, and many thousand official replies might be required. While some might argue that this is democratic, it happened with Downing Street e-petitions, and petition creators were often very upset with the speed, brevity and tone of the official replies (Wright, 2012a). As this was the main democratic output, the danger was that it decreased rather than strengthened trust in government and arguably made it harder for petitions to actually make an impact on policy because successful petitions might be competing with each other. Set the bar high, and it could disempower individual activists and make it so hard to get a response that people opt not to participate.

One of the ongoing moderation debates across all of the major e-petition systems has been how to handle apparently humorous petitions. For example, the Downing Street system initially had an open policy, surprisingly perceiving e-petitions as a 'light-hearted' forum, and the attitude that moderators 'shouldn't be kind of het up about these things. Fine, let people have their say'.[7] However, after receiving hundreds of humorous petitions it was decided 'to take a stand and say in truth this isn't a serious matter for government. It is potentially crowding out those people who are submitting petitions that are of more legitimate concern'.[8] The Leader of the House of Commons placed emphasis on the moderators stopping 'frivolous' petitions, which he said was 'the nature of the internet' (Chapman,

2011). Attempts to meet humorous petitions with humorous responses have come in for criticism either because it is a waste of taxpayer money, or because it belittles the democratic process. Examples include a petition to make the *Top Gear* presenter, Jeremy Clarkson, Prime Minister, which Downing Street responded to with its own humorous video,[9] and a petition to fund and begin building a Death Star (heavily facilitated by the huge 4Chan discussion forum[10]). The Chief of the Science and Space Branch in the White House Office of Management and Budget replied to the Death Star proposal, joking about the '$850 000 000 000 000 000' cost and questioning, 'Why would we spend countless taxpayer dollars on a Death Star with a fundamental flaw that can be exploited by a one-man starship?'[11]

While the main focus of this chapter is on formal, government-led e-petition systems, it should be noted that there are significant differences in moderation and agenda-setting between the different non-governmental organizations. The Australian organization, GetUp! places more emphasis on strategic decisions about where to invest campaign resources and a process of deliberation. For example, the centre chose to prioritize a same-sex marriage campaign, which was not listed in the top ten priorities from its membership survey, because it was considered an important issue to campaign on.[12] The organization has now removed its Campaign Ideas Forum (CIF), and has developed Communityrun.org as a platform to allow individuals to develop their own campaigns, separate from GetUp. Its position is summarized in a response to a campaign which it decided not to take up from the CIF:

> Just because a campaign suggestion receives a lot of votes, doesn't mean that GetUp will take it on as a campaign. It does mean GetUp staff, volunteers and interns will look into the issue and see if it fits with GetUp's values, fires up a movement of GetUp members, and that there is a good moment when GetUp members can help win a victory on the issue. A few people try to game the system by flooding certain campaign suggestions with votes from fake accounts, and so forth – but it's usually pretty obvious when that's happening. (cited in Morris, 2011)

The GetUp! approach can be compared with Avaaz (which shares some of the same co-founders as GetUp!), which is more user-led and undertakes a range of activities such as regular polling of supporters and testing response rates to e-mail messages to see if people are interested. One final example is change.org, a for-profit social activism group that makes revenue by allowing sponsored petitions, with the financial logic further changing the nature of agenda-setting and moderation.[13]

IMPACT

Research into the impact of e-petitions has produced mixed results. The first key debate has been over how to measure the impact of e-petitions. One perspective has been to focus on whether they have substantive impacts on public policy (Shulman, 2009 – focusing specifically on mass e-mail campaigns), while others argue that this is an unduly narrow analytical frame that captures neither the rationale for many organizations in using e-petitions, nor how e-petitions can sit alongside a much wider range of campaigning activities (Karpf, 2010). The second key debate is more normative: just how much influence should e-petitions be allowed to have within a democratically elected representative democracy? As outlined below, while participating through e-petitions might be easier, participation is still unequal, and ultimately the silent majority is far larger than the more vocal minority of people who have signed any single petition. The focus on numbers – receiving 100 000 signatures – often seems to be more about minimizing the number of responses that have to be made (House of Commons Reform Committee, 2009), and is not particularly helpful. For example, after 1.8 million people signed an anti-road-pricing petition,[14] Blair (2007) argued that: 'it's not possible, wise or healthy for politicians to try and sweep [the views] under the carpet' and argued that the response to the e-petition 'shows that my government is listening'. However, Blair (2007) was not promising to drop the policy, but to have a public debate with a view to convincing people of the need for change, because 'road charging is surely part of the answer'. Less than a year later, however, road pricing was quietly shelved; with many people crediting the e-petition. But is this enhancing democratic legitimacy and accountability? Road pricing was included in Labour's 2005 General Election manifesto, on which the party was democratically elected (Labour Party, 2005: 25). Yet when the policy was dropped, the then Transport Minister, Lord Adonis, is reported to have stated that this was because road pricing lacked a democratic mandate (Wright, 2009). Moreover, representative survey evidence indicated that there was public support (61 per cent) for road pricing, if the revenue was invested in public transport (IPSOS Mori, 2010). The concern is that e-petitions (and more direct forms of democracy in general) could lead to reactionary, lowest common denominator policy-making where tough policy choices are avoided or get dropped because of mob rule.[15]

In systems where fewer petitions are submitted, such as in Scotland and Wales, it is possible for elected representatives to review virtually all of the petitions (indeed, the pre-checking explicitly encourages this) and it is less a quantitative exercise and more about the judgement of the elected representatives, which has normative appeal for many (Engel, 2011).

Fair assessment of the impact of e-petitions is difficult because it is hard to disentangle what impact the e-petition had when many other activities are typically happening; government-led e-petitions can become a political football amongst politicians; and non-governmental e-petitions have a vested interest in arguing that their actions have been successful, not least because they want to encourage people to keep acting, and they are in competition with each other for donations and support.

Government-led e-petitions are often criticized for being little more than public relations exercises, designed to garner large e-mail databases that can be used to lobby people (Snider, 2013). The Queensland state-level e-petition system, for example, has been found to have minimal impact: 'little, if any, action is taken on individual petitions other than a ceremonial reading followed by the placement of a printed copy on the relevant minister's desk' (Belot, 2013). Criticisms have also been made of some non-governmental organizations for choosing e-petition topics that they believe will recruit new supporters (adding to their e-mail list), rather than it being considered the most socially important or impactful campaign (Wright, forthcoming a).

Starting with the impact of e-petitions on public policy, research to date has indicated several cases where there has been an apparent impact. Cotton's (2011: 38) analysis of Scottish e-petitions discovered that '12.7% of E-petitions were closed as a result of the issues raised being implemented, [indicating] that E-petitions do have the ability to affect policy formulation, that the Scottish Parliament takes E-petitioning seriously, and that E-petitions have the ability to become or change laws'. Cotton then looked at specific cases, such as a petition to provide cancer drugs on the National Health Service (NHS), submitted by the partner of a cancer sufferer who was self-funding a key treatment. This petition received a relatively modest 632 signatures. However, the Scottish system does not just focus on numbers, and the committee invited the petition creator and her husband to give evidence before the committee. On the basis of their evidence, the committee launched an inquiry, which led to the Better Cancer Care report and significant changes to the funding of cancer treatment. The committee's closing response is worth highlighting at length:

> I am sure that the committee would like to reflect not only on the positive actions of the Scottish Government but on the indispensable input of the petitioner, Tina McGeever, on behalf of her husband, the late Mike Gray. Without the petitioner and the energy of both individuals directly involved, we would not be seeing the real improvements that I am sure the petition will effect throughout Scotland in respect of patients accessing newly licensed medicines, in the process for considering objectively individual patient treatment

requests and in the arrangements for the combination of care that is available to patients. Finally, we should reflect on the fact that all of those real improvements for people throughout Scotland have been effected through the simple process of lodging a petition. The petitioner should take great pride in that. (Scottish Parliament Public Petitions Committee, 2011 cited in Cotton, 2011)

On the basis of his analysis, Cotton (2011: 48) concludes:

It is clear that the E-petition process has allowed individuals such as Tina McGeever, Rajiv Joshi, Deryck Beaumont, and hundreds of others to participate in policy formulation with the Scottish Parliament. It has facilitated public debate with the Parliament, and given a new outlet for citizens and groups to voice their grievances and concerns. Increasing public participation in the democratic process was one of the goals of the new Scottish Parliament, and the development and use of its E-petitioning system has fulfilled this function. The case studies have demonstrated the tangible changes in policy that E-petitioning can bring, and further cements its critical role in creating a participative Parliament.

Moving away from Scotland, Wright (2012a) cites a range of examples from Downing Street e-petitions, such as the dropping of the fledgling road pricing policy proposal, stopping hospitals from using revenue-producing premium phone lines for contacting them, and a national holiday to commemorate troops. Other successful petitions were not specifically policy-related, making it easier for the government to act, such as a pardon for the famous scientist and code-breaker, Alan Turing. Birrell (2012: 121–122) cites some examples from the Welsh Assembly, including a petition that led to increases in funding for disabled children, and one that led to the adoption of a mandatory charge for plastic bags. Similar to Scotland, petitions typically lead to debates and reports, which can then influence policy. Impactful examples cited in a Hansard Society report (2012) on the new direct.gov system were few in number and sketchier, although a petition to release in full all documents relating to the Hillsborough tragedy has been debated in Parliament and the government has committed to do this.[16] Non-governmental e-petition systems have also been widely credited with successful actions, such as stopping the closure of BBC Radio 6 Music (where the weight of the campaign was specifically cited in the BBC Trust's decision), and the reversing of a policy to sell off national forests.[17]

As shown above, there are a number of examples of e-petitions that have impacted upon policy. However, it must be recognized that the vast majority of e-petitions have little or no impact. While with the Downing Street model several thousand official responses were made, content analysis has shown that these were largely dismissing or correcting e-petitioners, with no evidence of policy impact (Wright, 2012a). Perhaps unsurprisingly, petition creators were often very negative about the official replies. While this is not a policy impact, it is a form of political engagement. Interestingly, many petition creators had a positive view about the impact of their petition, precisely because they did not expect it to have an impact and measured success with a lower, broader bar (Wright, forthcoming b). In fact, several petition creators felt that the government was more open and responsive privately than what was communicated in the official replies. In most descriptive models, such as We The People, direct.gov and Downing Street e-petitions, only a tiny fraction of petitions reach the threshold and the vast majority fall into a void where nothing happens. When petitions reach the threshold, we might expect a stronger response and policy impact – but the results have been mixed.

As noted above, a parliamentary committee decides whether a petition that receives 100 000 signatures is debated in the UK parliament, and Bochel (2013: 804) found that only five of the eight petitions that achieved 100 000 signatures had been debated in Parliament (they may have been debated subsequently). Moreover, the actual topic of the debates often did not match the text of the petition. In perhaps the most famous example, a petition asking for those convicted of participating in the 2011 London Riots to lose their welfare benefits was debated in Parliament for three hours, but the debate actually failed to cover the substantive issue of cutting benefits.[18]

BARRIERS TO PARTICIPATION

The legitimacy of e-petition systems rests, in part, on there being equality of access to participate. While divides in access to the Internet may be decreasing in many countries, there are still many people who cannot afford, or do not want, to be online. As Carman (2014: 168) has noted: 'Politicians and advocates need to be careful to avoid the "if we build it, they will come" perspective. Just because new, innovative methods of participation are made available to the public it does not mean that the playing field suddenly becomes level and that "critical citizens" will spontaneously shed their cynicism'. This divide, Carman argues, will be exacerbated if e-democratic innovations are given more influence over policy-making. While Carman does not make clear why he believes this, one can assume that it is because the higher incentive will further motivate the already politically active. Thomas and Streib's (2003) survey data support Carman's broader point: the digital divide is exacerbated if we look specifically at the people who visit government websites. As Bimber

(1999: 423) noted: 'Government officials who attempt to gauge public concerns by paying attention to citizens through e-mail will draw slightly different conclusions than they would from paying attention to traditional contacts through letters and phone', and history indicates that increases in access to information about politics has not led to increases in political engagement (Bimber, 1998: 139).

These arguments reflect the view that while e-petition platforms might make it easier for people to participate, this is not uniform. Conceiving participation as a journey, I have previously argued that access to the Internet is but one issue. For some, such as those without Internet access at home, participating in an e-petition would be an expedition (Wright, 2012b), but for those who participate in politics regularly it is akin to making their commute easier. Most e-petition systems are designed to make participation easier, and many of the widgets being built by developers enhance this (see the section on 'Priorities for future research', below).[19] If Carman's research is correct, it is questionable just how much impact they will have, and just how representative and equal participation in e-petitions actually is. However, research by Margetts et al. (2011) indicates that the size of the petition has a significant predictive value on decisions to participate.

CONCLUSION

While e-petitions remain controversial, empirical analysis suggests that they are popular with citizens and have had impacts on public policy. However, there are significant variations between e-petition platforms, and how they are designed and institutionalized is crucial. For example, well-resourced, substantive e-petition systems such as in Scotland have had clear impacts. However, even descriptive e-petition platforms, such as Downing Street and direct.gov, have had an influence; although there remain concerns about the normative desirability of this, not least because of inequality of access. A key difference is that the Scottish system is fully embedded into the parliamentary process, and petitions are taken forward by a panel of Members of the Scottish Parliament (MSPs) who understand how the system works. This also seems to make it easier to identify discrete examples where e-petitions have made a difference. Nevertheless, it must be recognized that the majority of e-petitions make no impact. However, petition creators were often still quite positive about their petitions, because they were using broader definitions of success.

PRIORITIES FOR FUTURE RESEARCH

E-petitions are in their infancy, and changing rapidly. As this chapter has highlighted, a number of important studies of e-petitions now exist, but there are a number of areas that should be prioritized in future research.

Transparency, Apps and Big Data

One relatively new field is apps and visualization widgets that can be used to help make sense of, and promote, e-petitions. The best example of this is featured on We The People,[20] which has developed an application programming interface (API) (and subsequent Hackathon at the White House) to allow people to develop the capacity of the platform.[21] There are numerous fascinating and potentially very powerful tools, including:

- A tool that overlaps petition data, such as the postcode where signers are from, onto geographic maps (Catherine D'Ignazio, MIT Center for Civic Media; Mick Thompson, Code For America). This can be embedded with further public data such as political leanings based on voting, and this can be broken down by day (Matt Loff, Visionistic).[22]
- A tool that extracts locations mentioned in petitions and then visualizes this (for example, the country focus for foreign policy petitions; Garrett Miller, Map Box).
- A tool that can analyse the sentiment in petitions (Yoni Ben-Meshulam, Opower).
- The petition.io dashboard, which includes a range of functions, such as tracking signature rates over time. This makes analysing what drives petition rates simpler (see Wright, 2012a for an example).
- A tool that enable people to embed petitions into blogs, making then easier to promote (Steve Grunwell).

These tools make it possible to do even more sophisticated analyses of e-petitions. Many may also be used by petition creators to promote their petition, and analysis of the success of this would be interesting, and fits into the broader open data agenda.

Policy Impact

While there has already been some research that has attempted to assess the policy impact of e-petitions, this has been relatively limited in scope, focused on a narrow number of cases, and has struggled to disentangle what role the e-petition played in successful broader campaigns. For example, the campaign to create a national holiday to commemorate troops was partially successful: a holiday was created, but not in the requested long gap between the August Bank Holiday and Christmas, and it was placed on a Saturday as a cost-saving measure. This also fitted with an existing government agenda. Surprisingly, very little has been written about We the People to date, and an assessment of its policy impact would be welcome.

Participants' Perspectives

Surprisingly few studies of e-petitions have focused on surveying the people who signed the petitions, to understand why they chose to act, what they thought about the process and how impactful the petition was. In particular, scholars analysing patterns of participation have identified 'super-participants': people who create and/or sign dozens of petitions (Wright, 2012; Jungherr and Jurgens, 2010; Cotton, 2011: 36). Closer analysis of these participants would be helpful to better understand what motivates them, whether they reflect the broader population, their links to formal organizations, and how they operate (for example, whether they write to journalists).

Non-Governmental E-Petitions

Detailed research to date has tended to focus on government and Parliament petition platforms. Further analysis of non-governmental e-petition systems would be of interest.

NOTES

1. For example, a right to petition the monarch dates back to Saxon times, and was formalized in the Magna Carta of 1406. Over time, the volume and support for petitions has ebbed and flowed, with a significant decline around the time of the First World War for several decades, but the system itself remained relatively unchanged (Leys, 1955). Both the UK and Australian parliaments moved in recent years to make the rules for submitting paper petitions less restrictive, though it would be fair to say that the changes are modest. The UK's approach is freer than the Australian one. See http://www.parliament.uk/documents/commons-information-office/P07.pdf and http://www.aph.gov.au/~/media/02%20Parliamentary%20Business/24%20Committees/243%20Reps%20Committees/Petitions/PDF/PetitionsBrochure.pdf (accessed 16 June 2014).
2. https://petitions.whitehouse.gov/how-why/terms-participation (accessed 15 June 2014).
3. https://petitions.whitehouse.gov/how-why/moderation-policy (accessed 15 June 2014).

4. Interview with senior Downing Street official.
5. Interview with senior Downing Street official.
6. Direct.gov e-petitions was even more extreme, with 22000 petitions received in six months (House of Commons Procedure Committee, 2011).
7. Interview with senior Downing Street official.
8. Interview with senior Downing Street official.
9. Video available at https://www.youtube.com/watch?v=cNy1w4DV5Hw (accessed 16 June 2014).
10. http://imgur.com/gallery/byAkC (accessed 17 June 2014).
11. https://petitions.whitehouse.gov/response/isnt-petition-response-youre-looking (accessed 16 June 2014).
12. Invited talk given by Sam McLean, National Director, GetUp! at the University of Melbourne, May 2014.
13. Though they note that they do not accept hateful sponsored petitions. See http://www.change.org/en-AU/about/advertising-guidelines (accessed 17 June 2014).
14. The proposed Road Pricing policy would have required the installation of monitoring devices to track where, and at what time, people drove, charging them a tax if they used the busiest roads at peak periods.
15. See, for example, a withering editorial in *The Guardian*: http://www.theguardian.com/commentisfree/2011/nov/17/e-petitions-the-peoples-voice (accessed 18 June 2014).
16. http://epetitions.direct.gov.uk/petitions/2199 (accessed 18 June 2014).
17. See, for example, the Facebook group: https://www.facebook.com/groups/saved6music/; and newspaper articles: http://www.theguardian.com/commentisfree/2014/feb/24/e-petitions-often-worse-than-useless and http://www.theguardian.com/media/2010/jul/05/bbc-6-music-saved.
18. http://www.bbc.com/news/uk-politics-15283837 (accessed 17 June 2014).
19. Though campaigners have noted that small changes such as hiding the names of the people who sign a petition can make it harder to build a campaign.
20. https://petitions.whitehouse.gov/how-why/api-gallery. The direct.gov platform also makes public a significant amount of data through its API, though the app community is less developed. See http://epetitions.direct.gov.uk/faq#question11 (accessed 15 June 2014).
21. Some video from the Hackathon is available: https://www.youtube.com/watch?feature=player_embedded&v=sjfsUzECqK0#t=0m52s (accessed 17 June 2014).
22. http://wtp.visionistinc.com/# (accessed 15 June 2014).

FURTHER READING

Karpf, D. (2010) 'Online political mobilization from the advocacy group's perspective: looking beyond clicktivism', *Policy and Internet*, 2: 7–41.
Margetts, H., John, P., Escher, T. and Reissfelder, S. (2011) 'Social information and political participation on the internet: an experiment', *European Political Science Review*, 3(3): 321–344.
Riehm, U., Böhle, K. and Lindner, R. (2011) 'Electronic petitioning and modernisation of petitioning systems in Europe', TAB Report No. 146.
Smith, G. (2009) *Democratic Innovations: Designing Institutions for Citizen Participation.* Cambridge: Cambridge University Press.
Wright, S. (2012) 'Assessing (e-)democratic innovations: "democratic goods" and Downing Street E-petitions', *Journal of Information Technology and Politics*, 9(4): 453–470.

REFERENCES

Belot, H. (2013) 'Petitions are in vogue, but do they make a difference?', *Citizen*. Available at: http://www.thecitizen.org.au/analysis/petitions-are-vogue-do-they-make-difference#sthash.hKVKGFSZ.dpuf (accessed 18 June 2014).

Bimber, B. (1998) 'The Internet and political transformation: populism, community, and accelerated pluralism', *Polity*, 31(1): 133–160.

Bimber, B. (1999) 'The Internet and citizen communication with government: does the medium matter?', *Political Communication*, 16(4): 409–428.

Birrell, D. (2012) *Comparing Devolved Governance*. Basingstoke: Palgrave Macmillan.

Blair, T. (2007) 'The e-petition shows that my government is listening', *Guardian*, 18 February. Available at: http://www.theguardian.com/commentisfree/2007/feb/18/uk.transport (accessed 18 June 2014).

Blumler, J. and Coleman, S. (2001) *Realising a Civic Commons in Cyberspace*. London: IPPR.

Bochel, C. (2012) 'Petitions: different dimensions of voice and influence in the Scottish Parliament and the National Assembly for Wales', *Social Policy and Administration*, 46(2): 142–160.

Bochel, C. (2013) 'Petition systems: contributing to representative democracy?', *Parliamentary Affairs*, 66(4): 798–815.

Carman, C.J. (2014) 'Barriers are barriers: asymmetric participation in the Scottish public petitions system', *Parliamentary Affairs*, 67(1): 151–171.

Chadwick, A. (2012) 'Web 2.0: new challenges for the study of e-democracy in an era of informational exuberance', in Coleman, S. and Shane, P. (eds), *Connecting Democracy: Online Consultation and the Flow of Political Communication*. Cambridge, MA: MIT Press, pp. 45–75.

Chapman, J. (2011). 'Government's e-petitions website crashes on its first day as debate over capital punishment heats up', *Daily Mail*. Available at: http://www.dailymail.co.uk/news/article-2022147/Capital-punishment-People-power-forces-MPs-vote-death-penalty.html (accessed 16 June 2014).

Coleman, S., Hall, N. and Howell, M. (2002) *Hearing Voices: The Experience of Online Public Consultations and Discussions in UK Governance*. London: Hansard Society.

Cotton, R.D. (2011) 'Political participation and e-petitioning: an analysis of the policy-making impact of the Scottish Parliament's e-petition system', unpublished thesis, University of Central Florida. Available at: http://etd.fcla.edu/CF/CFH0004083/Cotton_Ross_D_201112_BA.pdf (accessed 18 June 2014).

Engel, N. (2011) 'Why the e-petitions system isn't working', *Guardian*, 17 November. Available at: http://www.theguardian.com/commentisfree/2011/nov/16/e-petitions-system (accessed 18 June 2014).

Hansard Society (2012) *What Next for E-petitions?* London: Hansard Society.

House of Commons Procedure Committee (2011) 'Uncorrected transcript of oral evidence (HC1706)'. Available at: http://www.publications.parliament.uk/pa/cm201012/cmselect/cmproced/uc1706/uc170601.htm (accessed 16 June 2014).

House of Commons Reform Committee (2009) 'Reform of the House of Commons Select Committee – First Report: Rebuilding the House (Section 5: Involving the public)'. Available at: http://www.publications.parliament.uk/pa/cm200809/cmselect/cmref-hoc/1117/111709.htm (accessed 19 June 2014).

Ipsos MORI (2010) 'Support for road pricing if revenues used for public transport', research archive. Available at: http://www.ipsos-mori.com/researchpublications/researcharchive/poll.aspx?oItemId=234.

Jungherr, A. and Jurgens, P. (2010) 'The political click: political participation through e-petitions in Germany', *Policy and Internet*, 2(4): 131–165.

Karpf, D. (2010) 'Online political mobilization from the advocacy group's perspective: looking beyond clicktivism', *Policy and Internet*, 2: 7–41.

Labour Party (2005) *Forward Not Back: The Labour Party Manifesto 2005*. London: Labour Party.

Leys, C. (1955) 'Petitioning in the nineteenth and twentieth centuries', *Political Studies*, 3(1): 45–64.

Lindner, R. and Riehm, U. (2011) 'Broadening participation through e-petitions? An empirical study of petitions to the German Parliament', *Policy and Internet*, 3(1): 1–23.

Margetts, H., John, P., Escher, T. and Reissfelder, S. (2011) 'Social information and political participation on the internet: an experiment', *European Political Science Review*, 3(3): 321–344.

Morozov, E. (2009) 'The brave new world of slacktivism', *Foreign Policy*, 19 May. Available at: http://neteffect.foreignpolicy.com/posts/2009/05/19/the_brave_new_world_of_slacktivim (accessed 28 November 2014).

Morris, K. (2011) 'Inside GetUp and the new youth politics', *Quadrant Online*. Available at: http://quadrant.org.au/magazine/2011/3/inside-getup-and-the-new-youth-politics/ (accessed 17 June 2014).

Navarria, G. (2011) 'The Internet and representative democracy: a doomed marriage? Lessons learned from the Downing Street E-petition website and the case of the 2007 Road-Tax Petition', in Manoharan, A. and Holzer, M. (eds), *E-Governance and Civic Engagement: Factors and Determinants of E-democracy*, Hershey, PA: IGI Global, pp. 362–380.

Scottish Parliament (2010) 'Petitioning the Scottish Parliament: making your voice heard', Edinburgh. Available at: http://www.scottish.parliament.uk/PublicInformationdocuments/Petitioning-Eng-250712.pdf.

Scottish Parliament Public Petitions Committee (2011) 'Official Report: Public Petitions Committee (Session 3)'. Available at: http://www.scottish.parliament.uk/parliamentary-business/28862.aspx?r=6216&mode=pdf (accessed 21 January 2015).

Shulman, S. (2009) 'The case against mass e-mails: perverse incentives and low quality public participation in US federal rulemaking', *Policy and Internet*, 1(1): 23–53.

Smith, G. (2009) *Democratic Innovations: Designing Institutions for Citizen Participation*. Cambridge: Cambridge University Press.

Snider, J.H. (2013) 'Updating Americans' First Amendment right to petition their government', paper presented at the Harvard Law School Luncheon, 20 September.

Thomas, J.C. and Streib, G. (2003) 'The new face of government: citizen-initiated contacts in the era of e-government', *Journal of Public Administration Research and Theory*, 13(1): 83–102.

Thomson, B. (2009) 'Access and participation: aiming high', in Jeffery, C. and Mitchell, J. (eds), *The Scottish Parliament 1999–2009: The First Decade*. Edinburgh: Luath Press, pp. 43–48.

Wilhelm, A. (2000) *Democracy in the Digital Age: Challenges to Political Life in Cyberspace*. London: Routledge.

Wright, R. (2009) 'Adonis shelves road-price policy', *Financial Times*, 25 June. Available at: http://www.ft.com/intl/cms/s/0/0ebf2b32-6120-11de-aa12-00144feabdc0.html#axzz35k5SltF8 (accessed 26 June 2014).

Wright, S. (2006) 'Government-run online discussion fora: moderation, censorship and the shadow of control', *British Journal of Politics and International Relations*, 8(4): 550–568.

Wright, S. (2012a) 'Assessing (e-)democratic innovations: "democratic goods" and Downing Street E-petitions', *Journal of Information Technology and Politics*, 9(4): 453–470.

Wright, S. (2012b) 'Mapping the participatory journey in online consultations', in Shane, P. and Coleman, S. (eds), *Re-connecting Democracy*. Cambridge, MA: MIT Press, pp. 149–172.

Wright, S. (forthcoming a) 'Populism and Downing Street e-petitions: connective action, hybridity and the changing nature of organizing', *Political Communication*.

Wright, S. (forthcoming b) 'Perceptions of success in online political participation'.

10. Argumentation tools for digital politics: addressing the challenge of deliberation in democracies
Neil Benn

INTRODUCTION

This present volume brings together researchers in the areas of democratic theory, political communication, and information technology. Many at the intersection of these areas have continually made the case for using information and communication technology (ICT) to improve our democracy, governance, and political culture; see for example, notable contributions on the topics of online deliberation (Davies and Peña Gangadharan, 2009; Coleman and Moss, 2012) and e-participation (Tambouris et al., 2011, 2012). Many who aim to see the Internet and other digital technologies transform democracy and governance for the better have suggested that there is a need for radical new tools and techniques to address one particular challenge of the political process: helping citizens be better informed about the key political issues of the day and to better make sense of the inherent conflict of opinion in political debate (Coleman and Blumler, 2009; Macintosh et al., 2009). To this end, these scholars have recently turned to research and development in the area of argumentation technology – particularly computer-supported argument visualization (CSAV) technology – for new software tools to address this challenge.

One of the main aims of this chapter is briefly to survey current research on the use of argumentation technology to enhance the political process, particularly those aspects of the political process related to policy deliberation. It is important to emphasize at the beginning that this chapter is restricted narrowly to CSAV technology rather than ICT for digital politics more broadly, thus excluding such areas as e-voting and e-petitioning technology, even if such technologies can sometimes form part of a policy deliberation process. The chapter begins by examining the use and understanding of the term 'deliberation' from the perspective of CSAV research and development (R&D). The chapter then continues with a survey of the main trends in CSAV R&D with respect to their use in deliberative political processes. Based on this brief survey and critical analysis of current work, the chapter next identifies possible areas of future research and the priority research questions that should be the basis of this future research. Finally, the chapter lists supplementary reading material for the reader interested in further exploring the themes only briefly covered and surveyed here.

CSAV AS DELIBERATIVE TECHNOLOGY

This section has two main aims: first, to review the case that is often made for the use of CSAV technology as a tool for fostering deliberative democracy; and second, to provide a brief survey of the key research and development in this area. It should be noted here that it is outside the scope of this chapter to provide a detailed survey of all CSAV R&D, including all techniques and tools that have been or are currently being used. Such a survey would quickly become obsolete and thus useless for what is intended as a reference volume. Instead, this section identifies the key aspects of the intellectual lineage of CSAV research, which then prepares the reader for the next section on possible priority areas for future research that can build on key advances. Where useful, actual systems will be referenced for their historical importance.

CSAV as Deliberative Technology: Motivation and Rationale

At a time when trust in politicians and the political class is in apparent decline, the public is, perhaps as a consequence, demanding greater transparency and accountability within the political process. As the public seemingly grows more cynical, policy-makers are increasingly obliged to articulate more clearly the reasons for their policy decisions and to open up the deliberative process to a broader audience.

It should be noted that the terms 'deliberation' and 'deliberative' are primarily used here (and in most of the CSAV research surveyed in the next section) in the sense of what Coleman and Moss (2012) refer to as 'deliberative rationalism': that is, 'rational argumentation modeled upon the Habermasian notion of discourse ethics'. These authors and others (see, for example, Chadwick, 2012) rightly criticize this construction of 'deliberation' and 'deliberative' as a possibly 'romanticized' – to use Chadwick's characterization – model of political discourse. Nonetheless, most of the CSAV field has used this as a normative ideal against which R&D outcomes can be measured.

Hilbert (2009) presents CSAV as one of the building blocks of new approaches to democracy that try to bring greater transparency through the use of ICT. For him, argumentation is at the heart of democratic deliberation, and in his view CSAV can enhance this deliberative process by providing new information-structuring techniques; for example, through clarifying the discursive relations that exist between different viewpoints. Hilbert further suggests that these tools can enable citizens not only to be better informed about parliamentary deliberation but also to participate in the deliberative process as they communicate their viewpoints to their representatives. On this reading, CSAV has the ability to take citizens beyond merely eavesdropping to actively engaging in the political decision-making process.

Furthermore, effective democratic citizenship is not solely about the right to express one's own opinions but is also about the responsibility to consider and reflect on the opinions and arguments of others. One promise of CSAV technology is that it can better facilitate deliberation by helping citizens locate and link their opinions within a larger deliberative space consisting of the opinions of their fellow citizens. Fulfilling such a promise is particularly important in an age of successful Web 2.0 and social media technology such as Twitter and Facebook, which have undoubtedly had a positive democratic impact (as observed and promoted by writers such as Howard and Hussain, 2013; Lotan et al., 2011; Shirky, 2011), but potentially at a price. With the proliferation of publishing platforms and commenting spaces, Web 2.0 seems to have succeeded in giving people a voice to speak, but failed at getting people to listen to each other, particularly in contentious matters of public political debate.

CSAV as Deliberative Technology: Key Research Themes

CSAV, hypertext and solving 'wicked problems'
The earliest contributions to the topic of CSAV as deliberative technology come from the computing field of computer-supported collaborative work (CSCW), particularly those researchers interested in so-called group decision support systems who were operating within the subfield of hypertext. Hypertext researchers were inspired by the insight of scholars such as Horst Rittel in the field of policy sciences, who writing with Webber (Rittel and Webber, 1973) helped to establish a theory of social policy planning as a fundamentally argumentative process. That is, in matters of social policy, Rittel and Weber posit that it is not possible to prove – in a mathematical or purely scientific manner of logical proof – that a social policy solution or decision is the correct one; rather, one can only try to persuade, through argumentation, that a given solution or decision should be preferred given a set of goals, constraints and value preferences. These authors coined the term 'wicked problems' to distinguish the kinds of

problems that social policy planners deal with from the kinds of problems that scientists encounter.

In addition to his theoretical work, Rittel in collaboration with Werner Kunz (Kunz and Rittel, 1970), created the Issue-Based Information System (IBIS), a software tool and approach. Assuming a model of policy-making and planning as argumentative processes, in the IBIS approach all policy deliberations start with one or more key issues that need to be resolved. Participants in the deliberation then assume positions in response to these issues and arguments are then brought in that support or object to a particular position.

Building on this work, the early hypertext community envisioned that graphical hypertext could be used to record, in a structured and organized manner, the outputs of the argumentation and reasoning processes taken to solve these so-called 'wicked problems'. One of the first and most influential graphical hypertext systems that built directly on the insights of Rittel was the gIBIS system (Conklin and Begeman, 1988), which was explicitly framed as a hypertext tool for 'exploratory policy discussion'. Indeed, the gIBIS system was presented as a graphical incarnation of Kunz and Rittel's IBIS approach and tool. A more recent descendent of gIBIS is the Compendium project, which is an open-source software tool with an international user community that includes governmental and non-governmental organizations, educational and business institutions (Buckingham Shum et al., 2006). Renton and Macintosh (2007) describe their use of Compendium to encourage debate and deliberation by citizens on issues of public policy, specifically by providing graphic argument maps that serve as 'policy memories'. Other relatively recent CSAV R&D inspired by IBIS work and the CSCW and hypertext pioneers include Zeno (Gordon and Richter, 2002) and Cope_It (Karacapilidis et al., 2009). More recently still, a Web-based tool called PolicyCommons has been developed within a European Union (EU)-funded project for visualizing arguments in policy deliberation (Benn and Macintosh, 2012).

Figure 10.1 shows a series of visualizations from the PolicyCommons CSAV tool as described by Benn and Macintosh (2012). As the authors explain, the purpose of the visualizations is to enable users to browse the arguments made during deliberation about a policy, in this case policy related to copyright in the knowledge economy. In the first two visualizations, the authors have adapted the 'treemap' technique pioneered by Shneiderman (1992) and created 'issue maps' – borrowing the IBIS terminology of Kunz and Rittel (1970) – which use color-coded regions to depict issues raised within the policy deliberation. The different sizes of the regions indicate the comparative number of responses associated with each issue. In the second visualization in particular, each 'issue

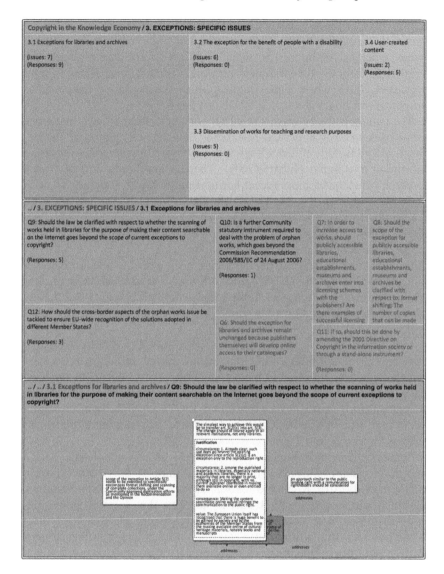

Source: Benn and Macintosh (2012).

Figure 10.1   A series of screenshots from the PolicyCommons Argument Visualization Tool showing Issue Map and Argument Network styles of visualization

region' (except those issues without responses) is clickable, so that users can click through to view the responses and arguments made on those issues which are of interest to them. Clicking on an issue generates the third visualization – the 'argument network' visualization – which is more typical of CSAV work. This visualization shows issues and responses as graphical text boxes connected by labeled lines.

The common theme of this research and these technologies that trace a line back to early hypertext research is that they all emphasize the key function and benefit of technology as a way of structuring and organizing information, echoing Hilbert (2009). There is another strand of CSAV research which has a more recent heritage in the Web 2.0 paradigm, with its focus on users as producers as well as consumers of content, and with connecting users in social networks.

CSAV, Web 2.0 and mass online deliberation
Although some of the research described in the previous section has specifically tried to address the challenge of group deliberation, the work has tended to focus on groups of limited size, where participants still mainly play the role of information consumers, typically contributing to the deliberative process indirectly via a facilitator.

In contrast, more recently there has been another strand of research exploring the use of CSAV for deliberation but now with a focus on mass online deliberation and on implementing important concepts of the Web 2.0 paradigm, with its emphasis on easier publishing of content (as in technology platforms such as Wordpress and YouTube), collaborative editing (as in platforms like Wikipedia), and social networking (as in platforms such as Facebook and Twitter).

Some of the more influential research in this strand includes the design and development of Cohere (Buckingham Shum, 2008), Deliberatorium (Gürkan et al., 2010; Klein and Iandoli, 2008) and Debategraph,[1] all three of which aim to provide Web-based platforms that support geographically dispersed user communities in controversial discussions online. Note, however, that all three also can trace a direct line of influence from Rittel's IBIS work. One of the most recent platforms which has adopted a non-IBIS approach to CSAV for deliberation is the ConsiderIt platform (Kriplean et al., 2012). These authors reject the more traditional IBIS-inspired approach of fine-grained information structuring on the grounds that it requires too high a level of training and commitment on the part of facilitators and users, thus creating a barrier to their widespread adoption. Kriplean et al. instead opt for an approach of 'more casually used interfaces' for deliberation, where users view and represent positions simply as pros and cons listed side by side in two columns.

Another CSAV tool that emphasizes the collaborative editing aspect of the Web 2.0 paradigm is Debatepedia.[2] However, its goal with respect to supporting online deliberation is distinct from the other platforms thus far introduced. That is, most of the technology platforms mentioned aim to represent arguments produced as part of a particular deliberative process that is directed towards solving a particular issue or set of issues. Debatepedia in contrast aims to be a reference resource providing summarized arguments on general issues of the day, which can then be incorporated as background material into a subsequent deliberative process. For example, the United Nations Foundation has funded the Climate Change portal on Debatepedia to summarize and outline the pros and cons with respect to addressing climate change in order that these summarized arguments can then inform social policy planning and deliberation. A similar goal of providing maps outlining debate on general issues of the day in order to inform subsequent political discussion is one of the motivations of a recent project headed by a pre-eminent sociologist, Bruno Latour. This project, called MACOSPOL (for Mapping Controversies on Science for Politics),[3] provides a platform for visualizing debates in an interactive map showing the most important stakeholders and their positions.

Furthermore, in this strand of CSAV that promotes more widespread, online user engagement, there is a class of technology that aims to provide novel ways of reflecting back a picture of a user's political viewpoints and where they are positioned with respect to other users. One very recent example of this is YourView,[4] which is an online platform for public political debate among citizens in Australia. The aim of the platform is to allow users to take a public stand on any major issue, ultimately with the goal of identifying the 'collective wisdom' of the Australian public and communicating these collective viewpoints to policy-makers. The platform then provides interactive visualization for users to see where they stand in the political landscape.

Another influential deliberative CSAV tool that has focused on 'closing the loop', to represent the collective views of the public on issues back to policy-makers, is the Parmenides platform (Cartwright and Atkinson, 2009). In Parmenides, users participating in a policy deliberation are presented with a particular policy proposal and its associated justification. The tool then guides the user through a structured survey where the user is asked to give agree/disagree responses to each part of the justification. Furthermore, because of this detailed level of argument representation, a policy analyst is able to clearly identify which elements of the policy justification are most or least acceptable for the sample of citizens participating in the survey.

Two similar platforms that have appeared at the European level are the EU Profiler[5] and Puzzled By Policy.[6] Part of an academic project that claimed independence from any party or political movement, the EU Profiler was a 'voting advice application' built for the 2009 European Parliament elections to help citizens understand the political landscape across Europe, specifically through helping them to answer questions such as: 'Where do the major parties stand on important issues?' and 'Where do I stand in relation to these parties?'. Each party's position was determined based on expert analysis of manifestos, public statements, and other documentation. Users were then invited to complete a survey which then generated a variety of visualizations to depict the proximity of each participant's views to the views of the various political parties as an indication of who best supports their views and who best supports their political profile. Puzzled By Policy has similar aims, but with a focus exclusively on the issue of immigration in the EU.

At the US level, there is one notable initiative of the US Department of State called Opinion Space.[7] This platform maps the opinions of participants on policy issues in a two-dimensional space, where position on the map is based on similarity of opinions: on the opinion map, participants are graphically depicted as closer to those they agree with and further away from those they disagree with.

PRIORITY AREAS FOR FUTURE RESEARCH

Although many advances have been made in CSAV research in general and CSAV applied to digital politics in particular, much still remains to be done if CSAV is to fulfill its potential to significantly impact the way we conduct our politics. Here I turn to identifying three priority areas of research that need to address key challenges: CSAV and information design, CSAV and the mass media, and evaluating CSAV as deliberative technology.

CSAV and Information Design

Kolb (2008) reminds us that making sense of arguments requires both having a sense of the detail as well as having a sense of the whole, which makes understanding complex arguments, whether in text or in graphical form, a challenging task. But, as Benn and Macintosh (2012) have indicated, whereas written language has evolved certain conventions for expressing arguments, no similar advancement has occurred in the design of visual languages for depicting argumentation. In particular, one of the main challenges facing tools for mapping and visualizing argument structures is the usability and readability of the visualizations themselves. Thus, one of the key challenges for CSAV research is to design visual languages for argument mapping, with appropriate visual cues for helping users to read and understand arguments, in much the same way that written language consists of many linguistic cues to help readers understand narrative and argumentative structure in order to make sense of a piece of text.

Indeed, generally speaking, most CSAV research has not systematically explored how the design of different elements of a visual language can represent the different elements of argumentation. Most tools tend to use a limited selection of visual features such as basic color textures and basic shapes. And even within the limited selection of visual features, most tend to adopt an even more limited visual metaphor of text-boxes connected via labeled or unlabeled arrows to depict connected elements of argumentation.

This limited use of visual language partly stems from the emphasis in CSAV until now (starting from the earliest Hypertext research) on the challenge of organizing and structuring information (decomposition, categorization, and linking of information units) rather than on issues of graphic design.

If CSAV technology is to attain widespread usage, research needs to focus more on the visual in argumentation visualization, and, to borrow and adapt a phrase from Sack (2011), perhaps to formulate an aesthetics of argumentation visualization, with the dual aim of understanding how this influences the way people consume information and the role it plays in sensemaking and managing information overload. This is perhaps the most urgent area for future research and development for CSAV, particularly for digital politics. To address this challenge, CSAV needs to be seen as an approach to information design and not just information structuring, where 'information design' is taken in the sense of Horn (1999); that is, as an area that combines cognitive science, interface design, visual communication and learning, and typography and graphic design. In this regard, CSAV research can learn from mass publishers such as newspapers and magazines who Horn has identified as leaders in the popularization of information design, recognizing how aesthetics affects whether readers will actually consume and use information.

Thus, some of the challenging research questions in CSAV from the perspective of information design are:

● What elements and conventions of visual language and graphics are most appropriate for depicting argumentation?
● What types of computer-generated graphical representations are best suited for communicating political debate?

- How can we create debate visualizations that are both visually appealing and understandable?

CSAV and the Mass Media

Research on the effects of technology on citizenship tends to focus on the impact of what is referred to as 'new' media (for example, social networking technology such as Facebook and Twitter). This research tends to ignore the effects that 'old' media such as TV and radio continue to have on civic life. Indeed, despite an increasingly diverse media ecosystem, TV and radio are still the primary media by which the public is informed about politics and by which they consume political debate. But rather than focus on the unhelpful division of old versus new media, future research should investigate the role that each plays in a mixed-media ecosystem. In particular, it is expected that new media will provide the long-sought-after interactivity to old media, where the views and reactions of the public at large are fed back to policy-makers who have traditionally used old media for one-way communication.

Until now, policy-makers have been confined to methods such as focus groups for eliciting detailed views from a sample of citizens. The trend will be towards tools such as Twitter that can capture these views *en masse* and reflect these back to policy-makers. Indeed, there has been a recently concluded European Commission-funded project, WeGov, which has investigated the use of popular social media rather than dedicated e-participation websites for mining public opinion and sentiment as a way of closing the loop between policy-makers and citizens (Wandhöfer et al., 2012). The WeGov project has been in part inspired by commercial analytics techniques employed by advertising and marketing agencies. Marketing and advertising departments of big commercial organizations already rely on social media tools to monitor consumer sentiment. The story is told how on 'Super Bowl Sunday', Nestlé's marketing team gathered in a situation room to monitor a 'map' of where social media chatter was most active, and the level of positive and negative sentiment towards their products in comparison to their competitors' products. In cases where there is negative sentiment, public relations (PR) people are able to intervene to try to address any problems. The marketing chief is quoted as saying that 'we have entered the age of conversation' between corporations and consumers (*The Economist*, 2013).

However, the current techniques of sentiment analysis and opinion mining need to move beyond simple good/bad or positive/negative binary feedback in order to show more nuanced aspects of public opinion. In particular, this will involve new techniques for mining aspects of public arguments such as value preferences.

The challenging research questions in this future CSAV area are:

- How can these new types of argument visualizations of political debates enhance existing media coverage of politics, political debates and elections?
- Can new media and techniques such as opinion mining and sentiment analysis add an interactive and reactive dimension to old media?

Evaluating CSAV as Deliberative Technology

Traditionally, evaluation is the one area in which CSAV R&D has been particularly deficient. Most evaluation studies have been related to the use of CSAV technology for educational purposes and to teach critical thinking skills, rather than the use of CSAV technology in deliberative settings. Even so, Braak et al. (2006) assessed most of the then available research into the effectiveness of argument visualization tools and found that many studies were not valid and generalizable, for reasons such as not defining reliable measures of the effectiveness of their tools, not creating control groups against which to compare experimental groups, and often not replicating the experiments beyond the narrow confines of the laboratory and extending them to other settings with other subject populations. Thus, Braak et al. argue that at best one can conclude that there is a slight trend towards CSAV tools improving critical thinking skills, but no stronger conclusion can be drawn from existing studies about the effectiveness of CSAV technology for its stated aims.

With respect to evaluating CSAV technology in deliberative settings, there have been a few small-scale studies, for example those conducted by Gürkan et al. (2010) for the Deliberatorium tool and Kriplean et al. (2012) for the ConsiderIt tool. In such studies, one common evaluation method is to automatically track user activity as they use the tool to participate in deliberation. Common metrics include logging the number of contributions submitted and the number of rebuttals made against others' contributions. In terms of deliberative outcomes, one common metric relates to the extent to which users shift their views during the deliberation, such shifting being considered as one key normative goal of deliberation.

Specifically for Gürkan et al. (2010), the aim was to evaluate the tool in terms of knowledge accumulation and organization as well as user participation and satisfaction. For Kriplean et al. (2012), they were most interested in the extent to which users considered alternative views and the extent to which users engaged with others' viewpoints. However, in both studies the authors recognize the limits of the experimental settings, acknowledging that, in the case of the ConsiderIt evaluations, 'without comparative data, we cannot say whether the pro/con list format caused people to consider both pros and cons more than they otherwise would have' (Kriplean et al., 2012).

Thus, it is apparent that if CSAV technology is to make any positive contribution to the deliberative process, there is an urgent need for empirical studies to ground many of the claims made on behalf of the technology and to inform future design of the technology.

The challenging research questions with respect to evaluating CSAV as deliberative technology are:

- Following Price (2012), can we define more robust evaluation frameworks, including more reliable measures for the effectiveness of CSAV tools, encompassing outcomes of deliberation as well as the discursive process itself?
- Following Chadwick (2012), can we design evaluation studies that consider the democratic value of other forms of political expression than textual deliberative discourse?

NOTES

1. Debategraph homepage, http://www.debategraph.org.
2. Debatepedia homepage, http://www.debatepedia.org/.
3. Mapping Controversies homepage, http://www.mappingcontroversies.net.
4. YourView homepage, http://www.yourview.org.au/.
5. EU Profiler homepage, http://www.euprofiler.eu.
6. Puzzled by Policy homepage, http://www.puzzledbypolicy.eu/.
7. Opinion Space homepage, http://www.state.gov/opinionspace.

FURTHER READING

Following is an alphabetized list of readings that the reader is invited to investigate for a more complete picture of the research issues and advances only briefly covered in this chapter.

Buckingham Shum and Hammond (1994). For the reader interested in a full excavation of the roots of hypertext groupware and how it led to the implementation of the first CSAV tools, this journal article by Buckingham Shum and Hammond is still perhaps, to date, the most extensive treatment of the history of the area.

Hilbert (2009). This journal article provides a thorough examination of the key challenges faced by CSAV technology when applied in the context of digital politics (the author uses the term 'e-democracy'). It is one of the rare articles that gives equal emphasis to the perspective of political communication and democratic theory, as well as the technological perspective.

Kirschner et al. (2003). Although written over a decade ago, this volume edited by Kirschner et al., entitled *Visualizing Argumentation*, still remains core reading in the field of CSAV. The reader is especially invited to read contributions in that volume by Simon Buckingham Shum ('The roots of computer supported argument visualization'), Tim van Gelder ('Enhancing deliberation through computer supported argument visualization') and Robert Horn ('Infrastructure for navigating interdisciplinary debates: critical decisions for representing argumentation').

Rittel and Webber (1973). The work of Horst Rittel is invoked time and again in the CSAV literature and is widely regarded as providing the intellectual roots, from a policy sciences perspective, for the understanding of the role that argumentation plays in social policy planning and deliberation. This article, written four decades ago, is perhaps the most definitive of his contributions.

Schneider et al. (2013). From a tools and technology perspective, Schneider et al. provide a comprehensive review of current CSAV research and development. The authors focus on argumentation tools that are Web-based and that have collaborative features. They are particularly interested in tools that are intended for use by the general public rather than tools that are exclusively about government consultation or policy-making. Their review of 37 argumentation tools compares them along a number of dimensions, including visualization, ease of use, collaboration, user engagement, and deliberative polling.

REFERENCES

Benn, N. and Macintosh, A. (2012). Policycommons – visualizing arguments in policy consultation. In Tambouris, E., Macintosh, A. and Sæbø, Ø. (eds), *Electronic Participation – 4th IFIP WG 8.5 International Conference, ePart 2012, Kristiansand, Norway, September 3–5, 2012. Proceedings*, Vol. 7444 of Lecture Notes in Computer Science. Kristiansand: Springer, pp. 61–72.

Braak, S.v.d., Oostendorp, H.v., Prakken, H. and Vreeswijk, G. (2006). A critical review of argument visualization tools: do users become better reasoners? In Grasso, F., Kibble, R. and Reed, C. (eds), *Workshop Notes of the ECAI-2006 Workshop on Computational Models of Natural Argument (CMNA VI)*, pp. 67–75, Riva del Garda, Italy.

Buckingham Shum, S. (2008). Cohere: towards Web 2.0 argumentation. In *2nd International Conference on Computational Models of Argument (COMMA '08)*, pp. 97–108, Toulouse: IOS Press.

Buckingham Shum, S. and Hammond, N. (1994). Argument-based design rationale: what use at what cost? *International Journal of Human–Computer Studies*, 40(4): 603–652.

Buckingham Shum, S., Selvin, A.M., Sierhuis, M., Conklin, J., Haley, C.B. and Nuseibeh, B. (2006). Hypermedia support for argumentation-based rationale: 15 years on from gibis and qoc. In Dutoit, A.H., McCall, R., Mistrik, I. and Paech, B. (eds), *Rationale Management in Software Engineering*, pp. 111–132. New York: Springer.

Cartwright, D. and Atkinson, K. (2009). Using computational argumentation to support e-participation. *IEEE Intelligent Systems*, 24(5): 42–52.

Chadwick, A. (2012). Web 2.0: new challenges for the study of e-democracy in an era of informational exuberance. In Coleman, S. and Shane, P.M. (eds), *Connecting Democracy: Online Consultation and the Flow of Political Communication*, pp. 45–73. Cambridge, MA: MIT Press.

Coleman, S. and Blumler, J.G. (2009). *The Internet and Democratic Citizenship*. New York: Cambridge University Press.

Coleman, S. and Moss, G. (2012). Under construction: the field of online deliberation research. *Journal of Information Technology and Politics*, 9(1): 1–15.

Conklin, J. and Begeman, M.L. (1988). Gibis: a hypertext tool for exploratory policy discussion. In *Proceedings of the 1988 ACM Conference on Computer-Supported Cooperative Work (CSCW '88)*, pp. 140–152. New York: ACM.

Davies, T. and Peña Gangadharan, S. (2009). *Online Deliberation: Design, Research and Practice*. Stanford, CA: CSLI Publications. *The Economist* (2013). Marketing: less guff, more puff. May 18.

Gordon, T.F. and Richter, G. (2002). Discourse support systems for deliberative democracy. In Lenk, K. and Traunmüller, R. (eds), *Electronic Government, First International Conference, EGOV02*, Vol. 2456 of Lecture Notes in Computer Science, pp. 248–255. Berlin and Heidelberg: Springer Verlag.

Gürkan, A., Iandoli, L., Klein, M. and Zollo, G. (2010). Mediating debate through on-line large-scale argumentation: evidence from the field. *Information Sciences*, 180: 3686–3702.

Hilbert, M. (2009). The maturing concept of e-democracy: from e-voting and online consultations to democratic value out of jumbled online chatter. *Journal of Information Technology and Politics*, 6: 87–110.

Horn, R.E. (1999). Information design: the emergence of a new profession. In Jacobson, R. (ed.), *Information Design*, pp. 15–33. Cambridge, MA: MIT Press.

Howard, P.N. and Hussain, M.M. (2013). *Democracy's Fourth Wave?: Digital Media and the Arab Spring*, Oxford Studies in Digital Politics. Oxford: Oxford University Press.

Karacapilidis, N.I., Tzagarakis, M., Karousos, N., Gkotsis, G., Kallistros, V., Christodoulou, S. and Mettouris, C. (2009). Tackling cognitively-complex collaboration with cope_it! *International Journal of Web-Based Learning and Teaching Technologies*, 4(3): 22–38.

Kirschner, P., Buckingham Shum, S. and Carr, C. (eds) (2003). *Visualizing Argumentation: Software Tools for Collaborative and Educational Sense-Making*, Computer-Supported Cooperative Work. London: Springer-Verlag.

Klein, M. and Iandoli, L. (2008). Supporting collaborative deliberation using a large-scale argumentation system: the MIT collaboratorium. In *Proceedings of the 11th Directions and Implications of Advanced Computing Symposium and 3rd International Conference on Online Deliberation (DIAC-2008/OD2008)*, pp. 5–12. Berkeley, CA.

Kolb, D. (2008). The revenge of the page. In Brusilovsky, P., and Davis, H., (eds) *HT '08: Proceedings of the 19th ACM Conference on Hypertext and Hypermedia (HT '08)*, pp. 89–96. Pittsburgh, PA. New York: ACM.

Kriplean, T., Morgan, J.T., Freelon, D., Borning, A. and Bennett, L. (2012). Supporting reflective public thought with considerit. In *Proceedings of the ACM 2012 Conference on Computer Supported Cooperative Work (CSCW '12)*, pp. 265–274. New York: ACM.

Kunz, W. and Rittel, H.W.J. (1970). Issues as elements of information systems. Technical Report Working Paper 131, Center for Planning and Development Research, University of California, Berkeley, CA.

Lotan, G., Graeff, E., Ananny, M., Gaffney, D., Pearce, I. and Danah Boyd (2011). The revolutions were tweeted: Information flows during the 2011 Tunisian and Egyptian revolutions. *International Journal of Communications*, 5: 1375–1405.

Macintosh, A., Gordon, T.F. and Renton, A. (2009). Providing argument support for e-participation. *Journal of Information Technology and Politics*, 6: 43–59.

Price, V. (2012). Playing politics: the experience of e-participation. In Coleman, S. and Shane, P.M. (eds), *Connecting Democracy: Online Consultation and the Flow of Political Communication*, pp. 125–148. Cambridge, MA: MIT Press.

Renton, A. and Macintosh, A. (2007). Computer supported argument maps as a policy memory. *Information Society*, 23(2): 125–133.

Rittel, H.W.J. and Webber, M.M. (1973). Dilemmas in a general theory of planning. *Policy Sciences*, 4: 155–169.

Sack, W. (2011). Aesthetics of information visualization. In Lovejoy, M., Paul, C. and Vesna, V. (eds), *Context Providers: Conditions of Meaning in Media Arts*, pp. 123–150. Bristol: Intellect Ltd.

Schneider, J., Groza, T. and Passant, A. (2013). A review of argumentation for the social semantic web. *Semantic Web*, 4(2): 159–218.

Shirky, C. (2011). The political power of social media. *Foreign Affairs*, 90(1): 28–41.

Shneiderman, B. (1992). Tree visualization with tree-maps: 2-d space-filling approach. *ACM Trans. Graph.*, 11: 92–99.

Tambouris, E., Macintosh, A. and de Bruijn, H. (2011). *Electronic Participation – Third IFIP WG 8.5 International Conference, ePart 2011, Delft. Proceedings. LNCS 6847*. Heidelberg: Springer.

Tambouris, E., Macintosh, A. and Sæbø, Ø. (eds) (2012). *Electronic Participation – 4th IFIP WG 8.5 International Conference, ePart 2012, Kristiansand, Norway, September 3-5, 2012. Proceedings*, volume 7444 of Lecture Notes in Computer Science. Springer.

Wandhöfer, T., Taylor, S., Alani, H., Joshi, S., Sizov, S., Walland, P., Thamm, M., Bleier, A. and Mutschke, P. (2012). Engaging politicians with citizens on social networking sites: the wegov toolbox. *International Journal of Electronic Government Research*, 8(3): 22–43.

PART III

COLLECTIVE ACTION AND CIVIC ENGAGEMENT

11. The logic of connective action: digital media and the personalization of contentious politics

W. Lance Bennett and Alexandra Segerberg

From the Arab Spring and *los indignados* in Spain, to Occupy Wall Street (and beyond), large-scale, sustained protests are using digital media in ways that go beyond sending and receiving messages. Some of these action formations contain relatively small roles for formal brick-and-mortar organizations. Others involve well-established advocacy organizations, in hybrid relations with other organizations, using technologies that enable personalized public engagement. Both stand in contrast to the more familiar organizationally managed and brokered action conventionally associated with social movement and issue advocacy. This chapter examines the organizational dynamics that emerge when communication becomes a prominent part of organizational structure. It argues that understanding such variations in large-scale action networks requires distinguishing between at least two logics that may be in play: the familiar logic of collective action associated with high levels of organizational resources and the formation of collective identities, and the less familiar logic of connective action based on personalized content sharing across media networks. In the former, introducing digital media does not change the core dynamics of the action. In the case of the latter, it does. Building on these distinctions, the chapter presents three ideal types of large-scale action networks that are becoming prominent in the contentious politics of the contemporary era.

With the world economy in crisis, the heads of the 20 leading economies held a series of meetings beginning in the fall of 2008 to coordinate financial rescue policies. Wherever the G20 leaders met, whether in Washington, London, St Andrews, Pittsburgh, Toronto, or Seoul, they were greeted by protests. In London, anti-capitalist, environmental direct activist, and non-governmental organization (NGO)-sponsored actions were coordinated across different days. The largest of these demonstrations was sponsored by a number of prominent NGOs including Oxfam, Friends of the Earth, Save the Children and World Vision. This loose coalition launched a Put People First (PPF) campaign promoting public mobilization against social and environmental harms of 'business-as-usual' solutions to the financial crisis. The website for the campaign carried the simple statement:

> Even before the banking collapse, the world suffered poverty, inequality and the threat of climate chaos. The world has followed a financial model that has created an economy fuelled by ever-increasing debt, both financial and environmental. Our future depends on creating an economy based on fair distribution of wealth, decent jobs for all and a low carbon future. (Put People First, 2009)

The centerpiece of this PPF campaign was a march of some 35 000 people through the streets of London a few days ahead of the G20 meeting, to give voice and show commitment to the campaign's simple theme.

The London PPF protest drew together a large and diverse protest with the emphasis on personal expression, but it still displayed what Tilly (2004, 2006) termed 'WUNC': worthiness, embodied by the endorsements by some 160 prominent civil society organizations and recognition of their demands by various prominent officials; unity, reflected in the orderliness of the event; numbers of participants, that made PPF the largest of a series of London G20 protests and the largest demonstration during the string of G20 meetings in different world locations; and commitment, reflected in the presence of delegations from some 20 nations who joined local citizens in spending much of the day listening to speakers in Hyde Park or attending religious services sponsored by Church-based development organizations.[1] The large volume of generally positive press coverage reflected all of these characteristics, and responses from heads of state to the demonstrators accentuated the worthiness of the event (Bennett and Segerberg, 2011).[2]

The protests continued as the G20 in 2010 issued a policy statement making it clear that debt reduction and austerity would be the centerpieces of a political program that could send shocks through economies from the United States and the UK, to Greece, Italy, and Spain, while pushing more decisive action on climate change onto the back burner. Public anger swept cities from Madison to Madrid, as citizens protested that their governments, no matter what their political stripe, offered no alternatives to the economic dictates of a so-called neoliberal economic regime that seemed to operate from corporate and financial power centers beyond popular accountability and, some argued, even beyond the control of states.

Some of these protests seemed to operate with surprisingly light involvement from conventional organizations. For example, in Spain '*los indignados*' (the indignant ones) mobilized in 2011 under the name of 15M for the date (May 15) of the mass mobilization that involved protests in some 60 cities. One of the most remarkable aspects of this sustained protest organization was its success at keeping political parties, unions, and other powerful political organizations out: indeed, they were targeted as part of the political problem. There were, of course, civil society organizations supporting 15M, but they generally stayed in the background to honor the personalized identity of the movement: the faces and voices of millions of ordinary people displaced by financial and political crises. The most visible organization consisted of the richly layered digital and interpersonal communication networks centering around the media hub of Democracia real YA![3] This network included links to more than 80 local Spanish city nodes, and a number of international solidarity networks. On the one hand, Democracia real YA! seemed to be a website, and on the other, it was a densely populated and effective organization. It makes sense to think of the core organization of the *indignados* as both of these and more, revealing the hybrid nature of digitally mediated organization (Chadwick, 2013).

Given its seemingly informal organization, the 15M mobilization surprised many observers by sustaining and even building strength over time, using a mix of online media and offline activities that included face-to-face organizing, encampments in city centers, and marches across the country. Throughout, the participants communicated a collective identity of being leaderless, signaling that labor unions, parties, and more radical movement groups should stay at the margins. A survey of 15M protesters by a team of Spanish researchers showed that the relationships between individuals and organizations differed in at least three ways from participants in an array of other more conventional movement protests, including a general strike, a regional protest, and a pro-life demonstration: (1) where strong majorities of participants in other protests recognized the involvement of key organizations with brick-and-mortar addresses, only 38 per cent of *indignados* did so; (2) only 13 per cent of the organizations cited by 15M participants offered any membership or affiliation possibilities, in contrast to large majorities who listed membership organizations as being important in the other demonstrations; and (3) the mean age range of organizations (such as parties and unions) listed in the comparison protests ranged from 10 to over 40 years, while the organizations cited in association with 15M were, on average, less than three years old (Anduiza et al., 2014). Despite, or perhaps because of, these interesting organizational differences, the ongoing series of 15M protests attracted participation from somewhere between 6 and 8 million people, a remarkable number in a nation of 40 million (RTVE, 2011).

Similar to PPF, the *indignados* achieved impressive levels of communication with outside publics both directly via images and messages spread virally across social networks, and indirectly when anonymous Twitter streams and YouTube videos were taken up as mainstream press sources. Their actions became daily news fare in Spain and abroad, with the protesters receiving generally positive coverage of their personal messages in local and national news; again defying familiar observations about the difficulty of gaining positive news coverage for collective actions that spill outside the bounds of institutions and take to the streets (Gitlin, 1980).[4] In addition to communicating concerns about jobs and the economy, the clear message was that people felt the democratic system had broken to the point that all parties and leaders were under the influence of banks and international financial powers. Despite avoiding association with familiar civil society organizations, lacking leaders, and displaying little conventional organization, *los indignados*, similar to PPF, achieved high levels of WUNC.

Two broad organizational patterns characterize these increasingly common digitally enabled action networks. Some cases, such as PPF, are coordinated behind the scenes by networks of established issue advocacy organizations that step back from branding the actions in terms of particular organizations, memberships, or conventional collective action frames. Instead, they cast a broader public engagement net using interactive digital media and easy-to-personalize action themes, often deploying batteries of social technologies to help citizens spread the word over their personal networks. The second pattern, typified by the *indignados* and the Occupy protests in the United States, entails technology platforms and applications taking the role of established political organizations. In this network mode, political demands and grievances are often shared in very personalized accounts that travel over social networking platforms, e-mail lists, and online coordinating platforms. For example, the easily personalized action frame, 'We are the 99 per cent', that emerged from the US Occupy protests in 2011 quickly traveled the world via personal stories and images shared on social networks such as Tumblr, Twitter, and Facebook.

Compared to many conventional social movement protests, with identifiable membership organizations leading the way under common banners and collective identity frames, these more personalized, digitally mediated collective action formations have frequently been larger; have scaled up more quickly; and have been flexible in tracking moving political targets and bridging different issues. Whether we look at PPF, Arab Spring, the *indignados*, or Occupy, we note surprising success in communicating simple political messages directly to outside publics using common digital technologies such as Facebook or Twitter. Those media feeds are often picked up as news sources by conventional journalism organizations.[5] In addition, these digitally mediated action networks often seem to be accorded higher levels of WUNC than their more conventional social movement counterparts. This observation is based on comparisons of more conventional anti-capitalist collective actions organized by movement groups, in contrast with both the organizationally enabled PPF protests and the crowd-enabled 15M mobilizations in Spain and the Occupy Wall Street protests, which quickly spread to thousands of other places. The differences between both types of digitally mediated action and more conventional organization-centered and brokered collective actions led us to see interesting differences in underlying organizational logics and in the role of communication as an organizing principle.

The rise of digitally networked action (DNA) has been met with some understandable skepticism about what really is so very new about it, mixed with concerns about what it means for the political capacities of organized dissent. We are interested in understanding how these more personalized varieties of collective action work: how they are organized, what sustains them, and when they are politically effective. We submit that convincingly addressing such questions requires recognizing the differing logics of action that underpin distinct kinds of collective action networks. This chapter thus develops a conceptual framework of such logics, on the basis of which further questions about DNA may then be tackled.

We propose that more fully understanding contemporary large-scale networks of contentious action involves distinguishing between at least two logics of action that may be in play: the familiar logic of collective action, and the less familiar logic of connective action. Doing so in turn allows us to discern three ideal action types, of which one is characterized by the familiar logic of collective action, and two other types involve more personalized action formations that differ in terms of whether formal organizations are more or less central in enabling a connective communication logic. A first step in understanding DNA, the DNA at the core of connective action, lies in defining personalized communication and its role along with digital media in the organization of what we call connective action.

PERSONAL ACTION FRAMES AND SOCIAL MEDIA NETWORKS

Structural fragmentation and individualization in many contemporary societies constitute an important backdrop to the present discussion. Various breakdowns in group memberships and institutional loyalties have trended in the more economically developed industrial democracies, resulting from pressures of economic globalization spanning a period from roughly the 1970s through to the end of the last century (Bennett, 1998; Putnam, 2000). These sweeping changes have produced a shift in social and political orientations among younger generations in the nations that we now term the post-industrial democracies (Inglehart, 1997). These individualized orientations result in engagement with politics as an expression of personal hopes, lifestyles, and grievances. When enabled by various kinds of communication technologies, the resulting DNAs in post-industrial democracies bear some remarkable similarities to action formations in decidedly undemocratic regimes such as those swept by the Arab Spring. In both contexts, large numbers of similarly disaffected individuals seized upon opportunities to organize collectively through access to various technologies (Howard and Hussain, 2011). Those connectivities fed in and out of the often intense face-to-face interactions going on in squares, encampments, mosques, and general assembly meetings.

In personalized action formations, the nominal issues may resemble older movement or party concerns in terms of topics (environment, rights, women's equality, and trade fairness) but the ideas and mechanisms for organizing action become more personalized than in cases where action is organized on the basis of social group identity, membership, or ideology. These multifaceted processes of individualization are articulated differently in different societies, but include the propensity to develop flexible political identifications based on personal lifestyles (Giddens, 1991; Inglehart, 1997; Bennett, 1998; Bauman, 2000; Beck and Beck-Gernsheim, 2002), with implications in collective action (McDonald, 2002; Micheletti, 2003; della Porta, 2005) and organizational participation (Putnam, 2000; Bimber et al., 2012). People may still join actions in large numbers, but the identity reference is more derived through inclusive and diverse large-scale personal expression rather than through common group or ideological identification.

This shift from group-based to individualized societies is accompanied by the emergence of flexible social 'weak tie' networks (Granovetter, 1973) that enable identity expression and the navigation of complex and changing social and political landscapes. Networks have always been part of society, to help people navigate life within groups or between groups, but the late modern society involves networks that become more central organizational forms that transcend groups and constitute core organizations in their own right (Castells, 2000). These networks are established and scaled through various sorts of digital technologies that are by no means value-neutral in enabling quite different kinds of communities to form and diverse actions to be organized, from auctions on eBay to protests in different cultural and social settings. Thus, the two elements of 'personalized communication' that we identify as particularly important in large-scale connective action formations are:

1. Political content in the form of easily personalized ideas such as PPF in the London 2009 protests, or 'We are the 99 per cent' in the later Occupy protests. These frames require little in the way of persuasion, reason, or reframing to bridge differences in how others may feel about a common problem. These personal action frames are inclusive of different personal reasons for contesting a situation that needs to be changed.
2. Various personal communication technologies that enable sharing these themes. Whether through texts, tweets, social network sharing, or posting YouTube mashups, the communication process itself often involves further personalization through the spreading of digital connections among friends or trusted others. Some more sophisticated custom coordinating platforms can resemble organizations that exist more online than off.

As we followed various world protests, we noticed a dazzling array of personal action frames that spread through social media. Both the acts of sharing these personal calls to action and the social technologies through which they spread help to explain both how events are communicated to external audiences and how the action itself is organized. Indeed, in the limiting case, the communication network becomes the organizational form of the political action (Earl and Kimport, 2011). We explore the range of differently organized forms of contention using personalized communication up to the point at which they enter the part of the range conventionally understood as social movements. This is the boundary zone within which what we refer to as connective action gives way to collective action.

The case of PPF occupies an interesting part of this range of contentious action because there were many conventional organizations involved in the mobilization, from Churches to social justice NGOs. Yet, visitors to the sophisticated, stand-alone, PPF coordinating platform (which served as an interesting kind of organization in itself) were not asked to pledge allegiance to specific political demands on the organizational agendas of the protest sponsors. Instead, visitors to the organizing site were met with an impressive array of social technologies, enabling them to communicate in their own terms with each other and with various political targets. The centerpiece of the PPF site was a prominent text box under an image of a megaphone that invited the visitor to 'Send Your Own Message to the G20'. Many of the messages to the G20 echoed the easy-to-personalize action frame of PPF, and they also revealed a broad range of personal thoughts about the crisis and possible solutions.

PPF as a personal action frame was easy to shape and share with friends near and far. It became a powerful example of what students of viral communication refer to as a meme: a symbolic packet that travels easily across large and diverse populations because it is easy to imitate, adapt personally, and share broadly with others. Memes are network-building and bridging units of social information transmission similar to genes in the biological sphere (Dawkins, 1989). They travel through personal appropriation, and then by imitation and personalized expression via social sharing in ways that help others to appropriate, imitate, and share in turn (Shifman, 2013). The simple PPF protest meme traveled interpersonally, echoing through newspapers, blogs, Facebook friend networks, Twitter streams, Flickr pages, and other sites on the Internet, leaving traces for years after the events.[6] Indeed, part of the meme traveled to Toronto more than a year later where the leading civil society groups gave the name 'People First' to their demonstrations. And many people in the large crowds in Seoul in the last G20 meeting of the series could be seen holding up red and white 'PPF' signs in both English and Korean (Weller, 2010).

Something similar happened in the case of the *indignados*, where protesters raised banners and chanted 'Shhh . . . the Greeks are sleeping', with reference to the crushing debt crisis and severe austerity measures facing that country. This idea swiftly traveled to Greece where Facebook networks agreed to set alarm clocks at the same time to wake up and demonstrate. Banners in Athens proclaimed: 'We've awakened! What time is it? Time for them to leave!' and 'Shhh . . . the Italians are sleeping' and 'Shhh . . . the French are sleeping'. These efforts to send personalized protest themes across national and cultural boundaries met with varying success, making for an important cautionary point: we want to stress that not all personal action frames travel equally well or equally far. The fact that these messages traveled more easily in Spain and Greece than in France or Italy is an interesting example pointing to the need to study failures as well as successes. Just being easy to personalize (for example, I am personally indignant about x, y, and z, and so I join with *los indignados*) does not ensure successful diffusion. Both political opportunities and conditions for social adoption may differ from situation to situation. For example, the limits in the Italian case may reflect an already established popular anti-government network centered on comedian–activist Beppe Grillo. The French case may involve the ironic efforts of established groups on the left to lead incipient solidarity protests with the *indignados*, and becoming too heavy-handed in suggesting messages and action programs.

Personal action frames do not spread automatically. People must show each other how they can appropriate, shape, and share themes. In this interactive process of personalization and sharing, communication networks may become scaled up and stabilized through the digital technologies people use to share ideas and relationships with others. These technologies and their use patterns often remain in place as organizational mechanisms. In the PPF and the *indignados* protests, the communication processes themselves represented important forms of organization.

In contrast to personal action frames, other calls to action more clearly require joining with established groups or ideologies. These more conventionally understood collective action frames are more likely to stop at the edges of communities, and may require resources beyond communication technologies to bridge the gaps or align different collective frames (Snow and Benford, 1988; Benford and Snow, 2000). For example, another set of protests in London at the start of the financial crisis was organized by a coalition of more radical groups under the name G20 Meltdown. Instead of mobilizing the expression of large-scale personal concerns, they demanded ending the so-called neoliberal economic policies of the G20, and some even called for the end to capitalism itself. Such demands typically come packaged with more demanding calls to join in particular repertoires of collective action. Whether those repertoires are violent or non-violent, they typically require adoption of shared ideas and behaviors. These anarcho-socialist demonstrations drew on familiar anti-capitalist slogans and calls to 'storm the banks' or 'eat the rich' while staging dramatic marches behind the four horsemen of the economic apocalypse riding from the gates of old London to the Bank of England. These more radical London events drew smaller turnouts (some 5000 for the Bank of England march and 2000 for a climate encampment), higher levels of violence, and generally negative press coverage (Bennett and Segerberg, 2011). While scoring high on commitment in terms of the personal costs of civil disobedience, and displaying unity around anti-capitalist collective action frames, these demonstrations lacked the attributions of public worthiness (for example, recognition from public officials, getting their messages into the news) and the numbers that gave PPF its higher levels of WUNC.

Collective action frames that place greater demands on individuals to share common identifications or political claims can also be regarded as memes, in the sense that slogans such as 'eat the rich' have rich histories of social transmission. This particular iconic phrase may possibly date to Rousseau's quip: 'When the people shall have nothing more to eat, they will eat the rich'. The crazy course of that meme's passage down through the ages includes its appearance on T-shirts in the 1960s and in rock songs of that title by Aerosmith and Motorhead, just to scratch the surface of its history of travel through time and space, reflecting the sequence of appropriation, personal expression, and sharing. One distinction between personal action and collective action memes seems to be that the latter require somewhat more elaborate packaging and ritualized action to reintroduce them into new contexts. For example, the organizers of the 'storm the banks' events staged an elaborate theatrical ritual with carnivalesque opportunities for creative expression as costumed demonstrators marched behind the Four Horsemen of the financial apocalypse.[7] At the same time, the G20 Meltdown discourse was rather closed, requiring adopters to make common cause with others. The Meltdown coalition had an online presence, but they did not offer easy means for participants to express themselves in their own voices (Bennett and Segerberg, 2011). This suggests that more demanding and exclusive collective action frames can also travel as memes, but more often they hit barriers at the intersections of social networks defined by established political organizations, ideologies, interests, class, gender, race, or ethnicity. These barriers often require resources beyond social technologies to overcome.

While the idea of memes may help to focus differences in transmission mechanisms involved in more personal versus collective framing of action, we will use the terms 'personal action frames' and 'collective action frames' as our general concepts. This conceptual pairing locates our work alongside analytical categories used by social movement scholars (Snow and Benford, 1988; Benford and Snow, 2000). As should be obvious, the differences we are sketching between personal and collective action frames are not about being online versus offline. All contentious action networks are in important ways embodied and enacted by people on the ground (Juris, 2008; Routledge and Cumbers, 2009). Moreover, most formal political organizations have discovered that the growing sophistication and ubiquity of social media can reduce the resource costs of public outreach and coordination, but these uses of media do not change the action dynamics by altering the fundamental principles of organizing collectivities. By contrast, digital media networking can change the organizational game, given the right interplay of technology, personal action frames, and, when organizations get in the game, their willingness to relax collective identification requirements in favor of personalized social networking among followers.

The logic of collective action that typifies the modern social order of hierarchical institutions and membership groups stresses the organizational dilemma of getting individuals to overcome resistance to joining actions where personal participation costs may outweigh marginal gains, particularly when people can ride on the efforts of others for free, and reap the benefits if those others win the day. In short, conventional collective action typically requires people to make more difficult choices and adopt more self-changing social identities than DNA based on personal action frames organized around social technologies. The spread of collective identifications typically requires more education, pressure, or socialization, which in turn makes higher demands on formal organization and resources such as money to pay rent for organization offices, to generate publicity, and to hire professional staff organizers (McAdam et al., 1996).[8] Digital media may help to reduce some costs in these processes, but they do not fundamentally change the action dynamics.

As noted above, the emerging alternative model that we call the logic of connective action applies increasingly to life in late modern societies in which formal organizations are losing their grip on individuals, and group ties are being replaced by large-scale, fluid social networks (Castells, 2000).[9] The organizational processes of social media play an important role in how these networks operate, and their logic does not require strong organizational control or the symbolic construction of a united 'we'. The logic of connective action, we suggest, entails a dynamic of its own and thus deserves analysis on its own analytical terms.

TWO LOGICS: COLLECTIVE AND CONNECTIVE ACTION

Social movements and contentious politics extend over many different kinds of phenomena and action (Melucci, 1996; McAdam et al., 2001; Tarrow, 2011). The talk about new forms of collective action may reflect ecologies of action that are increasingly complex (Chesters and Welsh, 2006). Multiple organizational forms operating within such ecologies may be hard to categorize, not least because they may morph over time or context, displaying hybridity of various kinds (Chadwick, 2013). In addition, protest and organizational work is occurring both online and off, using technologies of different capabilities, sometimes making the online/ offline distinction relevant, but more often not (Earl and Kimport, 2011; Bimber et al., 2012).

Some observers mark a turning point in patterns of contemporary contentious politics, which mix different styles of organization and communication, along with the intersection of different issues with the iconic union of 'teamsters and turtles' in the Battle of Seattle in 1999, during which burly union members marched alongside environmental activists wearing turtle costumes in battling a rising neoliberal trade regime that was seen as a threat to democratic control of both national economies and the world environment. Studies of such events show that there are still plenty of old-fashioned meetings, and issue brokering and coalition building, going on (Polletta, 2002). At the same time, however, there is increasing coordination of action by organizations and individuals using digital media to create networks, structure activities, and communicate their views directly to the world. This means that there is also an important degree of technology-enabled networking (Livingston and Asmolov, 2010) that makes highly personalized, socially mediated communication processes fundamental structuring elements in the organization of many forms of connective action.

How do we sort out what organizational processes contribute what qualities to collective and connective action networks? How do we identify the borders between fundamentally different types of action formations: that is, what are the differences between collective and connective action, and where are the hybrid overlaps? We propose a starting point for sorting out some of the complexity and overlap in the forms of action by distinguishing between two logics of action. The two logics are associated with distinct dynamics, and thus draw attention to different dimensions for analysis. It is important to separate them analytically as one is less familiar than the other, and this in turn constitutes an important stumbling block for the study of much contemporary political action that we term connective action.[10]

The more familiar action logic is the logic of collective action, which emphasizes the problems of getting individuals to contribute to the collective endeavor that typically involves seeking some sort of public good (for example, democratic reforms) that may be better attained through forging a common cause. The classical formulation of this problem was articulated by Olson (1965), but the implications of his general logic have reached far beyond the original formulation. Olson's intriguing observation was that people in fact cannot be expected to act together just because they share a common problem or goal. He held that in large groups in which individual contributions are less noticeable, rational individuals will free-ride on the efforts of others: it is more cost-efficient not to contribute if you can enjoy the good without contributing. Moreover, if not enough people join in creating the good, your efforts are wasted anyway. Either way, it is individually rational not to contribute, even if all agree that all would be better off if everyone did. This thinking fixes attention on the problematic dynamics attending the rational action of atomistic individuals, and at the same time makes resource-rich organizations a central concern. Both the solutions Olson discerned – coercion and selective incentives – implied organizations with substantial capacity to monitor, administer, and distribute such measures.

In this view, formal organizations with resources are essential to harnessing and coordinating individuals in common action. The early application of this logic to contentious collective action was most straightforwardly exemplified in resource mobilization theory (RMT), in which social movement scholars explicitly adopted Olson's framing of the collective action problem and its organization-centered solution. Part of a broader wave rejecting the idea of social movements as irrational behavior erupting out of social dysfunction, early RMT scholars accepted the problem of rational free-riders as a fundamental challenge and regarded organizations and their ability to mobilize resources as critical elements of social movement success. Classic formulations came from McCarthy and Zald (1973, 1977) who theorized the rise of external support and resources available to social movement organizations (SMOs), and focused attention on the professionalization of movement organizations and leaders in enabling more resource-intensive mobilization efforts.

The contemporary social movement field has moved well beyond the rational choice orientation of such earlier work. Indeed, important traditions developed independently of, or by rejecting, all or parts of the resource mobilization perspective and by proposing that we pay more attention to the role of identity, culture, emotion, social networks, political process, and opportunity structures (Melucci, 1996; McAdam et al., 2001; della Porta and Diani, 2006). We do not suggest that these later approaches cling to rational choice principles. We do, however, suggest that echoes of the modernist logic of collective action can still be found to play a background role even in work that is in other ways far removed from the rational choice orientation of Olson's original argument. This comes out in assumptions about the importance of particular forms of organizational coordination and identity in the attention given to organizations, resources, leaders, coalitions, brokering differences, cultural or epistemic communities, the importance of formulating collective action frames, and bridging of differences among those frames. Connective action networks may vary in terms of stability, scale, and coherence, but they are organized by different principles. Connective action networks are typically far more individualized and technologically organized sets of processes, that result in action without the requirement of collective identity framing or the levels of organizational resources required to respond effectively to opportunities.

One of the most widely adopted approaches that moved social movement research away from the rational choice roots toward a more expansive collective action logic is the analysis of collective action frames, which centers on the processes of negotiating common interpretations of collective identity linked to the contentious issues at hand (Snow et al., 1986; Snow and Benford, 1988; Hunt et al., 1994; Benford and Snow, 2000). Such framing work may help to mobilize individuals and ultimately lower resource costs by retaining their emotional commitment to action.

At the same time, the formulation of ideologically demanding, socially exclusive, or high-conflict collective frames also invites fractures, leading to an analytical focus on how organizations manage or fail to bridge these differences. Resolving these frame conflicts may require the mobilization of resources to bridge differences between groups that have different goals and ways of understanding their issues. Thus, while the evolution of different strands of social movement theory has moved away from economic collective action models, many still tend to emphasize the importance of organizations that have strong ties to members and followers, and the resulting ways in which collective identities are forged and fractured among coalitions of those organizations and their networks.

Sustainable and effective collective action from the perspective of the broader logic of collective action typically requires varying levels of organizational resource mobilization deployed in organizing, leadership, developing common action frames, and brokerage to bridge organizational differences. The opening or closing of political opportunities affects this resource calculus (Tarrow, 2011), but overall, large-scale action networks that reflect this collective action logic tend to be characterized in terms of numbers of distinct groups networking to bring members and affiliated participants into the action and to keep them there. On the individual level, collective action logic emphasizes the role of social network relationships and connections as informal preconditions for more centralized mobilization (for example, in forming and spreading action frames, and forging common identifications and relations of solidarity and trust). At the organizational level, the strategic work of brokering and bridging coalitions between organizations with different standpoints and constituencies becomes the central activity for analysis (see also Diani, forthcoming). Since the dynamics of action in networks characterized by this logic tends not to change significantly with digital media, it primarily invites analysis of how such tools help actors do what they were already doing (see also Bimber et al., 2009; Earl and Kimport, 2011).

Movements and action networks characterized by these variations on the logic of collective action are clearly visible in contemporary society. They have been joined by many other mobilizations that may superficially seem like movements, but on closer inspection lack many of the traditional defining characteristics. Efforts to push these kinds of organization into recognizable social movement categories diminish our capacity to understand one of the most interesting developments of our times: how fragmented, individualized populations, that are hard to reach and even harder to induce to share personally transforming collective identities, somehow find ways to mobilize protest networks from Wall Street to Madrid to Cairo. Indeed, when people are individualized in their social orientations, and thus structurally or psychologically unavailable to modernist forms of political movement organization, resource mobilization becomes increasingly costly and has diminishing returns. Organizing such populations to overcome free-riding and helping them to shape identities in common is not necessarily the most successful or effective logic for organizing collective action. When people who seek more personalized paths to concerted action are familiar with practices of social networking in everyday life, and when they have access to technologies from mobile phones to computers, they are already familiar with a different logic of organization: the logic of connective action.

The logic of connective action foregrounds a different set of dynamics from the ones just outlined. At the core of this logic is the recognition of digital media as organizing agents. Several collective action scholars have explored how digital communication technology alters the parameters of Olson's original theory of collective action. Lupia and Sin (2003) show how Olson's core assumption about weak individual commitment in large groups (free-riding) may play out differently under conditions of radically reduced communication costs. Bimber et al. (2005) in turn argue that public goods themselves may take on new theoretical definition as erstwhile free-riders find it easier to become participants in political networks that diminish the boundaries between public and private; boundaries that are blurred in part by the simultaneous public–private boundary crossing of ubiquitous social media.

Important for our purposes here is the underlying economic logic of digitally mediated social networks, as explained most fully by Benkler (2006). He proposes that participation becomes self-motivating as personally expressive content is shared with, and recognized by, others who in turn repeat these networked sharing activities. When these interpersonal networks are enabled by technology platforms of various designs that coordinate and scale the networks, the resulting actions can resemble collective action, yet without the same role played by formal organizations or transforming social identifications. In place of content that is distributed and relationships that are brokered by hierarchical organizations, social networking involves co-production and co-distribution, revealing a different economic and psychological logic: co-production and sharing based on personalized expression. This does not mean that all online communication works this way. Looking at most online newspapers, blogs, or political campaign sites makes it clear that the logic of the organization-centered brick-and-mortar world is often reproduced online, with little change in organizational logic beyond possible efficiency gains (Bimber and Davis, 2003; Foot and Schneider, 2006). Yet, many socially mediated networks do operate with an alternative logic that also helps to explain why people labor collectively for free to create such things as open source software, Wikipedia, WikiLeaks, and the free and open source software that powers many protest networks (Calderaro, 2011).

In this connective logic, taking public action or contributing to a common good becomes an act of personal expression and recognition or self-validation achieved by sharing ideas and actions in trusted relationships. Sometimes the people in these exchanges may be on the other side of the world, but they do not require a club, a party, or a shared ideological frame to make the connection. In place of the initial collective action problem of getting the individual to contribute, the starting point of connective action is the self-motivated (though not necessarily self-centered) sharing of already internalized or personalized ideas, plans, images, and resources with networks of others. This 'sharing' may take place in networking sites such as Facebook, or via more public media such as Twitter and YouTube through, for example, comments and re-tweets.[11] Action networks characterized by this logic may scale up rapidly through the combination of easily spreadable personal action frames and digital technology enabling such communication. This invites analytical attention to the network as an organizational structure in itself.

Technology-enabled networks of personalized communication involve more than just exchanging information or messages. The flexible, recombinant nature of DNA makes these web spheres and their offline extensions more than just communication systems. Such networks are flexible organizations in themselves, often enabling coordinated adjustments and rapid action aimed at often shifting political targets, even crossing geographic and temporal boundaries in the process. As Diani (forthcoming) argues, networks are not just precursors or building blocks of collective action: they are in themselves organizational structures that can transcend the elemental units of organizations and individuals.[12] As noted earlier, communication technologies do not change the action dynamics in large-scale networks characterized by the logic of collective action. In the networks characterized by connective action, they do.

The organizational structure of people and social technology emerges more clearly if we draw on the actor-network theory of Latour (2005) in recognizing digital networking mechanisms (for example, various social media and devices that run them) as potential network agents alongside human actors (that is, individuals and organizations). Such digital mechanisms may include organizational connectors (for example, web links), event coordination (for example, protest calendars), information sharing (for example, YouTube and Facebook), and multifunction networking platforms in which other networks become embedded (for example, links in Twitter and Facebook posts), along with various capacities of the devices that run them. These technologies not only create online meeting places and coordinate offline activities, but they also help to calibrate relationships by establishing levels of transparency, privacy, security, and interpersonal trust. It is also important that these digital traces may remain behind on the web to provide memory records or action repertoires that might be passed on via different mechanisms associated with more conventional collective action such as rituals or formal documentation.

The simple point here is that collective and connective logics are distinct logics of action (in terms of both identity and choice processes), and thus both deserve analysis on their own terms. Just as traditional collective action efforts can fail to result in sustained or effective movements, there is nothing preordained about the results of digitally mediated networking processes. More often than not, they fail badly. The transmission of personal expression across networks may or may not become scaled up, stable, or capable of various kinds of targeted action depending on the kinds of social technology designed and appropriated by participants, and the kinds of opportunities that may motivate anger or compassion across large numbers of individuals. Thus, the Occupy Wall Street protests that spread in a month from New York to more than 80 countries and 900 cities around the world might not have succeeded without the inspiring models of the Arab Spring or the *indignados* in Spain, or the worsening economic conditions that provoked anger among increasing numbers of displaced individuals. Yet, when the Occupy networks spread under the easy-to-personalize action frame of 'We are the 99 per cent', there were few identifiable established political organizations at the center of them. There was even a conscious effort to avoid designating leaders and official spokespeople. The most obvious organizational forms were the layers of social technologies and websites that carried news reported by participants and displayed tools for personalized networking. One of the sites was '15.10.11 united for #global change'.[13] Instead of the usual 'Who are we?' section of the website, #globalchange asked: 'Who are you?'.

Collective and connective action may co-occur in various formations within the same ecology of action. It is nonetheless possible to discern three clear ideal types of large-scale action networks. While one is primarily characterized by collective action logic, the other two are connective action networks distinguished by the role of formal organizations in facilitating personalized engagement. As noted above, conventional organizations play a less central role than social technologies in relatively crowd-enabled networks such as the *indignados* of Spain, the Arab Spring uprisings, or the Occupy protests that spread from Wall Street around the world. In contrast to these more technology-enabled networks, we have also observed hybrid networks (such as PPF) where conventional organizations operate in the background of protest and issue advocacy networks to enable personalized engagement. This hybrid form of organizationally enabled connective action sits along a continuum somewhere between the two ideal types of conventional organizationally brokered collective action and relatively more crowd-enabled connective action. The following section presents the details of this three-part typology. It also suggests that co-existence, layering, and movement across the types becomes an important part of the story.

A TYPOLOGY OF COLLECTIVE AND CONNECTIVE ACTION NETWORKS

We draw upon these distinct logics of action (and the hybrid form that reveals a tension between them) to develop a three-part typology of large-scale action networks that feature prominently in contemporary contentious politics. One type represents the brokered organizational networks characterized by the logic of collective action, while the others represent two significant variations on networks primarily characterized by the logic of connective action. All three models may explain differences between and dynamics within large-scale action networks in event-centered contention, such as protests and sequences of protests as in the examples we have already discussed. They may also apply to more stable issue advocacy networks that engage people in everyday life practices supporting causes outside of protest events, such as campaigns. The typology is intended as a broad generalization to help understand different dynamics. None of the types are exhaustive social movement models. Thus, this is not an attempt to capture, much less resolve, the many differences among those who study social movements. We simply want to highlight the rise of two forms of digitally networked connective action that differ from some common assumptions about collective action in social movements and, in particular, that rely on mediated networks for substantial aspects of their organization.

Figure 11.1 presents an overview of the two connective action network types and contrasts their organizational properties with more familiar collective action network organizational characteristics. The ideal collective action type at the right side in the figure describes large-scale action networks that depend on brokering organizations to carry the burden of facilitating cooperation and bridging differences when possible. As the anti-capitalist direct action groups in the G20 London summit protests exemplified, such organizations will tend to promote more exclusive collective action frames that require frame bridging if they are to grow. They may use digital media and social technologies more as means of mobilizing and managing participation and coordinating goals, rather than inviting personalized interpretations of problems and self-organization of action. In addition to a number of classic social movement accounts (for example, McAdam, 1986), several of the NGO networks discussed by Keck and Sikkink (1998) also accord with this category (Bennett, 2005).

At the other extreme, on the left side in the figure we place connective action networks that self-organize largely without central or lead organizational actors, using technologies as important organizational agents. We call this type crowd-enabled connective action. While some formal organizations may be present, they tend to remain at the periphery or may exist as much in online as in offline forms. In place of collective action frames, personal action frames become the transmission units across trusted social networks. The loose coordination of the *indignados* exemplifies this ideal type, with conventional organizations deliberately kept at the periphery as easily adapted personal action frames travel online and offline with the aid of technology platforms such as the Democracia real Ya! organization.[14]

In between the organizationally-brokered collective action networks and the crowd-enabled connective action network is the hybrid pattern introduced above. This middle type involves formal organizational actors stepping back from projecting strong agendas, political brands, and collective identities in favor of using resources to deploy social technologies enabling loose public networks to form around personalized action themes. The middle type may also encompass more informal organizational actors that develop some capacities of conventional organizations in terms of resource mobilization and coalition building without imposing strong brands and collective identities.[15] For example, many of the general assemblies in the Occupy protests became resource centers, with regular attendance, division of labor, allocation of money and food, and coordination of actions. At the same time, the larger communication networks that swirled around these protest nodes greatly expanded the impact of the network. The surrounding technology networks invited loose-tied participation that was often in tension with the face-to-face ethos of the assemblies, where more committed protesters spent long hours with dwindling numbers of peers debating on how to expand participation without diluting the levels of commitment and action that they deemed key to their value scheme. Thus, even as Occupy displayed some organizational development, it was defined by its self-organizing roots.

Networks in this hybrid model engage individuals in causes that might not be of such interest if stronger demands for membership or subscribing to collective demands accompanied the organizational offerings. Organizations facilitating these action networks typically deploy an array

CONNECTIVE ACTION
Crowd-Enabled Networks

- Little or no formal organizational coordination of action
- Large scale personal access to multilayered social technologies
- Communication content centers on emergent inclusive personal action frames
- Personal expression shared over social networks
- Crowd networks may shun involvement of existing formal organizations

CONNECTIVE ACTION
Organizationally Enabled Networks

- Loose organizational coordination of action
- Organizations provide social technology outlays – both custom and commercial
- Communication content centers on organizationally generated inclusive personal action frames
- Some organizational moderation of personal expression through social networks
- Organizations in the background in loosely linked networks

COLLECTIVE ACTION
Organizationally Brokered Networks

- Strong organizational coordination of action
- Social technologies used by organizations to manage participation and coordinate goals
- Communication content centers on collective action frames
- Organizational management of social networks – more emphasis on interpersonal networks to build relationships for collective action
- Organizations in the foreground as coalitions with differences bridged through high resource organization brokerage

Figure 11.1   Elements of collective and connective action networks

of custom-built (for example, 'send your message') and outsourced (for example, Twitter) communication technologies. This pattern fits the PPF demonstrations discussed earlier, where some 160 civil society organizations – including major NGOs such as Oxfam, Tearfund, Catholic Relief, and World Wildlife Fund – stepped back from their organizational brands to form a loose social network inviting publics to engage with each other and take action. They did this even as they negotiated with other organizations over such things as separate days for the protests (Bennett and Segerberg, 2011).

The formations in the middle type reflect the pressures that Bimber et al. (2005) observed in interest organizations that are suffering declining memberships and have had to develop looser, more entrepreneurial relations with followers. Beyond the ways in which particular organizations use social technologies to develop loose ties with followers, many organizations also develop loose ties with other organizations to form vast online networks sharing and bridging various causes. Although the scale and complexity of these networks differ from the focus of Granovetter's (1973) observations about the strength of weak ties in social networks, we associate this idea with the elements of connective action: the loose organizational linkages, technology deployments, and personal action frames. In observing the hybrid pattern of issue advocacy organizations facilitating personalized protest networks, we traced a number of economic justice and environmental networks, charting protests, campaigns, and issue networks in the UK, Germany, and Sweden (Bennett and Segerberg, 2013).[16] In each case, we found (with theoretically interesting variations) campaigns, protest events, and everyday issue advocacy networks that displayed similar organizational signatures: (1) familiar NGOs and other civil society organizations joining loosely together to provide something of a networking backbone; (2) for digital media networks engaging publics with contested political issues; yet with (3) remarkably few efforts to brand the issues around specific organizations, own the messages, or control the understandings of individual participants. The organizations had their political agendas on offer, to be sure, but as members of issue networks, put the public face on the individual citizen and provided social technologies to enable personal engagement through easy-to-share images and personal action frames.

The organizations that refrain from strongly branding their causes or policy agendas in this hybrid model do not necessarily give up their missions or agendas as name-brand public advocacy organizations. Instead, some organizations interested in mobilizing large and potentially 'WUNC-y' publics in an age of social networking are learning to shift among different organizational repertoires, morphing from being hierarchical, mission-driven NGOs in some settings to being facilitators in loosely linked public engagement networks in others. As noted by Chadwick (2007, 2013), organizational hybridity makes it difficult to apply fixed categories to many organizations as they variously shift from being issue advocacy NGOs to policy think tanks, to SMOs running campaigns or protests, to multi-issue organizations, to being networking hubs for connective action. In other words, depending on when, where, and how one observes an organization, it may appear differently as an NGO, SMO, INGO, TNGO, NGDO (non-governmental organization, social movement organization, international non-governmental organization, transnational non-governmental organization, non-governmental development organization), an interest advocacy group, a political networking hub, and so on. Indeed, one of the advantages of seeing the different logics at play in our typology is to move away from fixed categorization schemes, and observe actually occurring combinations of different types of action within complex protest ecologies, and shifts in dominant types in response to events and opportunities over time.

The real world is of course far messier than this three-type model. In some cases, we see action formations corresponding to our three models side by side in the same action space. The G20 London protest offered a rare case in which organizationally enabled and more conventional collective action were neatly separated over different days. More often, the different forms layer and overlap, perhaps with violence disrupting otherwise peaceful mobilizations as occurred in the Occupy Rome protests on 15 October 2011, and in a number of Occupy clashes with police in the United States. In still other action cycles, we see a movement from one model to another over time. In some relatively distributed networks, we observe a pattern of informal organizational resource-seeking, in which informal organizational resources and communication spaces are linked and shared (for example, re-tweeted), enabling emergent political concerns and goals to be nurtured without being co-opted by existing organizations and their already fixed political agendas. This pattern occurred in the crowd-enabled Twitter network that emerged around the 15th UN Climate Change Conference in Copenhagen. As the long tail of that network handed its participants off to the Twitter stream devoted to the next summit in Cancun, we saw an increase in links to organizations of various kinds, along with growing links to and among climate bloggers (Segerberg and Bennett, 2011). Such variations on different organizational forms offer intriguing opportunities for further analyses aimed at explaining whether mobilizations achieve various goals, and attain different levels of WUNC.

In these varying ways, personalized connective action networks cross paths (sometimes with individual organizations morphing in the process) with more conventional collective action networks centered on SMOs, interest organizations, and brand-conscious NGOs. As a result, while we argue that these networks are an organizational form in themselves, they are often hard to grasp and harder to analyze because they do not behave like formal organizations. Most formal organizations are centered (for example, located in physical space), hierarchical, bounded by mission and territory, and defined by relatively known and countable memberships (or in the case of political parties, known and reachable demographics). By contrast, many of today's issue and cause networks are relatively decentered (constituted by multiple organizations and many direct and cyber activists), distributed, or flattened organizationally as a result of these multiple centers, relatively unbounded, in the sense of crossing both geographical and issue borders, and dynamic in terms of the changing populations who may opt in and out of play as different engagement opportunities are presented (Bennett, 2003, 2005). Understanding how connective action engages or fails to engage diverse populations constitutes part of the analytical challenge ahead.

Compared to the vast number of theoretically grounded studies on social movement organizing, there is less theoretical work that helps to explain the range of collective action formations, running from relatively crowd-enabled to organizationally enabled connective action networks. While there are many descriptive and suggestive accounts of this kind of action, many of them insightful (for example, Castells, 2000; Rheingold, 2002), we are concerned that the organizational logic and underlying dynamic of such action is not well established. It is important to gain clearer understandings of how such networks function and what organizing principles explain their growing prominence in contentious politics.

CONCLUSION

DNA is emerging during a historic shift in late modern democracies in which, most notably, younger citizens are moving away from parties, broad reform movements, and ideologies. Individuals are relating differently to organized politics, and many organizations are finding that they must engage people differently: they are developing relationships to publics as affiliates rather than members, and offering them personal options in ways to engage and express themselves. This includes greater choice over contributing content, and introduces micro-organizational resources in terms of personal networks, content creation, and technology development skills. Collective action based on exclusive collective identifications and strongly tied networks continues to play a role in this political land-scape, but this has become joined by, interspersed with, and in some cases supplanted by personalized collective action formations in which digital media become integral organizational parts. Some of the resulting DNA networks turn out to be surprisingly nimble, demonstrating intriguing flexibility across various conditions, issues, and scales.

It has been tempting for some critics to dismiss participation in such networks as noise, particularly in reaction to sweeping proclamations by enthusiasts of the democratic and participatory power of digital media. Whether from digital enthusiasts or critics, hyperbole is unhelp-ful. Understanding the democratic potential and effectiveness of instances of connective and collective action requires careful analysis. At the same time, there is often considerably more going on in DNA than clicktivism or facile organizational outsourcing of social networking to various com-mercial sites.[17] The key point of our argument is that fully explaining and understanding such action and contention requires more than just adjust-ing the classic social movement collective action schemes. Connective action has a logic of its own, and thus attendant dynamics of its own. It deserves analysis on its own terms.

The linchpin of connective action is the formative element of 'sharing': the personalization that leads actions and content to be distributed widely across social networks. Communication technologies enable the growth and stabilization of network structures across these networks. Together, the technological agents that enable the constitutive role of sharing in these con-texts displace the centrality of the free-rider calculus and with it, by exten-sion, the dynamic that flows from it; most obviously, the logical centrality of the resource-rich organization. In its stead, connective action brings the action dynamics of recombinant networks into focus, a situation in which networks and communication become something more than mere precondi-tions and information. What we observe in these networks are applications of communication technologies that contribute an organizational principle that is different from notions of collective action based on core assumptions about the role of resources, networks, and collective identity. We call this different structuring principle the logic of connective action.

Developing ways to analyze connective action formations will give us more solid grounds for returning to the persistent questions of whether such action can be politically effective and sustained (Tilly, 2004; Gladwell, 2010; Morozov, 2011). Even as the contours of political action may be shifting, it is imperative to develop means of thinking meaningfully about the capacities of sustainability and effectiveness in relation to con-nective action and to gain a systematic understanding of how such action plays out in different contexts and conditions.

The string of G20 protests surrounding the world financial crisis illustrate that different organizational strategies played out in different political settings produce a wide range of results. The protests at the Pittsburgh and Toronto G20 summits of 2009 and 2010, respectively, were far more chaotic and displayed far less WUNC than those organized under the banner of PPF in London. Disrupted by police assaults and weak organizational coordination, the Pittsburgh protests displayed a cacophony of political messages that were poorly translated in the press and even became the butt of late-night comedy routines. The *Daily Show* sent a correspondent to Pittsburgh and reported on a spectrum of messages that included: a Free Tibet marching cymbal band; Palestinian peace advocates; placards condemning genocide in Darfur; hemp and marijuana awareness slogans; and denunciations of the beef industry; along with the more expected condemnations of globalization and capitalism. One protester carried a sign saying 'I protest everything', and another dressed as Batman stated that he was protesting the choice of Christian Bale to portray his movie hero. The correspondent concluded that the Pittsburgh protests lacked unity of focus, and turned for advice to some people who knew how to get the job done: members of the Tea Party. The *Daily Show* panel of Tea Party experts included a woman wearing a black Smith & Wesson holster that contained a wooden crucifix with an American flag attached. When asked what the Pittsburgh protesters were doing wrong, they all agreed that there was a message problem. One said, 'I still don't know what their message is', and another affirmed, 'Stay on message and believe what you say'. The *Daily Show* report cut back to show a phalanx of Darth Vader-suited riot police lined up against the protesters; according to the correspondent, the 'one single understandable talking point' in Pittsburgh (*Daily Show*, 2009). Humor aside, this example poses a sharp contrast to the more orderly London PPF protests that received positive press coverage of the main themes of economic and environmental justice (Bennett and Segerberg, 2011).

The challenge ahead is to understand when DNA becomes chaotic and unproductive, and when it attains higher levels of focus and sustained engagement over time. Our studies suggest that differing political capacities in networks depend, among other things, on whether: (1) in the case of organizationally enabled DNA, the network has a stable core of organizations sharing communication linkages and deploying high volumes of personal engagement mechanisms; or (2) in the case of crowd-enabled DNA, the digital networks are redundant and dense with pathways for individual networks to converge, enabling viral transmission of personally appealing action frames to occur.

Attention to connective action will neither explain all contentious politics nor replace the model of classic collective action that remains useful for analysing social movements. But it does shed light on an important mode of action making its mark in contentious politics today. A model focused primarily on the dynamics of classic collective action has difficulties accounting for important elements in the Arab spring, the *indignados*, the Occupy demonstrations, or the global protests against climate change. A better understanding of connective action promises to fill some of these gaps. Such understanding is essential if we are to attain a critical perspective on some of the prominent forms of public engagement in the digital age.

ACKNOWLEDGEMENT

The original version of this chapter was published as: W. Lance Bennett and Alexandra Segerberg (2012), The logic of connective action: digital media and the personalization of contentious politics. *Information, Communication and Society*, 15(5), 739–768. The authors are grateful for permission from Taylor & Francis (http://www.tandfonline.com) to reprint the article as this chapter. This version has been updated to reflect changes that appear in *The Logic of Connective Action* (Bennett and Segerberg, 2013).

NOTES

1. Simultaneous protests were held in other European cities with tens of thousands of demonstrators gathering in the streets of Berlin, Frankfurt, Vienna, Paris, and Rome.
2. US Vice President Joe Biden asked for patience from understandably upset citizens while leaders worked on solutions, and the British Prime Minister at the time, Gordon Brown, said: 'the action we want to take (at the G20) is designed to answer the questions that the protesters have today' (Vinocur and Barkin, 2009).
3. http://www.democraciarealya.es/.
4. Beyond the high volume of Spanish press coverage, the story of the *indignados* attracted world attention. BBC *World News* devoted no fewer than eight stories to this movement over the course of two months, including a feature on the march of one group across the country to Madrid, with many interviews and encounters in the words of the protesters themselves.
5. For example, our analyses of the US Occupy protests show that increased media attention to economic inequality in the USA was associated with the coverage of the Occupy protests (Bennett and Segerberg, 2013). While political elites were often reluctant to credit the occupiers with their new-found concern about inequality, they nonetheless seemed to find the public opinion and media climate conducive to addressing the long-neglected issue.
6. A Google search of 'put people first g20' more than two years after the London events produced nearly 1.5 million hits, with most of them relevant to the events and issues of the protests well into 75 search pages deep.

7. We would note, however, that carnivalesque or theatrical expressions may entail strategically depersonalized forms of expression in which individuals take on other personae that often have historically or dramatically scripted qualities. We thank Stefania Milan for this comment.

8. We are not arguing here that all contemporary analyses of collective action rely on resource mobilization explanations (although some do). Our point is that whether resource assumptions are in the foreground or the background, many collective action analyses typically rely on a set of defining assumptions centered on the importance of some degree of formal organization and some degree of strong collective identity that establishes common bonds among participants. These elements become more marginal in thinking about the organization of connective action.

9. While we focus primarily on cases in late modern, post-industrial democracies, we also attempt to develop theoretical propositions that may apply to other settings such as the Arab Spring, where authoritarian rule may also result in individualized populations that fall outside of sanctioned civil society organization, yet may have direct or indirect access to communication technologies such as mobile phones.

10. Routledge and Cumbers (2009) make a similar point in discussing horizontal and vertical models as useful heuristics for organizational logics in global justice networks (see also Robinson and Tormey, 2005; Juris, 2008).

11. We are indebted to Bob Boynton for pointing out that this sharing occurs both in trusted friends networks such as Facebook and in more public exchange opportunities among strangers of the sort that occur on YouTube, Twitter, or blogs. Understanding the dynamics and interrelationships among these different media networks and their intersections is an important direction for research.

12. We have developed methods for mapping networks and inventorying the types of digital media that enable actions and information to flow through them. Showing how networks are constituted in part by technology enables us to move across levels of action that are often difficult to theorize. Network technologies enable thinking about individuals, organizations, and networks in one broad framework. This approach thus revises the starting points of classic collective action models, which typically examine the relationships between individuals and organizations and between organizations. We expand this to include technologies that enable the formation of fluid action networks in which agency becomes shared or distributed across individual actors and organizations as networks reconfigure in response to changing issues and events (Bennett and Segerberg, 2013; Bennett et al., 2014).

13. http://www.15october.net (accessed 19 October 2011).

14. We wish to emphasize that there is much face-to-face organizing work going on in many of these networks, and that the daily agendas and decisions are importantly shaped offline. However, the connectivity and flow of action coordination occurs, importantly, online.

15. We thank an anonymous referee for highlighting this subtype.

16. Our empirical investigations focused primarily on two types of networks that display local, national, and transnational reach: networks to promote economic justice via more equitable North–South trade norms (fair trade) and networks for environmental and human protection from the effects of global warming (climate change). These networks display impressive levels of collective action and citizen engagement and they are likely to remain active into the foreseeable future. They often intersect by sharing campaigns in local, national, and transnational arenas. As such, these issue networks represent good cases for assessing the uses of digital technologies and different action frames (from personalized to collective) to engage and mobilize citizens, and to examine various related capacities and effects of those engagement efforts.

17. Technology is not neutral. The question of the degree to which various collectivities have both appropriated and become dependent on the limitations of commercial technology platforms such as Flickr, Facebook, Twitter, or YouTube is a matter of considerable importance. For now, suffice it to note that at least some of the technologies and

their networking capabilities are designed by activists for creating political networks and organizing action (Calderaro, 2011).

REFERENCES

Anduiza, Eva, Camilo Cristancho and José M. Sabucedo (2014). Mobilization through online social networks: the political protest of the *indignados* in Spain. *Information, Communication and Society*, 17(6), 750–764.

Bauman, Z. (2000). *Liquid Modernity*. Cambridge: Polity.

Beck, U. and Beck-Gernsheim, E. (2002). *Individualization: Institutionalized Individualism and its Social and Political Consequences*. London: SAGE.

Benford, R.D. and Snow, D.A. (2000). Framing processes and social movements: an overview and an assessment. *Annual Review of Sociology*, 26, 611–639.

Benkler, Y. (2006). *The Wealth of Networks: How Social Production Transforms Markets and Freedom*. New Haven, CT: Yale University Press.

Bennett,W.L. (1998). The uncivic culture: communication, identity, and the rise of lifestyle politics. Ithiel de Sola Pool Lecture, American Political Science Association, *PS: Political Science and Politics*, 31, 41–61.

Bennett,W.L. (2003). Communicating global activism: strengths and vulnerabilities of networked politics. *Information, Communication and Society*, 6(2), 143–168.

Bennett, W.L. (2005). Social movements beyond borders: organization, communication, and political capacity in two eras of transnational activism. In della Porta, D. and Tarrow, S. (eds) *Transnational Protest and Global Activism* (pp. 203–222). Boulder, CO: Rowman & Littlefield.

Bennett, W.L., Lang, S. and Segerberg, A. (2014). European issue publics online: the cases of climate change and fair trade. In Risse, Thomas (ed.) *European Public Spheres: Politics Is Back* (pp. 108–137). Cambridge: Cambridge University Press.

Bennett, W.L. and Segerberg, A. (2011). Digital media and the personalization of collective action: social technology and the organization of protests against the global economic crisis. *Information, Communication and Society*, 14, 770–799.

Bennett, W.L. and Segerberg, A. (2013). *The Logic of Connective Action: Digital Media and the Personalization of Contentious Politics*. New York: Cambridge University Press.

Bimber, B. and Davis, R. (2003). *Campaigning Online: The Internet in US Elections*. New York: Oxford University Press.

Bimber, B., Flanagin, A. and Stohl, C. (2005). Reconceptualizing collective action in the contemporary media environment. *Communication Theory*, 15, 389–413.

Bimber, B., Flanagin, A. and Stohl, C. (2012). *Collective Action in Organizations: Interaction and Engagement in an Era of Technological Change*, New York: Cambridge University Press.

Bimber, B., Stohl, C. and Flanagin, A. (2009). Technological change and the shifting nature of political organization. In Chadwick, A. and Howard, P. (eds), *Routledge Handbook of Internet Politics* (pp. 72–85). London: Routledge.

Calderaro, A. (2011). New political struggles in the network society: the case of free and open source software (FOSS) movement. Paper presented at ECPR General Conference, Reykjavik, 25–27 August.

Castells, M. (2000). *The Network Society*. 2nd edn. Oxford: Blackwell.

Chadwick, A. (2007). Digital network repertoires and organizational hybridity. *Political Communication*, 24(3), 283–301.

Chadwick, A. (2013). *The Hybrid Media System: Politics and Power*. New York: Oxford University Press.

Chesters, G. and Welsh, I. (2006). *Complexity and Social Movements: Multitudes at the End of Chaos*. London: Routledge.

Daily Show (2009). Tea partiers advise G20 protesters. *Daily Show*, October 1. Available at:

at: http://www.thedailyshow.com/watch/thu-october-1-2009/tea-partiers-advise-g20-prot esters (accessed October 6, 2010).

Dawkins, R. (1989). *The Selfish Gene*. Oxford: Oxford University Press.

della Porta, D. (2005). Multiple belongings, flexible identities and the construction of "another politics": between the European social forum and the local social fora. In della Porta, D. and Tarrow, S. (eds), *Transnational Protest and Global Activism* (pp. 175–202). Boulder, CO: Rowman and Littlefield.

della Porta, D. and Diani, M. (2006). *Social Movements: An Introduction*. 2nd edn. Malden, MA: Blackwell.

Diani, Mario (forthcoming). *The Cement of Civil Society: Civic Networks in Localities*. Cambridge: Cambridge University Press.

Earl, J. and Kimport, K. (2011). *Digitally Enabled Social Change: Online and Offline Activism in the Age of the Internet*. Cambridge, MA: MIT Press.

Foot, K. and Schneider, S. (2006). *Web Campaigning*. Cambridge, MA: MIT Press.

Giddens, A. (1991). *Modernity and Self-Identity: Self and Society in the Late Modern Age*. Stanford, CA: Stanford University Press.

Gitlin, T. (1980). *The Whole World is Watching: Mass Media in the Making and Unmaking of the New Left*. Berkeley, CA: University of California Press.

Gladwell, M. (2010). Small change: why the revolution will not be tweeted. *New Yorker*, October 4.

Granovetter, M. (1973). The strength of weak ties. *American Journal of Sociology*, 78, 1360–1380.

Howard, P. and Hussain, M. (2011). The role of digital media. *Journal of Democracy*, 22(3), 35–48.

Hunt, S., Benford, R.D. and Snow, D.A. (1994). Identity fields: framing processes and the social construction of movement identities. In Laraña, E., Johnston, H. and Gusfield, J.R. (eds), *New Social Movements: From Ideology to Identity* (pp. 185–208). Philadelphia, PA: Temple University Press.

Inglehart, R. (1997). *Modernization and Post-Modernization: Cultural, Economic and Political Change in 43 Societies*. Princeton, NJ: Princeton University Press.

Juris, J. (2008). *Networking Futures: The Movements against Corporate Globalization*. Durham, NC: Duke University Press.

Keck, M. and Sikkink, K. (1998). *Activists beyond Borders: Advocacy Networks in International Politics*. Ithaca, NY: Cornell University Press.

Latour, B. (2005). *Reassembling the Social: An Introduction to Actor-Network-Theory*. Oxford: Oxford University Press.

Livingston, S. and Asmolov, G. (2010). Networks and the future of foreign affairs reporting. *Journalism Studies*, 11(5), 745–760.

Lupia, A. and Sin, G. (2003). Which public goods are endangered? How evolving communication technologies affect 'the logic of collective action'. *Public Choice*, 117, 315–331.

McAdam, D. (1986). Recruitment to high-risk activism: the case of freedom summer. *American Journal of Sociology*, 92, 64–90.

McAdam, D., McCarthy, J.D. and Zald, M.N. (1996). Opportunities, mobilizing structures, and framing processes: toward a synthetic, comparative perspective on social movements. In McAdam, D., McCarthy, J.D. and Zald, M.N. (eds), *Comparative Perspectives on Social Movements: Political Opportunities, Mobilizing Structures, and Cultural Framings* (pp. 1–20). New York: Cambridge University Press.

McAdam, D., Tarrow, S. and Tilly, C. (2001). *Dynamics of Contention*. New York: Cambridge University Press.

McCarthy, J.D. and Zald, M.N. (1973). *The Trend of Social Movements in America: Professionalization and Resource Mobilization*. Morristown, NJ: General Learning Press.

McCarthy, J.D. and Zald, M.N. (1977). Resource mobilization and social movements: a partial theory. *American Journal of Sociology*, 82(6), 1212–1241.

McDonald, K. (2002). From solidarity to fluidarity: social movements beyond 'collective identity' – the case of globalization conflicts. *Social Movement Studies*, 1(2), 109–128.

Melucci, A. (1996). *Challenging Codes: Collective Action in the Information Age*. Cambridge: Cambridge University Press.

Micheletti, M. (2003). *Political Virtue and Shopping*. New York: Palgrave.

Morozov, E. (2011). *The Net Delusion: How Not to Liberate the World*. London: Allen Lane.

Olson, M. (1965). *The Logic of Collective Action: Public Goods and the Theory of Groups*. Cambridge, MA: Harvard University Press.

Polletta, F. (2002). *Freedom is an Endless Meeting: Democracy in American Social Movements*. Chicago, IL: University of Chicago Press.

Putnam, R. (2000). *Bowling Alone: The Collapse and Revival of American Community*. New York: Simon & Schuster.

Put People First (2009). http://www.putpeoplefirst.org.uk/ (accessed 6 July 2011).

Rheingold, H. (2002). *Smart Mobs: The Next Social Revolution*. Cambridge, MA: Perseus Pub.

Robinson, A. and Tormey, S. (2005). Horizontals, verticals and the conflicting logics of transformative politics. In el-Ojeili, C. and Hayden, P. (eds), *Confronting Globalization* (pp. 208–226). London: Palgrave.

Routledge, P. and Cumbers, A. (2009). *Global Justice Networks: Geographies of Transnational Solidarity*. Manchester: Manchester University Press.

RTVE (2011). Mas de seis millones de Espanoles han participado en el movimiento 15M. August 6. Available at: http://www.rtve.es/noticias/20110806/mas-seis-millones-espanoles-han-participado-movimiento-15m/452598.shtml (accessed September 18, 2011).

Segerberg, A. and Bennett, W.L. (2011). Social media and the organization of collective action: using Twitter to explore the ecologies of two climate change protests. *Communication Review*, 14(3), 197–215.

Shifman, L. (2013). *Memes in Digital Culture*. Cambridge, MA: MIT Press.

Snow, D.A. and Benford, R.D. (1988). Ideology, frame resonance, and participant mobilization. *International Social Movement Research*, 1, 197–217.

Snow, D.A., Rochford, B. Jr., Worden, S.K. and Benford, R.D. (1986). Frame alignment processes, micromobilization, and movement participation. *American Sociological Review*, 51, 464–481.

Tarrow, S. (2011). *Power in Movement: Social Movements in Contentious Politics*. 3rd edn. New York: Cambridge University Press.

Tilly, C. (2004). *Social Movements, 1768–2004*. Boulder, CO: Paradigm.

Tilly, C. (2006). WUNC. In Schnapp, J.T. and Tiews, M. (eds), *Crowds* (pp. 289–306). Stanford, CA: Stanford University Press.

Vinocur, N. and Barkin, N. (2009). G20 marches begin week of protests in Europe. Reuters. March 28. Available at: http://www.reuters.com/article/ 2009/03/28/us-g20-britain-march-idUSTRE52R0TP20090328 (accessed 9 July 2011).

Weller, B. (2010). G20 protests in Seoul. *Demotix*. Available at: http://www.demotix.com/photo/504262/g20-protests-seoul (accessed 9 July 2011).

12. Youth civic engagement

Chris Wells, Emily Vraga, Kjerstin Thorson,
Stephanie Edgerly and Leticia Bode

INTRODUCTION

For as long as there has been a study of digital politics, young citizens have occupied a special place in it. Why? Two major reasons stand out.

First, young people early on were recognized as 'digital natives', a term meant to capture something special about the relationship between youth and digital media (Prensky, 2001): a kind of electronic sixth sense to explain aptitudes for videocassette recorder (VCR) programming in the 1980s, website surfing in the 1990s, and social media development and use in the twenty-first century. For scholars of political engagement, one resulting assumption was that previously disengaged youth might be reached with a preferred medium and so brought back to civic life (for example, Delli Carpini, 2000). Originating as it did with perceptions of older people less well inclined to use technology, the concept of digital native has limitations in describing the practices and abilities of a large and very diverse group, as we shall shortly see. Yet the fact that the generation entering adulthood today is the first to have come of age completely immersed in digitally connected media is one worth paying attention to – although not simply because of their facility with technology.

A second reason for interest in the political uses to which young people have put digital media are high-profile examples in which they have been on the forefront of experimenting with digital communication in political life. Barack Obama's 2008 United States presidential campaign gained an early youth following, in part thanks to creative videos such as Obama Girl's 'I've Got a Crush on Obama', and pop star will.i.am's 'Yes We Can', which were shared widely among youth networks and media. Young people have also been prominent in protest politics, with the 2006 'MySpace protests' against harsh immigration policies in the United States inaugurating a series of youth-driven movements that culminated most spectacularly in the 2011 protests of the Egyptian revolution, Spanish *indignados*, Chilean students demanding increased access and more public investment in education, and Occupy Wall Street. In important ways these activities are not unique to this generation: political protest has often emerged among students and other young activists. Still, in each of these examples young people have been observed doing something interesting and new in the political realm, often innovating with digital tools as they go.

We would cite a further reason for the importance of following the fortunes of young citizens as their political involvements via digital media take shape. In a number of ways the global youth generation is unique, and uniquely troubled politically and economically. In the US and much of the developed world, 'Millennials' are markedly tolerant and liberal; but many are also unemployed and underemployed (Pew Research Center, 2010; Zogby, 2008). There is growing inequality among those in higher social classes and those further down, features of late modern societies generally, but often felt most acutely at the youth level. The emerging generation faces great challenges, and use of digital media to engage with public life will surely be at the heart of how they approach those challenges.

Thus, as changes, possibilities, and threats swirl around young citizens, we can expect to see youth continuing to be a focal point of activity and innovation. But we have reached a point where it may be useful to look back at what we have seen – and thought – about young people and digital politics over the past 15 years. The young people who were the subjects of the first articles about the fascinating nexus of young citizens, digital media and politics, are now approaching middle age. Even as Internet time churns ahead rapidly enough that 'generations' of distinct digital experience turn over ever couple of years, we take this chapter as an opportunity to look back on two decades of research on the possibilities that young people will increase their engagement in politics through or because of digital media. We train our lens on six areas of special interest when it comes to young citizens and digital politics:

- the changing bases of the civic identities of citizens in the industrialized democracies;
- how younger generations are consuming – or not consuming – news;
- the practices and patterns emerging as formal political campaigns attempt to reach young people through digital media;
- the role of socio-economic status, skill, and the definition of political engagement;
- how content creation and interaction in digital media enables, for some young citizens, a form of cultural engagement that pushes at the boundaries of the political; and
- what these changes imply for the study of political socialization and the practice of civic education.

Before proceeding, a note on what is meant by 'young' citizens: the literature on youth civic engagement is inconsistent in its definitions of the precise boundaries of youth. Work from education frequently includes teens up to the age of 18; research from national surveys typically includes those only 18 or older, typically up to age 24 or 30; other research includes some or all of these groups. This is in fact less of a problem than it might at first seem, for the reason that the overarching perspective of contemporary work is typically concerned with contextual effects on cohorts. Thus, for the purposes of this chapter we are comfortable adopting the term 'Millennial' to refer to a generation imperfectly defined as including members born in 1980 or later (Ng et al., 2010).

YOUNG CITIZENS AND THE CHANGING BASES OF CIVIC IDENTITY

What Digital Natives?

The notion that members of the current younger generation are somehow different from their elders, especially in their use of media, is prevalent and appealing, though research in recent years has shown it to often be oversimplified. For this reason, it is useful to begin with some close consideration of just how exceptional 'digital natives' are, as opposed to older 'digital immigrants' (Palfrey and Gasser, 2008).

Many older individuals likely have had personal experiences with a young person who quickly developed skills with digital devices, programming a VCR, configuring a computer, or setting up social media profiles (Jones and Czerniewicz, 2010). Combined with intense media attention to dozens of whizz-kid entrepreneurs, especially in the social media age, many members of the public have been receptive to a narrative of underlying intuitions and innate skills of the younger generation. But the notion that youth are fundamentally different in their digital skill development has been undercut by careful research, and shown to be more the result of inferences drawn from the most visible and interesting examples than social reality (Bennett et al., 2008). Especially, whereas a minority of young citizens are in possession of high-level digital skills, most are not; most of the sophisticated practices of digital content creation, editing, and sharing are not engaged in by most people, young or otherwise (Eynon and Malmberg, 2011; Hargittai and Walejko, 2008; Schlozman et al., 2010). Even youth heavy users of social networking sites have been shown to be surprisingly unaware and unconfident when asked to explore beyond the basic functionalities of those platforms (Livingstone, 2008: 406–407). And the finding that socio-economic status, especially parents' education, and types of Internet access are much more significant factors than age alone in determining a young person's opportunities for use, comfort with, and skill level in using digital media has been consistently replicated (Hargittai, 2010; Schlozman et al., 2010; Schradie, Chapter 5 in this volume; Smith et al., 2013; though see Blank, 2013).

Situating the Digital Citizen in Late Modern Society

This is not to say that we should dismiss claims that young people are doing special things with digital media in politics. Young citizens are using digital media more often, and for more aspects of their lives, than older citizens (Lenhart et al., 2010); it is just that the usual 'digital natives' focus on aptitude and skill are often overstated. Instead, recent scholarship has moved toward placing young people in a social-historical context: because as much as societies are experiencing great upheaval, its effects may be especially strong for citizens growing up in rapidly changing environments (Bennett et al., 2010; Zukin et al., 2006).

Several important social-structural changes should inform our understanding of young people's civic identities, relationships to politics, and use of digital media to those ends. First, the period preceding and encompassing the childhoods of contemporary young people has been one of marked economic change. Led by processes of economic globalization, this period saw the completion of the interpenetration of national economies, the rising power of transnational corporations, and the decline of unions; and correspondingly, a faltering working-class way of life for many in the rust belts of the United States and other developed nations. Concurrent with economic globalization was a shift toward network structures of organization, in associational contexts from global capital and international finance to civil society and local communities (Castells, 1996). Paired with new economic stresses, these changes meant a constellation of pressures that undermined the group-based associational life of the high modern society (Bennett, 1998). The resulting decline in participation in place-based, face-to-face community organizations (Putnam, 2000) means that many Millennials have had less exposure to the traditional interpersonal community interactions that were formative for older citizens. In their place, argue some scholars, is a networked individualism in which young people in particular are comfortable creating interest-based communities via online social networks (Rainie and Wellman, 2012).

Parallels can be seen in the media structures younger citizens have grown up alongside. By the time Millennials in the USA were teenagers, 63 percent of their households had cable television and many had video games; the three-network media sphere was already a thing of the past (Nielsen, 2009). And of course, it is the multi-channel, infinite-capacity Internet that truly makes possible space-independent networked individualism. Thus, whereas their parents and grandparents inhabited a world in which the evening newscast was being simultaneously seen by a third of US television watchers (and the others were enjoying similar fare), when Millennials got home from school they were greeted by a panoply of media choices, from MTV to the Golf Channel to Super Nintendo, to suit whatever particular interest (or ennui) gripped them at the moment.

All of this demonstrates that what is special about young people is about much more than the ability to create a YouTube mashup: members of the Millennial generation have grown up in a world in which potent labor unions, single-employer careers, long-term economic security, limited-channel media systems and exclusively space-based communities were historical artifacts, not lived experiences.

Some scholars have proposed that these changes engender a shift in the bases for civic identity for many young people. Following in the footsteps of sociologists such as Giddens (1991) and Beck (1999), these writers contend that because of their different experiences, recent generations of citizens have found ways of acting out their citizenship in ways that differ from their elders: Inglehart (1997) has termed the newer mode 'postmaterialist'; Bennett (1998) has observed the rise of 'lifestyle' politics; Schudson (1998) sees the rise of a 'rights-bearing' citizenship. What these perspectives share is a view of citizenship that entails a decreased experience of duty and obligation, decreased identification with and trust in parties and official leaders, and decreased inclination to participate in organized, bounded protests. In place of these old norms are rising demands for expression, individuality, personalization and flexibility in the acting out of civic identity, which may take the form of acts that can be practiced on a daily, lifestyle basis, such as becoming a vegetarian or making (at least occasional) conscious consumer choices, or non-political 'community' participation, such as volunteering (Zukin et al., 2006). From this perspective, both changes in civic participation and digital media uses are seen as products of young people's situatedness in a changing civic order and the particular technologies available (and developing) at that time (Bennett et al., 2010; Wells, 2013).

How do these changes play out in the actual experiences and activities of young people as they engage the political world with digital media? How can we understand the participation of citizens experimenting with tools we are only beginning to understand, who also had formative experiences – of community, of media, of civic life – quite unlike those of their elders? This attention to the larger contexts in which young citizens leverage digital media for political purposes yields many of the insights described below.

YOUTH AND NEWS

If we wish to understand the implications of economic, social, and media changes for young citizens' civic activity, news is an excellent first place to look. There can be little doubt that the habits of news media consumption among today's youth are shifting dramatically in comparison to previous generations. A recent news consumption survey in the US found that more than a third of older adults read news every day, but only 16 percent of 18–30-year-olds report the same. Sixty percent of older adults watch television news every day, but only 30 percent of young adults do (Patterson, 2007). While there has always been a usage gap between older and younger cohorts, this gap is growing. In 1963, 79 percent of US high school seniors reported reading about politics in the newspaper more than three days a week and 70 percent watched political news on television more than three days a week (data from Jennings et al., 2005, authors' analysis). The 2008 Future Voters Study found that only 4 percent of US adolescents read a print newspaper more than three days a week while 17 percent watched news on television more than three days a week (Thorson, 2013).

News consumption as a habit of a particular time and place is being replaced by the new media-enabled possibilities for news on demand. Online news use among young adults is growing; but there is little evidence that use of the Internet and mobile technologies for news is becoming a regular habit to replace the news routines of earlier generations. In Patterson's 2007 study, only a fifth of young adults in the US reported daily use of the Internet for news and less than half of those reported actively seeking out news content. The rest encountered news incidentally while they are online for other purposes. This trend has not changed recently, with frequent findings that whatever the medium of delivery, 'young people continue to spend less time with news' (Pew Research Center, 2012). A variety of observers have pointed out that consumption of news media content is increasingly driven by personal preference, enabling those with a great deal of interest in keeping up with the world to gain greater knowledge while those who do not enjoy the news may find it increasingly easier to avoid such content altogether (Bimber, 2003; Couldry et al., 2007; Prior, 2007).

A bright spot in concerns over transforming news media repertoires among youth has been the increased potential for peer-shared news and information content via social network platforms such as Facebook and

Twitter. New digital media tools used by news organizations and other content sites combined with the in-platform content sharing capabilities of services like Facebook and Twitter may make for a renaissance in the social delivery of news content. Research has shown that exposure to news content on sites like Facebook can lead to political learning, and is linked to the emergence of social capital (Bode, 2012). However, the extent to which such exposure occurs among younger citizens is not yet settled. In a survey of US college students, Baumgartner and Morris (2010) reported that less than half of respondents get news from social networking sites (SNS) even once a week and that use of SNS as a news source conferred no benefit to political knowledge. A September 2012 Pew Internet and American Life report characterized the role of SNS in politics as 'modest', noting that the vast majority of SNS users (84 percent) posted little or nothing about politics during the 2012 US presidential election season. Nearly 60 percent said their friends posted little or no political content as well (Pew Internet and American Life Project, 2012). A series of in-depth interviews in the run-up to the 2012 election revealed that many young adults feel social pressures to avoid engaging in the exchange of opinionated political content on Facebook, in great part arising from the complexities of the networked audiences for posts to that site (Thorson et al., 2014).

Not surprisingly, given this, polls regularly show younger citizens to be at the bottom of the heap when it comes to the standard measures of political knowledge and news quiz questions (Wattenberg, 2008). Yet, curiously, in the face of most classic accounts of citizen engagement in politics, which posit learning about politics through news as a necessary precursor to action, some research is unearthing evidence of a sort of decoupling of knowledge and action. Östman (2012), looking at a sample of Swedish youth, shows contributing content to online spaces to be predictive of civic participation – both on and offline – but negatively associated with knowledge. In some ways this finding comports with the theoretical view of an emerging citizenship as decreasingly structured by formal entry points into the political process, such as news and institutions, and increasingly acting out on an individually defined basis. More research is clearly needed to consider whether the role of knowledge in civic engagement is indeed undergoing a change as the nature of the media system, and young citizens' inclinations, evolve.

FORMAL POLITICS

This brings us to the question of how digital technologies are impacting the participation rates of young people. In the United States and other

Western countries, the most elemental act of formal political participation has long been the vote. Historically, young adult turnout has been quite low in comparison to other age cohorts. In the US case, those under 25 vote at a rate roughly 20 points lower than their older counterparts (Levine and Lopez, 2002), a trend that has endured across time and generations, suggesting that it is a characteristic of a stage of the life process more than one of any particular generation. Not surprisingly, the low – and, through 2000, declining – level of youth participation in formal politics has long been a source of great concern, a state of affairs that had many observers prepared to hope for a turnaround prompted by new communication technologies.

In this context, the 2008 US presidential election was an exciting landmark election for youth involvement. Youth turnout was an impressive 51.1 percent, representing an increase of 2 million voters under 30 compared to 2004, and the highest portion of young voters participating since 1992. The increases were particularly significant given that turnout among older generations slightly declined (see Figure 12.1; CIRCLE, 2008).

For this success, considerable credit went to the Obama campaign and its novel youth outreach campaign. This campaign had a number of important elements, including providing extensive organizing training to young people and making existing infrastructure available through university political groups across the nation. But the element that received some of the most attention was its digital media strategy. The story goes that digital media allowed the campaign to reach Millennials where they naturally are – on YouTube and Facebook – and through text messaging (Dreier, 2008). Even as the traditional campaign infrastructure reliant on landline telephones became useless when considering the under-30 generation, the Obama campaign reached into social media, allowing mobilized youth to reach out to one another, independent of the campaign itself. Indeed, McCain was widely criticized for failing to take full advantage of digital platforms in his campaign outreach (Dreier, 2008).

Still, the popular narrative over the role of digital media in mobilizing youth in 2008 must contend with some additional facts. First, the increase in youth turnout from 2004 to 2008 was a continuation of an upward trend that began in the beginning of the 2000s, and in fact the greater part of the rise took place between 2000 and 2004 (see Figure 12.1). Also, the increase in turnout between 2004 and 2008 came entirely from increases in non-white youth turnout (CIRCLE, 2008): white youth were no more likely to turn out in 2008 than they were in 2004. Further, though some expected the 2012 election to be an encore performance of 2008, it shaped up much differently. The digital tools were still employed by both campaigns, and to great accolades by the American punditry, but the result of their

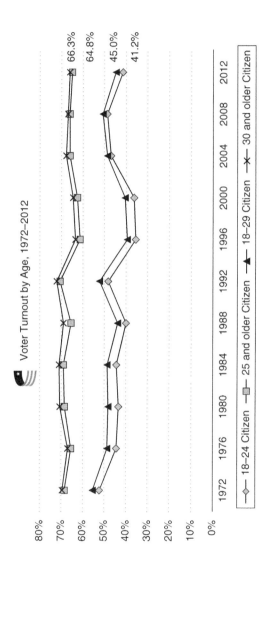

Voter Turnout by Age, 1972–2012

66.3%
64.8%
45.0%
41.2%

80%
70%
60%
50%
40%
30%
20%
10%
0%

1972 1976 1980 1984 1988 1992 1996 2000 2004 2008 2012

18–24 Citizen —◆— 25 and older Citizen —■— 18–29 Citizen —▲— 30 and older Citizen —✕—

Source: CIRCLE's tabulations from the CPS Nov. Voting and Registration Supplements, 1972–2012 (reprinted by permission).

Figure 12.1   Voter turnout in American presidential elections, 1972–2012 (by age)

targeting was less successful: only 45 percent of the under-30 demographic turned out to vote in 2012, down six points from 2008's high (CIRCLE, 2012). This was despite a major emphasis of both presidential campaigns on micro-targeting, relying on unprecedented integration of data across all aspects of the campaign to identify voters and reach them by the most effective means possible (McCoy, 2012; Issenberg, 2012).

This does not mean that digital technology played no role: as Garcia-Castañon et al. (2011) have shown, digital technology may have played a particularly important role in bringing young people of color into the election. Unlike older generations, technology access has relative parity among youth of different ethnicities (holding socio-economic class constant), many of whom regularly access social media via smartphones, and so were reachable by Obama's strategy.

But the story of a mass of youth being engaged online is too simple. For all his campaign's innovations with digital media, Obama was also a remarkable candidate with special appeal to young people. Further, his campaign engaged in significant shoe-leather outreach to youth on college campuses and in urban areas. And too often forgotten about the campaign are all the ways in which digital tools within Obama's campaign were leveraged to make traditional forms of outreach – phone calls, doorknocks, neighborly conversations – as effective as possible (Kreiss, 2012). The perception of Obama as a 'Facebook candidate' does not hold up to this evidence, and in fact the specific contribution of digital media to his election is quite difficult to tease out: research since 2008 has been equivocal as to the direct impacts of digital, and especially social, media in turning out young people in 2008 (Baumgartner and Morris, 2010; Hargittai and Shaw, 2013; Kushin and Yamamoto, 2010).

SOCIO-ECONOMIC STATUS, SKILLS AND ONLINE PRACTICES

Scholars of youth engagement have generally been moving away from searches for the direct effects of digital media, which have had mixed, if generally weakly positive results (Boulianne, 2009). Instead, the field is turning toward more nuanced perspectives in which technology is seen as one of many factors influencing an individual or group's likelihood of participating, sometimes directly, more often moderating or mediating, or being moderated or mediated by, other factors, often socio-demographic and attitudinal characteristics (Hargittai and Shaw, 2013). The emerging consensus is, 'it depends'. But the factors on which it depends are coming into better focus.

One factor on which the impact of digital media on youth engagement depends is socio-economic status (SES). The study of how SES impacts youth engagement has deep implications, foremost among them the question of whether digital media significantly changes the makeup of who becomes engaged. Because SES has long been a primary predictor of political participation, digital media may be poised to level the playing field if media use is more equitably distributed (Schlozman et al., 2010). Alternately, if digital media is primarily available to or adopted by those with high SES, inequalities might be expected to remain unchanged, or even be exacerbated (Margolis and Resnick, 2000; Schradie, Chapter 5 in this volume).

There is evidence for some optimism on these points. In particular, evidence is mounting that race, once a core structuring factor in SES, plays a different role in connecting youth to digital engagement and politics: African-American youth are somewhat more likely than their white counterparts to engage in higher-level digital content creation (Correa, 2010), and as likely to use the media that can connect them to politics (Garcia-Castañon et al., 2011). However, these sanguine notes must be taken in the context of an increasingly inequitable society, and one in which income inequality is having no trouble taking up the slack once played by race in keeping young people at different levels of opportunity.

Another factor, related to SES, is skill with digital media (Hargittai, 2010). Those with wider and deeper repertoires online do tend to participate more, both online and offline. Hargittai and Shaw (2013) found among a college student sample that those with higher levels of skill did report higher levels of civic engagement and certain forms of political engagement, such as petition-signing – but notably, not voting.

It is also crucial to note that interest in politics continues to play a major role in whether a young person's digital media use leads them to become politically engaged (Boulianne, 2009; Xenos and Moy, 2007; Xenos and Kyoung, 2008). Ultimately, younger citizens' lower interest in politics will always remain a barrier to equal engagement, insurmountable by any degree of digital media innovation (Mindich, 2004). Part of the importance of political interest derives from the fact that it leads young people into very different varieties of digital media use. 'Digital media' can refer to a variety of platforms, devices and uses, and some uses of digital media are more tightly linked to the emergence of political behavior than others (Shah et al., 2001; Valenzuela et al., 2012). Across a series of studies, Kahne et al. (2012; Kahne and Middaugh, 2012) have shown that non-political, interest-driven practices online (that is, taking part in online communities of interest such as hobbies or film) are predictive of political engagement, but time spent engaging in friendship-driven social activities is not. Pasek et al. (2009) similarly find young people's use of the Web for information seeking to be associated with civic engagement and political knowledge, while social networking site use is not.

Finally, one of the most active and contentious areas of debate concerning young citizens and engagement is drawing the lines around what, exactly, constitutes engagement. It is clear that a discussion of only news consumption and formal political participation is no longer adequate to the task of describing youth uses of digital media to participate in public life, and discussions over the levels of youth engagement depend increasingly on one's definition of engagement.

The result in the field of youth uses of digital media for political participation has been calls for more diversified conceptions of what political engagement is (Bakker and de Vreese, 2011; Dalton, 2009). No longer is it sufficient to examine youth rates of voting, contacting public officials, contributing money, and following conventional news: young people now inhabit a political communication sphere in which their options are much more varied, and include a host of opportunities to learn, share, and express ideas on topics ranging from 'lolcats' to Vladimir Putin's affection for Siberian tigers. Correspondingly, the sorts of activities in which young citizens engage are diversifying, and there is some evidence that one of the ways in which they differ from their elders is in perceiving a variety of online communicative forms as political action (Lariscy et al., 2011). That is, they sense a mode of influence available to them via online expressions to digital networks that are more resonant than formal, institutionalized actions. Earl and Kimport (2011) have especially led the field in documenting how young people are using online petitioning to communicate views on issues ranging from the formally political to entertainment-oriented concerns.

We should note that we are not describing a distinction between those who participate online and those who do so offline: though the question of whether an activity took place online or offline received considerable attention in the past, recent work on youth participation suggests that citizens are not dividing into camps, young or otherwise, that participate predominantly in one but not the other of these realms. Rather, the two forms are highly correlated (Conroy et al., 2012), and tend to behave quite similarly in analyses (Bakker and deVreese, 2011; Calenda and Mosca, 2007; Östman, 2012). Valenzuela et al. conclude: 'activism does not confine itself to separate online and offline spheres, but instead online interactions can aid offline forms of citizen participation' (Valenzuela et al., 2012: 311).

CULTURAL ENGAGEMENT

Digital media enable an assortment of activities that exist at the boundary of what is commonly accepted as 'the political'. And foremost in testing those boundaries has been work on what has variously been described as content production, user-generated content, engagement with participatory culture, cultural engagement, or interactivity.

According to Jenkins et al., participatory cultures are defined as having several essential properties:

1. Relatively low barriers to artistic expression and civic engagement.
2. Strong support for creating and sharing one's creations with others.
3. Some type of informal mentorship whereby what is known by the most experienced is passed along to novices.
4. Members who believe that their contributions matter.
5. Members who feel some degree of social connection with one another (at least, they care what other people think about what they have created) (Jenkins et al., 2006: 7).

What is special about the participatory cultures emergent in digital environments are the spaces and tools young people have at their fingertips. For example, digital communication technology now enables communities to transcend geography by existing (and thriving) solely within the digital ether (Benkler, 2006). Similarly, advancements in online video tools and social networking have made creating and sharing media more accessible (Bennett et al., 2010). That said, it is not 'better' technology that drives participation, but rather cultural practices that inspire individuals to use technology for acts of participation (Jenkins et al., 2006). Thus, a key civic benefit of participatory cultures is the potential for youth to develop skill sets that extend beyond personal expression to include the social skills needed to interact with a larger community.

Research on 'fandoms' has revealed communities formed around entertainment media, such as popular novels, television shows, and comics. It is important to make the distinction between 'fans' and 'fandoms'; where the former denotes individuals who have a passionate connection to a media franchise, and the latter consists of communities that feel strongly connected to both a franchise and other members (Jenkins et al., 2013: 166). In this light, fandoms are participatory cultures in which members hone communication skills and coordinate with others to accomplish goals. The Harry Potter Alliance is an organization based around the Harry Potter fandom that raises money for charitable causes. The founder, Andrew

Slack, describes his motivation for creating the alliance as similar to the power and corruption struggles that a young Harry Potter faced when he formed Dumbledore's Army (Jenkins, 2009). The Alliance builds from this connection in launching campaigns against genocide in Africa, in support of same-sex marriage rights, and for Haitian earthquake relief, to name just a few; mobilizing more than 100000 young people worldwide in the process (Jenkins et al., 2013). According to Slack, the campaigns appeal to youth and young adults who are 'passionate, enthusiastic, and idealistic', but do not consider themselves to be overly political or traditional activists (Jenkins, 2009). The Harry Potter Alliance provides these young people with a community that marries their enthusiasm with less intimating forms of activism.

Youth online engagement has also been demonstrated through civic gaming. Research in this area is interested in the potential of digital game exposure to foster the learning and practice of civic skills. This is an especially promising line of investigation given that a 2008 survey in the US found 97 percent of youth, aged 12 to 17, reported playing some type of video game (Lenhart et al., 2008). So, how might game use relate to participatory engagement? Jenkins et al. (2006) argue that gaming cultures can provide individuals with the core experiences of play, simulation, and performance, which are precursors to participation. Games require young people to make decisions, communicate their ideas effectively, and in many cases, work with multiple players to achieve an end goal. For example, The Sims is a life simulation video game series where players control the development of an area by making strategic decisions about population size, construction, money allocation, and more. While The Sims is a commercial product (more than 150 million copies have been sold worldwide; Sacco, 2011), it does provide players with an environment where valuable skills are developed and practiced. When the virtual city of Alphaville developed virtual social problems with prostitution and organized crime, 'residents' responded by using the democratic process of debating and voting to determine the best solutions (Jenkins, 2006). Similar arguments have also been made regarding the democratic merits of Second Life, Ultima Online, and World of Warcraft, where players manage the freedom to act individually with the rewards of collaboration and community building (Benkler, 2006; Gerson, 2007; Gordon and Koo, 2008). Coordinated acts within the virtual gaming world can resemble traditional acts of political participation, like when World of Warcraft players during the 2008 US presidential campaign organized a virtual march of avatars in support of candidate Ron Paul (GamePolitics.com, 2008). Examples such as these, well beyond the conventional territory of civic engagement, point to the ongoing need for both conceptual and empirical work to understand the involvement of young people through digital channels.

POLITICAL SOCIALIZATION AND CIVIC EDUCATION

One area working especially hard to come to terms with these changes is that of political socialization and civic education. There, scholars are increasingly moving away from the transmission model of socialization, which focused almost entirely on the role of parents in fostering political attitudes (for example, Carmines et al., 1987; Niemi and Jennings, 1991), to recognize the diversity of influences that shape how young adults become involved in the political process and develop their political identity (Lee et al., 2013; McDevitt and Chaffee, 2002; Wolak, 2009). Similarly, emergent research is highlighting the wide diversity of ways in which parents and children can co-orient themselves towards the political process, moving beyond studies of transmission versus trickle-up socialization; what we are seeing is that intra-family dynamics are more complex than simple transmission, and there are even cases in which youth seem to 'socialize' their parents (Vraga et al., 2014).

Although research has long suggested the mass media contribute to socialization by focusing attention on the political process and intersecting with parental, classroom, and peer discussions on the topic, the more complex media and online environment requires scholars to differentiate between forms of online engagement. Within this field, particular attention has been paid to youth engagement on social networking sites, which often promote civic and political engagement (Bode et al., 2014; Gil de Zuniga et al., 2012; Vitak et al., 2011), and lead youth to diverge from their parents in political activism. This important online engagement may have several roots: the influence of peer discussion on political and citizenship identities (Glynn et al., 2012; McDevitt and Kiousis, 2007), the development of norms and 'citizenship vocabularies' regarding appropriate political behavior and action (Thorson, 2012; Thorson et al., 2014); the potential for exposure to broader political viewpoints than in offline discussions (Kahne et al., 2012; Pew Internet and American Life Project, 2012), and the ability to engage in self-expression and identity formation (Bode et al., 2014; Shah et al., 2007).

This more complex media environment has also contributed to a renewed interest in the role that civic education can play in socialization. While earlier research found little impact of school on socialization, more recent evidence suggests that strong civic curricula that encourage children to actively debate and discuss issues in class can play a vital role in socialization processes, particularly in conjunction with media exposure (Flanagan et al., 2005; Hess, 2009). Similarly, researchers are increasingly emphasizing that schools can play an important role in helping youth develop the skills they need to interpret online information correctly (Kahne and Middaugh, 2012), as well as encouraging youth to both seek out and value exposure to diverse perspectives (Borah et al., 2013; Kahne et al., 2012), long recognized as important to promoting tolerance in the back-and-forth conflict inherent in the democratic process (Mutz, 2006).

Together, scholars have begun to develop a fairly robust measure of the forces expected to contribute to youth socialization into civic and political life. But despite the recognition that studying socialization is important because the orientations developed during youth tend to endure throughout the life cycle and shape engagement with politics (Fiorina, 1981; Green et al., 2002; Kroh and Selb, 2009; Thorson, 2012; Plutzer, 2002), scholars differ on a fundamental question: when are these orientations actually established? Are some ideological preferences genetically inherited, while others are socialized over time (Alford et al., 2005)? Do children develop stable partisan attitudes aged 5 to 8 years, before they enter grade school (van Deth et al., 2011), or is adolescence the key time to observe changes in partisan identity (Kroh, 2010; Wolak, 2009)? And does socialization occur gradually, in a linear fashion, as youth recognize their place in the political process, or does it occur in fits and starts during campaigns, building on the agenda-setting potential for the mass media to make political discussion salient (Kiousis et al., 2005; Valentino and Sears, 1998; Sears and Valentino, 1997)?

CONCLUSION

We hesitate to speculate about the future of Internet time, but perhaps the maturation of both the first digital generation and our first generation of research on digital media allows us to imagine ourselves near the 'end of the beginning' of youth digital politics. That is, though we are far from coherent answers to many of our questions about the implications of digital media on the political world (in a time of change as rapid as ours, giving any such answers would be foolhardy), we are accumulating bodies of evidence on a number of critical issues. What, then, is the relationship between digital politics and youth civic engagement? Our reading of the evidence leads us to a sort of paraphrase of Kranzberg's first law of technology (Kranzberg, 1986): digital media are sometimes good for youth engagement, occasionally bad, and almost always complicated.

The evidence that digital media are ushering in a straightforward sea change in participatory engagement is not there; but this does not mean that they are having no effect. Far from it: we have seen evidence that racial disparities in digital media use may be closing, with corresponding enhancements of engagement, even as disparities in income grow. New styles of engagement are also emerging that offer exciting new ways of thinking about how to participate in public life in a society running on the digital media operating system. And tremendous innovation is taking place in the field of civic education to help us better understand the processes that may aid young people in gaining valuable civic skills at a young age.

This is deeply important work that extends far beyond the halls of academia. Because, more than digital natives or anything else, the young people coming of age today are citizens who will need to deal with political and social problems of very great magnitude. Our field can and must continue to work to better understand how digital technologies can be used to improve meaningful participation and the solving of social problems.

After all, one of the most exciting things about research on young people's uses of digital media for political purposes is how often human beings, and the technologies they use, defy our expectations: it is more complicated than that. What we found time and again in our survey of the literature is that, first, almost all the 'logical' assumptions about youth uses of digital media depend on a host of factors; and second, understanding this field will always be impossible without a rich appreciation for the social-political contexts young people inhabit. But this is why we do research: to come to grips with these complexities and render them into something that occasionally resembles understanding. We have made the case that the field has made a good start in understanding how young citizens are situated to engage in digital politics. But there is much more research – and surprises, surely – ahead.

FURTHER READING

On changing patterns of citizenship in the face of social and technological changes, see Bennett et al. (2010). For a defining study of youth engagement in the early twenty-first century, Zukin et al. (2006). For the primary exposition of mixing media and participatory culture, Jenkins (2006). On web use skills and their implications for youth engagement, Hargittai (2010) is essential. Kahne's work, such as Kahne et al. (2012), leads in the area of civic education and digital media. Vitak et al. (2011) is a leading study of the massive social networking site Facebook, and its implications for engagement.

REFERENCES

Alford, J.R., C.L. Funk and J.R. Hibbing (2005), 'Are political orientations genetically transmitted?', *American Political Science Review*, 99 (2), 153–167.

Bakker, T.P. and C.H. de Vreese (2011), 'Good news for the future? Young people, Internet use, and political participation', *Communication Research*, 38 (4), 451–470.

Baumgartner, J.C. and J.S. Morris (2010), 'MyFaceTube politics: Social networking web sites and political engagement of young adults', *Social Science Computer Review*, 28 (1), 24–44.

Beck, Ulrich (1999), *World Risk Society*, Cambridge: Polity.

Benkler, Yochai (2006), *The Wealth of Networks: How Social Production Transforms Markets and Freedom*, New Haven, CT: Yale University Press.

Bennett, S. (2008), 'The "Digital Natives" debate: a critical review of the evidence', *British Journal of Educational Technology*, 39 (5), 775–786.

Bennett, W. Lance (1998), 'The uncivic culture: communication, identity, and the rise of lifestyle politics', *PS: Political Science and Politics*, 31 (4), 741–761.

Bennett, W. Lance, Deen Freelon and Chris Wells (2010), 'Changing citizen identify and the rise of a participatory media culture', in Lonnie R. Sherrod, Judith Torney-Purta and Constance A. Flanagan (eds), *Handbook of Research on Civic Engagement in Youth*, Hoboken, NJ: Wiley Publishing, pp. 393–423.

Bennett, Sue, Karl Maton and Lisa Kervin (2008), 'The digital natives debate: a critical review of the evidence', *British Journal of Educational Technology*, 39, 775–786.

Bimber, Bruce (2003), *Information and American Democracy: Technology in the Evolution of Political Power*, New York: Cambridge University Press.

Blank, Grant (2013), 'Who creates content?', *Information, Communication and Society*, 16 (4), 590–612.

Bode, Leticia (2012), 'Facebooking it to the polls: a study in social networking, social capital, and political behavior', *Journal of Information Technology and Politics*, 9 (4), 352–369.

Bode, Leticia, Emily K. Vraga, Porismita Borah and Dhavan V. Shah (2014), 'A new space for political behavior: political social networking and its democratic consequences', *Journal of Computer-Mediated Communication*, 19 (3), 414–429.

Borah, Porismita, Stephanie Edgerly, Emily K. Vraga and Dhavan V. Shah (2013), 'Hearing and talking to the other side: antecedents of cross-cutting exposure in adolescents', *Mass Communication and Society*, 16 (3), 391–416. DOI: 10.1080/15205436.2012.693568.

Boulianne, Shelley (2009), 'Does Internet use affect engagement? A meta-analysis of research', *Political Communication*, 26, 193–211.

Calenda, D. and L. Mosca (2007), 'The political use of the internet: some insights from two surveys of Italian students', *Information, Communication, and Society*, 10 (1), 29–47.

Carmines, E.G., J.P. McIver and J.A. Stimson (1987), 'Unrealized partisanship: a theory of dealignment', *Journal of Politics*, 49, 376–400.

Castells, Manuel (1996), *The Rise of the Network Society*, Malden, MA: Blackwell Publishers.

CIRCLE (2008), 'New census data confirm increase in youth voter turnout in 2008 election', available at http://www.civicyouth.org/new-census-data-confirm-increase-in-youth-voter-turnout-in-2008-election/ (accessed 28 August 2013).

CIRCLE (2012), 'Youth voting', available at http://www.civicyouth.org/quick-facts/youth-voting/ (accessed 28 August 2013).

Conroy, M., J.T. Feezell and M. Guerrero (2012), 'Facebook and political engagement: a study of online political group membership and offline political engagement', *Computers in Human Behavior*, 28 (5), 1535–1546. DOI:10.1016/j.chb.2012.03.012.

Correa, T. (2010), 'The participation divide among "online experts": experience, skills and psychological factors as predictors of college students' Web content creation', *Journal of Computer-Mediated Communication*, 16 (1), 71–92.

Couldry, Nick, Sonia Livingstone and Tim Markham (2007), *Media Consumption and Public Engagement: Beyond the Presumption of Attention*, London: Palgrave Macmillan.

Dalton, Russell (2009), *The Good Citizen: How a Younger Generation Is Reshaping American Politics*, Washington, DC: CQ Press.

Delli Carpini, Michael X. (2000), 'Gen.com: youth, civic engagement, and the new information environment', *Communication Abstracts*, 24 (3), 341–349.

Dreier, P. (2008), 'Obama's youth movement', available at http://www.huffingtonpost.com/peter-dreier/obamas-youth-movement_b_127169.html (accessed 28 August 2013).

Earl, Jennifer and Katrina Kimport (2011), *Digitally Enabled Social Change: Activism in the Internet Age*, Cambridge, MA: MIT Press.

Eynon, R. and L.E. Malmberg (2011), 'A typology of young people's Internet use: implications for education', *Computers and Education*, 56 (3), 585–595. DOI:10.1016/j.compedu.2010.09.020.

Fiorina, M.P. (1981), *Retrospective Voting in American National Elections*, New Haven, CT: Yale University Press.

Flanagan, C., L.S. Gallay, S. Gill, E. Gallay and N. Nti (2005), 'What does democracy mean? Correlates of adolescents' views', *Journal of Adolescent Research*, 20, 193–218.

GamePolitics.com (2008), 'Top 20 video game moments from presidential campaign', available at http://gamepolitics.com/2008/11/05/top-20-video-game-moments-presidential-campaign#.UjXWjsasim4 (accessed 1 July 2013).

Garcia-Castañon, M., A.D. Rank and M.A. Barreto (2011), 'Plugged in or tuned out? Youth, race, and Internet usage in the 2008 Election', *Journal of Political Marketing*, 10 (1–2), 115–138. DOI:10.1080/15377857.2011.540209.

Gerson, M. (2007), 'Where the avatars roam', available at http://www.washingtonpost.com/wp-dyn/content/article/2007/07/05/AR2007070501824.html (accessed 1 July 2013).

Giddens, Anthony (1991), *Modernity and Self-identity: Self and Society in the Late Modern Age*, Stanford, CA: Stanford University Press.

Gil de Zuniga, H., N. Jung and S. Valenzuela (2012), 'Social media use for news and individuals' social capital, civic engagement, and political participation', *Journal of Computer-Mediated Communication*, 17, 319–336. DOI:10.1111/j.1083-6101.2012.01574.x.

Glynn, C.J., M.E. Huge and C.A. Lunney (2012), 'The influence of perceived social norms on college students' intention to vote', *Political Communication*, 26, 48–64. DOI:10.1080/10584600802622860.

Gordon, E. and G. Koo (2008), 'Placeworld: using virtual worlds to foster civic engagement', *Space and Culture*, 11 (3), 204–221.

Green, Donald P., Bradley Palmquist and Eric Schickler (2002), *Partisan Hearts and Minds*, New Haven, CT: Yale University Press.

Hargittai, E. (2010), 'Digital na(t)ives? Variation in internet skills and uses among members of the "Net Generation"', *Sociological Inquiry*, 80, 92–113.

Hargittai, E. and Y.P. Hsieh (2012), 'Succinct survey measures of Web-use skills', *Social Science Computer Review*, 30 (1), 95–107. DOI:10.1177/0894439310397146.

Hargittai, E. and A. Shaw (2013), 'Digitally savvy citizenship: the role of Internet skills and engagement in young adults' political participation around the 2008 presidential election', *Journal of Broadcasting and Electronic Media*, 57 (2), 115–134. DOI:10.1080/08838151.2013.787079.

Hargittai, E. and G. Walejko (2008), 'The participation divide: content creation and sharing in the digital age', *Information Communication and Society*, 11 (2), 239–256.

Hess, Diana E. (2009), *Controversy in the Classroom*, New York: Routledge.

Inglehart, Ronald (1997), *Modernization and Postmodernization: Cultural, Economic, and Political Change in 43 Societies*, Princeton, NJ: Princeton University Press.

Issenberg, S. (2012), 'Obama's white whale: how the campaign's top-secret project Narwhal could change this race, and many to come', available at http://www.slate.com/articles/news_and_politics/victory_lab/2012/02/project_narwhal_how_a_top_secret_obama_campaign_program_could_change_the_2012_race_.html (accessed 15 July 2013).

Jenkins, Henry (2006), *Convergence Culture: Where Old and New Media Collide*, New York: NYU Press.

Jenkins, Henry (2009), 'How Dumbledore's Army is transforming the real word: an interview with the HP Alliance's Andrew Slack', available at http://henryjenkins.org/2009/07/how_dumbledores_army_is_transf.html (accessed 15 July 2013).

Jenkins, Henry, Sam Ford and Joshua Green (2013), *Spreadable Media*, New York: NYU Press.

Jenkins, Henry, Ravi Puroshotma, Katherine Clinton, Martha Weigel and Alison J. Robison (2006), 'Confronting the challenges of participatory culture: Media education for the 21st century', available at http://www.newmedialiteracies.org/wp-content/uploads/pdfs/NMLWhitePaper.pdf (accessed 15 July 2013).

Jennings, M.K., G.B. Markus, R.G. Niemi and L. Stoker (2005), 'Youth–parent socialization panel study, 1965–1997: four waves combined', Ann Arbor, MI: Inter-university Consortium for Political and Social Research. DOI:10.3886/ICPSR04037.v1.

Jones, C. and L. Czerniewicz (2010), 'Describing or debunking? The Net Generation and Digital Natives', *Journal of Computer Assisted Learning*, 26 (5), 317–320. DOI:10.1111/j.1365-2729.2010.00379.x.

Kahne, J. and E. Middaugh (2012), 'Digital media shapes youth participation in politics', *Phi Delta Kappan*, 94 (3), 52–56.

Kahne, J., E. Middaugh, N.J. Lee and J.T. Feezell (2012), 'Youth online activity and exposure to diverse perspectives', *New Media and Society*, 14 (3), 492–512. DOI:10.1177/1461444811420271.

Kiousis, S., M. McDevitt and X. Wu (2005), 'The genesis of civic awareness: agenda setting in political socialization', *Journal of Communication*, 55, 756–774. DOI: 10.1111/j.1460-2466.2005.tb03021.x.

Kranzberg, M. (1986), 'Technology and history: "Kranzberg's Laws"', *Technology and Culture*, 27 (3), 544–560. DOI:10.2307/3105385.

Kreiss, Daniel (2012), *Taking Our Country Back: The Crafting of Networked Politics from Howard Dean to Barack Obama*, Oxford: Oxford University Press.

Kroh, Martin (2010), 'The formative period of party identification: parental education in childhood and adolescence', paper presented at the annual meeting of the American Political Science Association, Washington, DC.

Kroh, M. and P. Selb (2009), 'Inheritance and the dynamics of party identification', *Political Behavior*, 31, 559–574. doi 10.1007/s11109-009-9084-2.

Kushin, M.J. and M. Yamamoto (2010), 'Did social media really matter? College students' use of online media and political decision making in the 2008 election', *Mass Communication and Society*, 13 (5), 608–630. DOI:10.1080/15205436.2010.516863.

Lariscy, R.W., S.F. Tinkham and K.D. Sweetser (2011), 'Kids these days: examining differences in political uses and gratifications, Internet political participation, political information efficacy, and cynicism on the basis of age', *American Behavioral Scientist*, 55 (6), 749–764. DOI:10.1177/0002764211398091.

Lee, Nam-Jin, Dhavan V. Shah and Jack M. McLeod (2013), 'Processes of political socialization: a communication mediation approach to youth civic engagement', *Communication Research*, 40 (5), 669–697. DOI:10.1177/0093650212436712.

Lenhart, A., J. Kahne, E. Middaugh, A. Macgill, C. Evans and J. Vitak (2008), 'Teens, video games, and civics', available at http://www.pewinternet.org/Reports/2008/Teens-Video-Games-and-Civics.aspx (accessed 12 June 2013).

Lenhart, A., K. Purcell, A. Smith and K. Zickuhr (2010), 'Social media and mobile Internet use among teens and young adults', available at http://pewresearch.org/pubs/1484/social-media-mobile-internet-use-teens-millennials-fewer-blog (accessed 12 June 2013).

Levine, P. and M.H. Lopez (2002), 'Youth voter turnout has declined, by any measure', available at http://civicyouth.org/research/products/Measuring_Youth_Voter_Turnout.pdf (accessed 18 June 2013).

Livingstone, S. (2008), 'Taking risky opportunities in youthful content creation: teenagers' use of social networking sites for intimacy, privacy and self-expression', *New Media and Society*, 10 (3), 393–411. DOI:10.1177/1461444808089415.

Margolis, Michael and David Resnick (2000), *Politics as Usual: The Cyberspace 'Revolution'*, Thousand Oaks, CA: Sage Publications.

McCoy, T. (2012), 'The creepiness factor: how Obama and Romney are getting to know you', available at http://www.theatlantic.com/politics/archive/2012/04/the-creepiness-factor-how-obama-and-romney-are-getting-to-know-you/255499/ (accessed 15 June 2013).

McDevitt, M. and S.H. Chaffee (2002), 'From top-down to trickle-up influence: Revisiting assumptions about the family in political socialization', *Political Communication*, 19, 281–301.

McDevitt, M. and S. Kiousis (2007), 'The red and blue of adolescence: origins of the compliant voter and the defiant activist', *American Behavioral Scientist*, 50, 1214–1230.

Mindich, David T.Z. (2004), *Tuned Out: Why Americans Under 40 Don't Follow the News*, New York: Oxford University Press.

Mutz, Diana C. (2006), *Hearing the Other Side: Deliberative versus Participatory Democracy*, Cambridge: Cambridge University Press.

Ng, E.S.W., L. Schweitzer and S.T. Lyons (2010), 'New generation, great expectations: a field study of the millennial generation', *Journal of Business Psychology*, 25, 281–292.

Niemi, R.G. and M.K. Jennings (1991), 'Issues and inheritance in the formation of party identification', *American Political Science Review*, 35, 970–988.

Östman, J. (2012), 'Information, expression, participation: how involvement in user-generated content relates to democratic engagement among young people', *New Media and Society*, 14 (6), 1004–1021. DOI:10.1177/1461444812438212.

Palfrey, John and Urs Gasser (2008), *Born Digital: Understanding the First Generation of Digital Natives*, New York: Basic Books.

Pasek, J., E. More and D. Romer (2009), 'Realizing the social Internet? Online social networking meets offline social capital', *Journal of Information Technology and Politics*, 6 (3–4), 197–215.

Patterson, Thomas E. (2007), *Young People and News*, Cambridge, MA: Shorenstein Center for Press and Politics.

Pew Internet and American Life Project (2012), 'Politics on social networking sites', available at http://pewinternet.org/~/media//Files/Reports/2012/PIP_PoliticalLifeonSocial NetworkingSites.pdf (accessed 2 August 2013).

Pew Research Center (2010), 'Millennials: a portrait of generation next', available at http://www.pewsocialtrends.org/files/2010/10/millennials-confident-connected-open-to-change.pdf (accessed 14 September 2013).

Pew Research Center (2012), 'In changing news landscape, even television is vulnerable', available at http://www.people-press.org/files/legacy-pdf/2012%20News%20Consumption%20Report.pdf (accessed 15 September 2013).

Plutzer, E. (2002), 'Becoming a habitual voter: inertia, resources, and growth in young adulthood', *American Political Science Review*, 96, 41–56.

Prensky, M. (2001), 'Digital natives, digital immigrants', *On the Horizon*, 9(5), 1–6. DOI:10.1108/10748120110424816.

Prior, Markus (2007), *Post-Broadcast Democracy: How Media Choice Increases Inequality in Political Involvement and Polarizes Elections*, Cambridge: Cambridge University Press.

Putnam, Robert (2000), *Bowling Alone: The Collapse and Revival of American Community*, New York: Simon & Schuster.

Rainie, Lee and Barry Wellman (2012), *Networked: The New Social Operating System*, Cambridge, MA: MIT Press.

Sacco, D. (2011), 'The Sims 3: Generations', available at http://www.mcvuk.com/retail-biz/recommended/read/the-sims-3-generations/02925 (accessed 19 June 2013).

Schlozman, K.L., S. Verba and H.E. Brady (2010), 'Weapon of the strong? Participatory inequality and the Internet', *Perspectives on Politics*, 8 (2), 487–509. DOI:10.1017/S1537592710001210.

Schudson, Michael (1998), *The Good Citizen: A History of American Civic Life*, New York: Martin Kessler Books.

Sears, D.O. and N.A. Valentino (1997), 'Politics matter: political events as catalysts for pre-adult socialization', *American Political Science Review*, 91, 45–65.

Shah, D.V., J. Cho, S. Nah, M.R. Gotlieb, H. Hwang, N.J. Lee, R.M. Scholl and D.M. McLeod (2007), 'Campaign ads, online messaging, and participation: extending the communication mediation model', *Journal of Communication*, 57, 676–703.

Shah, D.V., N. Kwak and R.L. Holbert (2001), '"Connecting" and "disconnecting" with civic life: patterns of Internet use and the production of social capital', *Political Communication*, 18 (2), 141–162.

Smith, J., Z. Skrbis and M. Western (2013), 'Beneath the "Digital Native" myth: understanding young Australians' online time use', *Journal of Sociology*, 49 (1), 97–118. DOI:10.1177/1440783311434856.

Thorson, Kjerstin (2012), 'What does it mean to be a good citizen? Citizenship vocabularies as resources for action', *Annals of the American Academy of Political and Social Science*, 644, 70–85.

Thorson, Kjerstin (2013), 'Finding gaps and building bridges: mapping youth citizenship', unpublished manuscript.

Thorson, Kjerstin, Emily K. Vraga and Neta Klinger-Vilenchik (2014), 'Don't push your opinions on me: young citizens and political etiquette on Facebook', in J.A. Hendricks and D. Schill (eds), *Presidential Campaigning and Social Media: An Analysis of the 2012 Campaign*, Boulder, CO: Oxford University Press, pp. 74–93.

Valentino, N.A. and D.O. Sears (1998), 'Event-driven political communication and the preadult socialization of partisanship', *Political Behavior*, 20 (2), 127–154. DOI:10.1023/A:1024880713245.

Valenzuela, S., A. Arriagada and A. Scherman (2012), 'The social media basis of youth protest behavior: the case of Chile', *Journal of Communication*, 62 (2), 299–314. DOI:10.1111/j.1460-2466.2012.01635.x.

Van Deth, J.W., S. Abendschon and M. Vollmar (2011), 'Children and politics: an empirical reassessment of early political socialization', *Political Psychology*, 32 (1), 147–174. DOI: 10.1111/j.1467-9221.2010.00798.x.

Vitak, J., P. Zube, A. Smock, C.T. Carr, N. Ellison and C. Lampe (2011), 'It's complicated: Facebook users' political participation in the 2008 election', *Cyberpsychology, Behavior, and Social Networking*, 14 (3), 107–114. DOI:10.1089/cyber.2009.0226.

Vraga, E.K., L. Bode, J. Yang, S. Edgerly, K. Thorson, C. Wells and D.V. Shah (2014), 'Political influence across generations: partisanship and candidate evaluations in the 2008 election', *Information, Communication and Society*, 17 (2), 184–202. DOI:10.1080/13691 18X.2013.872162.

Wattenberg, Martin P. (2008), *Is Voting for Young People?* New York: Pearson Longman.

Wells, C. (2013), 'Two eras of civic information and the evolving relationship between civil society organizations and young citizens', *New Media and Society*. DOI:10.1177/14614448 13487962.

Wolak, J. (2009), 'Explaining change in party identification in adolescence', *Electoral Studies*, 28, 573–583. DOI:10.1016/j.electstud.2009.05.020.

Xenos, M.A. and K. Kyoung (2008), 'Rocking the vote and more: An experimental study of the impact of youth political portals', *Journal of Information Technology and Politics*, 5 (2), 175–189. DOI:10.1080/19331680802291400.

Xenos, M.A. and P. Moy (2007), 'Direct and differential effects of the internet on political and civic engagement', *Journal of Communication*, 57, 704–718.

Zogby, J. (2008), *The Way We'll Be: The Zogby Report on the Transformation of the American Dream*. New York: Random House.

Zukin, Cliff, Scott Keeter, Molly Andolina, Krista Jenkins and Michael X. Delli Carpini (2006), *A New Engagement: Political Participation, Civic Life, and the Changing American Citizen*. New York: Oxford University Press.

13. Internet use and political engagement in youth
Yunhwan Kim and Erik Amnå

Under the central theme of this handbook, politics in the digital era, this chapter provides an overview of youth political engagement in relation to their individual level of everyday Internet use. For around the last two decades, the Internet has increased in popularity as a subject of academic inquiry in many areas, and for various reasons. In the field of politics, in particular, the Internet has become popular because it is considered as one of the most promising tools to encourage political engagement, especially among the younger generations (Delli Carpini, 2000; di Gennaro and Dutton, 2006; Gibson et al., 2005; Owen, 2006; Quintellier and Vissers, 2008). The debates revolving around Internet use in relation to politics vary considerably and include, for example, how to conceptualize human–technology interaction; who are (political) Internet users; Internet usage by political institutions and politicians; whether the Internet is useful in mobilizing people and closing social divides; whether the Internet contributes to change in the meta-level political topography that includes forms of democracy, and so forth. Accordingly, it is necessary specifically to narrow and clarify the focus of this chapter at the outset.

Three aspects should be noted. First, the chapter focuses on the age bracket that comprises youth. Therefore, the literature focusing on this age group was prioritized in our review of the literature. This does not always apply, however, largely because of the insufficiency of the studies that exclusively focus on youth. Second, this chapter mainly covers the specific connection between youth Internet use and their political engagement at the individual level. Therefore, although valuable and popular, studies of political Internet use at societal level are beyond the scope of this chapter. Third, the chapter focuses mainly on quantitative studies.

The overall structure of the chapter is as follows. Since research on youth Internet use and political engagement is a relatively recent development in comparison with other traditionally popular issues in the field, such as conventional political behaviors among adults, we aim to present both general background knowledge on the subject and specific issues in terms of research. In brief, the first more general section covers youth in the Internet era, and youth political engagement. The second more specific section includes an overview of agendas and findings from research on the Internet and youth political engagement, an empirical analysis driven by our own data that exemplifies the aforementioned issues, and suggestions for future studies. Therefore, our hope is that the readers of this chapter have both a bird's-eye view on, and a more specific critical look at, the topic of the Internet and political engagement among youth.

SECTION I

Contemporary Youth in the Internet Era

There are two general underlying reasons for regarding the topic of Internet use among youth today as significant. Most of all, in many societies where the Internet is fairly well disseminated, it is becoming more and more inseparable from youth life in general (Lenhart et al., 2007; Rideout et al., 2010; Shapiro and Margolin, 2014). It has for long been known that youth and adolescents are usually the fastest adopters and avid users of any new technology, and the Internet is no exception (di Gennaro and Dutton, 2006; Ekström et al., 2014; Lee et al., 2013; Quintellier and Vissers, 2008; Ward et al., 2003). As an illustration, Lenhart et al. (2010) reported, on the basis of survey data obtained from a representative American adolescent sample, that more than 90 percent of the adolescents were embedded in the Internet environment. Among young generations, this surprisingly widespread use of the Internet does not seem to be much influenced by other traditionally important discriminant social factors, such as economic climate or cultural norms. Taking a look at data from the European Union, for example, despite the wide range of differences between its member states on various socio-economic factors, youth Internet use between the countries was similar in terms of its universality and massive scale (Eurostat, 2008–2013). In addition, advances in technology allow Internet users to have a ubiquitous, easily affordable connection to the Internet, making it harder to separate the Internet from its users, which include youth and adolescents. Again, they are the ones who are most sensitive to and enthusiastic about a new form of technology. All of this indicates that the Internet is getting more and more critical to understanding the lives of a broad range of contemporary youth.

The second, more important, reason is that many aspects of youth life are indeed intertwined with changes derived from the Internet, which creates developmental contexts for youth today that are distinct from those faced by previous generations. In response to the increasing importance of the Internet in understanding youth life, many questions arise as to the relations between Internet use and various youth behaviors. For example, researchers have explored issues, such as novel behaviors, which have emerged in the online sphere, and also the effects of offline behaviors on online behaviors (and vice versa). Research of this kind is a unique line of inquiry that applies especially to contemporary young people, who have been referred to as the 'Net Generation' (Tapscott, 1998, 2008) or 'digital natives' (Prensky, 2001). They have been in the Internet environment prior to their formative years (Jones et al., 2010; Kennedy et al., 2007), which means that their life styles as youth and adolescents may be meaningfully different from those of former generations in many ways. The Internet is where they acquire information, how they construct their knowledge, how they interact with each other, how they build and maintain friendships, and how they spend their spare time. This vast change in life style has been enough to attract a wide range of researchers in various fields, including education, adjustment problems, mental health, and so on. It has implications for studies of youth political engagement as well, since how young people today go through the process of political socialization is not fully explicable on the basis of earlier knowledge.

Youth Political Engagement

Active political engagement of the public is an indicator of a healthy democratic society (Putnam, 2000). When it comes to young generations, it is also a precondition for the continuation and development of democracy. As such, properly understanding young people's political engagement has been regarded as an important social task. However, understanding of youth political engagement does not simply involve translation from understanding of adult political engagement, since it requires additional considerations: of things like restricted political rights, status and experiences, and the uniqueness of the youth development phase (Amnå and Zetterberg, 2010; Hirzalla and van Zoonen, 2011; Livingstone et al., 2005; Stepick et al., 2008). Indeed, such issues have contributed to our efforts to distinguish between understandings of youth political engagement and of adult political engagement.

Despite the long-lasting interest in and importance of political engagement among young generations, it was only around two decades ago that the topic became particularly prominent, and social concerns and research attention resurfaced. There was the social anxiety that political engagement among young generations was decreasing, which was interpreted as a crisis of democracy (Putnam, 2000). This pessimistic picture of youth political engagement was based on observations of decreasing voting turnout and party membership among the young, trends that are still ongoing (Mycock and Tonge, 2012; Pattie et al., 2004). Since these phenomena are easily observable, when coupled with warnings conveyed by experts in the field of politics they were enough to be a driving force for engagement in discourses on the civic engagement of young generations. The dominant perspectives of the time, derived from such discourses, tended to describe youth as passive in their political engagement.

However, a large number of studies have followed that do not share the pessimistic picture of youth political engagement illustrated above. Scholars on the optimistic side have pointed out that the picture of passive youth was obtained from a narrow focus on defining and observing political engagement in young generations (Ladd, 1999). They maintained that the political engagement of young generations is not as passive as described when alternative forms of political engagement are considered (Stolle and Hooghe, 2011). The alternative forms are things that are less overt, including everyday behaviors such as voluntary work (Flanagan, 2013), societally discriminatory consumer behavior (Micheletti, 2003), and online activity (Bennett et al., 2011; Loader, 2007). Such behaviors have lower thresholds than those applying to conventional, institutionalized indicators of political engagement, such as voting and party membership. Indeed, studies based on this perspective have illustrated that many young people are politically active, or at least, not notoriously passive as described earlier (Fisher, 2012; Livingstone et al., 2005; Norris, 2002; Sloam, 2012; Zukin et al., 2006). Not only does this alternative perspective make for a more optimistic outlook on contemporary youth political engagement, but it also seems to be more relevant to understanding younger age groups such as youth and adolescents. Some stereotypical forms of political engagement are not yet legally available to the young (for example, voting), and/or are not conventionally encouraged among people who have not yet reached adulthood (for example, party membership) (Stepick et al., 2008). In essence, the alternative perspective is one of a positive outlook on youth political engagement, in that it considers more developmentally appropriate political indicators.

Although quite a recent development, still at an initial stage of empirical investigation, a further school of thought that views youth in a partly more optimistic light has emerged (for example, Amnå and Ekman, 2014; Ekman and Amnå, 2012; Hansard Society, 2013). By contrast with the alternative perspective above, it shifts attention to different indicators of political engagement, and suggests a change in the way we should conceptualize political engagement among young people today. In particular, it points out that youth have been described as passive because the focus has been exclusively on manifest, behavioral aspects of youth engagement. Rather, these scholars suggest that the cognitive aspects of youth political engagement, such as political interest, also deserve to receive attention, and that by considering both manifest, behavioral aspects and latent, cognitive aspects, it is possible to obtain a better graded and more realistic picture of youth political and civic activity (Amnå and Ekman, 2014; Ekman and Amnå, 2012). Central to this contention is the concept of standby youth, referring to young people who show high interest but not substantial behavioral participation, who were considered passive when the focus was exclusively on behavioral indicators (Amnå, 2010). Indeed, a few recent studies that have tested this contention have confirmed that this type of political orientation does exist, both among adolescents (Amnå and Ekman, 2014), and among adults including youth (Hansard Society, 2013). Further, standby youth show levels of competence similar to those of active youth (who present behavioral participation) on most indicators of political engagement, and who therefore seem to be potential resources for healthy democracy (Amnå and Ekman, 2014). By showing that the latent aspect of political engagement can serve as an important marker for differentiating political orientations among contemporary youth, these findings indicate that it is worth focusing on so far neglected variability among younger people to obtain a clearer understanding of their political activities.

In sum, research on the political engagement of the young was sparked by the anxiety in society aroused by decreasing levels of participation in some overt forms of political actions. This view was challenged later by an alternative perspective, with a substantial amount of empirical backing, that considered a wider range of developmentally more appropriate political and civic behaviors among youth and adolescents. Further, the most recent perspective, although not yet as substantially grounded as the alternative perspective above, also challenges pessimistic evaluations of young people's political engagement, in particular by emphasizing the need to pay more careful attention to less overt sides of political engagement. Regardless of which contention is more or less convincing, a common message from all of these studies is that one key to understanding political engagement among contemporary youth lies in careful consideration of their developmental stage and living context.

SECTION II

Individual Internet Use and Political Engagement in Youth

This section reviews specific research findings with regard to the relation between youth Internet use and their political engagement. Although the history of research on this subject matter is relatively short, there are at least three issues that many scholars have tried to clarify and reach agreement on: (1) whether Internet use affects political engagement overall; (2) which types of Internet use are more or less related to political engagement than others; and (3) why and how Internet use is related to political engagement. Here, we present the debates and research agendas concerning each of these topics, followed by accounts of current understanding and/or agreement on them. Although our goal is to discuss young people's Internet use and political engagement, given the insufficiency of the studies focusing exclusively on this population, some studies exploring older populations had to be included. However, in an effort to minimize the literature irrelevant to understanding young citizens today, we include them only where it is unavoidable.

The effects of Internet use on political engagement[1]
Our initial and fundamental inquiry about the Internet when it comes to political engagement was whether the use of this new technology would have a negative or positive impact on civic lives in general. The contrasting perspectives that support either a negative or a positive impact are designated as the 'dystopian' and the 'utopian', respectively (Bakker and de Vreese, 2011; Pasek et al., 2009).

Scholars with a 'dystopian' perspective have given several reasons why people's Internet use might decrease political engagement (Kraut et al., 1998; Nie and Erbring, 2000; Putnam, 1995, 2000). The dystopians anticipated that the Internet would be used mostly for entertainment purposes, rather than for any political or civic purpose. Therefore, time spent on the Internet (for entertainment purposes) would occupy the spare time that otherwise might have been devoted to political engagement (the so-called 'time replacement hypothesis'). They also anticipated that the virtual world created in the online sphere would somewhat isolate people from their reality and decrease substantive communication between people.

By contrast, scholars with a 'utopian' perspective have a brighter view on the effects of people's Internet use on political engagement (Bennett, 2000; Delli Carpini, 2000; di Gennaro and Dutton, 2006; Polat, 2005). The utopians focus on the potentials that the Internet has to offer, such as faster and easier information acquisition, and the interactive and asynchronous communication that is a precondition for, or at least a booster of, political engagement. Further, these scholars do not see the virtual world and online communication as unreal, but rather as substantive. Accordingly, they see Internet use as having the potential not only to encourage already engaged citizens but also to engage not-yet-engaged citizens.

In this regard, perhaps the most concise answer can be obtained from Boulianne's (2009) study. Boulianne gave a good summary of the ongoing debates on this issue and presented empirical evidence obtained from a series of meta-analyses. In brief, the author conducted meta-analyses on the relation between Internet use and political engagement on the basis of the findings of 38 studies that used samples from the United States, and focused on offline political and civic behaviors. The analyses mainly addressed three aspects of the relationship: its direction, its significance, and its effect size. The overall results indicate that only a small proportion of findings on the relationship between Internet use and political and civic engagement are negative (16 percent). By contrast, most of the findings support a positive relationship between Internet use and political and civic engagement (77 percent). Accordingly, considering both the negative and the positive perspectives, the latter is generally more convincing. What is noteworthy, however, is that the magnitude of the positive effect was found to be not as substantial as the utopians had hoped (the estimated effect size being 0.07), and the mechanisms underlying the effect did not seem to be uniform, which left many tasks to be resolved in future studies (Boulianne, 2009).

A related question with regard to the relation between Internet use and overall political engagement is whether the effects of Internet use on political engagement are uniquely attributable to Internet use per se, or whether these effects largely reflect the indirect impacts of other factors that have been consistently found to explain political engagement, such as traditional media use and political interest. These two factors, among others, provide grounds for suspicion that the relation between Internet use and political participation, even if it turns out to be statistically significant, may be spurious. Specifically, it might be that avid traditional media users are also active Internet users, or that people who are interested in politics are also those who are active Internet users (self-selection bias). Empirical evidence on this issue is available in the literature, in studies that have examined the relative influences of traditional media and the Internet, and/or in studies that have used potential confounding factors as control variables (for example, Bakker and de Vreese, 2011; Ekström and Östman, 2013; Gil de Zúñiga et al., 2009; Lee et al., 2013; Östman, 2013; Pasek et al., 2006; Quintellier and Vissers, 2008; Velasquez and LaRose, 2014). In most cases, the additional portions of political engagement that are explained by Internet use are significant after the effects of traditional media use and/or political interest are taken into account (with a positive direction of influence). These findings therefore indicate that Internet use, overall, seems to have certain unique and positive impacts on political engagement.

Which types of Internet use are related to political engagement?
Despite some evidence that Internet use has positive effects on political engagement, a challenge still remained. The effects found were not large. Therefore, in an effort to obtain a clearer picture of the relation between Internet use and political engagement, the question of which aspects of Internet use are related to political engagement naturally followed. People do not use the Internet uniformly; rather, patterns of Internet use between individuals, and even of any one individual, are quite heterogeneous (Bakker and de Vreese, 2011; Östman, 2012; Pasek et al., 2006; Pasek et al., 2009; Polat, 2005; Quintellier and Vissers, 2008; Shah et al., 2001). Moreover, patterns and ranges of Internet use have been evolving and expanding, making it more complex than before to define and operationalize what it means to use the Internet. Indeed, one of the potential reasons for the less-than-expected positive effect of Internet use on political engagement was that omnibus measures of Internet usage, such as time spent on the Internet, are too crude to properly capture specific aspects of Internet use (Baumgartner and Morris, 2010; Ekström et al., 2014; Quintellier and Vissers, 2008). Use of such crude measures inevitably precludes the uncovering of genuine effects of Internet use on political engagement. Therefore, consideration of how to operationalize Internet use was necessary in order to clarify its relationship with political engagement.

There are several ways to decompose Internet usage conceptually and analytically into chunks or pieces. For example, overall usage may be broken down according to specific website or service uses, or types of activities, or methods of access, and so on. One widely used way of proceeding is to categorize Internet use according to type of activity. This method seems to be more popular than others since it is less affected by superficial-level rapid changes in the Internet sphere (recall how many websites and Internet services have a ridiculously short life span), and it can, to some extent, reflect users' qualitative characteristics, such as the motives underlying their Internet use (compared with methods that only reflect quantitative characteristics).

One particular approach that provides a theoretical basis for classifying types of Internet use is the 'uses-and-gratification' approach (Blumler and Katz, 1974; McQuail, 1994; Ruggiero, 2000). A type-based classification is made, which has the potential to reflect the internal characteristics of Internet users. The theoretical approach suggests that individuals actively choose which types of media content they seek, and subscribe on the basis of what satisfies them. Hence, the pattern of media use is expected to vary to the extent that what is satisfying differs between people. This approach has long served as a theoretical underpinning of research on various types of media, and it has also been applied specifically to the Internet

(Baumgartner and Morris, 2010; Ruggiero, 2000). The types of Internet use suggested in the literature so far are: informational, interactional/communicative/social, creative, expressive, entertainment/recreational, consumptive, finance-managerial, and participatory (Ekström and Östman, 2013; Pasek et al., 2009; Eynon and Malmberg, 2011; Livingstone et al., 2005). Given the advantages of type-based approaches to the media (mentioned above), and the reality that many studies are based on types, our review hereafter is based on types of activities on the Internet.

Research on Internet use in relation to political engagement initially focused on online activities that are rather specifically associated with traditional political engagement. The Internet initially received research attention as a supplementary medium for information and communication that has somewhat distinctive features compared with the traditional media, such as being up to date and having universal accessibility (Östman, 2012). It is not surprising, therefore, that informational Internet use (such as news consumption on the Internet and visitation to websites on political issues or for campaigning) has received the greatest research attention. In addition, interactional Internet use regarding politics (such as circulating political information to acquaintances, and participating in online political forums) has also received a great deal of attention. When the focus has been on these two specific kinds of Internet use, research findings have been more consistent and convincing than when an omnibus measure of Internet use has been used. More political Internet use is related to increased civic engagement. That is, the people who seek and access more political information online and those who discuss more societal and political issues online are those who report higher levels of political engagement (Bakker and de Vreese, 2011; Quintellier and Vissers, 2008). This indicates that politically driven Internet use is related to actual political engagement.

Later, however, as the range of possible experiences on the Internet expanded, the focus of research was extended correspondingly, from rather specific political Internet use to more variegated and general types of use (Ekström and Östman, 2013; Östman, 2012). This research stream, however, may not be typical, since the focus of many research topics usually switches from a general area to more specific sub-areas, rather than vice versa. This non-conventional research direction was taken on the basis of the realization that some experiences underlying general Internet use seem to be closely related to those underlying civic engagement. In this regard, especially relevant activities, which have received great research attention, are creative Internet use, such as running a blog, or sharing original content online. Indeed, this type of Internet use has been found to relate significantly to political engagement, both cross-sectionally

(Gil de Zúñiga et al., 2009; Quintellier and Vissers, 2008; Wei and Yan, 2010; Östman, 2013) and longitudinally (Ekström and Östman, 2013). All this indicates that, with the evolving and expanding potential of the Internet, the opportunity to reignite political engagement through, or mediated by, Internet use seems to expand correspondingly. Thus, if we restrict our focus to politics-specific Internet use, we may miss some promising opportunities that the Internet can provide.

Theoretical explanations for political engagement[2]
The most interesting and important questions concern why and how Internet use is related to political engagement, and under which conditions it works best. Although empirical evidence substantiating underlying mechanisms is still scanty, a handful of attempts have been made to construct theoretical explanations. The currently available explanations in the literature focus on some rather distinctive aspects, which make them complementary rather than competing.

Despite the paucity of theorizing in the literature overall, there is a well-established theoretical model, known as the citizen communication mediation model (McLeod et al., 2001; Shah et al., 2005). The structure, and naming, of the model is quite straightforward. It postulates that information obtained on an issue from the media sparks interaction and communication on the issue among people, who subsequently encourage political participation (Shah et al., 2005). Using US national panel survey data, Shah et al. (2005) provided empirical evidence that the effects of media use, including Internet use, on political participation are largely indirect, both cross-sectionally and longitudinally. In so doing, the authors showed, in particular, that communication between citizens, as measured by interactive civic messaging and interpersonal political discussion, plays a significant mediational role. Also, the model has been empirically supported by studies of youth that have focused on both traditional and online media channels (Lee et al., 2013), and also by studies that have focused specifically on Internet use (Ekström and Östman, 2013) in a longitudinal context. By showing the strong, but largely indirect, effects of the media on political participation, this model not only suggests a specific mechanism via which Internet use is connected to civic participation, but also emphasizes that youth play an active role in their own political socialization (Shah et al., 2005).

Another theoretical approach is based on a psychological perspective. Specifically, some scholars have focused on how the Internet use can influence political participation by psychologically empowering its users (Flanagin et al., 2010; Leung, 2009). The process of psychological empowerment has received attention mostly with regard to creative Internet use.

For example, Östman (2012) regarded the creative Internet use that is characterized by 'expressivity, performance and collaboration' (p. 1005) as a potential precursor of, or driving force for, civic engagement. These characteristics explain how creative Internet use may help to build psychological empowerment in adolescents. In creative online activities, adolescents have ample chances to spread their opinions to the public, and realize that their voices are heard in society and that they can affect that society. Although empirical evidence directly supporting the mechanism as a whole is scarce in the literature, there is some empirical evidence in support of all its different parts. That is, overall, more creative Internet use has been found to relate to a higher level of political engagement (Ekström and Östman, 2013; Gil de Zúñiga et al., 2009; Östman, 2012; Wei and Yan, 2010). In addition, a substantial number of studies have shown that young people with higher political self-efficacy are the ones who participate in more political activities (for example, Caprara et al., 2009; Velasquez and LaRose, 2014). There is no direct evidence in the literature that links creative Internet use in youth to political efficacy, but it has been shown that youth who report more creative Internet activities are also those who have higher levels of political self-efficacy (see An Empirical Illustration section below).

Another theoretical effort, made in the literature, to elaborate a link between Internet use and political participation among the young draws attention to the conditions that moderate the link. For example, Xenos and Moy (2007) focused on political interest as a potential factor that might have contingent effects. The authors found that the effects of Internet use (online news consumption) were connected to political behaviors, but more strongly among adults who reported higher levels of political interest than among those who reported lower levels. Later, Xenos et al. (2014) expanded this line of study, based on a youth sample, by exploring additional potential moderators of the relations between Internet use and political participation such as political discourse at home and civic education at school.

Although attempts to establish theoretical frameworks are still in their initial stages, with very little empirical evidence, they convey important messages. The first two lines of research we have considered highlight underlying interpersonal and intrapersonal mechanisms with regard to how Internet use is connected to political engagement. The third line indicates that the effects of Internet use on political engagement are not uniform, but rather heterogeneous. They vary according to many other politics-related factors.

An Empirical Illustration

We show here some empirical data that can illustrate some of the points reviewed above. We first derive some relevant information from a recently published work by Amnå and Ekman (2014) regarding political engagement among contemporary youth. Then, there is additional analysis of the same data set, which addresses the issue of young people's Internet use. Overall, since the purpose of this section is to exemplify empirically the threads in previous findings, for each analysis below, we first remind the reader of the issues in the literature reviewed above, and then present some results.

Studies of youth political engagement have been concerned with how best to capture youth participatory reality by reflecting upon better indicators of their political engagement. In the most recent development, Amnå and Ekman (2014) showed that, by focusing simultaneously on both manifest and latent aspects of political engagement, it is possible to obtain a more nuanced picture of the political orientation of youth. Specifically, using measures of both political participation and political interest, they clustered 863 Swedish adolescents (M_{age} = 16.6, 51.4 percent girls) into four groups: active, standby, unengaged, and disillusioned (Table 13.1).

The focal point, emphasized by Amnå and Ekman (2014), is that there is a cognitively highly engaged group of youth, although they are behaviorally reserved, namely standby youth. Further, the authors provided comparisons of various political indicators[3] between the four groups, showing that standby youth are qualitatively different from members of the two genuinely passive youth groups (unengaged and disillusioned youth), but are closer to active youth (see Table 13.2).

When it comes to Internet use among adolescents, there are many previous reports that they are deeply immersed in the Internet in many aspects of their lives. In line with this, we observed that our adolescents spent a substantial amount of their time on the Internet. Specifically, when they were asked how much of their spare time they spent on the Internet, out of a group of 861 adolescents, 39.2 percent reported that it was more than three hours per day. By comparison, only 5.2 percent reported that they spent less than half an hour per day.

Next, we preliminarily observed the relation between Internet use and political engagement. One argument, presented above, is that an omnibus measure of Internet use is not relevant to young people today given their massive and heterogeneous Internet use. Relatedly, the relation is more evident when a more specific measure of Internet use is adopted. This tendency was also observed in our analysis (see Table 13.3). Zero-order correlations between Internet use and political variables showed that

Table 13.1 Results of a 4 (citizenship orientation) × 2 (gender) MANOVA examining differences on measures used in the cluster analysis (z-scores)

	Active (n = 51)	Standby (n = 401)	Unengaged (n = 226)	Disillusioned (n = 185)	F-values (d.f.)	η^2
			M (SD)			
Participation	3.29_a (1.31)	0.03_b (0.56)	-0.44_c (0.18)	-0.44_c (0.19)	$793.85_{(3,859)}$ ***	0.74
Interest	0.71_a (1.10)	0.73_a (0.59)	-0.30_b (0.24)	-1.37_c (0.44)	$689.34_{(3,859)}$ ***	0.71

Notes:
Multivariate F-test (Wilks's λ), $F(6, 1716) = 690.44$, $p < 0.001$, $\lambda^2 = 0.71$.
Within each row, means with different subscripts differ significantly at $p < 0.05$ in Tukey's HSD post hoc comparisons.
$* p < 0.05$; $** p < 0.01$; $*** p < 0.001$.

Source: Reprinted with permission from Amnå and Ekman (2014).

Table 13.2 Results of a 4 (citizenship orientation) x 2 (gender) MANOVA examining citizenship orientation group differences on measures of citizenship competences (z-scores)

	M (SD)				F-values (d.f.)	η^2
	Active ($n = 51$)	Standby ($n = 401$)	Unengaged ($n = 226$)	Disillusioned ($n = 185$)		
Political efficacy	0.73_a (1.01)	0.36_b (0.84)	-0.24_c (0.84)	-0.67_d (1.01)	$75.55_{(3760)}$***	0.23
Trust in institutions	0.08_{ab} (1.06)	0.30_a (0.85)	-0.02_b (0.95)	-0.62_c (1.02)	$436.36_{(3760)}$***	0.13
Social trust	-0.17_{ab} (0.93)	0.12_a (0.97)	0.04_a (1.01)	-0.24_b (1.03)	$6.92_{(3760)}$***	0.03
Feelings about politics	0.74_a (0.99)	0.58_a (0.70)	-0.25_b (0.72)	-1.14_c (0.65)	$265.98_{(3760)}$***	0.51
Satisfaction with democracy	-0.44_b (1.24)	0.15_a (0.95)	0.12_a (0.82)	-0.35_b (1.10)	$17.43_{(3760)}$***	0.06
Ambitions	1.14_a (1.24)	0.33_b (0.91)	-0.31_c (0.71)	-0.66_d (0.83)	$88.21_{(3760)}$***	0.26
News consumption	0.50_a (0.89)	0.32_b (0.93)	-0.17_b (0.84)	-0.62_c (0.99)	$45.97_{(3760)}$***	0.15
Knowledge	0.06_{ab} (1.10)	0.23_a (0.98)	-0.17_b (0.96)	-0.30_c (0.93)	$15.51_{(3760)}$***	0.06

Notes:
Multivariate F-test (Wilks's λ), $F(24, 2184) = 34.87$, $p < 0.001$, $\eta^2 = 0.27$.
Within each row, means with different subscripts differ significantly at $p < 0.05$ in Tukey's HSD post hoc comparisons.
Means sharing any one of the subscripts do not differ. For example, ab does not differ significantly from a or b, but ab differs from c.
* $p < 0.05$; ** $p < 0.01$; *** $p < 0.001$.

Source: Reprinted with permission from Amnå and Ekman (2014).

Table 13.3 Zero-order correlations between political engagement and Internet use

	1	2	3	4	5	6
1. Political participation	–					
2. Political interest	0.31***	–				
3. Time spent on the Internet	0.04	−0.06	–			
4. Informational Internet use	0.26***	0.38***	0.16***	–		
5. Interactional Internet use	0.08*	0.02	0.28***	0.16***	–	
6. Creative Internet use	0.28***	−0.01	0.25***	0.30***	0.36***	–
7. Entertainment Internet use	0.03	−0.12***	0.38***	0.31***	0.20***	0.39***

Note: * $p < 0.05$; ** $p < 0.01$; *** $p < 0.001$.

the omnibus measure of Internet use (time spent on the Internet) was related neither to political participation nor to political interest. However, some significant relations were found when it came to specific uses of the Internet (informational, interactional, creative, and entertainment). Specifically, informational Internet use was positively related to both political participation and interest. In addition, both interactional and creative Internet use were positively related to political participation. Entertainment Internet use was also related to political interest, not surprisingly in a negative direction. In a nutshell, the patterns found in this preliminary analysis are largely consistent with the literature reviewed above, in terms of both the issue of measurement and the specific relation found between Internet use and political engagement.

Finally, we merged the two findings presented above. That is, we observed differences in adolescents' Internet use according to their group membership in terms of political engagement. To this end, we conducted a series of analyses of covariance, with adolescent group membership entered as an independent variable, each type of Internet use as dependent variables, and overall time spent on the Internet as a covariate (Table 13.4). Significant results were found for three types of Internet use: informational, creative and entertainment. Overall, despite slight deviation, politically more engaged youth (either active or standby) tended to report more informational and creative Internet use and less entertainment Internet use than less engaged youth (either unengaged or disillusioned). The results are largely consistent with the main messages from the two findings presented above. That is, some differences between the groups of youth classified by their political engagement were found in their specified use of the Internet. Also, standby youth differed from the other two groups of genuinely passive youth (unengaged and disillusioned), especially in terms of the cognitive type of Internet use (informational Internet use), rather than with regard to the behavioral type of Internet use (creative Internet use).

Future Research Directions

Suggested future directions are twofold, theoretical and methodological, but they are not mutually exclusive. First, theoretically, a clear goal must be to construct more and precise frameworks that can explain the link between Internet use and political participation. At the beginning of this chapter, we pointed out that studies of individual Internet use and political engagement among young people are less prevalent than those in other traditionally popular areas. On the other hand, it is also true that studies in the political arena have been growing rapidly. In comparison

Table 13.4 Results of ANCOVAs examining citizenship orientation group differences on measures of Internet use

	Estimated marginal means (SE)				F-values (d.f.)	η^2
	Active (n = 51)	Standby (n = 401)	Unengaged (n = 226)	Disillusioned (n = 185)		
Informational Internet use	3.68_a (0.12)	3.25_b (0.04)	2.78_c (0.06)	2.51_d (0.06)	$49.678_{(3850)}$***	.149
Interactional Internet use	3.95 (0.14)	3.74 (0.05)	3.76 (0.06)	3.68 (0.07)	$1.112_{(3851)}$	.004
Creative Internet use	2.52_a (0.11)	1.63_b (0.04)	1.59_b (0.05)	1.62_b (0.06)	$20.904_{(3849)}$***	.069
Entertainment Internet use	3.21_{ab} (0.13)	2.96_b (0.05)	2.94_b (0.06)	3.28_a (0.07)	$6.206_{(3851)}$***	.021

Notes:
Within each row, means with different subscripts differ significantly at $p < 0.05$ in Bonferroni post hoc comparisons. Means sharing any one of the subscripts do not differ. For example, ab does not differ significantly from a or b.
* $p < 0.05$; ** $p < 0.01$; *** $p < 0.001$.

with the increase in empirical evidence investigating the relation between Internet use and political engagement, an increase in efforts devoted to developing theoretical understanding is more difficult to detect. Few of the efforts reviewed above are well founded, although they complement each other and are being further elaborated. However, theoretical build-up in the literature overall is still at a preliminary level, leaving many potentially related factors and mechanisms unexplored, which precludes the grasping of any integrated understanding. In making such efforts, we also note that what has been less discussed in this chapter, but is crucial to future research, would be to clarify theoretically the very concepts of political engagement and political participation, so as to develop a consistent and inter-subjectively efficient terminology (see van Deth, 2014). Another potential avenue would be to develop also what youth Internet use and political networking may imply when it comes to both the citizen norms and the ideals of democracy, for example whether 'new forms of networked young citizenship, more compatible for the times and contemporary youth culture, may be more fruitful for both understanding contemporary developments and also for future democratic governance' (Loader et al., 2014: 149).

The value of elaborating specific mechanisms cannot be emphasized enough, especially when it comes to the young people of today. When the Internet was first popularized, technological aspects (for example, whether one is capable of using the Internet and related services) and social-structural aspects (for example, whether the Internet is affordable for all people regardless of background) were treated as the significant barriers to actualization of the political potential of Internet use. However, these concerns are fading since technological development seems to have overcome them easily. Today, Internet connectivity is widespread, universally ubiquitous, and easily affordable in most developed countries. It is even possible to create one's own webpage or blog with just one click. This indicates that young people today, if they wish to, can fully enjoy the political opportunities that derive from the Internet. Accordingly, in light of the decreasing social-structural and external difficulties in realizing the political potentials of the Internet, elucidating the internal mechanisms and conditions of the impacts of Internet use on political engagement can be said to be becoming more important. This is in line with the contemporary stances of researchers in the fields of media and youth political socialization, in that scholars support the claims that the Internet per se hardly has deterministic impacts on human behaviors, and that young people are their own agents of political socialization (Amnå et al., 2009; Lee et al., 2013).

Second, two methodological points in relation to improvement are worth mentioning: one about measurement, the other about study design. When it comes to measurement, as reviewed above, a simple omnibus measure (such as time spent on the Internet) is unable to capture much variability in Internet use, hence effectively precluding the detection of nuanced relationships between Internet use and political engagement. This can be (and has been) easily overcome by using more specific measures of Internet use. However, in terms of how to specify Internet use, there are still some challenges to overcome. Although this chapter has been based on types of Internet use, it is also true that other specifications can be more relevant according to the research questions that scholars aim to answer. Yet, in comparison with types of Internet use, other specifications have not become well established. In addition, it is inherently difficult to develop a measure that is both generalizable and specific, especially when it comes to Internet use. This is because changes in the Internet environment keep occurring at a rapid pace, both in superficial respects (such as the advent and extinction of specific websites and online services) and in fundamental respects (such as the advent of Web 2.0; O'Reilly, 2005). Accordingly, the challenge is to find a way to simultaneously track changes in the Internet sphere and to capture efficiently the generalizable fundamental aspects beyond the superficial changes. For example, in line with this concern, Ekström et al. (2014) tried to adopt theories of space and public orientation to specify Internet use among adolescents. This kind of effort should be continued.

With regard to study design, it should be noted that the majority of the studies conducted so far have had a cross-sectional design. As is commonly pointed out, findings based on a cross-sectional design are not enough for any conclusions regarding causality to be drawn. And, as is often the case, the issue of causality is deeply related to the most important and exciting parts of arguments revolving around the relation between Internet use and political engagement. Relatedly, studies of mechanisms and conditions can be convincing when based on a longitudinal design. In addition, especially when it comes to youth and adolescents, a study becomes substantially more meaningful when it can inform about developments over time. However, it is only recently that studies based on a longitudinal or experimental design have been reported in the literature. Thus, more studies based on study designs that can substantiate causal claims are called for.

CONCLUSION

When the Internet became the object of academic inquiry in the field of politics, it sparked competing arguments between pessimists and optimists regarding the potential that Internet use has for the revitalization of the younger generation's political engagement. Now, around two decades later, our initial understanding is that the Internet can be used positively for the purpose of furthering engagement. At the same time, however, in order for this purpose to be fully met, more efforts to further elucidate the relation between young people's Internet use and political engagement are called for.

NOTES

1. We point out that most of the literature reviewed here on conceptual debates and meta-analyses in relation to 'dystopian' versus 'utopian' perspectives do not exclusively refer to youth.
2. We note again that the focus of this chapter is specifically on the link between Internet use and political engagement at the individual level. More fundamental and/or broader conceptual and philosophical discourses are not considered here.
3. Due to limitations of space, we are not able to provide more detailed explanations of each measure, although we believe that their meanings are self-explanatory. For further details, see Amnå and Ekman (2014).

REFERENCES

Amnå, E. (2010). Active, passive, or standby citizens? Latent and manifest political participa-tion. In Amnå, E. (ed.), *New Forms of Citizen Participation: Normative Implications* (pp. 191–203). Baden-Baden: Nomos.
Amnå, E. and Ekman, J. (2014). Standby citizens: diverse faces of political passivity. *European Political Science Review*, 6(2), 261–281.
Amnå, E., Ekström, M., Kerr, M. and Stattin, H. (2009). Political socialization and human agency: the development of civic engagement from adolescence to adulthood. *Statsvetenskaplig Tidskrift*, 111(1), 27–40.
Amnå, E. and Zetterberg, P. (2010). A political science perspective on socialization research: young Nordic citizens in a comparative light. In Sherrod, L.R., Torney-Purta, J. and Flanagan, C.A. (eds), *Handbook of Research on Civic Engagement in Youth* (pp. 43–66). Hoboken, NJ: John Wiley & Sons.
Bakker, T.P. and de Vreese, C.H. (2011). Good news for the future? Young people, Internet use, and political participation. *Communication Research*, 38, 451–470.
Baumgartner, J.C. and Morris, J.S. (2010). MyFaceTube politics: social networking web sites and political engagement of young adults. *Social Science Computer Review*, 28(1), 24–44.
Bennett, W.L. (2000). Introduction: communication and civic engagement in comparative perspective. *Political Communication*, 17(4), 307–312.
Bennett, W.L., Wells, C. and Freelon, D. (2011). Communicating civic engagement: con-trasting models of citizenship in the youth web sphere. *Journal of Communication*, 61(5), 835–856.
Blumler, J. and Katz, E. (1974). *The Uses of Mass Communication: Current Perspectives on Gratification Research*. Beverly Hills, CA: Sage.
Boulianne, S. (2009). Does Internet use affect engagement? Meta-analysis of research. *Political Communication*, 26, 193–211.
Caprara, G.V., Vecchione, M., Capanna, C. and Mebane, M. (2009). Perceived political

self-efficacy: theory, assessment, and applications. *European Journal of Social Psychology*, 39, 1002–1020.

Delli Carpini, M.X. (2000). Gen.com: youth, civic engagement, and the new information environment. *Political Communication*, 17(4), 341–349.

Di Gennaro, C. and Dutton, W. (2006). The Internet and the public: online and offline political participation in the United Kingdom. *Parliamentary Affairs*, 59(2), 299–313.

Ekman, J. and Amnå, E. (2012). Political participation and civic engagement: towards a new typology. *Human Affairs*, 22(3), 283–300.

Ekström, M., Olsson, T. and Shehata, A. (2014). Spaces for public orientation? Longitudinal effects of internet use in adolescence. *Information, Communication and Society*, 17(2), 168–183.

Ekström, M. and Östman, J. (2013). Information, interaction, and creative production: the effects of three forms of internet use on youth democratic engagement. *Communication Research*. Advance online publication. doi:10.1177/0093650213476295.

Eurostat (2008–2013). Information society statistics. http://ec.europa.eu/eurostat/statistics-explained/index.php/Information_society_statistics#Households_and_individuals.

Eynon, R. and Malmberg, L. (2011). A typology of young people's internet use: implications for education. *Computers and Education*, 56, 585–595.

Fisher, D.R. (2012). Youth political participation: bridging activism and electoral politics. *Annual Review of Sociology*, 38(1), 119–137.

Flanagan, C.A. (2013). *Teenage Citizens: The Political Theories of the Young*. Cambridge, MA: Harvard University Press.

Flanagin, A.J., Flanagin, C. and Flanagin, J. (2010). Technical code and the social construction of the Internet. *New Media and Society*, 12, 179–196.

Gibson, R.K., Lusoli, W. and Ward, S. (2005). Online participation in the UK: testing a 'contextualised' model of internet effects. *British Journal of Politics and International Relations*, 7, 561–583.

Gil de Zúñiga, H., Puig-I-Abril, E. and Rojas, H. (2009). Weblogs, traditional sources online and political participation: an assessment of how the internet is changing the political environment. *New Media and Society*, 11(4), 553–574.

Hansard Society (2013). *Audit of Political Engagement 10*. London: Hansard Society.

Hirzalla, F. and van Zoonen, L. (2011). Beyond the online/offline divide: how youth's online and offline civic activities converge. *Social Science Computer Review*, 29, 481–498.

http://epp.eurostat.ec.europa.eu/portal/page/portal/information_society/data/database.

Jones, C., Ramanau, R., Cross, S. and Healing, G. (2010). Net generation or digital natives: is there a distinct new generation entering university? *Computers and Education*, 54, 722–732.

Kennedy, G., Dalgarno, B., Gray, K., Judd, T., Waycott, J., Bennett, S., . . . Churchward, A. (2007). The net generation are not bit users of Web 2.0 technologies: preliminary findings. In *ICT: Providing Choices for Learners and Learning. Proceedings Ascilite Singapore 2007*. http://www.ascilite.org.au/conferences/singapore07/procs/kennedy.pdf.

Kraut, R., Patterson, M., Lundmark, V., Kiesler, S., Mukopadhyay, T. and Scherlis, W. (1998). Internet paradox: a social technology that reduces social involvement and psychological well-being? *American Psychologist*, 53(9), 1017–1031.

Ladd, E.C. (1999). *The Ladd Report*. New York: Free Press.

Lee, N.J., Shah, D.V. and McLeod, J.M. (2013). Processes of political socialization: a communication mediation approach to youth civic engagement. *Communication Research*, 40(5), 669–697.

Lenhart, A., Madden, M., Macgill, A.R. and Smith, A. (2007). *Teens and Social Media*. Washington, DC: Pew Internet and American Life Project. Retrieved July 2014, from http://www.pewinternet.org/files/old-media//Files/Reports/2007/PIP_Teens_Social_Media_Final.pdf.pdf.

Lenhart, A., Purcell, K., Smith, A. and Zickuhr, K. (2010). *Social Media and Mobile Internet Use among Teens and Young Adults*. Washington, DC: Pew Internet and American Life Project. Retrieved July 2014, from http://www.pewinternet.org/files/

old-media//Files/Reports/2010/PIP_Social_Media_and_Young_Adults_Report_Final_ with_toplines.pdf.

Leung, L. (2009). User-generated content on the internet: an examination of gratifications, civic engagement and psychological empowerment. *New Media and Society*, 11(8), 1327–1347.

Livingstone, S., Bober, M. and Helsper, E.J. (2005). Active participation or just more information? Young people's take-up of opportunities to act and interact on the Internet. *Information, Communication and Society*, 8(3), 287–314.

Loader, B.D. (2007). Introduction. Young citizens in the digital age: disaffected or displaced? In Loader, B.D. (ed.), *Young Citizens in the Digital Age: Political Engagement, Young People and New Media* (pp. 11–19). London: Routledge.

Loader, B.D., Vromen, A. and Xenos, M.A. (2014). The networked young citizen: social media, political participation and civic engagement. *Information, Communication and Society*, 17(2), 143–150.

McLeod, J.M., Zubric, J., Keum, H., Deshpande, S., Cho, J., Stein, S., et al. (2001). Reflecting and connecting: testing a communication mediation model of civic participation. Paper presented to the annual convention of the Association for Education in Journalism and Mass Communication, August, Washington, DC.

McQuail, D. (1994). The rise of media of mass communication. In McQuail, D. (ed.), *Mass Communication Theory: An Introduction* (pp. 1–29). London: Sage.

Micheletti, M. (2003). *Political Virtue and Shopping: Individuals, Consumerism, and Collective Action*. New York: Palgrave Macmillan.

Mycock, A. and Tonge, J. (2012). The party politics of youth citizenship and democratic engagement. *Parliamentary Affairs*, 65(1), 138–161.

Nie, N.H. and Erbring, L. (2000). *Internet and Society: A Preliminary Report*. Stanford, CA: Stanford Institute for the Quantitative Study of Society.

Norris, P. (2002). *Democratic Phoenix: Reinventing Political Activism*. Cambridge: Cambridge University Press.

O'Reilly, T. (2005). What is Web 2.0: design patterns and business models for the next generation of software. Retrived July 2014 from http://oreilly.com/web2/archive/what-is-web-20.html.

Östman, J. (2012). Information, expression, participation: how involvement in user generated content relates to democratic engagement among young people. *New Media and Society*, 14(6), 1004–1021.

Owen, D. (2006). The Internet and youth civic engagement in the United States. In Oates, S., Owen, D. and Gibson, R.K. (eds), *The Internet and Politics* (pp. 20–38). New York: Routledge.

Pasek, J., Kenski, K., Romer, D. and Jamieson, K.H. (2006). America's youth and community engagement: how use of mass media is related to civic activity and political awareness in 14- to 22-year-olds. *Communication Research*, 33(3), 115–135.

Pasek, J., More, E. and Romer, D. (2009). Realizing the social internet? Online social networking meets offline civic engagement. *Journal of Information Technology and Politics*, 6, 197–215.

Pattie, C., Seyd, P. and Whiteley, P. (2004). *Citizenship in Britain: Values, Participation and Democracy*. Cambridge: Cambridge University Press.

Polat, R.K. (2005). The Internet and political participation: exploring the explanatory links. *European Journal of Communication*, 20(4), 435–459.

Prensky, M. (2001). Digital natives, digital immigrants Part 1. *On the Horizon*, 9(5), 1–6.

Putnam, R.D. (1995). Tuning in, tuning out: the strange disappearance of social capital in America. *PS: Political Science and Politics*, 28(4), 664–683.

Putnam, R.D. (2000). *Bowling Alone: The Collapse and Revival of American Community*. New York: Simon & Schuster.

Quintelier, E. and Vissers, S. (2008). The effect of internet use on political participation: an analysis of survey results for 16-year-olds in Belgium. *Social Science Computer Review*, 26, 411–427.

Rideout, V.J., Foehr, U.G. and Roberts, D.F. (2010). *Generation M2: Media in the Lives of 8- to 18-Year-Olds*. A Kaiser Family Foundation Study. Menlo Park, CA: Henry J. Kaiser Family Foundation.

Ruggiero, T.E. (2000). Uses and gratifications theory in the 21st century. *Mass Communication and Society*, 3(1), 3–37.

Shah, D.V., Cho, J., Eveland, W.P. and Kwak, N. (2005). Information and expression in a digital age: modeling internet effects on civic participation. *Communication Research*, 32, 531–565.

Shah, D.V., Kwak, N. and Holbert, R.L. (2001). 'Connecting' and 'disconnecting' with civic life: patterns of internet use and the production of social capital. *Political Communication*, 18, 141–162.

Shapiro, L.A.S. and Margolin, G. (2014). Growing up wired: social networking sites and adolescent psychosocial development. *Clinical Child and Family Psychology Review*, 17, 1–18.

Sloam, J. (2012). Introduction: youth, citizenship and politics. *Parliamentary Affairs*, 65(1), 4–12.

Stepick, A., Stepick, C.D. and Labissiere, C.Y. (2008). South Florida's immigrant youth and civic engagement. *Applied Developmental Science*, 12(2), 57–65.

Stolle, D. and Hooghe, M. (2011). Shifting inequalities: patterns of exclusion and inclusion in emerging forms of political participation. *European Societies*, 13(1), 119–142.

Tapscott, D. (1998). *Growing Up Digital: The Rise of the Net Generation*. New York: McGraw-Hill.

Tapscott, D. (2008). *Grown Up Digital: How the Net Generation is Changing Your World*. New York: McGraw-Hill.

Van Deth, J.W. (2014). A conceptual map of political participation. *Acta Politica*. Advance online publication. doi: 10.1057/ap.2014.6.

Velasquez, A. and LaRose, R. (2014). Youth collective activism through social media: the role of collective efficacy. *New Media and Society*. Advance online publication. doi: 10.1177/1461444813518391.

Ward, S., Gibson, R. and Lusoli, W. (2003). Online participation and mobilisation in Britain: hype, hope and reality. *Parliamentary Affairs*, 56, 652–668.

Wei, L. and Yan, Y. (2010). Knowledge production and political participation: reconsidering the knowledge gap theory in the Web 2.0 environment. Paper presented at the international conference on Information Management and Evaluation, March, Cape Town, South Africa.

Xenos, M. and Moy, P. (2007). Direct and differential effects of the Internet on political and civic engagement. *Journal of Communication*, 57, 704–718.

Xenos, M.A., Vromen, A. and Loader, B. (2014). The great equalizer? Patterns of social media use and youth political engagement in three advanced democracies. *Information, Communication and Society*, 17(2), 151–167.

Zukin, C., Keeter, S., Andolina, M., Jenkins, K. and Delli Carpini, M.X. (2006). *A New Engagement? Political Participation, Civil Life, and the Changing American Citizen*. New York: Oxford University Press.

PART IV

POLITICAL TALK

14. Everyday political talk in the Internet-based public sphere
Todd Graham

Ever since the advent of the Internet, political communication scholars have debated its potential to facilitate and support public deliberation as a means of revitalizing and extending the public sphere (Coleman and Blumler, 2009; Dahlberg, 2001a; Dahlgren, 2005; Papacharissi, 2002; Sunstein, 2002). In a time of growing cynicism and disillusionment towards politics, we have seen across Western democracies erosions of trust and engagement in political and media systems (see, for example, Brants, 2012; Coleman, 2012; Coleman and Blumler, 2009). As Blumler and Coleman (2001) argue, the current political communication structures that make up the public sphere are poorly serving certain democratic values, such as 'opportunities for committed advocacy, rounded dialogue, sustained deliberation, and especially the provision of incentives for citizens to learn, choose, and become involved in, rather than merely to follow and kibbitz [chat] over, the political process' (p. 8). The belief that the Internet may play a significant role in reducing some of this deliberative deficit has generated significant interest in the possible benefits and drawbacks of online communication. Much of the debate has focused on the medium's potential in offering communicative spaces that transcend the limitations of time, space and access (and the traditional mass media) whereby open communication, deliberation and exchange of information among the public can prosper.

Following the initial enthusiasm over the possibilities of a more interactive and deliberative electorate, along with the cyber-pessimist response, a growing body of rich empirical research into online deliberation has arisen in its wake. In search of online deliberation, scholars have conducted a broad range of investigations, developing several prominent directions in the field. One popular line of research has been the study of informal political talk through the lens of public sphere ideals. Drawing from the work of Jürgen Habermas and other deliberative democratic theorists (for example, Barber, 1984; Bohman, 1996; Dryzek, 2000), researchers have analysed and assessed the extent to which various online forums and communicative practices approximate deliberative ideals.

Why should scholars bother studying everyday political talk online?

247

The focus on such talk is based on the belief that at the heart of civic culture should be a talkative public and that the Internet affords citizens the communicative space necessary to rehearse and debate the pressing political and societal issues of the day. It is through ongoing participation in informal political talk whereby citizens become aware of other opinions, discover the important issues of the day, test new ideas, and develop and clarify their preferences. It is such talk, which takes place over time and across different spaces, that prepares citizens, the public sphere and the political system at large for political action. Thus, understanding political talk that occurs in these spaces is necessary because of its links with the deliberative system in general and other forms of political engagement specifically.

The aim of this chapter is to detail and discuss this growing body of research and its significance. First, I begin by discussing what scholars mean by political talk and why it is thought to be essential for (a more deliberative) democracy. Following this, the major findings to date are set out, focusing specifically on three of the most common features of political talk investigated by scholars in the field. I discuss scholarly disagreement and offer my thoughts and critical reflection on the topic. Finally, the chapter ends with several recommendations for future research into informal political talk in the Internet-based public sphere.

WHY EVERYDAY POLITICAL TALK MATTERS

One of the most common questions I am asked when explaining my research to students is: why study everyday political talk? As one of my former students reacted, 'It's just talk, right? It's not as though it really leads to anything meaningful'. Many of us are even taught to avoid discussing politics.[1] So, then, why does everyday political talk matter? In order to address this question, we first need to situate the role of political talk within the broader notion of the public sphere.

The concept of the public sphere has become synonymous with the work of Habermas, (1989[1962]) initially developed in *The Structural Transformation of the Public Sphere*. His evolving theory (1984 [1981], 1987 [1985], 1990 [1983], 1996 [1992], 2006), which is situated within the broader theory of deliberative democracy, provides, as Dahlberg (2004) argues, one of the richest and most systematically developed critical theories of the public sphere. It is conceptualized as the realm of social life – separate from political and economic interests – where the exchange of information, ideas and positions on the discovery and questions of the common good take place. The public sphere 'springs into being' when private citizens come together freely to debate openly the political and social issues of the day. Central to the concept is the promotion and cultivation of rational-critical discourse through the active reasoning of the public. It is through the ongoing development of such discourse that public opinion is formed, which in turn guides the political system.

In contemporary societies, however, the public sphere is highly complex and opinion formation occurs in a variety of interacting publics across a multitude of spaces at varying levels. As Habermas (1996 [1992]) himself states:

> It represents a highly complex network that branches out into a multitude of overlapping international, national, regional, local, and subcultural arena. Functional specifications, thematic foci, policy fields, and so forth, provide the points of reference for a substantive differentiation of public spheres that are, however, still accessible to laypersons (for example, popular science and literary publics, religious and artistic publics, feminist and 'alternative' publics, publics concerned with health-care issues, social welfare, or environmental policy). Moreover, the public sphere is differentiated into levels according to the density of communication, organizational complexity, and range – from the episodic publics found in taverns, coffee houses, or on the streets, through the occasional or 'arranged' publics of particular presentations and events, such as theatre performances, rock concerts, party assemblies, or church congresses; up to the abstract public sphere of isolated readers, listeners, and viewers scattered across larger geographic areas, or even around the globe, and brought together only through mass media. (pp. 373–374)

Even with such complexity the public sphere, or rather the network of public spheres, can be conceptualized via four fundamental conditions.[2] First, it needs active citizens who engage not only in institutional forms of political participation, such as voting, but more importantly in public deliberation on the relevant societal and political issues of the day. Second, it requires autonomous communicative spaces (free from both state and commercial influence) whereby citizens can engage freely and openly in public deliberation. Third, it requires the mass media. Journalism's role is not only to fuel public debate by providing information and keeping a critical eye on government and corporate affairs, but also to encourage, facilitate and act as a platform for it. Finally, there is deliberation, the guiding communicative form of the public sphere, which stipulates the structural, procedural and dispositional requirements of the process.

Now, we should make the distinction between (formal) deliberation on one hand and everyday political talk on the other. For many scholars, deliberation is a normative concept, which is guided by the principle of rationality based on a set of norms and rules oriented towards the common good aimed at achieving a rationally motivated consensus. However, this kind of deliberation seems inappropriate for the everyday commutative spaces of the public sphere. Such spaces are not bound to any formal agendas or outcomes, and political talk that emerges in these spaces is often spontaneous and tends to lack any direct purpose outside the purpose of talk for talk's sake. Unlike deliberation within public decision-making bodies, everyday political talk is not necessarily aimed at decision-making or other forms of political action, but rather is often expressive in nature (Mansbridge, 1999: 212).

However, everyday political talk is not meaningless simply because it does not typically lead to immediate or direct political action. On the contrary, there is a growing body of evidence that suggests talking about politics can increase levels of political knowledge, civic engagement, exposure and tolerance to differencing perspectives, and facilitate preference change (Bennett et al., 2000; Coleman and Blumler, 2009; Conover et al., 2001; Eveland, 2004; Kim et al., 1999; McClurg, 2003; Monnoyer-Smith, 2006; Price and Cappella, 2002). It is through such talk whereby citizens achieve mutual understanding about each other and the political and societal problems (and solutions) they face. It is the web of informal political conversations, conducted over time and across and between the multitude of levels and spaces, which fosters public opinion, preparing citizens and the political system at large for political action.[3]

ANALYSING AND ASSESSING ONLINE POLITICAL TALK

Over the past two decades, the field of online deliberation has developed into one of the central areas of interest among research on the public sphere and deliberative democracy. It now covers a variety of research agendas, which include comparisons between face-to-face and online deliberation (Wojcieszak et al., 2009; Baek et al., 2012); the use of online consultations (Albrecht, 2006; Åström and Grönlund, 2012; Coleman, 2004; Fishkin, 2009; Karlsson, 2012; Kies, 2010; Winkler, 2005); moderation and the design of forums (Bendor et al., 2012; Edwards, 2002; Wright, 2009; Wright and Street, 2007); the extent to which forums facilitate contact between opposing perspectives (Brundidge, 2010; Stromer-Galley, 2003; Wojcieszak and Mutz, 2009); and the effects of online deliberation on civic engagement (Price and Cappella, 2002). One of the most popular lines of research, however, has been the study of informal political talk through the lens of deliberative ideals.

Researchers have investigated informal political talk in a variety of online forums, which include Usenet newsgroups (Davis, 2005; Hill and Hughes, 1998; Papacharissi, 2004; Schneider, 1997; Wilhelm, 1999;

Zhang et al., 2013); news media-sponsored forums such as newspapers (Graham, 2010b; Schultz, 2000; Strandberg, 2008; Tanner, 2001; Tsaliki, 2002); forums hosted by political parties and governments, excluding e-consultations (Dunne, 2009; Graham and Witschge, 2003; Hagemann, 2002; Jankowski and Van Os, 2004; Winkler, 2005); online deliberative initiatives (Dahlberg, 2001b); comparisons between different types (Brants, 2002; Graham, 2011; Jensen, 2003); 'third spaces', that is, nonpolitical forums (Graham, 2008, 2010a, 2012a; Graham and Wright, 2014); other platforms such as chat (Stromer-Galley and Martinson, 2009), blogs (Koop and Jansen, 2009) and readers' comments (Graham, 2012b; Ruiz et al., 2011); and social media network sites such as Facebook and YouTube (Halpern and Gibbs, 2013; Robertson et al., 2010).[4] Studies here focus on measuring the deliberativeness of political talk as a means of determining the extent to which the Internet is conducive to (particular) conditions of deliberation. Namely, researchers construct a set of criteria that embody the ideal of deliberation, which are then operationalized into measurable concepts and employed in an empirical analysis. Habermas's work – especially his theory of communicative rationality and discourse ethics – has been highly influential in this process. However, there has been a lack of consistency among researchers regarding the conditions used for evaluation, which is partly due to different interpretations of Habermas's work.[5] This has made making comparisons between findings difficult, along with the different contexts and types of forums analysed. Despite this, the criteria used can be categorized into four elements of deliberation: the communicative form, dispositional requirements, norms (for example, of equality and diversity) and outcomes.[6]

First, scholars have been consumed by measuring the quality of rational-critical debate (see, for example, Graham, 2008; Graham and Witschge, 2003; Jensen, 2003; Kies, 2010; Stromer-Galley, 2007; Wilhelm, 1999). Researchers have focused on various elements such as gauging the level and quality of rationality, critical reflection, types of argumentation and the use of supporting evidence. There has also been a focus on coherence (thematic consistency) of online debates and the extent to which arguments are grounded in the common good as opposed to particular interests. Second, researchers have investigated the dispositional requirements of listening and understanding. As Barber (1984) maintains, deliberation requires 'listening as well as speaking, feeling as well as thinking and acting as well as reflecting' (p. 178). Studies have analysed the level of reciprocity: the extent to which participants read and reply to each other's posts. Some have investigated deeper levels of understanding such as reflexivity and empathy. Third, another popular element has been the (social) norms of deliberation. Online debates have been assessed for discursive equality (for example, distribution of participation, acts of inequality and mutual respect), diversity of opinions and the level of sincerity. Finally, scholars have attempted to analyse the outcomes of online debates by measuring the level of continuity (extended debate), commissive speech acts (acknowledging the better argument) and convergence of opinions. I will now take a closer look at three of the most commonly used conditions, which are reciprocity, discursive equality and diversity of opinions.

Reciprocity

Reciprocity has been one of the most common conditions operationalized by scholars. Although there are some variations in its conceptualization (as discussed below), simply put, it requires participants to listen and respond to each other. As Schneider (1997) states, 'Reciprocity is an important consideration in assessing the public sphere because it indicates the degree to which participants are actually interacting with each other, and working on identifying their own interests with those of the group, as opposed to talking past each other or engaging in simple bargaining or persuasion' (p. 105).

Early commentators questioned whether the Internet would facilitate or impede reciprocal exchange. Kolb (1996), for example, argued that the rhythm of the Internet is ideal for Habermasian dialogue. The asynchronous nature of forums affords participants the time needed to read and react to arguments of other participants. Other scholars were less than optimistic. Streck (1998), for example, argued that the Internet 'elevates the right to speak above all others, and all but eliminates the responsibility to listen' (pp. 45–46). Similarly, Schultz (2000) maintained, 'A new discipline is required since the Internet involves a great temptation to publish and communicate too much, which consequently weakens the overall significance and excludes many people just because they cannot keep up and cannot get through the dense communicative jungle' (p. 219).

Early empirical research into the level of reciprocal exchange indicated that online discussion forums tended to foster shouting matches. With less than one in five messages representing a reply, Wilhelm (1999) concluded that rather than listening participants of Usenet newsgroups used the forums to amplify their own views. Davis (2005) similarly concluded that 'people often talk past one another when they are not verbally attacking each other' (p. 67). Research by Hagemann (2002), Jankowski and Van Os (2004) and Strandberg (2008) all revealed similar findings. Most of the empirical evidence, however, suggests a different story: political talk online tends to be reciprocal (Brants, 2002; Dahlberg, 2001b; Jensen, 2003; Papacharissi, 2004; Schneider, 1997; Tsaliki, 2002; Winkler, 2005).

My own research, which has covered a variety of forum types from news media message boards to political talk in third spaces, has revealed that participants tend to engage with each other rather than talk past one another (Graham, 2010a, 2010b, 2011, 2012a; Graham and Witschge, 2003; Graham and Wright, 2014). For example, my analysis of over 3000 posts from the reader comment sections of the *Guardian* found that more than half of the posts were engaged in reciprocal exchange, which was typically rational and critical in nature (Graham, 2012b).

The conflicting findings can partly be explained by differing conceptualizations of reciprocity. For some scholars, it took on a broader meaning and included the dispositional requirement of reflexivity. These studies reported a lower level of reciprocity, which is no surprise given the (more demanding) requirements (for example, Hagemann, 2002; Jankowski and Van Os, 2004). In other cases, the concept was attached to the communicative form (rational-critical debate) and was too narrow in focus. For example, Strandberg (2008) assessed reciprocity by measuring the level of messages coded as questions, thereby excluding other types of responses. Moreover, many of the studies counted messages as replies based solely on the structural markers of the forum (for example, posts embedded in other posts) without reading the content of the posts. Consequently, it is unclear whether participants were actually listening. Furthermore, most studies do not take the context of the thread into account, coding simply at the level of the post. This is problematic because, for example, all messages in a thread could be replies, but replies to the original seed post, falling well short of the deliberative ideal (see Graham, 2008).

Discursive Equality

Discursive equality, not to be mistaken for inclusion (that is, access to the debate), refers to the normative claim that all participants are equal. It requires the rules and guidelines that coordinate and maintain the process of deliberation to not privilege one individual or group of individuals over another; participants to have an equal opportunity to express their views and question the position of others; and participants to respect and recognize each other as having an equal voice and standing within this process. Past studies have typically focused on two aspects: analysing the formal and informal rules, and management of those rules for acts of exclusion and inequality; and identifying communicative exclusions and inequalities that occur in communicative practices. The former has been applied in many e-consultation studies (and those that focus on design), and the latter has been the primary focus for analysing informal political talk.

For some early commentators, the Internet's ability to allow for anonymity and the breakdown of social cues was seen as liberating. As Agre (2002) explains, 'Conventional markers of social difference (gender, ethnicity, age, rank) are likewise held to be invisible, and consequently it is contended that the ideas in an online message are evaluated without the prejudices that afflict face-to-face interaction' (p. 314). However, these same liberating characteristics were viewed by others in a less promising light. For example, Barber (1998) questioned whether deliberation within the public sphere could be 'rekindled on the net, where identities can be concealed and where flaming and other forms of incivility are regularly practiced' (p. 269). Issues concerning deception and flaming gained much attention.[7]

In the past, discursive equality has been examined from two angles: distribution of voice and substantial equality. The most common measurement has been the equal distribution of voice indicator. Researchers measure the number of participants along with their share of the postings, thereby determining the concentration of participation. Forums that maintain a distribution of voice skewed towards a small group of frequent posters are considered to be discursively unequal because they threaten the participatory opportunities of other participants. One of the most common findings has been that online forums typically feature highly active minorities of content creators (Brants, 2002; Dahlberg, 2001b; Dunne, 2009; Graham, 2010a, 2010b; Jankowski and Van Os, 2004; Jensen, 2003; Koop and Jansen, 2009; Robertson et al., 2010; Schultz, 2000; Winkler, 2005). For example, Schneider's (1997) analysis of newsgroups found that only 5 per cent of participants accounted for 80 per cent of the messages, indicating substantial inequalities in the rate and distribution of participation.

The problem with many of these studies, as Scott Wright and I have argued elsewhere (Graham and Wright, 2014), is that they tend to make assumptions about the behaviour of active minorities (super participants): that they dictate the agenda and style of debate. However, most studies do not examine in any great detail (if at all) their communicative practices (outside the frequency of participation). On those rare occasions when researchers have analysed the communicative practices of active minorities, they have found their impact to be largely positive (Albrecht, 2006; Kies, 2010). For example, our research found that contrary to conceived wisdom, most did not attempt to stop other users from posting, or attack them, but rather they performed a range of positive roles including helping other users, summarizing longer threads, being empathetic towards others' problems, and engaging in (largely) rational-critical debate (Graham and Wright, 2014).

Substantial equality requires that participants respect and recognize each other as having an equal voice and standing in the deliberative process. One common approach has been to analyse forums for acts of exclusion and inequality through identifying instances of aggressive and abusive posting behaviour. Hill and Hughes (1998), for example, examined newsgroups for flaming: 'personal, ad hominem attacks that focus on the individual poster not the ideas of the message' (p. 52). They found that more than a third of all threads ended in 'flame-fests'. Studies conducted by Davis (2005), Jankowski and Van Os (2004), Wilhelm (1999) and Zhang et al. (2013) all report similar findings, pointing to anonymity, and the lack of norms and moderation as the contributing factors to such behaviour (see also Strandberg, 2008).

However, we need to be careful when using these findings as an indicator of the state of informal political talk online. First, these studies focus almost exclusively on newsgroups (and very partisan ones at that), which have a reputation of being attack-orientated (Wright, 2012a). Many of these studies too were conducted during the late 1990s, a time when Net-etiquette and communicative norms were just developing (and the number of Internet users was significantly lower). Moreover, once we move away from partisan-based newsgroups, a somewhat different picture emerges. For example, my own research of reader comment sections and reality TV, news media and government-sponsored forums found that in all cases degrading exchanges – to lower in character, quality, esteem or rank another participant and/or participant's position – represented less than 15 per cent of posts (Graham, 2010a, 2010b, 2012a, 2012b; Graham and Witschge, 2003; Graham and Wright, 2014). These findings are supported by other studies of various forum types (Hagemann, 2002; Halpern and Gibbs, 2013; Jensen, 2003; Papacharissi, 2004; Rowe, 2014; Ruiz et al., 2011; Winkler, 2005).

Diversity of Opinion

One common concern since the arrival of the Internet has been the fragmentation of the public sphere. The most prominent advocate of this position is Cass Sunstein (2002). He claims that because the Internet eliminates geographical boundaries and makes it easier for individuals with similar views to find one another, that this will lead to echo-chambers of like-minded individuals, thus fostering the polarization of opinions and widening the gap between extreme positions. Do Internet forums foster homogeneity or diversity of opinions?

As we might expect, the findings reveal a somewhat mixed account. Wilhelm's (1999) study, for example, concluded that newsgroups represented 'communities of interest, virtual gathering places in which those people who share a common interest can discuss issues without substantial transaction or logistical costs' (p. 171). Similar findings were reported by Davis (2005) and Hill and Hughes (1998). Conversely, several other studies of newsgroups and/or news media message boards found that such spaces hosted a diversity of opinions (Graham, 2011; Schneider, 1997; Strandberg, 2008; Tsaliki, 2002). For example, Stromer-Galley's (2003) interviews with participants revealed that people not only meet and engage with different points of view online, but also actively seek it out.

One of the problems with the debate surrounding the diversity of opinion (and the empirical evidence) is that almost all the commentary and studies focus on political communicative spaces. However, informal political talk is not bound to these spaces. For example, my analysis of the *Big Brother* and *Wife Swap* forums found that such spaces hosted not only political talk (accounting for 22 per cent and 32 per cent of the postings, respectively), but also a diversity of opinions (Graham, 2010a, 2012a; see also Wojcieszak and Mutz, 2009). Participants of third spaces are not there to talk politics. Therefore, when it does emerge, the chances are for greater diversity of opinions (Graham and Harju, 2011). Thus, the fragmentation theory makes little sense once we move beyond the political communicative landscape. It is beyond such spaces where political discussion grounded in diversity is more likely to be found.

Another problem with this debate is that it puts too much emphasis on the exchange of dissimilar perspectives when analysing individual cases, thereby neglecting the deliberative system; that is, there is room for online spaces that foster both diversity and uniformity of opinions. Contact with opposing perspectives is certainly necessary, but it does not always need to happen at the level of the forum. It can also occur between various discourses that develop online in the so-called enclaves of like-minded individuals. Moreover, these spaces allow minority discourses to develop. For example, some theorists have argued that political talk in such spaces allows for the development of in-group strategies and narratives, thus increasing the chance that some discourses are expressed publicly (see, for example, Barber, 1984; Fraser, 1990; Mansbridge, 1999).

CONCLUSION

More than a decade ago, Lincoln Dahlberg (2004) reflected critically on the state of online deliberation research, calling for a new agenda that would move 'beyond the first phase' of empirical analysis. Since then we have seen not only an increase in the number of studies, but also the development of new approaches and the fine-tuning of past ones. As a result, the field has blossomed, expanding our knowledge of the phenomena and developing into one of the main areas of interest among research on the public sphere and deliberative democracy.

However, with the rise of social media and the participatory culture that has followed in its wake, the need to develop innovative ways of studying everyday political talk is apparent. Although the field has grown, much of the research still, for example, focuses on case studies of discussion forums; research which explores political talk on new social media networks, and more importantly, across networks is scant. Regarding the latter, we need to start investigating the flow and clash of discourses that take place online (and offline, and between the two), examining the connections and relationships between different discursive spaces.[8] This type of research is already on its way, though usually not grounded in deliberative ideals. For example, there has been innovative work on the flow of discourse between the Twittersphere and mainstream news media (see for example, Broersma and Graham, 2012). One possible way forward is to analyse how a particular issue develops and flows across several networks, tracking and examining the overflow and clash of discourses between them. This type of research would provide us with much-needed insight into the (online) deliberative system.

Another priority should be to explore the extent to which online political talk contributes to meaningful political action. As Coleman and Moss (2012) have argued, 'for most online deliberation researchers it seems as if the political process ends when civic talk stops' (p. 11). Does engaging in political talk within such spaces support a movement towards participation in the formal political process? What we need, to take online deliberation research forward, is longitudinal and ethnographic studies which focus on how political talk (in both political and non-political spaces) transfers into participation in the political process and/or collective action in the public sphere.

As discussed above, there also is a clear need to investigate and explore online political talk in third spaces (see also Coleman and Blumler, 2009; Wright, 2012a, 2012b). As Kees Brants (2002) had already argued over a decade ago, 'politics online is e-verywhere', and this certainly includes everyday political talk. By focusing exclusively on political discussion forums, we are left with an incomplete picture or, worse, a distorted one. Moreover, these spaces are meaningful because they open up windows for researchers to explore and begin to understand the ways in which citizenry is intertwined with aspects and practices of everyday life.

In addition to better understanding the frequency and quality of political talk that emerges in third spaces, researchers should investigate what triggers such talk in everyday conversation and to what extent such spaces foster political action.[9] Regarding the former, how do people in the course of everyday conversation make connections to formal politics? Regarding the latter, are the political discussions that take place in such spaces characterized by argument and debate (as often found in political forums), or do people utilize the potential of third spaces to provide support, networking and community building, leading to various forms of political action?

NOTES

1. See for example Eliasoph's (1998) research on why people avoid political talk in (face-to-face) public settings.
2. See Graham (2009: 8–11) for a more comprehensive account of these conditions.
3. See Goodin (2008) and Mansbridge (1999) for a discussion on the deliberative system; that is, viewing deliberation as a broader process, which is spread throughout time and space.
4. Note that both Strandberg (2008) and Tsaliki (2002) examined a mix of forum types (not intended for comparison), but forums hosted by newspapers were the dominant type. Winkler (2005) analysed both consultations and a forum hosted by the European Union. Jankowski and Van Os's (2004) study included a government-sponsored forum where local politicians participated. Davis (2005) also investigated chats and blogs.
5. See Dahlberg (2004) for a critical account of past frameworks used for empirical analysis.
6. See Graham (2008), Kies (2010) and Stromer-Galley (2007) for three fruitful sets of deliberative measures.
7. One of the most common criticisms lodged against readers' comments by journalists is that they tend to foster abusive and aggressive posting behaviour (flaming). As Santana (2014: 19) points out, this is often blamed on anonymity: 'the pervasiveness of the incivility' that supposedly plagues readers' comments has reached 'fever pitch' among 'a rising chorus of journalists and industry observers' calling 'for the end of anonymous comments'. In response, many news organizations have recently stopped anonymous comments, while others restrict the number of stories opened to comments.
8. See also Wright's (2012b) argument.
9. See research by Graham and Harju (2011) and Graham et al. (forthcoming).

FURTHER READING

Coleman, S. and Blumler, J.G. (2009). *The Internet and Democratic Citizenship: Theory, Practice and Policy*. Cambridge, UK: Cambridge University Press.
Dahlberg, L. (2001a). The Internet and democratic discourse: exploring the prospects of online deliberative forums extending the public sphere. *Information, Communication and Society*, 4, 615–633. DOI:10.1080/13691180110097030.
Graham, T. (2008). Needles in a haystack: a new approach for identifying and assessing political talk in nonpolitical discussion forums. *Javnost – The Public*, 15(2), 5–24.
Graham, T. and Harju, A. (2011). Reality TV as a trigger of everyday political talk in the net-based public sphere. *European Journal of Communication*, 26, 18–32. DOI:10.1177/0267323110394858.
Graham, T. and Wright, S. (2014). Discursive equality and everyday talk online: the impact of 'super-participants'. *Journal of Computer-Mediated Communication*, 19, 624–642. DOI:10.1111/jcc4.12016.
Mansbridge, J. (1999). Everyday talk in the deliberative system. In Macedo, S. (ed.),

Deliberative Politics: Essays on Democracy and Disagreement (pp. 211–239). Oxford: Oxford University Press.

Wright, S. (2012a). Politics as usual? Revolution, normalization and a new agenda for online deliberation. *New Media and Society*, 14, 244–261. doi:10.1177/1461444811410679.

Wright, S. (2012b). From 'third place' to 'third space': everyday political talk in non-political online spaces. *Javnost – The Public*, 19(3), 5–20.

REFERENCES

Agre, P.E. (2002). Real-time politics: the Internet and the political process. *Information Society*, 18, 311–331. DOI:10.1080/01972240290075174.

Albrecht, S. (2006). Whose voice is heard in online deliberation? A study of participation and representation in political debates on the Internet. *Information, Communication and Society*, 9, 62–82. doi:10.1080/13691180500519548.

Åström, J. and Grönlund, Å. (2012). Online consultations in local government: what works, when and how. In Coleman, S. and Shane, P.M. (eds), *Connecting Democracy: Online Consultation and the Flow of Political Communication* (pp. 75–96). Cambridge, MA: MIT Press.

Baek, Y.M., Wojcieszak, M. and Delli Carpini, M.X. (2012). Online versus face-to-face deliberation: Who? Why? What? With what effects?. *New Media and Society*, 14, 363–383. DOI:10.1177/1461444811413191.

Barber, B.R. (1984). *Strong Democracy: Participatory Politics for a New Age*. Berkeley, CA: University of California Press.

Barber, B.R. (1998). *A Passion for Democracy: American Essays*. Princeton, NJ: Princeton University Press.

Bendor, R., Lyons, S.H. and Robinson, J. (2012). What's there not to 'like'? The technical affordances of sustainability deliberations on Facebook. *eJournal of eDemocracy and Open Government*, 4, 67–88.

Bennett, S.E., Flickinger, R.S. and Rhine, S.L. (2000). Political talk over here, over there, over time. *British Journal of Political Science*, 30, 99–119.

Blumler, J.G. and Coleman, S. (2001). *Realising Democracy Online: A Civic Commons in Cyberspace*. IPPR. Retrieved from http://www.ippr.org/publication/55/1230/realising-democracy-online-a-civic-commons-in-cyberspace.

Bohman, J. (1996). *Public Deliberation: Pluralism, Complexity and Democracy*. Cambridge, MA: MIT Press.

Brants, K. (2002). Politics is e-verywhere. *Communications: The European Journal of Communication Research*, 27, 171–188. DOI:10.1515/comm.27.2.171.

Brants, K. (2012). Trust, cynicism, and responsiveness: the uneasy situation of journalism in democracy. In Peters, C. and Broersma, M. (eds), *Rethinking Journalism: Trust and Participation in a Transformed News Landscape* (pp. 15–27). London: Routledge.

Broersma, M. and Graham, T. (2012). Social media as beat: tweets as news source during the 2010 British and Dutch elections. *Journalism Practice*, 6, 403–419. DOI:10.1080/1751278 6.2012.663626.

Brundidge, J. (2010). Encountering 'difference' in the contemporary public sphere: the contribution of the Internet to the heterogeneity of political discussion networks. *Journal of Communication*, 60, 680–700. DOI:10.1111/j.1460-2466.2010.01509.x.

Coleman, S. (2004). Connecting parliament to the public via the Internet: two case studies of online consultations. *Information, Communication and Society*, 7, 1–22. DOI:10.1080/136 9118042000208870.

Coleman, S. (2012). Believing the news: from sinking trust to atrophied efficacy. *European Journal of Communication*, 27, 25–35. DOI: 10.1177/0267323112438806.

Coleman, S. and Blumler, J.G. (2009). *The Internet and Democratic Citizenship: Theory, Practice and Policy*. Cambridge: Cambridge University Press.

Coleman, S. and Moss, G. (2012). Under construction: the field of online deliberation research. *Journal of Information Technology and Politics*, 9, 1–15. DOI:10.1080/1933168 1.2011.635957.

Conover, P.J., Searing, D.D. and Crewe, I. (2001). The deliberative potential of political discussion. *British Journal of Political Science*, 31, 21–62.

Dahlberg, L. (2001a). The Internet and democratic discourse: exploring the prospects of online deliberative forums extending the public sphere. *Information, Communication and Society*, 4, 615–633. DOI:10.1080/13691180110097030.

Dahlberg, L. (2001b). Extending the public sphere through cyberspace: the case of Minnesota e-democracy. *First Monday*, 6(3). doi:10.5210/fm.v6i3.838.

Dahlberg, L. (2004). Net-public sphere research: beyond the 'first phase'. *Javnost – The Public*, 11(1), 5–22.

Dahlgren, P. (2005). The Internet, public spheres, and political communication: dispersion and deliberation. *Political Communication*, 22, 147–162. doi:10.1080/10584600590933160.

Davis, R. (2005). *Politics Online: Blogs, Chatrooms, and Discussion Groups in American Democracy*. London: Routledge.

Dryzek, J.S. (2000). *Deliberative Democracy and Beyond: Liberals, Critics, Contestations*. Oxford: Oxford University Press.

Dunne, K. (2009). Cross cutting discussion: a form of online discussion discovered within local political online forums. *Information Polity*, 14, 219–232. DOI:10.3233/ IP-2009-0177.

Edwards, A.R. (2002). The moderator as an emerging democratic intermediary: the role of the moderator in Internet discussions about public issues. *Information Polity*, 7, 3–20.

Eliasoph, N. (1998). *Avoiding Politics: How Americans Produce Apathy in Everyday Life*. Cambridge: Cambridge University Press.

Eveland, W.P., Jr. (2004). The effect of political discussion in producing informed citizens: the roles of information, motivation, and elaboration. *Political Communication*, 21, 177–193. DOI:10.1080/10584600490443877.

Fishkin, J.S. (2009). *When the People Speak: Deliberative Democracy and Public Consultation*. Oxford: Oxford University Press.

Fraser, N. (1990). Rethinking the public sphere: a contribution to the critique of actual existing Democracy. *Social Text*, 25–26, 56–80. doi:10.2307/466240.

Goodin, R.E. (2008). *Innovating Democracy: Democratic Theory and Practice after the Deliberative Turn*. Cambridge: Cambridge University Press.

Graham, T. (2008). Needles in a haystack: a new approach for identifying and assessing political talk in nonpolitical discussion forums. *Javnost – The Public*, 15(2), 5–24.

Graham, T. (2009). What's *Wife Swap* got to do with it? Talking politics in the net-based public sphere. Doctoral dissertation. Retrieved from Universiteit van Amsterdam Digital Repository (314852).

Graham, T. (2010a). Talking politics online within spaces of popular culture: the case of the *Big Brother* forum. *Javnost – The Public*, 17(4), 25–42.

Graham, T. (2010b). The use of expressives in online political talk: impeding or facilitating the normative goals of deliberation?. In Tambouris, E., Macintosh, A. and Glassey, O. (eds), *Electronic Participation* (pp. 26–41). Berlin: Springer.

Graham, T. (2011). What's reality television got to do with it? Talking politics in the net-based public sphere. In Brants, K. and Voltmer, K. (eds), *Political Communication in Postmodern Democracy: Challenging the Primacy of Politics* (pp. 248–264). Basingstoke: Palgrave Macmillan.

Graham, T. (2012a). Beyond 'political' communicative spaces: talking politics on the *Wife Swap* discussion forum. *Journal of Information Technology and Politics*, 9, 31–45. DOI:10 .1080/19331681.2012.635961.

Graham, T. (2012b). Talking back, but is anyone listening? Journalism and comment fields. In Peters, C. and Broersma, M. (eds), *Rethinking Journalism: Trust and Participation in a Transformed News Landscape* (pp. 114–127). London: Routledge.

Graham, T. and Harju, A. (2011). Reality TV as a trigger of everyday political talk in the net-based public sphere. *European Journal of Communication*, 26, 18–32. DOI:10.1177/0267323110394858.

Graham, T., Jackson, D. and Wright, S. (forthcoming). From everyday conversation to political action: talking austerity in online 'third spaces'. *European Journal of Communication*.

Graham, T. and Witschge, T. (2003). In search of online deliberation: towards a new method for examining the quality of online discussions. *Communications: The European Journal of Communication Research*, 28, 173–204. DOI:10.1515/comm.2003.012.

Graham, T. and Wright, S. (2014). Discursive equality and everyday talk online: the impact of 'super-participants'. *Journal of Computer-Mediated Communication*, 19, 624–642. DOI:10.1111/jcc4.12016.

Habermas, J. (1984 [1981]). *The Theory of Communicative Action. Volume One. Reason and the rationalization of Society*. McCarthy, T. (trans.). Boston, MA: Beacon Press.

Habermas, J. (1987 [1985]). *The Theory of Communicative Action Volume Two. Lifeworld and System: A Critique of Functionalist Reason*. McCarthy, T. (trans.). Boston, MA: Beacon Press.

Habermas, J. (1989 [1962]). *The Structural Transformation of the Public Sphere: An Inquiry into a Category of Bourgeois Society*. Berger, T. and Lawrence, F. (trans.). Cambridge, MA: Polity Press.

Habermas, J. (1990 [1983]). *Moral Consciousness and Communicative Action*. Lenhardt, C. and Nicholsen, S.W. (trans.). Cambridge, MA: MIT Press.

Habermas, J. (1996 [1992]). *Between Facts and Norms: Contributions to a Discourse Theory of Law and Democracy*. Rehg, W. (trans.). Cambridge, MA: MIT Press.

Habermas, J. (2006). Political communication in media society: does democracy still enjoy an epistemic dimension? The impact of normative theory on empirical research. *Communication Theory*, 16, 411–426. DOI:10.1111/j.1468-2885.2006.00280.x.

Hagemann, C. (2002). Participation in and contents of two political discussion lists on the Internet. *Javnost – The Public*, 9(2), 61–76.

Halpern, D. and Gibbs, J. (2013). Social media as a catalyst for online deliberation? Exploring the affordances of Facebook and YouTube for political expression. *Computers in Human Behavior*, 29, 1159–1168. DOI:10.1016/j.chb.2012.10.008.

Hill, K.A. and Hughes, J.E. (1998). *Cyberpolitics: Citizen Activism in the Age of Internet*. New York: Rowman & Littlefield.

Jankowski, N.W. and Van Os, R. (2004). Internet-based political discourse: a case study of electronic democracy in Hoogeveen. In Shane, P.M. (ed.), *Democracy Online: The Prospects for Democratic Renewal through the Internet* (pp. 181–194). New York: Taylor & Francis.

Jensen, J.L. (2003). Public spheres on the Internet: anarchic or government sponsored – a comparison. *Scandinavian Political Studies*, 26, 349–374. DOI:10.1111/j.1467-9477.2003.00093.x.

Karlsson, M. (2012). Understanding divergent patterns of political discussion in online forums – evidence from the European Citizens' Consultation. *Journal of Information Technology and Politics*, 9, 64–81. DOI:10.1080/19331681.2012.635965.

Kies, R. (2010). *Promises and Limits of Web-Deliberation*. Basingstoke: Palgrave Macmillan.

Kim, J., Wyatt, R.O. and Katz, E. (1999). News, talk, opinion, participation: the part played by conversation in deliberative democracy. *Political Communication*, 16, 361–385. DOI:10.1080/105846099198541.

Kolb, D. (1996). Discourses across links. In Ess, C. (ed.), *Philosophical Perspectives on Computer-Mediated Communication* (pp. 15–26). Albany, NY: State University of New York Press.

Koop, R. and Jansen, H.J. (2009). Political blogs and blogrolls in Canada: forums for democratic deliberation?. *Social Science Computer Review*, 27, 155–173. doi:10.1177/0894439308326297.

Mansbridge, J. (1999). Everyday talk in the deliberative system. In Macedo, S. (ed.), *Deliberative Politics: Essays on Democracy and Disagreement* (pp. 211–239). Oxford: Oxford University Press.

McClurg, S.D. (2003). Social networks and political participation: the role of social interaction in explaining political participation. *Political Research Quarterly*, 56, 449–464. DOI:10.1177/106591290305600407.

Monnoyer-Smith, L. (2006). Citizen's deliberation on the Internet: an exploratory study. *International Journal of Electronic Government Research*, 2(3), 58–74. DOI:10.4018/jegr.2006070103.

Papacharissi, Z. (2002). The virtual sphere: the Internet as a public sphere. *New Media and Society*, 4, 9–27. doi:10.1177/14614440222226244.

Papacharissi, Z. (2004). Democracy online: civility, politeness, and the democratic potential of online political discussion groups. *New Media and Society*, 6, 259–283. DOI:10.1177/1461444804041444.

Price, V. and Cappella, J.N. (2002). Online deliberation and its influence: the electronic dialogue project in campaign 2000. *IT and Society*, 1(1), 303–329.

Robertson, S.P., Vatrapu, R. and Medina, R. (2010). Off the wall political discourse: Facebook use in the 2008 US presidential election. *Information Polity*, 15, 11–31. DOI:10.3233/IP-2010-0196.

Rowe, I. (2014). Civility 2.0: A comparative analysis of incivility in online political discussion. *Information, Communication and Society*. DOI:10.1080/1369118X.2014.940365.

Ruiz, C., Domingo, D., Micó, J.L., Díaz-Noci, J., Meso, K. and Masip, P. (2011). Public sphere 2.0? The democratic qualities of citizen debates in online newspapers. *International Journal of Press/Politics*, 16, 463–487. DOI:10.1177/1940161211415849.

Santana, A.D. (2014). Virtuous or vitriolic: the effect of anonymity on civility in online newspaper reader comment boards. *Journalism Practice*, 8, 18–33. DOI:10.1080/1751278 6.2013.813194.

Schneider, S.M. (1997). Expanding the public sphere through computer mediated communication: political discussion about abortion in a Usenet newsgroup. Doctoral dissertation. MIT, Cambridge, MA.

Schultz, T. (2000). Mass media and the concept of interactivity: an exploratory study of online forums and reader email. *Media, Culture and Society*, 22, 205–221. DOI:10.1177/016344300022002005.

Strandberg, K. (2008). Public deliberation goes on-line? An analysis of citizens' political discussions on the Internet prior to the Finnish parliamentary elections in 2007. *Javnost – The Public*, 15(1), 71–90.

Streck, J.M. (1998). Pulling the plug on electronic town meetings: participatory democracy and the reality of the Usenet (pp. 18–48). In Toulouse, C. and Luke, T.W. (eds), *The Politics of Cyberspace*. New York: Routledge.

Stromer-Galley, J. (2003). Diversity of political conversation on the Internet: users' perspectives. *Journal of Computer-Mediated Communication*, 8(3). DOI: 10.1111/j.1083-6101.2003.tb00215.x.

Stromer-Galley, J. (2007). Measuring deliberation's content: a coding scheme. *Journal of Public Deliberation*, 3(1). Retrieved from http://www.publicdeliberation.net/jpd/vol3/iss1/art12/.

Stromer-Galley, J. and Martinson, A.M. (2009). Coherence in political computer-mediated communication: analyzing topic relevance and drift in chat. *Discourse and Communication*, 3, 195–216. DOI:10.1177/1750481309102452.

Sunstein, C.R. (2002). *Republic.com*. Princeton, NJ: Princeton University Press.

Tanner, E. (2001). Chilean conversations: Internet forum participation debate Augusto Pinochet's detention. *Journal of Communication*, 51, 383–403. DOI:10.1111/j.1460-2466.2001.tb02886.x.

Tsaliki, L. (2002). Online forums and the enlargement of public space: research findings from a European project. *Javnost – The Public*, 9(2), 95–112.

Wilhelm, A.G. (1999). Virtual sounding boards: how deliberative is online political discussion?. In Hague, B.N. and Loader, B.D. (eds), *Digital Democracy: Discourse and Decision Making in the Information Age* (pp. 154–178). London: Routledge.

Winkler, R. (2005). *Europeans Have a Say: Online Debates and Consultations in the EU.*

Vienna: Austrian Federal Ministry for Education. Retrieved from http://epub.oeaw.ac.at/ita/ita-projektberichte/e2-2a34.pdf.

Wojcieszak, M.E., Baek, Y.M. and Delli Carpini, M.X. (2009). What is really going on? Structure underlying face-to-face and online deliberation. *Information, Communication and Society*, 12, 1080–1102. DOI:10.1080/13691180902725768.

Wojcieszak, M.E. and Mutz, D.C. (2009). Online groups and political discourse: do online discussion spaces facilitate exposure to political disagreement?. *Journal of Communication*, 59, 40–59. DOI:10.1111/j.1460-2466.2008.01403.x.

Wright, S. (2009). The role of the moderator: problems and possibilities for government-run online discussion forums. In Davies, T. and Gangadharan, S.P. (eds), *Online Deliberation: Design, Research, and Practice* (pp. 233–242). Stanford, CA: CSLI Publications.

Wright, S. (2012a). Politics as usual? Revolution, normalization and a new agenda for online deliberation. *New Media and Society*, 14, 244–261. DOI:10.1177/1461444811410679.

Wright, S. (2012b). From 'third place' to 'third space': everyday political talk in non-political online spaces. *Javnost – The Public*, 19(3), 5–20.

Wright, S. and Street, J. (2007). Democracy, deliberation and design: the case of online discussion forums. *New Media and Society*, 9, 849–869. DOI:10.1177/1461444807081230.

Zhang, W., Cao, X. and Tran, M.N. (2013). The structural features and the deliberative quality of online discussions. *Telematics and Informatics*, 30, 74–86. DOI:10.1016/j.tele.2012.06.001.

15. Creating spaces for online deliberation
Christopher Birchall and Stephen Coleman

WHY ONLINE DELIBERATION?

Contemporary political democracy is faced with two formidable challenges. Firstly, there is the problem of underinformed, unconfident citizens who find it difficult to make up their minds on many of the important policy issues that face society. They rarely talk about politics because they think that nobody in authority will take any notice of them; and they are seldom listened to because they rarely talk about politics. We could compel such people to vote on issues, regardless of whether they feel able to form a competent judgement; we can offer them opportunities to follow parties and leaders which serve as containers of composite values and preferences; or we might leave them to disengage from politics, allowing those who feel confident that they are well informed to make decisions for them. While such minimal terms of political engagement would be compatible with a highly parsimonious model of democracy, they would fall short of the norms of citizenship as formulated by participatory democrats. Secondly, there is the problem of dogmatic and inflexible citizens who have made up their minds on nearly all issues, often in accordance with an overarching ideological bias, and are open to neither new information nor ethical influence to change their rigidly held values and preferences. Such people satisfy the normative democratic requirement of being willing to enter the political fray, but the quality of their engagement tends to be inconsistent with the democratic principle of intellectual openness and adaptability. Neither citizens who cannot make up their minds, nor citizens who have finally and forever made up their minds, are ideal inhabitants of a healthy democracy.

Arguments and practical proposals for democratic deliberation respond to both of these challenges. Including the least confident or vocal members of society in something approaching a public conversation, while encouraging the permanently certain to encounter a wider range of perspectives and information, can only be good for democratic politics. Public deliberation fills a conspicuous vacuum in the public sphere in which self-referential political and media elites have often seemed to crowd out the voices of the citizenry.

The principles of deliberation are well known: all propositions should be on the table for inclusive and uncensored discussion; arguments for and against must be open to public scrutiny; those who deliberate must be regarded as equals (at least, in the context of the deliberative moment) and must listen with attention and respect to all arguments, evidence and experiential narratives; and, ideally at least, deliberative judgements should be based on the force of the strongest argument rather than narrow interests, blind commitments or appeals to external authority (Habermas, 1994; Dryzek, 2000; Gastil, 2000; Steiner, 2012). There are several other conditions that theorists might want to add to the list of deliberative requirements, with some setting the bar so high that it sometimes seems as if deliberation could only ever work in small-scale, experimental environments. Other scholars argue that even if full-blown deliberative democracy is too ambitious an objective, the creation of 'a more deliberative democracy' (Coleman and Blumler, 2009) would at least be preferable to the current situation in which the diverse testimonies of civic experience are drowned out by the relentless outpouring of sensational media headlines.

The case for democratic deliberation, in contrast to the mainly aggregative forms of decision-making associated with voting and mass parties, has gained momentum in recent years, partly in response to the two challenges discussed above, and partly because democratic legitimacy in a more culturally egalitarian era is ever more dependent upon the strength of communicative relationships between government and governed. Governments, parliaments, local authorities and parties, as suppliers of proposed solutions to social problems, can no longer depend upon popular deference, but are under increasing pressure to acknowledge the experience and expertise that lies beyond them, often within local neighbourhoods or communities of practice. Such inputs cannot be collected through the ballot box, which is a crude mechanism for capturing the rationale and multidimensionality of the public will (Coleman, 2013).

By enabling lots of different people to have the space and confidence to form, rehearse and articulate their views, and encouraging people to develop hitherto incomplete or inconsistent arguments, deliberation at its best helps people to acknowledge the political reality that it is sometimes politically preferable to engage in effective compromise than to remain isolated and impotent. By inviting citizens to account for their views rather than simply counting their bundled preferences, democratic outcomes might be more likely to reflect the values and experiences of citizens, stand a chance of being implemented with public support and be regarded as fair. However, establishing spaces, processes and cultural habits that are likely to result in meaningful, inclusive and consequential deliberation has proved to be a difficult challenge. Most citizens know where to go to vote when elections come around, and many have at least a clue about where to go to complain when elected representatives let them down in between elections. But where do citizens go to deliberate about the issues, policies and global forces that affect them? Deliberation has tended to invoke images of market squares, coffee houses and modern community centres, buzzing with civic dialogue; but how might these romantically quaint metaphors of deliberative space be reinvented as twenty-first-century arenas of democratic talk?

For some democratic theorists, the emergence of the Internet offered a potential solution to this problem. From the outset of the World Wide Web as a public network in the mid-1990s, theorists in search of contemporary space for deliberation and online enthusiasts in search of a democratizing role for the Internet gravitated towards visions of e-democracy: the potential of online space as an environment for a new kind of more inclusive and deliberative political practice. Millions of conversations and interactions of various kinds are going on all the time within online spaces that are now a routine domain of everyday interaction for a vast proportion – though not all – of the global population. But deliberative spaces do not form themselves. They are the consequences of intentionality and design. This chapter focuses on the problematics of designing space for online deliberation. Our aim is to consider what has been learned from research about the ways in which tools, protocols, structures and interfaces affect the quality of democratic deliberation. We then turn to the implications of these factors for future research regarding the promotion and evaluation of online deliberation.

PRINCIPLES OF DELIBERATIVE QUALITY

There is a theoretical distinction to be made between political deliberation, which seeks to encapsulate the benefits of focused, purposeful and honest talk, and everyday talk about politics, which is often fragmented, purposeless, uninformed and unequal. Whereas the latter 'is not always self-conscious, reflective or considered' (Mansbridge, 1999: 211), the quality of deliberative practice lies in its commitment to a process of shared reflection that eschews mere competitive self-interest and embedded injustice. In reality, the theoretical distinction between deliberation and everyday talk is less obvious; there can be greater or lesser degrees of the former within the latter.

Several commentators have observed that what passes for political debate online tends to be far from deliberative; that most online political exchanges seem to be partisan, prejudiced and uncivil; and that this raises significant doubts about the potential relationship between the Internet and more deliberative democracy (Hill and Hughes, 1999; Wilhelm, 2000; Morozov, 2012). A weakness of these studies is that they have tended to be based upon limited cases, such as fora in which members of the same party gather together to reinforce their collective values, or random exchanges between friends on social media sites. To dismiss arguments for online deliberation on the grounds that most online political talk is shallow, angry or uninformed is to miss the point of trying to design spaces that attempt to reduce the anti-deliberative influences of conversational homophily and group herding. The case for online deliberation rests on the assumption that it is a means of enhancing the quality of public debate and that such enhancement is unlikely to happen without well-planned design.

However, there is a temptation for scholars to 'discover' online deliberation by adopting the circular perspective that deliberation only occurs when people talk to one another in ways anticipated and facilitated by deliberative theorists. As Coleman and Moss (2012) have argued:

> Most researchers . . . continue to speak and write as if deliberation and the capacities it presupposes are naturally occurring and universal rather than constructed and contingent. Holding on to an essentialist conception of liberal citizenship, they fail to consider the extent to which the deliberative citizen is 'formed and normed', in Ivison's (1997: 41) evocative phrase, and to which they contribute to the construction of the object of their own research.

Rather than thinking of deliberation as an objective or formulaic practice in which one kind of technical platform can serve the needs of all citizens and all of the vast range of subjects they might want to discuss, it makes sense to acknowledge that different social groups behave differently in varying online spaces. Several important studies have identified determinants of online deliberative behaviour that preclude essentialism and recognize that there is no single way to realize the quality of deliberative outcomes (Freelon, 2010; Dahlgren, 2005; Pickard, 2008).

A first key factor determining deliberative outcomes, online or offline, is that most people prefer to talk to other people when they feel secure and comfortable rather than intimidated or under pressure. This accounts for the well-established finding that in both offline and online contexts people discuss politics with like-minded others and feel more comfortable in environments where their points of view and modes of expression are unlikely to be fundamentally challenged (McPherson et al., 2001; Nahon and Hemsley, 2014). The attraction of homophilic political communication presents a challenge to democracy, as the most likely effect of exchanging ideas with people who share one's views is to make such beliefs seem obviously right and to distance and marginalize alternative perspectives (Sunstein, 2002). A key mark of deliberative quality is the extent to which people find themselves in situations where they are compelled to justify their values and preferences; where, indeed, they might come to question or even change their original positions. Self-questioning and preference-shifting are strong empirical effects of high-quality deliberation. Of course, questioning one's opinions can be uncomfortable and all too often deliberative quality is realized at the expense of decreased participation in politics (Mutz, 2006). A well-designed deliberative online environment would allow people to feel safe in disclosing their views to strangers, while exposing them to perspectives that they would not usually encounter. As with the design of any public space, the aim should be to expose participants to the worldliness of politics without crushing personal dispositions. In the case of online deliberation, this entails an effective balance between the normative requirements of rational-critical interaction and the social practices and customs that people adopt as part of their personal performance of citizenship. In this regard Freelon's (2010) framework for exploring the ways in which distinctive 'democratic styles' lead people to deliberate in different ways provides a useful way of thinking about the pluralistic design of deliberative space. As he puts it, 'Rather than simply analyzing online forums in terms of the extent to which they adhere to a singular set of deliberative standards, scholars [should] bring to bear on their data an understanding that different kinds of public spheres exist'. Freelon argues that people come to public discussion with various ideas about what it means to perform as citizens. Liberal-individualists, he argues, are mainly interested in self-expression and self-actualization, while communitarians are mainly interested in strengthening collective ties, and classic deliberators are motivated by a search for the best argument. According to Freelon, both liberal-individualist and communitarian modes of discussion can incorporate elements of deliberation, but this calls for careful design to make it happen. That is to say, even in the absence of citizens who meet the normative requirements of fully fledged deliberators, the design of discursive environments can encourage degrees of deliberative outcomes. Taking this insight into account, designers of spaces for online deliberative talk might aim to create interfaces and protocols that allow discussants to pursue their own 'democratic styles', while being gently encouraged to interact with others committed to different styles. The important point here is that designers should acknowledge the nuances of cultural practice and expressive habit that frame deliberative interaction, rather than expecting such habits and practices to bend to the rigours of deliberative theory.

A second factor likely to affect deliberative quality is the subject matter being discussed. Some political topics are likely to arouse passions more than others (Coe et al., 2014). Karlsson (2010) analysed 28 online discussion forums, each sharing the same platform design, but in which contributors discussed different topics related to European Union (EU) policy. Significant variation was observed in levels of deliberative participation per visitor between the respective forums, suggesting that different discussion topics may make people more or less likely to participate in online deliberation. Just as citizens are often more likely to deliberate when in a comfortable environment, one might assume that they are more likely to deliberate about topics that make them feel safe, informed and relatively invulnerable to hostile feedback. Interestingly, Karlsson's study found that forums with the highest proportions of deliberative content were the ones that generated the most user engagement. Indeed, his conclusion that 'deliberation is more likely to be successful if the issue of deliberation is surrounded by a high level of engagement and conflicted opinions rather than being an issue that renders participants indifferent or is surrounded by a high level of consensus regarding the topics under investigation' is very promising from a democratic perspective. It suggests that contributors are more likely to put in the effort required for deliberation (as opposed to ranting) when they are exposed to a subject that they find not only engaging, but also intellectually challenging. Perhaps, then, an important requirement of a deliberative system is that it makes topics attractive and challenging to participants, particularly when they are outside of the target participants' usual areas of interest or comfort. Indeed, it might be that taking people beyond their ideological comfort zones is more likely to trigger deliberative activity than pandering to an imagined popular desire to avoid agonistic contestation.

Of course, it is not only the willingness of contributors to participate that matters, but also their ability to do so effectively. Designing spaces for online deliberation that compensate for structural inequalities offline (such as class, gender or ethnic inequality) can sometimes result in greater equality of voice between discussion participants. Monnoyer-Smith and Wojcik (2012) describe how online spaces can be designed in a fashion 'that welcomes women, the less informed, and the socioculturally deprived', thereby restructuring, but not eliminating, some of the unjust power structures that might be expected to prevail in the offline world. In short, design could play a role in helping people to discuss a diverse range of sometimes complex or sensitive subjects as well as in broadening the range of voices taking part.

A third factor likely to influence deliberative quality is the relationship (actual and perceived) between spaces in which people are invited to deliberate and institutions of power that are likely to be making decisions related to what is being discussed. A deliberative space discussing a proposed national policy might have clear links to the government, parliament or political party that has proposed it. If such institutions are involved in the discussion as sponsors, participants or respondents, this could have either positive or negative impacts upon deliberative outcomes. If participants' trust in the institution is high – if they believe that it is really listening to what they have to say, is minded to take their views and experiences into account before making a policy decision and is genuinely willing to offer honest feedback – this may well enhance the quality of deliberation. After all, people are more likely to engage in the hard work of deliberating if they believe that their efforts will have real-world consequences. Alternatively, if a governmental, legislative or corporate institution is deemed untrustworthy and people believe that a deliberative exercise is merely tokenistic or, worse still, an exercise in surveillance or data-gathering, this would surely diminish deliberative quality. In such circumstances, participants might decide to use the occasion to merely reaffirm their original positions or voice their scepticism towards the process. Some forms of online public deliberation are intentionally autonomous, refusing to be connected to any dominant political interest, especially government. These tend to entail lateral exchange of views between citizens, either for mainly epistemic ends or as a prelude to civic mobilization. Wright's (2009) study of discussions in online 'third spaces' – which he defines as 'online discussion spaces with a primarily non-political focus, but where political talk emerges within conversations' – found that such venues enable people to rehearse their own identities and encounter (often inadvertently) other perspectives and values. This, he suggests, may provide a crucial foundation for democratic deliberation. Indeed, there is evidence to suggest that peer-to-peer policy deliberation is often not regarded by participants as 'mere talk', but as a means of shaping policy by influencing public opinion, which in turn will put pressure upon elite decision-makers (Coleman et al., 2011 show how online protesters against the Iraq War had much more confidence in their capacity to influence fellow citizens than government per se). In this sense, effective deliberation in third spaces may be a valuable entry point to the informal political sphere. Within such informal contexts people learn to develop the quality of their arguments and gain the confidence to take more institutionally related collective action when necessary.

Taking these three factors into consideration can help deliberative practitioners to design spaces and interfaces that reflect the structural features of normatively effective deliberation. While some features of online deliberative quality call for the replication of offline practices that have proven to be effective, other features are distinctive to the online context. As Pingree et al. (2009) suggest, offline deliberative theory and practice may not be directly applicable to online environments and designers should 'strive to take advantage of the unique design flexibility of the online discussion environment'. DeCindio (2012) urges designers of online deliberative spaces to consider three key factors: the social grouping of people who are expected to deliberate (which she calls the *gemeinschaft* dimension); the social contract between developers, administrators and contributors (the *gesellschaft* dimension); and the technologies to be used in consolidating these relationships. Most of the research literature on online deliberation has tended to focus upon the first of these considerations: who deliberates and how their preferences change or stay the same. The other two considerations – developing the appropriate technological functionalities to facilitate deliberation, and devising rules and moderation structures that are most likely to generate productive deliberative outcomes – have received rather less research attention. It is to these considerations that we now turn.

DELIBERATIVE DESIGN: SOME TECHNICAL CONSIDERATIONS

A multitude of niches exist online in which conversation occurs with a greater or lesser degree of deliberative quality. Some attract user groups whose views are partisan; others attract participants whose views are more reflexive, reciprocal and cross-cutting. Some harbour highly deliberative political discussions almost by accident (Graham, 2012), while others generate deliberative content despite the design of the space. Occasionally, elements of deliberation emerge amongst the character-limited conversations on Twitter (Thimm et al., 2014; Upadhyay, 2014), while other sites devote considerable resources to the design of tools to facilitate public debate, but fail utterly (the UK government's 'Spending Challenge' is a case in point). Some deliberative success stories result from participants feeling safe and at home within a community, while other online sites, such as many of those established for official policy consultations, aim to attract politically disengaged citizens to specially designed spaces, outside of the familiar environments in which they might usually express themselves. We consider below five technical factors that have been identified by online deliberation researchers as being significant for effective design.

Engendering Substantive Debate

Creating the right environment for online deliberation to take place entails something of a balancing act. On the one hand, motivating people to participate in political talk with strangers often involves appealing to their passions; on the other hand, ensuring that debate is constructive often entails suppressing those same passions and encouraging some degree of dispassionate rationality. Scholars have given considerable thought to ways of engendering such a balance (Barton, 2005; Schlosberg et al., 2007). Here we discuss the extent to which designs for online deliberation have addressed the need to balance participant commitment and the informational foundations of thoughtful interaction.

Unchat, which was created by Noveck in 2003, is an experimental real-time discussion tool for small-group deliberation. It features 'speed bumps', designed to force users to encounter relevant information prior to participating in debate. Transcripts are provided to help latecomers to catch up with previous discussion. Like Unchat, the Deme interface (Davies et al., 2009) attempts to foster informed debate by providing access to relevant background information as well as features to enhance participant collaboration, including document-centred discussion and the sharing of files and links. The Deliberative Community Networks (OpenDCN) project (DeCindio, 2012) builds on these and other previous projects by including an 'informed discussion' tool that allows participants to upload their own background information in a wide array of formats. Participants use built-in templates to supply their own datasets or links to external datasets. In this way, they are able to offer their own interpretations of evidence, thereby transcending the rather artificial distinction between background information and deliberative practice. Implicit here is the principle of generating reciprocal interactions amongst participants, removing barriers between agenda-setting initiators of deliberative exercises and deliberative publics.

Real-Time or Asynchronicity

Some advocates of online deliberation claim that carefully designed interfaces for synchronous conversation can replicate the vivacity of face-to-face interactions. For example, Noveck's Unchat and Fishkin's tools for online deliberative polling gather dispersed people together online, as if they were in a single place at the same time. However, it is also acknowledged that synchronous conversations are difficult to schedule for large numbers of participants, so may need to be constrained by rules limiting group size and contribution frequency (Tucey, 2010; Cavalier et al., 2009). Such an approach sacrifices inclusive spontaneity for the sake of deliberative quality. Other scholars argue that asynchronous deliberation makes it more convenient for people to participate on their own terms and leads to more reflective outcomes because users have more time to think before committing themselves to a position. The majority of deliberative tools and models in recent years have been asynchronous (Macintosh, 2004), but these give rise to their own particular challenges. Entering into a large-scale asynchronous discussion that has already started presents users with a need to process, understand and organize the content that has emerged before they arrived. In the case of a large-scale discussion comprising thousands of threads and messages, this can prove to be a time-consuming challenge. As Pingree et al. (2009: 310) put it, 'The Problem of Scale manifests as a difficulty in keeping up with all messages being sent' while the 'Problem of Memory and Mental Organisation' arises from the limitations of human memory in assimilating argumentative material. Designers have sought to alleviate these problems by designing interface features that diminish the disadvantages faced by latecomers to a discussion. For example, OpenDCN seeks to optimize interactivity between participants by organizing content in such a way that specific individuals and arguments can be easily located within the overall discussion. Nested posts and replies help participants to visualize arguments, identify authors and find appropriate locations for their own contributions. Social rating features, such as 'likes' and 'recommends', organize the content further. Such features help, but can at times run counter to the principle of deliberation which expects everyone to be open to all arguments. By allowing users to rate the most popular comments, they are failing to reflect the quality of reasoning behind particular contributions, thereby shifting debate to the surface level of existing preferences (Buckingham Shum et al., 2014).

Visualizing the Arguments

The challenge of levelling the point of entry to deliberation, so that all participants are exposed not only to background information and each other's positions, but also to the core questions motivating the debate, is particularly necessary in the case of policy-related public deliberation, where it is of paramount importance that all contributors acknowledge a common agenda (Coleman and Blumler, 2009). Macintosh (2008) has argued that more complex discussion platforms are necessary to facilitate 'access to and analysis of factual information', 'preference formation' and 'community building'; systems that generate and present community knowledge as well as just information. In pursuit of such ends, Macintosh, like a number of other deliberative theorists, turned to argument visualization (AV). AV systems seek to provide not just spaces for people to pursue arguments, but also a way of making visible the flow of argumentation through graphical representations depicting the collision and convergence of arguments. AV's roots are in electronic collaborative theory which dates back more than 40 years to the creation of systems designed to support legal and political decision-making (Kunz and Rittel, 1970; Conklin and

Begeman, 1987). Expanding upon the Issue-Based Information System (IBIS) of Rittel and Webber (1973), AV formally structures conversations, the flows and components of which are used to create 'maps' of the arguments and evidence. This allows users to locate places within the debate where they feel that they can add value. Examples of AV include the Compendium, a tool used to structure and represent the content of public planning meetings (Okada et al., 2008), and DebateGraph which has been used in a number of governmental and third-sector-initiated deliberative consultations. Pingree's Decision Structured Deliberation system (DSD) and the Deliberatorium from Massachusetts Institute of Technology (MIT) (Klein, 2010) have taken AV a step further, utilizing Web 2.0 features such as ratings and filtering. The Deliberatorium provides participants with a personal homepage, which includes watchlists to help them to keep up with conversations that might be of particular interest to them. Such systems are yet to have a widespread impact on the norms of online participation, but may well in the future become useful facilitators of deliberative consultation (Klein, 2012).

Moderating the Discussion

Designing for online deliberation is not simply a matter of coming up with ever more sophisticated technical tools. Some qualities of deliberation depend upon more basic communicative interventions, such as moderation and facilitation. Wright and Street (2007) found that the social contract between contributors and administrators is a vital dimension to the success of deliberative spaces (see also Coleman and Gøtze, 2001; Noveck, 2003; Wright, 2006, 2009). The ways in which rules and protocols of a discussion space are maintained, contributors are encouraged to interact and discussion outcomes are encouraged can make the difference between friendly, sharing interaction and a breakdown in trust and civility. There is now considerable research evidence to suggest that open and uncontrolled discussion between large groups of people who do not know one another often results in reduced deliberative quality, measured in terms of rational content and contributor interaction (Sobieraj and Berry, 2011).

Moderation practices can be particularly sensitive in the case of governmental platforms where the management and structuring of discussion can be seen as a form of censorship. Wright (2009) has shown how discussion moderation can be vital in turning random position-stating into more focused and productive discourse. He describes two models of moderation: content moderation, in which humans (and also possibly automated programs) pre-moderate content against predefined criteria; and interactive moderation, in which the moderator acts as a facilitator, giving feedback, supplying resources and directing the conversation in productive ways. The latter can be seen in the Deliberatorium (Klein, 2010), in which the moderators have a 'part education and part quality control' role and can communicate with contributors to help them to produce acceptable posts. A recent study of journalists' involvement in online discussions generated by their stories shows that the presence of an 'official' or qualified voice in such debates often results in a more civil conversation (Meyer and Carey, 2014; Lewis et al., 2014).

An example of content moderation can be seen in the AV-based E-Liberate system which was built around the use of Robert's Rules of Order, 'a set of directives that designated an orderly process for equitable decision making in face-to-face meetings' (Schuler, 2009). However, this feature has not always been popular with users, who felt that their free expression was being constrained by overly formal rules. In response, the designers incorporated an 'auto pilot' feature into the system, allowing users to express themselves without constraint, but only when they considered that moderation was impeding their conversation. Similarly, designers of the Unchat system (Noveck, 2003) included a flexible moderation tool in which moderators are elected from amongst the discussion participants, who have the right to depose them if they disagree with their decisions.

Participant Authentication

Whether or not discussion participants are required to provide authentication before entering a deliberative space is a further pressing question for deliberative design. The case for requiring user authentication is that strong identities are more likely to contribute to trusting relationships between participants.

As DeCindio and Peraboni (2011) observe:

> Our long-standing experience managing the Milan Community Network and several related projects suggests that, in order to create a trustworthy social environment that encourage government officers and representatives to undertake online dialogue with citizens, this weak form of identification is not adequate: the online identity should, as much as possible, reflect the offline identity.

Authentication methods vary in strength, from postal confirmation of offline addresses used by banks and government departments, to weaker ones where email addresses or pseudonyms are all that is required to identify a participant. Marx (1999) identified graduated levels of identity that users could be asked to provide, ranging from their name and/or location to a pseudonym only traceable via an intermediary. In an experimental situation, Rhee and Kim (2009) found that when contributors to a discussion were required to reveal social identity cues this resulted in them being more attentive to messages and more likely to elaborate their arguments at a higher cognitive level than in a control group of anonymous discussants. However, authentication introduces barriers to participation (particularly for members of marginalized communities) and there is surely a case for distinguishing between weak authentication required for comment-posters on a political blog, and strong authentication required for contributors to a consequential exercise in policy deliberation.

SOME THOUGHTS ABOUT FUTURE RESEARCH

We began by referring to two types of citizen: the unconfidently undecided and the overconfidently dogmatic. Most deliberative practice has been geared towards helping the latter to be more flexible in formulating their preferences. By encouraging holders of hard preferences to justify their positions explicitly and publically and exposing them to counter-arguments, often stemming from radically different experience, some online deliberative exercises have proved to be a force for greater democratic understanding. A key research question here relates to the durability of such preference shifts. Do people adopt more open-minded outlooks during and shortly after exposure to other perspectives, but then return to ideological intransigence once the deliberative air has cleared? If so, might there be ways of sustaining such democratic outcomes beyond one-off mini-deliberations? Much thought has been devoted to designing spaces for time-limited deliberative events, but what about the possibility of establishing ongoing online deliberative institutions within which citizens might acquire enduring habits of democratic communication?

Several researchers have attempted to move beyond the notion of deliberation, both online and offline, as a discrete event. They argue that deliberative norms can best be realized in a scaled-up fashion: as macro rather than micro deliberation. Parkinson and Mansbridge's (2012) innovative notion of a 'deliberative system' in which there is division of labour and functions between individuals and institutions, each playing distinctive roles in the generation of deliberative outcomes, could have important implications for online deliberative design. If, instead of online spaces having to provide for all the complex norms of deliberation, they were to be seen as one element within a democratic media ecosystem, it would be possible to focus upon those aspects of public discussion that are best supported by digital technologies, leaving other elements to be provided elsewhere, such as television or newspaper content or local, face-to-face meetings. The practical, political and technical conditions and implications of the institutional interaction that could sustain a deliberative system have yet to be explored in any depth. The role of digital technologies within a macro-social order committed to democratic deliberation gives rise to much more complex problems than the relatively simple communicative challenge of creating isolated silos of high-level deliberation. The three principles of online deliberative quality considered above could be valuable in thinking through deliberation at a systemic level. The principle of encouraging cross-cutting debate, in which citizens encounter strangers and unsought-for perspectives, is a key precondition for normatively successful deliberation, but runs counter to the institutional structure of contemporary politics, whereby activists cluster together in partisan formations. There has been little research conducted on ways of enabling mass political parties to deliberate, whether internally, with the public or with one another. Indeed, much online deliberative experimentation has proceeded as if parties were irrelevant and preference formation and expression could be reconfigured at the micro level. Freelon's acknowledgement of divergent democratic styles is helpful here in opening up space for a more pluralistic sociology of discursive motivation.

There is space for more imaginative research on ways of supporting and empowering the second (possibly larger) group of disengaged citizens mentioned at the beginning of this chapter: those who are the least confident, informed and vocal. Such research might involve the development of hybrid spaces of deliberation, in which mass-media audiences are encouraged to go online and participate in debates triggered by television stories and images. Graham's work on the ways in which audiences of popular cultural content often use their viewing experience as a basis for broader social deliberation is highly promising in this regard. Might it be that the least politically confident or engaged people in contemporary society are unlikely to be attracted to the kinds of innovative Web-based spaces in which most deliberative innovation has occurred? The current popularity of social media platforms may well offer a more appropriate space for introducing elements of democratic deliberation. Most of the design innovations highlighted in this chapter have tended to work (when they do work) as niche products, operating within realms of specific consultative environments, rather than reaching out to the general public. While many researchers have been analysing the communicative dynamics of Facebook, Twitter, Weibo and YouTube, online deliberation scholars have been slow to do so (one exception here is Freelon, forthcoming). Perhaps that is because these massive social networks pose formidable challenges for the scoping of deliberative projects: in a world of global access to online media how does one generate a community of use that is open enough to be representative but controlled enough to connect a local or expert community to a local or expert discussion?

Here lies a major research challenge. In an age of seemingly endless choice in information source and participatory space, how can spaces be designed to encourage people to step outside of their comfort zones, to listen to opposing opinion about difficult topics, and to do so in spaces where efficacy might ensue? Many different niches exist on the Web in which conversation occurs with a greater or lesser degree of deliberative quality. Such digital niches are formed through complex combinations of social and technical dimensions that lead to varied conditions for effective deliberation. The challenge for designers of deliberative spaces is to translate the successful characteristics of these deliberative niches into more broadly inclusive spaces, shaped by interface design techniques and regulatory protocols that combine sensitivity to democratic normativity and an acknowledgement of cultural practice.

REFERENCES

Barton, Matthew D. (2005). 'The future of rational-critical debate in online public spheres'. *Computers and Composition* 22(2): 177–190.
Buckingham Shum, Simon, Anna De Liddo and Mark Klein (2014). 'DCLA meet CIDA: Collective Intelligence Deliberation Analytics'. 4th International Conference on Learning Analytics and Knowledge, Indianapolis, IN.
Cavalier, Robert, Miso Kim and Zachary S. Zeiss (2009). 'Deliberative democracy, online discussion, and Project PICOLA'. In T. Davies and S.P. Gangadharan (eds), *Online Deliberation: Design, Research and Practice*. San Francisco, CA: CSLI Publications.
Coe, Kevin, Kate Kenski and Stephen A. Rains (2014). 'Online and uncivil? Patterns and determinants of incivility in newspaper website comments'. *Journal of Communication* 64(4): 658–679.
Coleman, Stephen (2013). *How Voters Feel*. New York: Cambridge University Press.
Coleman, Stephen and Jay G. Blumler (2009). *The Internet and Democratic Citizenship: Theory, Practice and Policy*, Vol. 1. Cambridge: Cambridge University Press.
Coleman, Stephen and Gøtze, John (2001). *Bowling Together: Online Public Engagement in Policy Deliberation*. London: Hansard Society.
Coleman, Stephen, David E. Morrison and Simeon Yates (2011). 'The mediation of political disconnection'. In Kees Brants and Katrin Voltmer (eds), *Political Communication in Postmodern Democracy*. Basingstoke: Palgrave.
Coleman, Stephen and Giles Moss (2012). 'Under construction: the field of online deliberation research'. *Journal of Information Technology and Politics* 9(1): 1–15.
Conklin, Jeff and Michael L. Begeman (1987). 'gIBIS: a hypertext tool for team design deliberation'. *Proceedings of the ACM conference on Hypertext*. ACM.
Dahlgren, Peter (2005). 'The Internet, public spheres, and political communication: dispersion and deliberation'. *Political Communication* 22(2): 147–162.
Davies, Todd, Brendan O'Connor, Alex Cochran, Jonathan J. Effrat, Andrew Parker, Benjamin Newman and Aaron Tam (2009). 'An online environment for democratic deliberation: motivations, principles, and design'. In T. Davies and S.P. Gangadharan (eds), *Online Deliberation: Design, Research and Practice*. San Francisco, CA: CSLI Publications.

De Cindio, Fiorella (2012). 'Guidelines for designing deliberative digital habitats: learning from e-participation for open data initiatives'. *Journal of Community Informatics* 8: 2.

De Cindio, Fiorella and Cristian Peraboni (2011). 'Building digital participation hives: toward a local public sphere'. In M. Foth, L. Forlano, C. Satchell and M. Gibbs (eds), *From Social Butterfly to Engaged Citizen; Urban Informatics, Social Media, Ubiquitous Computing, and Mobile Technology to Support Citizen Engagement*. Cambridge, MA: MIT Press.

Dryzek, John S. (2000). *Deliberative Democracy and Beyond: Liberals, Critics, Contestations.* Oxford: Oxford University Press.

Freelon, Deen G. (2010). 'Analyzing online political discussion using three models of democratic communication'. *New Media and Society* 12(7): 1172–1192.

Freelon, Deen (forthcoming). 'Discourse architecture, ideology and democratic norms in online political discussion'. *New Media and Society.*

Gastil, John (2000). *By Popular Demand: Revitalizing Representative Democracy through Deliberative Elections.* Oakland, CA: University of California Press.

Graham, Todd (2012). 'Beyond 'political' communicative spaces: talking politics on the Wife Swap discussion forum'. *Journal of Information Technology and Politics* 9(1): 31–45.

Habermas, Jürgen (1994). 'Three normative models of democracy'. *Constellations* 1(1): 1–10.

Hill, Kevin A. and John E. Hughes (1999). *Cyberpolitics: Citizen Activism in the Age of the Internet.* Lanham, MD: Rowman & Littlefield Publishers.

Ivison, Duncan (1997). *The Self at Liberty: Political Argument and the Arts of Government.* Ithaca, NY: Cornell University Press.

Karlsson, Martin (2010). 'What does it take to make online deliberation happen?: A comparative analysis of 28 online discussion forums'. In F. De Cindio, A. Macintosh and C. Peraboni (eds), *From e-Participation to Online Deliberation, Proceedings of the Fourth International Conference on Online Deliberation, OD2010.* Leeds.

Klein, Mark (2010). 'Using metrics to enable large-scale deliberation'. In *Collective Intelligence in Organizations: A Workshop of the ACM Group 2010 Conference.*

Klein, Mark (2012). 'Enabling large-scale deliberation using attention-mediation metrics'. *Computer Supported Cooperative Work* 21: 449–473.

Kunz, Werner and Horst, W.J. Rittel (1970). 'Issues as elements of information systems'. Working Paper #131, Institut fur Grundlagen der Planung I.A., University of Stuttgart.

Lewis, Seth C., Avery E. Holton and Mark Coddington (2014). 'Reciprocal journalism: a concept of mutual exchange between journalists and audiences'. *Journalism Practice* 8(2): 229–241.

Macintosh, Ann (2004). 'Characterizing e-participation in policy-making'. *Proceedings of the 37th Annual Hawaii International Conference on System Sciences, 2004.* IEEE.

Macintosh, Ann (2008). 'The emergence of digital governance'. *Significance* 5(4): 176–178.

Mansbridge, Jane (1999). 'Everyday talk in the deliberative system'. In S. Macedo (ed.), *Deliberative Politics.* Oxford: Oxford University Press.

Marx, Gary T. (1999). 'What's in a Name? Some reflections on the sociology of anonymity'. *Information Society* 15(2): 99–112.

McPherson, Miller, Lynn Smith-Lovin and James M. Cook (2001). 'Birds of a feather: homophily in social networks'. *Annual Review of Sociology* 27: 415–444.

Meyer, Hans K. and Michael Clay Carey (2014). 'In moderation: examining how journalists' attitudes toward online comments affect the creation of community'. *Journalism Practice* 8(2): 213–228.

Monnoyer-Smith, Laurence and Stéphanie Wojcik (2012). 'Technology and the quality of public deliberation: a comparison between on and offline participation'. *International Journal of Electronic Governance* 5(1): 24–49.

Morozov, Evgeny (2012). *The Net Delusion: The Dark Side of Internet Freedom.* New York: Public Affairs.

Mutz, Diana C. (2006). *Hearing the Other Side: Deliberative versus Participatory Democracy.* New York: Cambridge University Press.

Nahon, Karine and Jeff Hemsley (2014). 'Homophily in the guise of cross-linking political blogs and content'. *American Behavioral Scientist.* DOI: 1177/0002764214527090.

Noveck, Beth Simone (2003). 'Designing deliberative democracy in cyberspace: the role of the cyber-lawyer'. *Boston University Journal of Science and Technology Law* 9: 1.
Okada, Alexandra, Simon J. Buckingham Shum and Tony Sherborne (2008). *Knowledge Cartography: Software Tools and Mapping Techniques*. Amsterdam: Springer.
Parkinson, John and Jane Mansbridge (eds) (2012). *Deliberative Systems: Deliberative Democracy at the Large Scale*. New York: Cambridge University Press.
Pickard, Victor W. (2008). 'Cooptation and cooperation: institutional exemplars of democratic internet technology'. *New Media and Society* 10(4): 625–645.
Pingree, Raymond J., T. Davies and S.P. Gangadharan (2009). 'Decision structure: a new approach to three problems in deliberation'. In T. Davies and S.P. Gangadharan (eds), *Online Deliberation: Design, Research and Practice*. San Francisco, CA: CSLI Publications.
Rhee, June W. and Eun-mee Kim (2009). 'Deliberation on the net: lessons from a field experiment'. In T. Davies and S.P. Gangadharan (eds), *Online Deliberation: Design, Research and Practice*. San Francisco, CA: CSLI Publications.
Rittel, Horst W.J. and Melvin M. Webber (1973). 'Dilemmas in a general theory of planning'. *Policy Sciences* 4(2): 155–169.
Schlosberg, David, Stephen Zavestoski and Stuart W. Shulman (2007). 'Democracy and e-rulemaking: web-based technologies, and the potential for deliberation'. eRulemaking Research Group, Paper 1, University of Massachusetts – Amherst.
Schuler, Douglas (2009). 'Online civic deliberation with E-Liberate'. In T. Davies and S.P. Gangadharan (eds), *Online Deliberation: Design, Research and Practice*. San Francisco, CA: CSLI Publications.
Sobieraj, Sarah and Jeffrey M. Berry (2011). 'From incivility to outrage: political discourse in blogs, talk radio, and cable news'. *Political Communication* 28(1): 19–41.
Steiner, Jürg (2012). *The Foundations of Deliberative Democracy: Empirical Research and Normative Implications*. New York: Cambridge University Press.
Sunstein, Cass R. (2002). 'The law of group polarization'. *Journal of Political Philosophy* 10(2): 175–195.
Thimm, Caja, Mark Dang-Anh and Jessica Einspänner (2014). 'Mediatized politics – structures and strategies of discursive participation and online deliberation on Twitter'. In Andreas Hepp and Friedrich Krotz (eds), *Mediatized Worlds: Culture and Society in a Media Age*. London: Palgrave.
Tucey, Cindy Boyles (2010). 'Online vs. face-to-face deliberation on the global warming and stem cell issues'. Western Political Science Association 2010 Annual Meeting Paper.
Upadhyay, Meenakshi (2014). 'Political deliberation on Twitter: is Twitter emerging as an opinion leader?', International Conference on People, Politics and Media (ICPPM), 25–26 April, Jagran Lakecity University.
Wilhelm, Anthony G. (2000). *Democracy in the Digital Age: Challenges to Political Life in Cyberspace*. New York: Routledge.
Wright, Scott (2006). 'Government – run online discussion fora: moderation, censorship and the shadow of control'. *British Journal of Politics and International Relations* 8(4): 550–568.
Wright, Scott (2009). 'The role of the moderator: problems and possibilities for government-run online discussion forums'. In T. Davies and S.P. Gangadharan (eds), *Online Deliberation: Design, Research and Practice*. San Francisco, CA: CSLI Publications.
Wright, Scott and John Street (2007). 'Democracy, deliberation and design: the case of online discussion forums'. *New Media and Society* 9(5): 849–869.

16. Computational approaches to online political expression: rediscovering a 'science of the social'

Dhavan V. Shah, Kathleen Bartzen Culver, Alexander Hanna, Timothy Macafee and JungHwan Yang

It is a curious fact that the empirical study of political talk, particularly online exchanges, is increasingly traced back to the social interactionism of nineteenth-century French sociologist Gabriel Tarde. As Terry Clark (1969) and Elihu Katz (2006) remind us, Tarde argued for conversation's place at the center of sociological inquiry, articulating a complex theory of 'inter-mental activity' concerning how people influence one another. In so doing, he developed the concepts that later became known as the two-step flow of communication and opinion leadership, among other propositions of interpersonal influence (Lazarsfeld et al., 1944; Berelson et al., 1954; Katz and Lazarsfeld, 1955). As Tarde writes about the late nineteenth century (1969 [1898]: 313), 'newspapers have transformed . . . the conversations of individuals, even those who do not read papers but who, talking to those who do, are forced to follow the groove of their borrowed thoughts. One pen suffices to set off a million tongues'. His thesis still holds true, with televised events such as the first 2012 presidential debate in the USA generating more than 10 million tweets in just a few hours, including many retweets of major accounts (Hanna et al., 2013).

Tarde was particularly concerned with the relationship between mass communication and interpersonal conversation for the formation of publics and their opinions. Katz (2006: 267) describes this mediated process as follows: 'To the press, he assigned the role of creating a public . . . The press, then, sets an agenda for the conversation of the cafes. Opinions are clarified and crystallized in these conversations, and then translated into actions in the world of politics'. The central tenets of this ordered model – press, conversation, opinion, and action – are supported by research on multi-step flow (Rogers and Shoemaker, 1971), opinion leadership (Shah and Scheufele, 2006), and communication mediation (Lee et al., 2013). In digital media environments, the sources of information and sites of conversation have begun to converge – amplified and reinforced within seemingly polarized ecologies – suggesting important directions for research on opinion and action in networked societies.

The Laws of Imitation (*Les lois de l'imitation*) (1903 [1890]), Tarde's most widely known work in English, speaks to communication and social influence within such settings. It also marks him as the 'founding father of innovation diffusion research' (Kinnunen, 1996), charting processes of communicative invention, reproduction, and opposition. Methodologically, his calls for attention to observable interpersonal interactions, particularly within conversational processes, presage the approaches central to computational social science, where each interaction 'leaves digital traces that can be compiled into comprehensive pictures of both individual and group behavior, with the potential to transform our understanding of our lives, organizations, and societies' (Lazer et al., 2009: 721). The technological affordances of social media permit tracking of message creation and expression within a network, as well as reception and diffusion through systems (Namkoong et al., 2010; Han et al., 2011). Tarde's insights about 'invention' and 'imitation' provide ways to study online talk as it intersects with deliberative democracy, informed opinion, and participatory citizenship (Schudson, 1978; Barber, 1984; Habermas, 1984; Kim et al., 1999; Price and Cappella, 2002; Mutz, 2006).

SITUATING POLITICAL TALK

Tarde, often presented as the foil to Emile Durkheim's efforts to distinguish the study of society from that of human psychology,[1] was profoundly concerned with the interplay of mental and social forces, particularly as seen in the locus of conversational exchanges. As Berelson et al. wrote in *Voting* (1954: 300), Tarde 'was convinced that opinions are really formed through the day-to-day exchanges of comments and observations which goes on among people . . . by the very process of talking to one another, the vague dispositions which people have are crystallized, step by step, into specific attitudes, acts, and votes'. From this perspective, conversation is not simply a site of networked information exchange, but also an opportunity for the composition and clarification of one's own views (see Pingree, 2007).

In the present day, this position seems all the more correct, especially in online environments, where political expression and conversation are increasingly common and visible. These environments also lend themselves to the sort of large-scale, highly detailed interactional analysis that Tarde's approach advocated. As Bruno Latour (2010) recently recognized, 'it is indeed striking that at this very moment, the fast expanding fields of "data visualisation," "computational social science," or "biological networks" (Lazer et al., 2009; Wimsatt, 2007) are tracing, before our eyes, just the sort of data Tarde would have acclaimed'. 'Big data' provide a way to understand everyday political talk online, its triggers, content, and structures.

Such close analyses of political talk, whether face-to-face or online, have typically been restricted to ethnographic or content analytic research. Works such as William Gamson's (1992) classic study, *Talking Politics*, used analysis of small-group discussions around affirmative action, nuclear power, the Arab–Israeli conflict, and US industry to counter the conventional wisdom of an uninformed and inactive electorate. Taking a similarly granular approach, Papacharissi (2004) examined the quality of online political talk by studying the level of civility in 287 discussion threads drawn randomly from 147 political newsgroups, concluding that discussions were civil but heated. Somewhat similarly, Walsh (2012) used participant observation of 37 reoccurring groups from 27 distinct communities across the state of Wisconsin to show how class- and place-based identity is linked with perceptions of relative deprivation.

Complementing this work on the actual content of conversations, research has also employed cross-sectional and panel survey methods to examine the causes and consequences of political talk (Huckfeldt and Sprague, 1995). Most notably, work by Jack McLeod et al. has examined the role of interpersonal conversation for community activism and political engagement (McLeod et al., 1999). Emphasizing the mediating roles of news and talk for participation in public life, communication behaviors are thought to shape and amplify the impact of background characteristics on citizens' engagement in democratic societies (McLeod et al., 1996). Various inquiries about the roles of media and conversation have coalesced into 'communication mediation models', which conclude that news consumption and political talk largely channel the effects of demographics, ideology, and social structure on outcome orientations and participatory responses (Sotirovic and McLeod, 2004; Cho et al., 2009).

This process has been further specified in the form of a 'citizen communication mediation model' (Shah et al., 2005; Shah et al., 2007), which theorizes and finds, consistent with Tarde's framework, that media influences are strong, but largely indirect, shaping opinion and action through effects on face-to-face and online discussion about news. Informational use of media, particularly newspaper and online news use, is found to stimulate expression and discussion through interpersonal and computer-mediated political talk, channeling effects on to engagement. The power of digital pathways to participation – from both conventional and online news sources through digital messaging – was particularly strong among the youngest generational groups (Lee et al., 2013).

POLITICAL TALK VIA DIGITAL MEDIA

Face-to-face and online talk share many virtues. Both spur compositional processing, mental elaboration, attitude crystallization, cross-cutting exposure, media reflection, knowledge gain, and mutual understanding (Shah et al., 2005; Mutz, 2006; Lee et al., 2013; Valenzuela et al., 2012). But political conversation via digital platforms may have certain advantages over face-to-face talk, as well. First, digital media often provide a source of political information and a sphere for political expression, readily facilitating their interplay and participatory consequences (Dahlgren, 2000, 2005). Second, the functionality of many online media emphasize discursive elements such as walls posts, online chat, photo sharing, and social networking, highlighting opportunities for public-spirited talk (Bennett et al., 2011; Boyd and Ellison, 2007). Third, generating user-created content, whether images, videos, or posts, may demand deeper forms of reflective and compositional processing (Freelon, 2010; Ekström and Östman, 2013). Regardless of the setting, 'conversation provides people with the opportunity to think through their "idea elements" and reduce cognitive inconsistency' (Kim et al., 1999: 363).

Less is known about the specific consequences of online political talk, such as whether deeper dialogue contributes to the enhancement of opinion quality and the development of efficacy (Kim et al., 1999). There is also some question as to the value of measuring behaviors in the digital world using analogue tools such as survey instruments, which rely on self-reported measures of network composition, structural heterogeneity, and discussion frequency (see Sotirovic and McLeod, 2001; Eveland and Hively, 2009; Kwak et al., 2005), rather than precise measures of content-specific message expression, reception, response, and repetition (see Han et al., 2011; Namkoong et al., 2010). In addition, it may be important to look across different conversational settings, such as messaging platforms (Shah et al., 2007; Hardy and Scheufele, 2005), political blogs (Gil de Zúñiga et al., 2009), and social networking sites (Bode et al., 2014), all of which have been linked to civic engagement and political action.

Social networking sites bring together the most powerful features of online interaction, messaging in real time and asynchronously, posting to smaller circles and to full networks, exchanging information and providing emotional support, creating new content and sharing the ideas of others, starting a group and joining those created by others; all seemingly strengthening social ties, albeit in unique registers to differing degrees (Boyd and Ellison, 2007; Pasek et al., 2009; Gil de Zúñiga et al., 2012). One recent study examines the consequences of political social networking behaviors, such as displaying a political preference on a profile, becoming a fan or a friend of a politician, joining a cause or political group, and using a news or politics application (Bode et al., 2014). It finds that such uses shape political participation above and beyond the effects of offline talk and online expression.

SOCIAL NETWORKING AROUND POLITICS

There is little doubt that social media provide a forum for the discussion of political issues, a system through which to recruit individuals to participate in pertinent political issues, and a means to find people who share similar political opinions (Papacharissi, 2010). In this sense, political engagement through social media may represent a shift from top-down political communication and participation to one that is interpersonal or bottom-up (Thackery and Hunter, 2010). It is also characterized by feedback mechanisms to elites, providing opinion leaders with an opportunity to shape the thinking of political and media elites if they can 'set off a million tongues'. Social media have certainly increased candidates' digital exposure, permitting communication with volunteers, donors, and constituents (Gueorguieva, 2008).

Alongside Facebook (Vitak et al., 2011) and YouTube (Dylko et al., 2012), Twitter has emerged as a major location of political interaction (Hanna et al., 2013; Tumasjan et al., 2010). Much of this work emphasizes the use of social media channels by politicians during elections (Bruns and Highfield, 2013). Studies of members of the US Congress have found that they use Twitter primarily for information dissemination, with little retweeting or hashtag use, suggesting elite efforts to lead opinion rather than echoing the ideas of others (Golbeck et al., 2010; Gainous and Wagner, 2014). Efforts to examine whether Twitter can predict election results are more mixed. Based on a sentiment analysis of more than 100,000 messages referencing a political party or politician in the 2009 German federal election, Tumasjan et al. (2010) found that Twitter 'is used extensively for political deliberation and that the mere number of party mentions accurately reflects the election result'. However, an alternative analysis of the same election found no link and argued that the predictive power of the prior study was a consequence of the 'arbitrary choices of the authors' (Jungherr et al., 2012: 229).

Regardless of its predictive power, it is widely acknowledged that political talk on Twitter is triggered by media happenings and news coverage (Hanna et al., 2013; Graham and Hajru, 2011). Only a handful of studies have examined the types of political expression that occur within social networking sites, with even fewer considering how citizens self-organize under certain banners or hashtags (that is, user-generated keywords organized around the # symbol). Given that 'hashtags are used to bundle together tweets on a unified, common topic, and that the senders of these messages are directly engaging with one another', they provide a potentially powerful way to track everyday political expression (Bruns and Burgess, 2011: 5). Hashtags can become identified with specific issue positions and protest efforts. Along these lines, Segerberg and Bennett (2011) analyzed hashtag usage around the 2009 United Nations Climate Change Conference, observing the strategic deployment of certain terms to organize action.

As noted above, Twitter also allows for the tracking of 'inventive' message expression and 'imitation' by others in the network as indicators of their recirculation and diffusion through systems (Tarde, 1903 [1890]; Rogers and Shoemaker, 1971). At the most basic level, this can be understood in terms of tweets and retweets (Boyd et al., 2010), although more sophisticated analyses have examined the flow of influence within social networking systems. These studies conclude that Twitter is an excellent medium for message propagation, finding that 37.1 percent of message flows spread more than three degrees of separation away from the original sender (Ye and Wu, 2010). Of course, not all issues are created equal. Tracing the diffusion of hashtags on Twitter, Romero et al. (2011: 695) find significant variation across topics, with hashtags on politically controversial topics 'particularly persistent, with repeated exposures continuing to have unusually large marginal effects on adoption'. Influence in online social networks is not simply a function of size of follower networks, at least as measured in retweets or user mentions (Cha et al., 2010). In fact, a recent analysis of 74 million diffusion events by 1.6 million Twitter users questioned the emphasis on online elites as influencers, concluding that 'word-of-mouth information spreads via many small cascades, mostly triggered by ordinary individuals' (Bakshy et al., 2011: 73).

COMPUTATIONAL APPROACHES

As these studies suggest, the next phase of research on political talk online may benefit from moving beyond ethnographic, content-analytic, or survey-based assessments of these phenomena, instead exploring the value of computational approaches to questions of issue attention, social influence, opinion formation, and political mobilization. The use of 'big data' may provide unique insights into how political elites and their followers deploy particular language around controversial political issues, and how these forms of expression splinter into polarized factions, coalesce into action, and get recirculated through networks.

Using such approaches to help illuminate how particular events or controversies spur different types of political talk and social activism, we explore two news controversies that erupted in 2012 in the United States: (1) Rush Limbaugh's statements about Sandra Fluke, a law student and women's rights activist who rose to national prominence when advocating for a contraception mandate on health care plans; and (2) the shooting of Trayvon Martin, an unarmed African-American teen killed by neighborhood watch volunteer George Zimmerman, who claimed self-defense under Florida's 'stand your ground' law.

The data used to examine these cases began with a purposive sample of politically active users. We identified the most prominent political user accounts in five categories: political advocacy groups; politicians and candidates; political party operatives; journalists and pundits; and political satirists and celebrities. Recency and frequency of activity, size of follower network, ideological diversity, and political prominence were considered when creating lists within each category. From this process, we compiled a final list of 165 political elites, including more than 40 candidates for national office, 40 major advocacy groups, and more than 50 journalists and pundits. We call these the 'top-level' users (see supplemental Online Appendix, available at http://bit.ly/shah-etal-2014).

We then collected follower lists for each of these users and drew a random sample of 80 of their followers, called the 'second-level' users. This resulted in 13,200 user accounts that we tracked. In practice, given the nature of Twitter user accounts, this number shrank slightly because some users' accounts were suspended or deleted. In fact, we began with the assumption that political elites gain and lose followers given their shifting prominence and visibility. Accordingly, we re-collected follower information at three distinct points during the US presidential election cycle: the initial collection in late 2011, mid-June 2012, and September 2012. Follower information was collected using the RESTful Twitter application programming interface (API) (https://dev.twitter.com/docs/api).

Using an account with expanded ('whitelisted') access to Twitter data, we gathered tweets from these users using the Streaming Twitter API (https://dev.twitter.com/docs/streaming-apis).[2] This collection gave us the following information for each user we specified: tweets created by the user, tweets which were retweeted by the user, replies to any tweet created by the user, retweets of any tweet created by the user, and 'manual' replies to the user created without using Twitter's 'Reply' button.

This resulted in a collection of more than 431 million tweets between December 30, 2011, and January 22, 2013, the day after President Obama's second inauguration. Given this sampling methodology, it is inevitable that we would pick up some 'elite' users among our follower

Table 16.1   Follower and following counts at level 1 and 2 of collection

		Level 1	Level 2
Followers			
	Mean	275,483	3,832
	Median	36,324	398
Following			
	Mean	10,551	2,243
	Median	578	689
Ratio			
	Mean	26.1	1.7
	Median	62.8	0.6

sample. For example, our second-level sample included four users with more than 1 million followers: rhythm and blues singer John Legend, Whole Foods, the *Huffington Post*, and National Public Radio (NPR)'s Scott Simon. Still, the vast majority of those included at this level were not political elites.

Table 16.1 shows the means and medians of these two levels, comparing number of follower accounts, number of accounts following, and the ratio of these two numbers. Given the timing of the cases considered in this chapter, this analysis centers on the first wave of data collection. Table 16.1 compares the two levels. Focusing on median values, which avoid inflated means due to outliers, we see that level 1 users (political elites) have nearly 100 times more followers than level 2 users (follower network). Level 1 users also follow fewer users than level 2 users, resulting in a ratio of followers to following that differs dramatically between our elite and follower groups. While it may be true that follower counts do not denote influence, per se (Cha et al., 2010), this certainly differentiates our elite users from those who followed them.

UNDERSTANDING THE CONTROVERSIES

To analyze these data, we turn to computational methods, namely computer-aided content analysis of keywords and hashtags, along with social network mapping of these online tokens. This chapter will proceed as follows: First, we examine the frequency of the appearance of particular keywords. We then identify which level of users are using these keywords. Next, we focus on the co-occurrence of hashtags as an indicator of public sentiment and political organizing, tracking these among elites and their followers. Finally, we map networks of message retweeting to understand the social structure of everyday political talk. Before moving to this computational analysis, we first provide some context on the two cases, the Sandra Fluke–Rush Limbaugh controversy and the Trayvon Martin–George Zimmerman shooting.

Sandra Fluke: A Washington, DC Conflict

Georgetown law student and women's health activist Sandra Fluke landed in the center of US national controversy – and a Twitter firestorm – in February 2012 shortly after testifying in a hearing staged by Congressional Democrats. An original hearing before the House Oversight and Government Reform Committee focused on the Affordable Care Act ('Obamacare') and provisions mandating coverage of contraceptives. Committee Chairman Darrell Issa (Republican, California) refused to allow Fluke to speak in a three-hour hearing concerning the contraceptive mandate during which only men testified, all opposing this policy.

Led by former House Speaker Nancy Pelosi (Democrat, California), Congressional Democrats convened an unofficial hearing February 23, 2012, to take Fluke's testimony. She spoke about the high cost of contraceptives and adverse effects on women's health, drawing praise from Democrats. Her appearance at the unofficial hearing drew derision shortly afterward, beginning with attacks from the conservative blogosphere, such as, 'Sex-Crazed Co-Eds Going Broke Buying Birth Control, Student Tells Pelosi Hearing Touting Freebie Mandate', from *CNS News*. The issue bubbled in those circles for five days before coming to a boil with conservative talk show radio host Rush Limbaugh. On his February 29, 2012, show, Limbaugh called Fluke a 'slut' and a 'prostitute', saying she wanted to be paid to have sex.

Swift and critical response to the statement came immediately from the left, and built over days to statements from Republicans calling the remarks inappropriate, including House Speaker John Boehner (Republican, Ohio) and National Republican Senatorial Committee Vice-Chair Carly Fiorina. Limbaugh was initially unmoved, saying on his March 1 show, 'If we are going to pay for your contraceptives, thus pay for you to have sex, we want something for it, and I'll tell you what it is: We want you to post the videos online so we can all watch'.

His intransigence proved costly. Responding to online outrage directed at the program's sponsors, advertisers began pulling their spots from the show on March 2, with Sleep Train Mattress Centers announcing its decision on Twitter: 'We don't condone negative comments directed toward any group. In response, we are currently pulling our ads from Rush with

Rush Limbaugh'. As online outrage and pressure on advertisers grew, Limbaugh apologized on his March 3 show: 'My choice of words was not the best, and in the attempt to be humorous, I created a national stir'; though some found the *mea culpa* lacking. The apology and a subsequent clarification did nothing to stem the sponsor losses, with some outlets reporting nearly 100 advertisers backing out of Limbaugh's show, an effect that stretched into 2014, more than two years later.[3] Pressure on advertisers was a clear trend on Twitter, as users called on Sleep Train, Netflix and others to pull their spots from the show.

Trayvon Martin: A Grassroots Firestorm

The Florida shooting of unarmed African-American teen Trayvon Martin in February 2012 slowly built into a national conversation on race and the US justice system, discourses that resurfaced to frame the August 2014 killing of Michael Brown in Ferguson, Missouri. Martin, 17, was shot and killed by George Zimmerman, a neighborhood watch volunteer, in Sanford, Florida, the night of February 26. Zimmerman was taken into custody immediately following the shooting but was released after claiming he acted in self-defense. Florida is a so-called 'stand your ground' state, providing legal protection for individuals who use deadly force to defend themselves when they feel their lives are in danger outside of their homes.

About ten days after Zimmerman's release, Martin's parents posted a petition to Change.org, urging Zimmerman's arrest, and the first mainstream news coverage was published the following day, when the parents filed suit to get records in the case. National media attention followed on March 13 to 15. A week later, the Sanford police chief stepped down from the investigation and a special prosecutor was assigned to the case. Scrutiny of the case ranged from the actions of the police and prosecutors to the media's inattention to the story. The controversy was charged both racially and politically, showing considerable polarization. In-person protests and demonstrations accompanied intense social media attention to the case in the early weeks.

Attention quickly narrowed to Martin's attire and questions of whether youths wearing 'hoodie' sweatshirts are inherently menacing. On March 23, speaking on *Fox and Friends*, Geraldo Rivera said, 'I think the hoodie is as much responsible for Trayvon Martin's death as George Zimmerman was'. Conservative blogs published excerpts from Martin's Twitter feed, including references to marijuana use (Graeff et al., 2014). Shortly after, a hacker broke into Martin's e-mail and social networking accounts and released personal information, which was subsequently covered by mainstream media. At the same time, Zimmerman's legal team set up a social media presence for their client, just as he earned scrutiny for posts about Mexicans made on his abandoned MySpace account. Nearly a month later, on April 11, 2012, the prosecutor announced second-degree murder charges against Zimmerman.

EXPLORING POLITICAL TALK ONLINE

Keyword Volume

To analyze the relative volume of keywords on Twitter, we graphed their concentration relative to all other content. Each line graph represents the proportion of the tweets within that level mentioning a keyword or hashtag. This allowed some insight into the intensity of political talk on these topics at these different levels and its flow between elites and their followers.

As can be seen in Figure 16.1, our analysis of the top-trending keywords on Twitter concerning the Sandra Fluke–Rush Limbaugh controversy

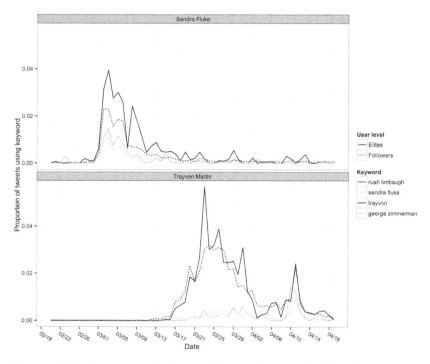

Figure 16.1   Proportional volume of keyword use for Sandra Fluke and Trayvon Martin cases

indicates a political conversation dominated by elites. The level 1 pool of candidates, party operatives, advocacy groups, media and pundits had a far greater within-level proportion of tweets, as volume relative to all other content, with the keywords 'Sandra Fluke' or 'Rush Limbaugh' than the randomly selected sample of their followers at level 2. This is particularly true for the keyword 'Rush Limbaugh', where elite attention appears to spur considerable attention among their followers.

In contrast, the Trayvon Martin shooting was initially discussed among the follower networks with somewhat more intensity than at the elite level. When the story initially gained attention in national news outlets between March 13 and 15, 2012, Twitter activity soon followed, with use of the keywords 'Trayvon' and 'George Zimmerman' proportionally higher for followers relative to the political and media elites. This was true through its first major spike of activity on March 20. Starting the next day and continuing through its peak on March 23, the date of both the special prosecutor's appointment and President Obama's public statement on the case, a greater proportion of elite posts mentioned Trayvon, suggesting a shift toward indexing. In contrast to the Sandra Fluke case, followers in aggregate appear to have begun the discourse on Trayvon absent elites, before major media outlets had started to pay attention to the issue.

Hashtag Clusters

Of course, keywords are only the starting point for any tracking over time of language use among elites and followers. To understand the specific uses and associations of different hashtags, we tracked the most prominent hashtags used to identify each case, and then conducted a principal component analysis of the use of particular hashtags by monitored accounts (see Table 16.2 and Table 16.3). We then verified these meaning clusters by comparing hashtag use relative to the other content of the tweet. That is, human coders verified whether these hashtags are used in consistent fashion in a random sample of actual tweets, and whether this use was consistent with the interpretation. This combination of scale building and content verification allowed us to capture the associations among hashtags quantitatively and qualitatively.

For the Fluke–Limbaugh case, we picked the hashtags that reflected support for Sandra Fluke and Rush Limbaugh, as well as the hashtags that represented calls to action. Principal component analysis in Table 16.2 suggests that there are two hashtag factors: a broad set discussing and supporting actors on both sides of the controversy (for example, #standwithsandra, #sandrafluke, #istandwithrush, #standwithrush, and #slutgate) and a distinct set calling for activism against Limbaugh (for example,

Table 16.2 *Principal component analysis of hashtag use in Sandra Fluke case*

	Factor 1	Factor 2	Communality
#boycottrush	0.87		0.76
#stoprush	0.87		0.76
#flushrush	0.78		0.62
#standwithrush		0.80	0.65
#standwithsandra	0.40	0.58	0.50
#istandwithrush		0.43	0.19
#sandrafluke		0.35	0.13
#slutgate		0.28	0.14

Note: Standardized loadings over 0.25 of varimax rotation.

Table 16.3 *Principal component analysis of hashtag use in Trayvon Martin case*

	Factor 1	Factor 2	Communality
#trayvon	0.92		0.89
#trayvonmartin	0.88		0.81
#zimmerman	0.60		0.37
#georgezimmerman	0.53		0.30
#iamtrayvon		0.82	0.68
#millionhoodiemarch		0.70	0.52

Note: Standardized loadings over 0.25 of varimax rotation.

#stoprush, #boycottrush, and #flushrush). Notably, the activism factor is more pronounced, with three clear loadings. The second factor, focusing on the actors and expressions of support for them, is more mixed, with relatively low communality estimates for a number of the hashtags, suggesting low fragmentation and considerable contestation within this cluster.

Human coders checked hashtag contexts by looking at 1,023 randomly selected tweets related to Sandra Fluke, a subset of which contained the relevant hashtags. The results suggest that tweets using 'activism' hashtags are overwhelmingly opposed to Limbaugh, and 37 percent of those make explicit calls for boycotting Limbaugh's radio program. In addition, most of the tweets that use #istandwithrush, #standwithrush, and #slutgate hashtags express sentiment positive toward Limbaugh or critical of Fluke, whereas 82 percent of tweets using #standwithsandra are supportive of Fluke. Based on this analysis and the principal component analysis, we distinguished the pro-Fluke, represented by #standwithsandra and #sandrafluke, from the pro-Limbaugh, represented by #istandwithrush, #standwithrush, and #slutgate, with these two distinct from the call to action against Rush represented by #stoprush, #boycottrush, and #flushrush.

A parallel process was used for the Trayvon Martin case. We began by selecting a number of hashtags that expressed key ideas, such as a dominant set possibly reflecting polarization and another set possibly expressing the idea of solidarity. Principal component analysis in Table 16.3 reveals that the hashtags can be grouped in two factors: hashtags debating the actors at the center of the controversy, Martin and Zimmerman (for example, #trayvon, #trayvonmartin, #zimmerman, and #georgezimmerman) and hashtags related to solidarity expression (for example, #iamtrayvon and #millionhoodiemarch). Again, the relatively low commonality estimates for the George Zimmerman hashtags within the actor factor suggest a more complex picture.

To confirm this classification, human coders checked hashtag contexts by looking at 1980 randomly selected tweets related to Trayvon Martin, a subset of which contained the relevant hashtags. Each hashtag was coded for whether it was an expression of support for Martin, opposition to Martin, or a call to action. Results suggest that 84 percent of tweets that use #trayvon and #trayvonmartin show overall support for Martin. For the hashtags #zimmerman and #georgezimmerman, 51 percent of tweets support Trayvon Martin, and 40 percent of them support Zimmerman, reflecting greater ambivalence. As we expected, a majority of the tweets that use solidarity hashtags support Martin over Zimmerman, with 24 percent of tweets that use #millionhoodiemarch and #iamtrayvon explicitly expressing calls to action. Since we can see clear distinctions in the use of hashtags to solicit support for Martin and Zimmerman, we grouped #trayvon and #trayvonmartin as pro-Trayvon and #zimmerman and #georgezimmerman as an alternate grouping, given that it was more mixed. We also built a solidarity grouping using #millionhoodiemarch and #iamtrayvon, a type of call to action.

Expression Intensity

Returning to the case of Sandra Fluke, when the volume of top relevant tweets is separated into the expression of particular hashtags, the elite group generates a greater proportion of tweets related to the controversy on all keywords and hashtags except the activism grouping of #stoprush, #boycottrush, and #flushrush (see Figure 16.2). These hashtags emerged March 2, the day after Limbaugh doubled down on his Fluke comments

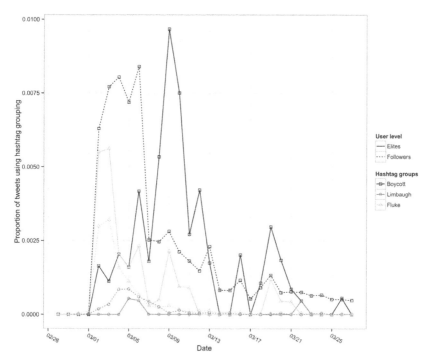

Figure 16.2   Proportional volume of hashtag clusters for Sandra Fluke case

and Sleep Train announced its withdrawal of advertisements. This call to action trailed off more quickly among the followers than the elites, who discussed the developing story, carrying the theme in about 1 percent of all tweets well into the next week. Proportionally, the elite group tweeted more about the case, with spikes in their activity marked by spikes among followers shortly afterward in most other ways.

These patterns may point to a number of possible influences. The Fluke case was a particularly Washington, DC-centric affair, beginning with Congressional hearings and quickly emerging as a flashpoint in the presidential campaign. The Washington, DC insider nature of the elite panel makes the group more likely to focus a greater proportion of Twitter activity on issues in their own arena. We also suspect Limbaugh's national profile may have led to quick and expansive focus among level 1 elites, perhaps indicated by the far greater use of Limbaugh's name than Fluke's (nearly 4.5 percent of tweets to 1.5 percent of tweets at the peak on March 3, 2012). Conversation among elites may also have been influenced by efforts from the left to highlight the case as an example of the 'war on

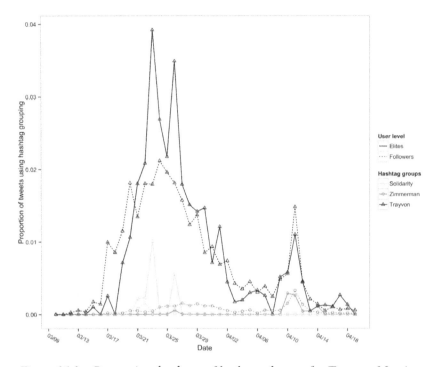

Figure 16.3   Proportional volume of hashtag clusters for Trayvon Martin case

women' and fundraise on the point. In contrast, it is notable that the followers were ahead of the elites in pointing toward a boycott, suggesting grassroots leadership on this matter.

Turning to the Trayvon Martin case, parallel analysis finds that among elites and their followers, #trayvon and #trayvonmartin were the most frequently used hashtags, followed by #georgezimmerman and #zimmerman (see Figure 16.3). In addition, a set of call-to-action hashtags appeared, including #iamtrayvon and #millionhoodiemarch, but was used far less frequently (less than 1 percent at its peak). It is surprising, then, that reporters reviewed the importance and effects of Internet activism through social media tools for this case (Graeff et al., 2014).

As an example of everyday political talk, the timing and volume of activity in the Martin case differed from what we found with Limbaugh and Fluke. Where elites quickly and clearly led and dominated the conversation surrounding Fluke, the Martin case shows non-elite followers picking up the story earlier and discussing it with more intensity before elites reacted and came to lead the conversation. Again, without a more detailed content analysis and message tracking – issues we return to below – we can only speculate on the reasons for this. It seems likely that the difference reflects early grassroots efforts to urge media and political actors to notice the case and direct attention to it. Followers began tweeting the keyword 'Trayvon' by March 8 and began circulating the hashtags #trayvon and #trayvonmartin by March 11, in the days following Martin's parents posting their petition on Change.org. The elite group first tweeted these hashtags on March 15, after national media coverage had begun.

However, once the elite group began discussing the case on Twitter, those users paid far more attention proportionally than the follower networks, coming close to double the percentage of tweets with #trayvon and #trayvonmartin on the March 23 spike. Although the elites spiked higher in tweets related to the case, their attention dropped off more quickly. The non-elite followers displayed more sustained conversation about the case, with drops that were less precipitous than those observed among the elites. Notably, the call to action implicit in the #iamtrayvon and #millionhoodiemarch hashtags were proportionally a very small part of the online conversation among elites and even less so among followers, indicating that this grassroots effort was somewhat smaller than the one behind the Rush Limbaugh boycott.

Network Mapping

For each of the topics, we generated a retweet network based upon tweets that included the major keywords and hashtags. Retweets are tweets in which users share content produced by another user, a near perfect reflection of Tarde's 'invention' and 'imitation'. Visualizations of the retweet network in both cases show clear distinctions between core groups discussing the issues. The graph depicts users and their centrality within the retweet network, with each node (bubble) in the network representing a user; the size of the node denotes how much that person retweets and is retweeted within the network; and the edges denoting a retweet. These visualizations represent the largest connected components of activity for each case. Users outside these components are not represented because of minimal interaction with this core group.[4]

Upon visual inspection of the Fluke network, this case shows two core clusters speaking almost entirely separately from each other (see Figure 16.4). One cluster is focused heavily on the boycott-focused hashtags of #stoprush, #boycottrush and #flushrush, represented in black. Retweets containing these hashtags clearly dominate that core, showing a group of users heavily focused on calls to action. This cluster has fewer instances of the grouping of #standwithsandra and #sandrafluke (denoted

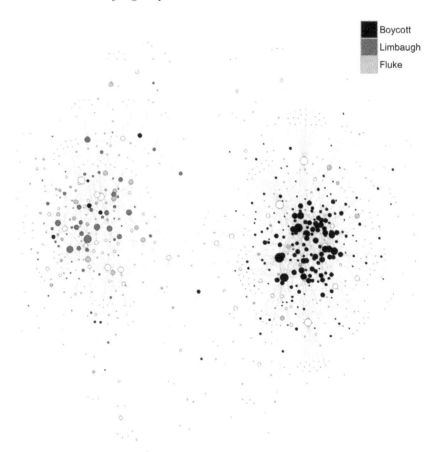

Figure 16.4   Retweet network for Sandra Fluke case with major hashtags

with light gray), as well as activity with keywords without hashtag (white). The calls to boycott are denser at the center of the network than the non-activist tweets, indicating activity focused more squarely on a set of central users. The activity among the nodes representing discussion of the actors in the event appears more disperse and less hierarchical.

The contrasting core of users tweeting about the Fluke case has a retweet network far less focused on political action. The most dominant hashtag use is the cluster for #standwithrush, #istandwithrush and #slutgate (dark gray). This network cluster also has relatively heavy use of the hashtag grouping of #standwithsandra and #sandrafluke, also apparent from our factor analysis, suggesting a spirited discussion about the

*Figure 16.5   Retweet network for Trayvon Martin case with major
             hashtags*

various players. This cluster was considerably less dense and active than
the boycott-dominated core.

The network in the Trayvon Martin case also split into two distinct
clusters (see Figure 16.5). Both were far less activist in their hashtag
use and focused almost entirely around the use of #trayvon and #tray-
vonmartin (black), with contending characterizations of the case and its
players. Each node represents a user who used this hashtag or retweeted
a message containing the hashtag. The grouping for #zimmerman and
#georgezimmerman (dark gray) appears in one cluster but with nowhere
near the dominance of the Trayvon-focused grouping. The solidarity-
focused grouping of #iamtrayvon and #millionhoodiemarch (light gray)

is virtually invisible in the network, showing almost no retweet activity among the hashtags.

The two clusters, both relying on the centrality of the Trayvon hashtags, have substantially more between-cluster interaction than we found in the Fluke case. Hashtag use within the network does not focus on calls to action though further analysis may reveal these in tweet content, rather than hashtag use. Overall, the cluster with the Zimmerman-focused grouping (dark gray) is less dense than the cluster containing the Trayvon-focused grouping.

As observed with the use of keywords and the clustering of hashtags, these two cases functioned quite differently, despite the fact that both addressed controversies regarding inequalities. Both apparently featured social influence by elites and followers – suggesting both a two-step flow and a two-way flow of communication – though who appears to exercise influence differs by topic and stages in each controversy's evolution, especially as it related to calls to action. The visualizations of these networks confirm these differences, and suggest that measuring message propagation, information diffusion, and interpersonal influence across these fluid levels is the future of research on everyday political talk.

LOOKING FORWARD

Rather than reflecting on the specifics of these cases, which were mainly deployed to illustrate the most basic potential of computational approaches to understanding online political talk, we close by using them to offer some broader suggestions about the future of online politics research and the implications of big data for this coming age of computational social science. In doing so, we look to the past, returning to the work of Gabriel Tarde that began this chapter. The digital traces contained in a single tweet provide a myriad of directions for research on interpersonal interaction and social influence linked directly to Tarde's main theoretical contributions. Yet the contemporary environment also complicates this linear formulation of elite influence through the press on to the constituted public, setting the conversational agenda and thereby distilling opinion into action. Rather, as we have observed here, within the feedback loops provided by the new information environment and social networking technologies, elite agendas appear to be influenced by their followers under certain circumstances, substantially complicating the framework Tarde proposed.

It may be that the most useful insights from Tarde are not derived from his early formulation of two-step flow or communication mediation, but instead from his assertion about the types of data that should be collected as part of a 'science of the social'. His approach, its full potential unrealized at the time, refuses the schism between individual and society as we have come to understand them after Durkheim. Moreover, technology and methodology has caught up to Tarde's theory, for as Latour writes (2010: 160):

> What we are witnessing, thanks to the digital medium, is a fabulous extension of [Tarde's] principle of traceability. It been put in motion not only for scientific statements, but also for opinions, rumors, political disputes, individual acts of buying and bidding, social affiliations . . . What has previously been possible for only scientific activity – that we could have our cake (the aggregates) and eat it too (the individual contributors) – is now possible for most events leaving digital traces, archived in digital databanks.

As this suggests, computational approaches that stress syntactical coding such as natural language processing may prove critical to understanding the nature of social influence and interpersonal dynamics in everyday political talk (Bird et al., 2009; Jurafsky and Martin, 2009; Shah et al., 2002; Han et al., 2011).[5] In addition, the growth of supervised and unsupervised machine learning tools focusing on the statistical co-occurrence of words or phrases may also prove powerful for managing large volumes of political messages (Blei and Lafferty, 2009; Hopkins and King, 2010). Such language processing allows for the coding of vast amounts of communications with considerable subtlety and qualification.

Complementing these language-processing approaches are tools for tracking their movement through social networks. Efforts to examine the role of audience size, pass-along value, and conversational ability on social influence now have ways of examining these questions at a large scale (Cha et al., 2010) and can do so alongside measures of whether the message is deemed interesting or elicits positive feelings (Bakshy et al., 2011), blending the structural and the psychological. Indeed, social networking sites like Twitter and Facebook provide 'a setting where many different kinds of information spread in a shared environment' (Romero et al., 2011: 695), permitting new insights regarding invention and expression of ideas and their imitation and diffusion with communication networks. These methods will provide new vistas onto political talk as it increasingly occurs through online channels, calling into question the accepted wisdom about message flows and offering new accounts of emergent forms of civic and political expression.

NOTES

1. As Terry Clark writes in the introduction of *Gabriel Tarde On Communication and Social Influence: Selected Papers* (1969: 25): 'Durkheim refused to accept that sociological principles should be grounded in psychology. Sociology as a distinctive science, he held, must take as its object of study social facts; and these social facts must find their causes as well as their consequences in other distinctly social facts'.
2. A brief tutorial of how to collect Twitter data using a simple Python script can be found at http://badhessian.org/2012/10/collecting-real-time-twitter-data-with-the-streaming-api/.
3. Online activists have maintained detailed databases of Rush Limbaugh sponsors, tracking continued supporters (who remain targets of a boycott), and noting which brands have pulled advertisements. At the time of writing in Fall 2014, these campaigners claim more than 2,700 local and national advertisers have withdrawn support (http://stoprush.net/rush_limbaugh_sponsor_list.php#current_a).
4. The Fluke network overall comprises 13,365 users, with this largest component including 870. For Trayvon, the overall network comprises 13,365, with 856 users in the largest component. The vast majority of users in both are 'isolates', nodes not tied with any other users.
5. Two main lines of modeling that have been developed do not rely so heavily on word counts and the manual curation of specific words: language modeling and statistical modeling (Monroe and Schrodt, 2008). Language modeling attempts to leverage as much information as it can out of a single document; it attempts to identify parts of speech in a given document and allows us to see the who, what, when, where, and how of a message.

REFERENCES

Bakshy, E., Hofman, J.M., Mason, W.A. and Watts, D.J. (2011). Everyone's an influencer: quantifying influence on twitter. In *Proceedings of the Fourth ACM International Conference on Web Search and Data Mining*, February (pp. 65–74). ACM.
Barber, B. (1984). *Strong Democracy: Participatory Politics for a New Age*. Berkeley, CA: University of California Press.
Bennett, W.L., Wells, C. and Freelon, D. (2011). Communicating civic engagement: contrasting models of citizenship in the youth web sphere. *Journal of Communication*, 61, 835–856.
Berelson, B., Lazarsfeld, P.F. and McPhee, W.N. (1954). *Voting: A Study of Opinion Formation in a Presidential Campaign*. Chicago, IL: University of Chicago Press.
Bird, S., Klein, E. and Loper, E. (2009). *Natural Language Processing with Python*. Sebastopol, CA: O'Reilly Media.
Blei, D.M. and Lafferty, J.D. (2009). Topic models. *Text Mining: Classification, Clustering, and Applications*, 10, 71.
Bode, L., Vraga, E K., Borah, P. and Shah, D.V. (2014). A new space for political behavior: political social networking and its democratic consequences. *Journal of Computer-Mediated Communication*, 19(3), 414–429.
Boyd, D.M. and Ellison, N.B. (2007). Social network sites: definition, history, and scholarship. *Journal of Computer-Mediated Communication*, 13, 210–230.
Boyd, D., Golder, S. and Lotan, G. (2010). Tweet, tweet, retweet: conversational aspects of retweeting on twitter. In *2010 43rd Hawaii International Conference on System Sciences (HICSS)* (pp. 1–10). January, IEEE.
Bruns, A. and Burgess, J.E. (2011). The use of Twitter hashtags in the formation of ad hoc publics. Paper presented at the 6th European Consortium for Political Research General Conference, August 25–27, University of Iceland, Reykjavik.
Bruns, A. and Highfield, T. (2013). Political networks on twitter: tweeting the Queensland state election. *Information, Communication and Society*, 16(5), 667–691.

Cha, M., Haddadi, H., Benevenuto, F. and Gummadi, P.K. (2010). Measuring user influence in Twitter: the million follower fallacy. *ICWSM*, 10, 10–17.

Cho, J. Shah, D.V., McLeod, J.M., McLeod, D.M., Scholl, R.M. and Gotlieb, M.R. (2009). Campaigns, reflection, and deliberation: advancing an O-S-R-O-R model of communication effects. *Communication Theory*, 19, 66–88.

Clark, Terry (ed.) (1969). *Gabriel Tarde On Communication and Social Influence: Selected Papers*. Chicago, IL: University of Chicago Press.

Dahlgren, P. (2000). The Internet and the democratization of civic culture. *Political Communication*, 17, 335–340.

Dahlgren, P. (2005). The Internet, public spheres, and political communication: dispersion and deliberation. *Political Communication*, 22(2), 147–162.

Dylko, I.B., Beam, M.A., Landreville, K.D. and Geidner, N. (2012). Filtering 2008 US presidential election news on YouTube by elites and nonelites: an examination of the democratizing potential of the internet. *New Media and Society*, 14(5), 832–849.

Ekström, M. and Östman, J. (2013). Information, interaction, and creative production: the effects of three forms of Internet use on youth democratic engagement. *Communication Research*, online pre-publication, 0093650213476295, 1–22.

Eveland, W.P. and Hively, M.H. (2009). Political discussion frequency, network size, and 'heterogeneity' of discussion as predictors of political knowledge and participation. *Journal of Communication*, 59(2), 205–224.

Freelon, D.G. (2010). Analyzing online political discussion using three models of democratic communication. *New Media and Society*, 12(7), 1172–1190.

Gainous, J. and Wagner, K.M. (2014). *Tweeting to Power: The Social Media Revolution in American Politics*. New York: Oxford University Press.

Gamson, W.A. (1992). *Talking Politics*. New York: Cambridge University Press.

Gil de Zúñiga, H.G., Jung, N. and Valenzuela, S. (2012). Social media use for news and individuals' social capital, civic engagement and political participation. *Journal of Computer-Mediated Communication*, 17(3), 319–336.

Gil de Zúñiga, H.G., Puig-I-Abril, E. and Rojas, H. (2009). Weblogs, traditional sources online and political participation: an assessment of how the internet is changing the political environment. *New Media and Society*, 11(4), 553–574.

Golbeck, J., Grimes, J. and Rogers, A. (2010). Twitter use by the UC Congress. *Journal of the American Society for Information Science and Technology*, 61(8), 1612.

Graeff, E., Stempeck, M. and Zuckerman, E. (2014). The battle for 'Trayvon Martin': Mapping a media controversy online and off-line. *First Monday*, 19(2). http://firstmonday.org/ojs/index.php/fm/article/view/4947.

Graham, T. and Hajru, A. (2011). Reality TV as a trigger of everyday political talk in the net-based public sphere. *European Journal of Communication*, 26(1), 18–32.

Gueorguieva, V. (2008). Voters, MySpace, and YouTube the impact of alternative communication channels on the 2006 election cycle and beyond. *Social Science Computer Review*, 26(3), 288–300.

Habermas, J. (1984). *The Theory of Communicative Action: Vol. 1. Reason and the Rationalization of Society*. McCarthy, T. (trans.). Boston, MA: Beacon.

Han, J.Y., Shah, D.V., Kim, E., Namkoong, K., Lee, S.Y., Moon, T.J., Cleland, R., McTavish, F.M. and Gustafson, D.H. (2011). Empathic exchanges in online cancer support groups: distinguishing message expression and reception effects. *Health Communication*, 26(2), 185–197.

Hanna, A., Wells, C., Maurer, P., Friedland, L., Shah, D. and Matthes, J. (2013, October). Partisan alignments and political polarization online: a computational approach to understanding the French and US presidential elections. In *Proceedings of the 2nd Workshop on Politics, Elections and Data* (pp. 15–22). ACM.

Hardy, B.W. and Scheufele, D.A. (2005). Examining differential gains from Internet use: comparing the moderating role of talk and online interactions. *Journal of Communication*, 55(1), 71–84.

Hopkins, D.J. and King, G. (2010). A method of automated nonparametric content analysis for social science. *American Journal of Political Science*, 54(1), 229–247.

Huckfeldt, R. and Sprague, J. (1995). *Citizens, Politics, and Social Communication: Information and Influence in an Election Campaign.* New York: Cambridge University Press.

Jungherr, A., Jürgens, P. and Schoen, H. (2012). Why the Pirate Party won the German election of 2009 or the trouble with predictions: a response to Tumasjan, A., Sprenger, T.O., Sander, P.G. and Welpe, I.M. 'Predicting Elections with Twitter: What 140 Characters Reveal about Political Sentiment'. *Social Science Computer Review*, 30(2), 229–234.

Jurafsky, D. and Martin, J.H. (2009). *Speech and Language Processing: An Introduction to Natural Language Processing, Speech Recognition and Computational Linguistics*, 2nd edn. Upper Saddle River, NJ: Prentice Hall.

Katz, E. (2006). Rediscovering Gabriel Tarde. *Political Communication*, 23(3), 263–270.

Katz, E. and Lazarsfeld, P. (1955). *Personal Influence: The Part Played by People in the Flow of Mass Communications.* Glencoe, IL: Free Press.

Kim, J., Wyatt, R.O. and Katz, E. (1999). News, talk, opinion, participation: the part played by conversation in deliberative democracy. *Political Communication*, 16, 361–385.

Kinnunen, J. (1996). Gabriel Tarde as a founding father of innovation diffusion research. *Acta Sociologica*, 39(4), 431–442.

Kwak, N., Williams, A., Wang, X. and Lee, H. (2005). Talking politics and engaging politics: an examination of the interactive relationships between structural features of political talk and discussion engagement. *Communication Research*, 32, 87–111.

Latour, B. (2010). Tarde's idea of quantification. In Candea, M. (ed.), *The Social After Gabriel Tarde: Debates and Assessments.* London: Routledge.

Lazarsfeld, P.F., Berelson, B. and Gaudet, H. (1944). *The People's Choice: How the Voter Makes Up His Mind in a Presidential Campaign.* New York: Duell, Sloan & Pearce.

Lazer, D., Pentland, A.S., Adamic, L., Aral, S., Barabasi, A.L., Brewer, D., . . . and Van Alstyne, M. (2009). Life in the network: the coming age of computational social science. *Science*, 323(5915), 721.

Lee, N.J., Shah, D.V. and McLeod, J.M. (2013). Processes of political socialization a communication mediation approach to youth civic engagement. *Communication Research*, 40(5), 669–697.

McLeod, J.M., Daily, K., Guo, Z., Eveland, W.P., Bayer, J., Yang, S. and Wang, H. (1996). Community integration, local media use, and democratic processes. *Communication Research*, 23(2), 179–209.

McLeod, J.M., Scheufele, D.A. and Moy, P. (1999). Community, communication, and participation: the role of mass media and interpersonal discussion in local political participation. *Political Communication*, 16, 315–336.

Monroe, B.L. and Schrodt, P.A. (2008). Introduction to the special issue: the statistical analysis of political text. *Political Analysis*, 16(4), 351–355.

Mutz, D.C. (2006). *Hearing the Other Side: Deliberative Versus Participatory Democracy.* New York: Cambridge University Press.

Namkoong, K., Shah, D.V., Han, J.Y., Kim, S.C., Yoo, W., Fan, D., . . . and Gustafson, D.H. (2010). Expression and reception of treatment information in breast cancer support groups: how health self-efficacy moderates effects on emotional well-being. *Patient Education and Counseling*, 81, S41–S47.

Papacharissi, Z. (2004). Democracy online: civility, politeness, and the democratic potential of online political discussion groups. *New Media and Society*, 6(2), 259–283.

Papacharissi, Z. (ed.). (2010). *A Networked Self: Identity, Community, and Culture on Social Network Sites.* New York: Routledge.

Pasek, J., More, E. and Romer, D. (2009). Realizing the social Internet? Online social networking meets offline civic engagement. *Journal of Information Technology and Politics*, 6(3–4), 197–215.

Pingree, R.J. (2007). How messages affect their senders: a more general model of message effects and implications for deliberation. *Communication Theory*, 17(4), 439–461.

Price, V. and Cappella, J.N. (2002). Online deliberation and its influence: the electronic dialogue project in campaign 2000. *IT and Society*, 1(1), 303–329.

Rogers, E.M. and Shoemaker, F.F. (1971). *Communication of Innovations: A Cross-Cultural Approach*. New York: Free Press.

Romero, D.M., Meeder, B. and Kleinberg, J. (2011, March). Differences in the mechanics of information diffusion across topics: idioms, political hashtags, and complex contagion on Twitter. In *Proceedings of the 20th International Conference on World Wide Web* (pp. 695–704). ACM.

Schudson, M. (1978). The ideal of conversation in the study of mass media. *Communication Research*, 5(3), 320–329.

Segerberg, A. and Bennett, W.L. (2011). Social media and the organization of collective action: using Twitter to explore the ecologies of two climate change protests. *Communication Review*, 14(3), 197–215.

Shah, D.V., Cho, J., Eveland, W.P. and Kwak, N. (2005). Information and expression in a digital age: modeling Internet effects on civic participation. *Communication Research*, 32, 531–565.

Shah, D.V., Cho, J., Nah, S., Gotlieb, M.R., Hwang, H., Lee, N., Scholl, R.M. and McLeod, D.M. (2007). Campaign ads, online messaging, and participation: extending the communication mediation model. *Journal of Communication*, 57, 676–703.

Shah, D.V. and Scheufele, D.A. (2006). Explicating opinion leadership: nonpolitical dispositions, information consumption, and civic participation. *Political Communication*, 23(1), 1–22.

Shah, D.V., Watts, M.D., Domke, D. and Fan, D.P. (2002). News framing and cueing of issue regimes: explaining Clinton's public approval in spite of scandal. *Public Opinion Quarterly*, 66(3), 339–370.

Sotirovic, M. and McLeod, J.M. (2001). Values, communication behavior, and political participation. *Political Communication*, 18, 273–300.

Sotirovic, M. and McLeod, J.M. (2004). Knowledge as understanding: the information processing approach to political learning. In Kaid, L. (ed.), *Handbook of Political Communication Research*. Mahwah, NJ: Lawrence Erlbaum Associates.

Tarde, G. (1969 [1898]). *Gabriel Tarde On Communication and Social Influence: Selected Papers*, Vol. 334. Chicago, IL: University of Chicago Press.

Tarde, G. (1903 [1890]). *The Laws of Imitation*, trans. E.C. Parsons. New York: Henry, Holt.

Thackeray, R. and Hunter, M. (2010). Empowering youth: use of technology in advocacy to affect social change. *Journal of Computer-Mediated Communication*, 15(4), 575–591.

Tumasjan, A., Sprenger, T.O., Sandner, P.G. and Welpe, I.M. (2010). Predicting elections with Twitter: what 140 characters reveal about political sentiment. *ICWSM*, 10, 178–185.

Valenzuela, S., Kim, Y. and de Zúñiga, H.G. (2012). Social networks that matter: exploring the role of political discussion for online political participation. *International Journal of Public Opinion Research*, 24(2), 163–184.

Vitak, J., Zube, P., Smock, A., Carr, C., Ellison, N. and Lampe, C. (2011). It's complicated: Facebook users' political participation in the 2008 election. *Cyberpsychology, Behavior and Social Networking*, 14(3), 107–114.

Walsh, K.C. (2012). Putting inequality in its place: rural consciousness and the power of perspective. *American Political Science Review*, 106(03), 517–532.

Wimsatt, W.C. (2007). *Re-engineering Philosophy for Limited Beings: Piecewise Approximations to Reality*. Cambridge, MA: Harvard University Press.

Ye, S. and Wu, S.F. (2010). Measuring message propagation and social influence on Twitter. com. In IEEE (ed.), *Social Informatics*. Berlin and Heidelberg: Springer.

17. Two-screen politics: evidence, theory and challenges
Nick Anstead and Ben O'Loughlin

INTRODUCTION

For many members of the audience, the experience of consuming traditional media has changed. No longer is it an event that takes place wholly in the privacy and relative seclusion of the living room. The development of social media platforms as well as relatively accessible communication hardware has given audiences the ability to publically debate, question and pontificate about what they are watching in real time to people not physically proximate. This new phenomenon has been given a number of names. In the marketing world in particular it has become known as 'two-screen viewing', reflecting the pattern of simultaneous television watching while reading and creating online content on a smartphone, laptop or tablet (Kelly, 2011; Moses, 2012). Alternatively, the UK's TV Licensing Authority referred to this new behaviour as 'chatterboxing' (TV Licensing, 2012). A Pew research report published in the USA referred to the 'connected viewer' (Smith and Boyles, 2012). Focusing on political broadcasting, our own work has noted the development of what we term the 'Viewertariat', arguing that the ability to comment on media (once largely the sole preserve of the professional journalistic class, often termed the 'Commentariat') has been decentralized to a much wider swathe of the population (Anstead and O'Loughlin, 2011).

Whatever term we choose to describe it, there is no doubt that the advent of social media is changing the way we watch traditional media, especially television. This has important consequences for audiences, who are able both to express their own opinions and to monitor the opinions of their fellow audience members in real time. For scholars in the field it has at least two important ramifications. In the first instance, social media has the potential to provide us with vast amounts of data for our own research endeavours, and even create new forms of 'live research' (Elmer, 2013). However, this raises important questions as to the usefulness and validity of the datasets concerned. What exactly can we extrapolate from them? Second, and as significant, the data generated by two-screen viewing is likely to be used by other political actors – including government, politicians, broadcasters, advertisers and corporations – for their own purposes (Anstead and O'Loughlin, 2014). This allows them to modify their own use of social media and apply strategic communications techniques to manage and steer public conversations, thereby 'polluting' the social media datasets scholars work with. Thus we need to work to understand exactly how these data and their use influences pre-existing institutions.

For political scientists and especially those interested in media and communications, two-screen politics has particular resonance, since television has long been a central focus in the sub-field. In recent years, some scholars and commentators interested in the internet have played something of a revisionist role, but even then the arguments they made tended to be based on at least the implicit assumption that 'new' media would be different to television: while television is portrayed as passive and isolating, the internet was presented as a mobilizing public space (Castells, 2009; Trippi, 2008). As such, the interaction and the merging of the televisual and social media have profound consequences for these areas of the discipline and the questions that they address.

This chapter sketches out some of the key issues that this raises, for both scholars and practitioners. We argue that two-screen viewing matters both because of its quantitative penetration and because, perhaps more significantly, the idea of the connected viewers fits into a number of broader trends in the study of contemporary media. This presents a challenge for traditional theories attempting to explain the role of the media in political life. We conclude by offering an overview of the challenges, both research-based and practical, that emerge in this new landscape.

DOES TWO-SCREEN VIEWING MATTER? DATA AND EXISTING LITERATURE

One simple way of understanding the scale of two-screen viewing is through the social media data that is being produced by viewers. Certainly, the number of updates published on Twitter during important political and broadcast events has been considered newsworthy. For example, during the first UK prime ministerial debate in 2010, the politics-focused Twitter monitoring firm Tweetminster claimed that 184 396 debate-related tweets had been published during the hour and a half long broadcast (Tweetminster.com, 2010), a figure which was widely reported in UK news media at the time. In the USA, it was claimed that the 2012 presidential election night saw the highest number of event-focused tweets ever published, with more than 20 million election-related status updates appearing (Taylor, 2012). However, these figures need to be treated with some care. Generally, the figures for social media updates are considerably smaller than the overall viewership of the broadcast (for example the first UK prime ministerial debate was viewed by 8.4 million people). Furthermore, the overall number of tweets may prove deceptive if we are seeking to assess the number of people actually publishing social media content, since individuals can publish multiple comments (analysts can control for this by counting unique users, but journalists may focus on such headline figures). This creates a long-tail effect, where a very limited number of people are responsible for a huge bulk of updates (Anstead and O'Loughlin, 2011; Bruns et al., 2012; Bruns et al., 2013).

An alternative way of understanding the extent of two-screen viewing is by using survey research. There are a number of pieces of data which offer some insight into the extent of two-screen viewing among the overall population. In the UK, the television regulator Ofcom has undertaken survey research that has found 54 per cent of the population has used the internet while watching television (Ofcom, 2013). In the US, research carried out by the Pew Research Center's Internet and American Life Project found that 52 per cent of the US population had watched the television while using an internet-enabled device (Smith and Boyles, 2012).

Results such as these clearly suggest that two-screen viewing has become a reasonably common practice. However, it is still important to raise a caveat on these data. In particular, it remains important to ask exactly what people are doing online when their televisions are also on. The research by Pew, for example, included evidence not just of engagement with the broadcast content, but also of disengagement. Certainly, sizeable proportions of the sample were, for example, commenting on the programme they were watching (2 per cent of all cell phone owners / 19 per cent of smartphone owners) or checking if statements made in the programme were true (6 per cent / 37 per cent). However, the largest group were actually using their phones and mobile computing devices to pass the time during commercial breaks (17 per cent / 58 per cent) (Smith and Boyles, 2012). A similar study undertaken by Google, Ipsos and Sterling in the USA supported this distinction. While it found that 77 per cent of people watch TV with another device, only 22 per cent of people actually engage in simultaneous use of the second device that is complementary to the other media source (Google et al., 2012).

It can also be argued that the rise of two-screen viewing is significant because it is representative of wider shifts in media practices and research. It is a good example of what has been termed media convergence (Jenkins, 2008) or hybridity (Chadwick, 2011, 2013). Although they draw on quite different traditions, these two ideas share one theme in common, arguing that the hard distinction between old and new media is no longer sustainable, nor the zero-sum idea that the success of one is to the detriment of the other. Rather, the key question becomes how the two co-exist and interact with each other. Two-screen viewing is a good example of this, as viewers are developing new ways of engaging with quite traditional forms of broadcasting. Programme makers must adapt to this. We analysed Twitter responses to an episode in 2009 of the BBC's flagship political panel show, *Question Time*, and found several new patterns of interaction (Anstead and O'Loughlin, 2011). In the first instance, it was members of the viewing audience that started this engagement, creating their own hashtag for discussing programme content. Noting this development, the show's producers themselves went on to fully embrace Twitter as a part of the programme's format, not only publicizing the hashtag in the broadcast, but also going so far as to add an additional panel member who would respond to the broadcast on Twitter (BBC News Online, 2012). This neatly encapsulates both the challenges and the opportunities that exist for traditional media in the hybrid information system.

A second strand of contemporary thought highly relevant to understanding broader interest in two-screen viewing is the widespread interest in so-called 'big data' methods for research (Anderson, 2008; Boyd and Crawford, 2012; Mayer-Schonberger and Cukier, 2013). This exists not just among scholars, but also relates to techniques being developed by government and the private sector. As Mayor-Schonberger and Cukier argue, the idea of big data defies easy definition (2013: 6). However, it is clearly a multifaceted development, encompassing the production, storage and analysis of vast datasets. Writing in *Wired* magazine in 2008, editor Chris Anderson described the big data era as a time when there would be 'Sensors everywhere. Infinite storage. Clouds of processors. Our ability to capture, warehouse and understand massive amounts of data is changing science, medicine, business, and technology' (Anderson, 2008).

Two-screen viewing is part of this historical shift in communication and society. Harvesting social media is one of the prime environments for the creation of big datasets (Manovich, 2011). Indeed, the connected viewer seems rather akin to Anderson's idea of the ubiquitous sensor, responding in real time to the broadcast stimulus. Additionally, it is important to note developments in techniques deployed for the analysis of big datasets, especially natural language processing technologies. Here computers are used to read and attribute numeric values to harvested data, especially focusing on what is commented on and the extent to which it contains positive or negative sentiment. Such techniques are now employed by a number of consultancy firms working in the marketing space, who claim to be able to analyse public reaction to products, brands and events in real time using social media data (Anstead and O'Loughlin, 2012). Big data techniques are used by media organizations to understand their audiences. For instance the BBC's Global News Audiences team use social media analytics to understand reactions to their content, but also to identify political attitudes among those audiences. O'Loughlin carried out analysis with them during the London 2012 Olympic Games, harvesting data in Arabic, Russian, Persian and English to explore how audience conversations during the Games were inflected by attitudes to gender, religion and nationhood (Gillespie et al., 2013). A number of academic teams have also conducted big data studies of politics. Axel Bruns et al. have used Twitter data to analyse interactions between English- and Arabic-speaking users during the 2011 Arab Spring (Bruns et al., 2013), how citizens and authorities use social media to negotiate their responses to natural disasters (Bruns et al., 2012; Shaw et al., 2013), and how audience responses to the Eurovision Song Contest are structured around nation and ethnicity (Highfield et al., 2013). All were major television events that triggered audiences to take on differing roles as witnesses, spectators, participants, activists, cheerleaders, and so on, but whose responses fed back into events and could then be analysed and reflected upon by public authorities, broadcasters and research teams.

The manner in which organizations make use of such data is a significant argument in favour of the importance of two-screen viewing. Even if we remain cautious about the validity of the data being gathered and the claims that can be made based on its analysis, there can be no doubt that these data are being harvested and the findings are being published. There is also no doubt that these techniques are being used by political parties and campaign teams to monitor the reputation of their leaders and policies (Anstead and O'Loughlin, 2012). As such, it becomes very important to attempt to offer some context for the use of two-screen viewing and to cast a critical gaze on it.

THEORETICAL QUESTIONS

The development of two-screen viewing raises a number of important theoretical questions for political and media theorists. These are especially relevant when we consider debates about democracy and new media, and the potential that new modes of communication have for increasing either the quality or the quantity of political participation and discussion.

One important question relates to the quality of political content published on social media. There is a sizeable literature of studies investigating whether it measures up to the standards of debate set by normative models of democracy. The pre-eminent idea in this debate is Jurgen Habermas's deliberative democratic model, in which citizens engage in rational debate to arrive at a consensus-based outcome and become socialized into rational debating practices in the process (Habermas, 1991). However, as a number of scholars have pointed out, the tone of discussion on the internet seems far removed from this normative ideal, with name-calling, aggression and irrelevancy being far more common currency, as well as significant quantities of misogyny, homophobia and racism (for example, Davis, 2005).

This is not the whole story, though. We know that citizens are using social media to post content relevant to political broadcasting online and debating that in real time. This might involve commenting on events that are happening in the broadcast, adding additional information or questioning editorial decisions that broadcasters have made (Ampofo et al., 2011; Anstead and O'Loughlin, 2011; Bennett, 2012). Often these comments are made with humour or sarcasm, exposing the relatively narrow definition of deliberation found in the Habermasian model of democracy. Freelon (2010: 1177) argues scholars should move away from analysing 'online deliberation' and instead find ways to analyse 'online political discussion' more broadly; the Habermasian model is just one of several that could fruitfully be applied. An alternative normative ideal that perhaps better describes the nature of the two-screen political debate on social media is John Dryzek's (2002) model of discursive democracy, which is better able to accommodate the inclusion of humour, as well as appeals that go beyond the rational and also appeal to others' emotions. Ultimately, we need frameworks to analyse how the rational-deliberative, the expressive and other communicative modalities operate at once.

Two-screen viewing also challenges how political scientists understand the significance of television. Certainly, historically, it has been common in media studies to draw a sharp distinction between the producers of content and the audience. This dualism has also had a strong influence on the history of political communication and especially discussions on the relationship between the evolution of the medium and democracy. Some scholars have a tendency to see television as an inherently passive medium, which both circumscribes citizens' involvement and infantilizes political debate (Postman, 2006; Putnam, 1995, 2001). As such, the age of television politics is often compared unfavourably to older forms of participation which took place through direct engagement in the public space. In contrast – at least in the eyes of the most optimistic commentators – the internet is frequently cast as a medium which promotes activity, mobilization and debate, allowing citizens to engage with each other (Castells, 2009; Trippi, 2008).

As we suggest above in the discussion of hybridization, it is not quite that simple. For one thing, studies of audience engagement with television before the arrival of the internet show that people were far from being the passive dupes depicted in pessimistic scholarship; rather, many citizens actively construct the meaning of broadcast content themselves in critical and reflexive ways, comparing television content to that consumed through radio or newspapers or with their direct personal experience of political, economic and social affairs (Ang, 1994; Gillespie, 1995). This hybrid media consumption continues with the arrival of two-screen viewing. One cannot neatly segment the experience of television viewing from social media consumption. However, it is also possible to argue that two-screen viewing is changing the nature of the audience experience. Wohn and Na (2011) argue that it is creating the idea of 'group viewing', where television watching is becoming an activity that is less restricted to the private space and can, for those who so choose, become a pseudo-public activity. To paraphrase Putnam, we are no longer watching alone. We can watch together.

The final significant theoretical issues relate to power and hierarchy. In particular, we need to ask whether the emergence of two-screen viewing has the potential to change the way power is distributed between media actors and the public. On the face of it, the obvious historical reading would suggest a redistribution of power: historically, the content of broadcast media was controlled by a relatively small and elite group. Now, a wider body of citizens has the ability to respond to this content in real time, which in turn might give them power to debate and interpret content, and through this put pressure on media producers and influence the types of content that will be presented to audiences. However, we should be cautious about such a simple reading. First, the power of interpretation has always been in the possession of audiences. Second, it is not necessarily clear that agenda-setting power is being redistributed in any meaningful manner. We have been here before. Scholars and pundits keenly debated at the time whether bloggers were setting news agendas during the 2004 US presidential race (Hindman, 2008: 110–112). Writing more broadly about the internet and government, Davis has argued that a form of 'fat democracy' is emerging, meaning that while power has been dispersed to a wider elite (now including figures who are prominent on social media, as well as traditionally powerful actors from the world of politics and the old media), the 'average' citizen is now further removed from political decision-makers than ever (Davis, 2010). This is because they lack either the skills or the inclination to engage with the political discussions that the 'influencers' are involved in. As such, the power to steer political debate has been redistributed to some extent, but only marginally, and many people are now actually more removed from the decision-making centre.

CHALLENGES

There are a number of challenges that two-screen viewing raises, for a variety of actors. These cover a range of different issues: methodological, normative and institutional. We suggest four of particular significance: how technology and analysis techniques are going to develop in the future; how media organizations respond to the challenge of two-screen viewing; how the data generated from two-screen viewing is institutionalized; and how the digital divide will manifest itself in two-screen viewing and how this is understood and tackled.

Challenge 1: The Future of Methodology and Analysis

While the data cited above suggest that two-screen viewing is already a presence in conversations about media, it must also be remembered that the technology and methods being employed remain at an embryonic level. In Chapter 25 in this volume, Freelon draws attention to the progress made in the field of social computing. There, the analysts' goal is to improve tools and techniques. Politics is merely one case study that allows them to test and refine those tools. Those working in political communication, where the goal is rather to build convincing theories of the role of communication in political behaviour and the operations of power, should not only work with their computing science colleagues, but also learn some programming themselves so as to bring their contextual knowledge to the gathering and processing of political data itself, Freelon argues. It is likely that in the next few years tools and techniques in political science will continue to develop rapidly as more scholars make these connections. This applies both to data production and processing techniques. At the moment, the information actually available about the individuals publishing content is limited in certain important respects. For example, while some research has focused on linking social media content with specific geographical locations, such as a 'hate map' made by examining the origin points of tweets from the United States that used racially offensive language (Zook et al., 2013), in reality only a small number of tweets are geotagged; that is, have their geographical point of origin embedded in their metadata. One recent study suggested that as few as 1.6 per cent of all published tweets actually contained geo-location data (Leetaru et al., 2013).

In the future, it may be that more people are willing to publish these kinds of personal details with their social media updates. However, it may also be that technology develops to cope with these limitations. For example, there are alternative ways for pinpointing the location of the tweeter, at least approximately. Indeed, Leetaru et al. go on to suggest a number of other methods that might be used, based on analysing the information that users put in their biographies or have posted elsewhere on their Twitter feed (for some of the dilemmas that emerge when we try to do this, see Freelon, 2013). Certainly, we may see increasingly inventive ways of segmenting and categorizing data, going beyond Twitter. The platform has proved an attractive tool for those seeking to analyse real-time comments largely because the data are readily accessible and the ease of posting makes it a genuinely real-time environment. However, the lack of demographic information on users is a significant limitation if scholars require it. In contrast, Facebook, while in some respects a more private platform, contains huge amounts of information on individual users: where they live, their educational background and their religion and politics, for example. Furthermore, it has been shown that an individual's biographical information published on Facebook can be used as a very accurate predictor of their psychological disposition (Kosinski et al., 2013). Similar techniques could possibly be used to further segment and scale groups expressing a particular sentiment.

The applications of such technologies for the analysis of two-screen viewing data could have profound ramifications for political communication. We argue that they can revitalize some classic methods and approaches. In particular, they echo the use of focus groups since the 1990s to measure the response of very specific segments of the electorate to particular political messages (Gould, 2011; Schier, 2000). Increased and more effective use of metadata could potentially allow parties and political consultants to much more accurately measure social media reactions in real time among more targeted parts of the electorate. However, the potential for the development of such tools also creates important ethical questions. It is far from clear that members of the public really understand the processes which their data are being subjected to. Nor is it clear that they have given their informed consent for their data to be harvested and processed in such a manner, even if the Terms of Service Agreements they have signed to join a social network explicitly sanction such use of the data (Debatin et al., 2009).

Challenge 2: How Will Traditional Broadcasters Respond to Two-Screen Politics?

As the techniques to more closely monitor the audience in real time develop, this will present a challenge to traditional broadcasters. Some are attempting to integrate audience comments on social media into their programmes, as indicated by the BBC *Question Time* example cited above. The wider entertainment industry certainly sees social media data as providing an additional metric that can be used in conjunction with viewing figures. For example, the US ratings monitoring company Nielsen has started to measure and publish figures on online activity that occurs while viewers are consuming traditional media (Nielsen, 2012). However, a broader question is how effectively broadcasters will function in an environment where members of the audience are able to rapidly question and challenge the content they are being presented with. Certainly, research has shown that audiences are now able to vocalize their criticisms of programmes using social media, questioning the data they are being presented with or directly arguing against the positions taken by guests on a programme (Ampofo et al., 2011; Bennett, 2012).

We might consider this an extension of older forums that allowed the audience to express displeasure with broadcasters, such as newspaper letters pages and Usenet. Additionally, we might – from a normative perspective – consider the increase in space for debate and dissent an overall improvement to the nature of the media–political environment. However, there is an important difference that broadcasters, other media organizations and especially regulators will need to consider in the future. This is that new dynamics can emerge through network effects, which create patterns of communication and interaction that challenge the informal norms and formal laws that sustain the texture of public debate. For instance, social media significantly lower the barriers to the mobilization of a large number of people to take direct action in response to an event or content. The best illustration of this occurred not with television but in the newspaper industry, when *Daily Mail* journalist Jan Moir published an article containing a homophobic commentary on the death of Irish pop star Stephen Gately. Following a Twitter campaign involving a number of celebrities, the Press Complaints Commission received 25 000 separate complaints on the column – more than the total number of complaints they had received in the previous five years (Brook, 2009). While the complaint was ultimately dismissed, this example highlights the potential challenge to regulators faced by responding to the demands of a networked audience, where outrage can spread rapidly and easily be mobilized. While the anger of the complainers in the Gately–Moir example may seem justified, it is not hard to imagine less righteous and more reactionary causes being mobilized in the same manner, with the potential to both intimidate media providers and overwhelm regulators.

Challenge 3: How Will Big Data be Institutionalized by Political Authorities?

The way in which broadcasters adapt to this change is part of the broader question of how and whether social media audience data should become institutionalized by existing political institutions. Certainly, there is much interest in this from government, who hope that data from social media can be harvested in order to respond more rapidly to public demands and provide better services. According to this logic, social media can provide what Hood and Margetts term 'detectors', 'the instruments that governments use for taking information' (Hood and Margetts, 2007: 3; see also Margetts, 2009). Social media monitoring therefore has the potential to augment more traditional means used by government to understand the preferences of the public (such as opinion polls) and the success of public services (such as league tables or numbers of complaints).

However, there is a potentially darker side to this monitoring as well, with governments around the world becoming increasingly interested in so-called horizon monitoring. Here, big data and predictive analytics are used to predict the occurrence of particular events, especially events that pose risks to public health, economic stability or social order. In June 2013, for example, it was revealed that the UK Department for Environment, Food and Rural Affairs (Defra) was employing horizon scanning software developed in the private sector to analyse social media, and identify and rapidly respond to agricultural diseases, animal food shortage and even public protests against badger culls (Wheeler, 2013). Certainly, the recent revelations that the US government has been monitoring massive quantities of social media data under the project codename of Prism demonstrates that governments are interested in the data being generated by citizens and will not always be open and transparent about what they are doing with it. If this approach to technology and security is to be effective, data are needed on the commentary and behaviour of all citizens in order to spot 'deviant' or 'risky' patterns among a few, generating actionable intelligence or, as it is called in marketing, 'insight'.

Challenge 4: Two-screen Viewing and the Digital Divide

Even if data from social media were successfully institutionalized in such a way that they could be analysed and used to inform government decision-making or the content of political broadcasting, an important concern remains. It is worth asking who is not commenting online, and thus which voices in society are under-represented. The problem is an old one, perhaps best expressed in Elmer Schattschneider's (1960) claim that

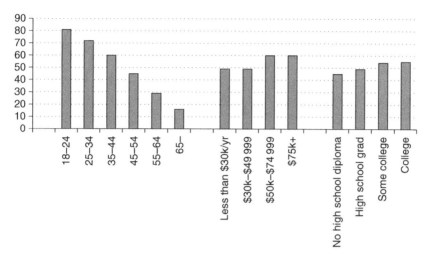

Source: Smith and Boyles (2012).

Figure 17.1 Percentage of people who use their mobile phones while watching television in the United States

the 'The flaw in the pluralist heaven is that the heavenly chorus sings with a strong upper-class accent'.

Figure 17.1 gives us some indication as to the demographics of people who actually said that they used an internet-enabled device while watching television, and suggests certain structural imbalances in the makeup of that group. Younger people are significantly more likely to engage in the practice. Those with higher levels of income and education are also more likely to be using the internet while having the television on.

This reflects the broader problem of the digital divide. Originally, this was conceived as simply being a binary problem. People were online because they had access to a computer and an internet connection, while others lacked the resources or the inclination for this (for a classic early statement of this kind, see Irving, 1999). While this remains quite a reductionist definition, it should be noted that such hardware and connectivity divisions may still play a role in the penetration of two-screen viewing today, given that young people and the wealthy are more likely to have access to equipment, such as smartphones and tablets, that is most useful to real-time commenting while watching television (Smith, 2013). More recent definitions of the digital divide are also relevant though, because they have moved away from a simple online–offline binary distinction, and instead recognize that there are important differences in skills between internet users (van Deursen and van Dijk, 2011). This is reflected in their ability to create content and engage in the wider networks of social participation that form around television broadcasts. It is also becoming clear that many users are simply not interested in using the internet, and some stop using it because of boredom or fears about privacy rather than financial expense. In the UK, for example, it appears that 18 per cent of the population have no intention of going online (Dutton and Blank, 2014).

Ironically, the digital divide becomes a bigger problem the more effectively social media is used by broadcasters, governments and other political actors. This is especially true if these institutions find mechanisms that make them more responsive to the wishes of their audiences, because there will be a group of citizens still not taking part in that debate.

CONCLUSIONS

As ever when considering the development of new media, the emergence of two-screen viewing forces us to ponder what is really new in the contemporary media environment, as well as the complex dichotomy that exists between continuity and change. Certainly, there has been a technological shift. For viewers seeking to produce their own content, new devices (notably tablets and smartphones) as well as new online platforms (such as Twitter and Facebook) provide unprecedented opportunities for reaction to media events at a previously unheard of scale and pace (Merrin, 2014). The traditional idea that the audience are the passive receivers of content is considerably problematized by these developments. Similarly, two-screen viewing creates opportunities for new forms of behaviour and interaction. Citizens can now annotate programmes or troll them, and certainly feel part of a larger collective in which they can monitor the views and reactions of their peers.

In the area of analysis, new big data analytics tools and storage facilitate new ways of processing, understanding and organizing this information. An emerging industry of monitoring firms promises new streams of data, which will enable us to better understand online publics. Whether these promises are fulfilled remains to be seen, but certainly, for media organizations, new metrics are emerging through which they can measure the reactions of their audiences. How will they integrate and harness social media to revitalize their traditional formats? Similarly, how will politicians, parties and government cope in an environment where their mediated relations with the public might be more direct and rapid, and potentially bi- and multidirectional?

But it is also important to remember what is not new in this equation. Viewers were never the passive caricatures that they are sometimes presented as, and were always capable of taking quite different readings from a political broadcast. Now they simply have the forum to articulate and share their interpretations. Additionally, many of the challenges and problems thrown up by the development of two-screen viewing are far from new. How to institutionalize the voices of citizens in political debate, or how to overcome the structural inequality embedded in the digital divide, are recurring problems in democracy.

Therefore the best way to understand two-screen viewing and address the challenges it presents is through this combination of change and continuity. Many of the old questions about power, influence, hierarchy, authority and order still apply, but to answer them we must get to grips with new developments such as two-screen viewing.

FURTHER READING

Ampofo, L., Anstead, N. and O'Loughlin, B. (2011). Trust, confidence, credibility: citizen responses on Twitter to opinion polls during the 2010 UK general election. *Information, Communication and Society*, 14(6), 850–871. DOI: 10.1080/1369118X.2011.587882.

Anstead, N. and O'Loughlin, B. (2011). The emerging viewertariat and BBC *Question Time*: television debate and real-time commenting online. *International Journal of Press-Politics*, 16(4), 440–462. DOI: 10.1177/1940161211415519.

Anstead, N. and O'Loughlin, B. (2014). Social media analysis and public opinion: the 2010 UK general election. *Journal of Computer-Mediated Communication*. Online First. DOI: 10.1111/jcc4.12102.

Boyd, D. and Crawford, K. (2012). Critical questions for big data: provocations for a cultural, technological, and scholarly phenomenon. *Information, Communication and Society*, 15(5), 662–679. DOI: 10.1080/1369118X.2012.678878.

Chadwick, A. (2013). *The Hybrid Media System: Politics and Power*. Oxford: Oxford University Press.

Elmer, G. (2013). Live research: Twittering an election debate. *New Media and Society*, 15(1), 18–30. DOI: 10.1177/1461444812457328.

REFERENCES

Ampofo, L., Anstead, N. and O'Loughlin, B. (2011). Trust, confidence, credibility: citizen responses on Twitter to opinion polls during the 2010 UK general election. Information, Communication and Society, 14(6), 850–871. DOI: 10.1080/1369118X.2011.587882.

Anderson, C. (2008). The end of theory? Wired.com, 23 June. Retrieved 27 June 2013, from http://www.wired.com/science/discoveries/magazine/16-07/pb_theory.

Ang, I. (1994). Desperately Seeking the Audience. London: Routledge.

Anstead, N. and O'Loughlin, B. (2011). The emerging viewertariat and BBC Question Time: television debate and real-time commenting online. International Journal of Press-Politics, 16(4), 440–462. DOI: 10.1177/1940161211415519.

Anstead, N. and O'Loughlin, B. (2012). Semantic Polling: The Ethics of Online

Public Opinion. London: London School of Economics Department of Media and Communications Media Policy Project.

Anstead, N. and O'Loughlin, B. (2014). Social media analysis and public opinion: the 2010 UK general election. Journal of Computer-Mediated Communication. Online First. DOI: 10.1111/jcc4.12102.

BBC News Online. (2012). Question Time to have Twitter panellist. 26 September. Retrieved 27 June 2013, from http://www.bbc.co.uk/news/entertainment-arts-19726751.

Bennett, L. (2012). Transformations through Twitter: the England riots, television viewership and negotiations of power through media convergence. Participations, 9(2), 511–525.

Boyd, D. and Crawford, K. (2012). Critical questions for big data: provocations for a cultural, technological, and scholarly phenomenon. Information, Communication and Society, 15(5), 662–679. DOI: 10.1080/1369118X.2012.678878.

Brook, S. (2009, 19th October). Jan Moir: more than 22,000 complain to PCC over Stephen Gately piece. Guardian. Retrieved 4 July, 2013, from http://www.guardian.co.uk/media/2009/oct/19/jan-moir-complain-stephen-gately.

Bruns, A., Burgess, J., Crawford, K. and Shaw, F. (2012). #qldfloods and @QPSMedia: crisis communication on Twitter in the 2011 South East Queensland floods. Brisbane: ARC Centre of Excellence for Creative Industries and Innovation.

Bruns, A., Highfield, T. and Burgess, J. (2013). The Arab Spring and social media audiences: English and Arabic Twitter users and their networks. American Behavioral Scientist, 57(7), 871–898.

Castells, M. (2009). Communication Power. Oxford, UK and New York, USA: Oxford University Press.

Chadwick, A. (2011). The political information cycle in a hybrid news system: the British Prime Minister and the 'Bullygate' affair. International Journal of Press/Politics, 16(1), 1–27. DOI: 10.1177/1940161210384730.

Chadwick, A. (2013). The Hybrid Media System: Politics and Power. Oxford: Oxford University Press.

Davis, A. (2010). Political Communication and Social Theory. London, UK and New York, USA: Routledge.

Davis, R. (2005). Politics Online: Blogs, Chatrooms and Discussion Groups in American Democracy. London, UK and New York, USA: Routledge.

Debatin, B., Lovejoy, J.P., Horn, A.-K. and Hughes, B.N. (2009). Facebook and online privacy: attitudes, behaviors, and unintended consequences. Journal of Computer-Mediated Communication, 15(1), 83–108. DOI: 10.1111/j.1083-6101.2009.01494.x.

Dryzek, J.S. (2002). Deliberative Democracy and Beyond: Liberals, Critics, Contestations. Oxford: Oxford University Press.

Dutton, W.H. and Blank, G. (2014). Cultures of the Internet: The Internet in Britain. Oxford: Oxford Internet Institute.

Elmer, G. (2013). Live research: Twittering an election debate. New Media and Society, 15(1), 18–30. DOI: 10.1177/1461444812457328.

Freelon, D.G. (2010). Analyzing online political discussion using three models of democratic communication. New Media and Society, 12(7), 1172–1190.

Freelon, D.G. (2013). Twitter geolocation and its limitations. 12 May. dfreelon.org. Retrieved 19 February 2014, from http://dfreelon.org/2013/05/12/twitter-geolocation-and-its-limitations/.

Gillespie, M. (1995). Television, Ethnicity and Cultural Change. London: Routledge.

Gillespie, M., O'Loughlin, B. and Proctor, R. (2013). Tweeting the Olympics: report. London.

Google, Ipsos and Sterling (2012). The New Multi-screen World: Understanding Cross-Platform Consumer Behaviour. New York: Google.

Gould, P. (2011). The Unfinished Revolution: How New Labour Changed British Politics for Ever. London: Abacus.

Habermas, J. (1991). The Structural Transformation of the Public Sphere: An Inquiry into a Category of Bourgeois Society. Boston, MA: MIT Press.

Highfield, T., Harrington, S. and Bruns, A. (2013). Twitter as a technology for audiencing and fandom: the Eurovision phenomenon. Information, Communication and Society, 16(3), 315–339. DOI: 10.1080/1369118X.2012.756053.

Hindman, M. (2008). The Myth of Digital Democracy. Princeton, NJ: Princeton University Press.

Hood, C. and Margetts, H. (2007). The Tools of Government in the Digital Age, new edn. Basingstoke, UK and New York, USA: Palgrave Macmillan.

Irving, L. (1999). Falling Through the Net: Defining the Digital Divide. Washington, DC: US Department of Commerce.

Jenkins, H. (2008). Convergence Culture: Where Old and New Media Collide. New York: New York University Press.

Kelly, S. (2011). BBC Click: The phenomenon of two screen viewing. BBC Click. London: BBC.

Kosinski, M., Stillwell, D. and Graepel, T. (2013). Private traits and attributes are predictable from digital records of human behavior. Proceedings of the National Academy of Sciences. DOI: 10.1073/pnas.1218772110.

Leetaru, K., Wang, S., Cao, G., Padmanabhan, A. and Shook, E. (2013). Mapping the global Twitter heartbeat: the geography of Twitter. First Monday, 18(5). Retrieved from http://firstmonday.org/ojs/index.php/fm/article/viewArticle/4366/. DOI: 10.5210/fm.v18i5.4366.

Manovich, L. (2011). Trending: the promises and the challenges of big social data. In Gold, M.K. (ed.), Debates in the Digital Humanities (pp. 460–475). Minneapolis, MN: University of Minnesota Press.

Margetts, H.Z. (2009). The Internet and public policy. Policy and Internet, 1(1), 1–21. DOI: 10.2202/1944-2866.1029.

Mayer-Schonberger, V. and Cukier, K. (2013). Big Data: A Revolution That Will Transform How We Live, Work and Think. London: John Murray.

Merrin, W. (2014). Media Studies 2.0. London,UK and New York, USA: Routledge.

Moses, L. (2012). Data-point: two-screen viewing. Adweek, 7 November. Retrieved 27 June 2013, from http://www.adweek.com/news/technology/data-points-two-screen-viewing-145014.

Nielsen (2012). US Media trends by demographic. Nielsen Research, 27 April. Retrieved 28 June 2013, from http://www.nielsen.com/us/en/newswire/2012/report-u-s-media-trends-by-demographic.html.

Ofcom (2013). Summary: UK audience attitudes to the broadcast media. Retrieved 14 May 2013, from http://stakeholders.ofcom.org.uk/market-data-research/other/tv-research/attitudes-broadcast-media/?utm_source=updates&utm_medium=email&utm_campaign=uk-media-attitudes.

Postman, N. (2006). Amusing Ourselves to Death: Public Discourse in the Age of Show Business. London, UK and New York, USA: Penguin.

Putnam, R.D. (1995). Tuning in, tuning out: the strange disappearance of social capital in America. PS: Political Science and Politics, 28(4), 664–683. DOI: 10.2307/420517.

Putnam, R.D. (2001). Bowling Alone: The Collapse and Revival of American Community. New York: Simon & Schuster.

Schattschneider, E.E. (1960). The Semisovereign People: A Realist's View of Democracy in America. New York, USA and London, UK: Holt, Rinehart & Winston.

Schier, S.E. (2000). By Invitation Only: The Rise of Exclusive Politics in the United States. Pittsburgh, PA: University of Pittsburgh Press.

Shaw, F., Burgess, J., Crawford, K. and Bruns, A. (2013). Sharing news, making sense, saying thanks: patterns of talk on Twitter during the Queensland floods. Australian Journal of Communication, 40(1), 23.

Smith, A. (2013). Smartphone ownership 2013. Washington, DC: Pew Internet and American Life Project.

Smith, A. and Boyles, J. L. (2012). The rise of the 'connected viewer'. Washington, DC: Pew Internet and American Life Project.

Taylor, C. (2012). Election night hits record high: 20 million tweets. Mashable.com, 7 November. Retrieved 23 June 2013, from http://mashable.com/2012/11/06/election-night-twitter-record/.

Trippi, J. (2008). The Revolution Will Not Be Televised. New York: William Morrow.

TV Licensing (2012). Telescope: a look at the nation's changing viewing habits. London: TV Licensing.

Tweetminster.com. (2010). The Leaders' Debate. Tweetminster.com webblog, 16 April. Retrieved 27 June 2013, from http://tweetminster.tumblr.com/post/524329305/the-leaders-debate.

van Deursen, A. and van Dijk, J. (2011). Internet skills and the digital divide. New Media and Society, 13(6), 893–911. DOI: 10.1177/1461444810386774.

Wheeler, B. (2013). Whitehall chiefs scan Twitter to head off badger protests. 29 June. Retrieved 4 July 2013, from http://www.bbc.co.uk/news/uk-politics-22984367.

Wohn, D.Y. and Na, E.-K. (2011). Tweeting about TV: sharing television viewing experiences via social media message streams. First Monday, 16(3). Retrieved from http://firstmonday.org/ojs/index.php/fm/article/view/3368. DOI: 10.5210/fm.v16i3.3368.

Zook, M., Graham, M., Shelton, T., Stephens, M. and Poorthuis, A. (2013). The geography of hate. Floating Sheep, 13 May. Retrieved 28 June 2013, from http://www.floatingsheep.org/2013/05/hatemap.html.

PART V

JOURNALISM

18. From news blogs to news on Twitter: gatewatching and collaborative news curation
Axel Bruns and Tim Highfield

Online engagement with news and current events, and especially with political developments, has a history which is almost as long as that of computer-mediated communication itself; even early online communities such as the famous bulletin board system Whole Earth 'Lectronic Link (WELL) included spaces for the discussion of news and politics by their users (Rheingold, 1993). Such activities were significantly simplified and popularized with the advent of the World Wide Web, and especially with the development of what are commonly referred to as 'Web 2.0' technologies: online platforms which were built around the active contribution of content and commentary by users, and which enabled a substantially larger number of users to publish their thoughts on virtually any topic, including the news of the day.

A major early catalyst for the use of such platforms in reporting and discussing the news occurred with the establishment of the first Independent Media Center in the lead-up to the controversial 1999 meeting of the World Trade Organization (WTO) in Seattle (Meikle, 2002; Hyde, 2002; Bruns, 2005). Activists who anticipated a largely one-sided coverage of the event by US and international media – painting anti-globalization and alternative-globalization protesters simplistically as rioters and disruptors of the event – used Web 2.0 technology to develop their own, alternative online media channels, through which they reported on the activist meetings and conferences accompanying the official WTO summit, and covered the clashes between demonstrators and riot police from a bottom-up perspective. Operated using open source technologies and borrowed equipment, Indymedia (as it became known) provided a vision of the WTO summit which was in stark contrast to the mainstream media coverage, and set the scene for the largely adversarial relationship between mainstream and alternative media which would dominate the following years; the first Independent Media Center (IMC) in Seattle also became the template for several hundred loosely networked IMCs which would emerge in countries around the world over subsequent years (Platon and Deuze, 2003).

While the success of Indymedia highlighted the possibilities of using comparatively cheap, lightweight web technology to offer a credible challenge to the agenda-setting authority of mainstream media organizations, it also presents a somewhat misleading picture of the everyday news engagement practices which would develop over the coming years; at first, especially in the US, but gradually also throughout other developed and well-connected nations. Seattle acted as a catalyst which attracted a critical mass of media-savvy activists who were able to engage in the first-hand reporting of summit events, but such preconditions are not necessarily met in everyday practice; the 'citizen reporters' who came to Seattle for a week to drive Indymedia's coverage eventually had to return to their own lives, which offered considerably less opportunity for the quasi-journalistic activities of researching and reporting new stories. In the aftermath of the Seattle event, therefore, the central focus both in Indymedia and in the wider movement which it had helped to invigorate shifted from reporting 'new' news to commenting on mainstream news, even at the same time as the term 'citizen journalism' began to enjoy increasing popularity (Singer, 2006; Deuze, 2009).

'Citizen journalism' is doubly misleading, however: on the one hand, it implies that professional journalists are not also citizens, and in final consequence denies the possibility that they may have entered their chosen profession in pursuit of civic ideals at least as much as in pursuit of a career. On the other, it claims that the online participants described as citizen journalists are indeed engaging in the full range of journalistic practices, including original reporting, and subscribe to the central ideals of professional journalism, including objective and disinterested coverage of issues and events. Neither of these are necessarily met for citizen journalists, however: for practical reasons related to available time and resources, citizen journalists often act as a second tier in the news process, commenting on and critiquing mainstream news content but generating few genuinely new stories of their own, and they do so very often specifically because they feel that their perspectives, their causes, are not accurately represented in such mainstream reporting.

This chapter, then, explores the role of this second tier of independent news blogs as it developed in the years following Seattle, and outlines the practice of gatewatching as a key element of news bloggers' activities. We critique perceptions of the news blogosphere as an echo chamber or filter bubble whose discussions about current events are detached from journalistic coverage, and demonstrate instead the close interconnections between independent news bloggers and professional journalists in the wider media ecology. Finally, we sketch the gradual transition and broadening of gatewatching practices in the news blogosphere towards the collaborative curation of news sharing in contemporary social media spaces, and outline the further research questions which emerge from such transformations of the flows of news and discussion.

NEWS BLOGGING AS GATEWATCHING

If (not least for practical reasons) the early citizen journalism of the Indymedia movement transformed to citizen commenting in subsequent years, then the emerging web technology and online cultural practice of blogs and blogging became its chief tool. Especially after the establishment of blog hosting platforms such as Blogger and Wordpress, blogs provided an even cheaper and more lightweight technology for online publishing than the Indymedia platform had been, and importantly they also did so for individual users, rather than mainly for activist groups. At times, compared to the arrival of cheap printing presses and their impact on the publishing and spread of political pamphlets during the American struggle for independence, blogging technology enabled anyone with an interest in news and politics, with a political grievance or an activist cause, to become their own publisher and share their thoughts on virtually any topic with a potentially worldwide audience (Bruns, 2006).

While the majority of such news and politics blogs may have attracted only minuscule audiences, some such sites rose to much more substantial popularity, to the point where they were able to affect the political process or prompt meaningful and impactful debate about the coverage of specific issues by mainstream media organizations. Several such events are by now well rehearsed in the literature about news and political blogging: the dogged pursuit of former US House Senate Leader Trent Lott by a loose coalition of political bloggers, over pro-segregationist remarks at a birthday dinner for Congress veteran Strom Thurmond, gradually led to increased media questioning and, eventually, to Lott's resignation; the Washington, DC, political insider blog Drudge Report claimed credit for being the first to break the Clinton–Lewinsky scandal; the detailed debunking of mainstream media claims about US President George W. Bush's service records by news bloggers served to prematurely end respected news anchor Dan Rather's career. News and politics blogs had arrived as players in US politics (and over time did so in other nations as well): by 2004, the major parties' presidential nomination conventions began to formally accredit leading bloggers alongside mainstream media representatives.

Again, however, we should not ignore the vast number of lesser news blogs over a focus on the handful of highly influential, widely read sites.

The latter eventually rose to their positions of prominence only because of the activities of the wider blogosphere: they became visible because many other, more minor bloggers linked to their posts, thereby demonstrating the dynamics of attention in what Benkler (2006: 3) has described as a new 'networked Information economy'. The common practice in blogging, as well as in other forms of social media, of linking to and thereby sharing the posts of others provides an important amount of amplification of these posts, enabling them to be widely visible well beyond their original point of publication; it is through such amplification processes, for example, that the Lott, Lewinsky or Rather stories were kept alive and eventually achieved domestic and international notoriety. (More recent controversies, by contrast, have tended to see a heightened level of Twitter and Facebook use over a shorter time frame to achieve similar impacts; so much so that the term 'social media shitstorm' to describe such intense moments of user-driven political crisis has entered recent dictionaries and scholarly literature; cf. Schmidt, 2013.)

This demonstrates a fundamental practice in news and political blogging, operating both within the blogosphere itself and in its intersections with the wider news mediasphere: the practice of gatewatching (Bruns, 2005). Because of their considerable operational constraints, usually resulting in a focus on citizen commenting over citizen journalism, news blogs cannot possibly act as gatekeepers in the way that mainstream news organizations have long positioned themselves; in an always-on, digitally determined media ecology, in fact, even those news organizations are now generally no longer able to control the news gates and claim major scoops for themselves. News bloggers, therefore, engage not in gatekeeping, but gatewatching: they observe and monitor what information passes through the gates of other news organizations, as well as those of original news sources (government and non-governmental organizations, companies, research institutes, and so on) and fellow commenters, and highlight, evaluate and discuss such material if it relates to their own fields of interest. Such gatewatching is a practice particularly well suited to online publication: it realizes the opportunities which emerge from the possibility of including direct hyperlinks to off-site material in one's own post, enabling news bloggers to engage in a conversation-at-a-distance with the material they link to.

What emerges from this everyday, widespread practice of gatewatching and commenting is a distributed process of tracking, discussing and, in aggregation, curating the news as it is published by other sites (Bruns, 2011; Bruns and Highfield, 2012). Overall, the news and political blogosphere acts as a second-order filtering mechanism, subsequent to the internal processes of mainstream news organizations, but also observing, critiquing and, where necessary, correcting them. While this has been seen by some, and especially by a number of mainstream journalists, as a parasitic relationship in which armchair journalists presume to improve the work of the professionals (see for example, *Australian*, 2007), in reality the relationship has become much more symbiotic, especially as major news bloggers have become mainstream media commentators in their own right, and as professional journalists have started their own blogs. Some professionals have acknowledged that, in the words of blogger-journalist Dan Gillmor (2003: vi), 'my readers know more than I do', especially about specialist topics which non-specialist journalists do not have the time or resources to explore in full detail; some news bloggers bring considerable energy and knowledge to bear on the task of critiquing mainstream reports, and of curating several such reports into a more comprehensive whole than any one news report could manage to achieve in its own right.

THE ECHO CHAMBER DEBATE

The quality of such gatewatching processes, and of their collaboratively curated outcomes, necessarily also depends on the diversity of the resources which are being gatewatched. One of the most persistent debates about news blogging amongst scholars and practitioners has been about this point, painting the news blogosphere at one extreme as a crucial new driver of the re-democratization of an otherwise oligarchical mass media system, telling truth to power (Benkler, 2006; Shirky, 2008; Castells, 2009), and at the other as a hermetically sealed echo chamber in which established nostrums circulate endlessly, continually amplified by bloggers linking to like-minded others (Sunstein, 2007; Morozov, 2011; Pariser, 2011).

Reality is likely to be found in between these two equally unrealistic caricatures of news blogging. It is true that attention in the blogosphere, as well as in the overall World Wide Web, is unevenly distributed; a handful of leading websites receive considerably more attention than the vast majority of sites. Such 'long-tail' patterns exist for most other media forms as well, and indeed constitute a fundamental constant for the distribution of public attention; online, they are also supported and reinforced by the operations of search engines such as Google. However, to claim on this basis that there is a 'Googlearchy' (Hindman, 2009) which fatally undermines the ability of web technologies (including blogs) to give voice to alternative perspectives, and generates an echo chamber instead, is to misunderstand public attention as singular rather than plural.

Long-tail distributions of attention can be calculated for the entire web, or for the entire blogosphere, but are comparatively meaningless from such highly aggregated perspectives. Any one blogger, or any one Internet user, has a specific range of interests, and will search for and link to material which relates to these interests; their distribution of attention represents a personalized long tail of sources which is different – subtly, or entirely – from that of any other user. Even if news bloggers were to predominantly link to like-minded other bloggers (and there are reasons to question this assumption, as they may also link to bloggers whom they seek to criticize or ridicule), then, like-minded does not mean identically minded, and in making such links bloggers broaden the potential range of views and opinions their readers might be exposed to.

Further, this distribution of attention is context- and time-specific, especially for blogs which deal with news and political events. For the French and Australian political blogospheres, for example, Highfield (2011) has shown that the coverage of a news story or political issue will develop across several categories of blogs, depending on the type of issue. A story pertaining to a particular issue, such as economic policy, will be discussed in detail by specialist economics bloggers, whose topic of interest for their sites is reflected in the issue at hand. These specialists' posts are then cited in further coverage of the issue by the 'A-list' blogs – the prominent sites in terms of readership and reputation – which cover in their posts a wider range of topics than the specialists. Within political blogospheres, for instance, the A-list blogs comment on politics in general, crossing multiple themes, fields and audiences, as opposed to focusing on a single subject. The A-list coverage then serves as both a filter and an amplifier: by drawing on analyses from the specialist bloggers, these blogs provide a more rounded and detailed, curated response to the issue, while also promoting the work of other blogs, and other media organizations, by highlighting their posts for a potentially different audience. For the coverage of general, everyday politics, this flow may be reversed, with the A-list leading the story and later adding thematic analysis from the specialists as a connection to their topics of interest emerges. In the response to both specialist issues and politics in general, though, the respective roles of topical experts and information curators are clear within the posts of these different types of blogs, and the follow-up coverage from A-list, specialist and more minor blogs further develops these patterns as bloggers link and respond to one another's posts. Nuernbergk (2013) shows similar patterns for the German political blogosphere's coverage of the 2007 G8 summit and surrounding protests in Heiligendamm (see also Moe, 2010; Ackland and Shorish, 2009; Nahon et al., 2011 for further approaches to these questions).

What emerges from this, then, is a picture not of a single echo chamber, a single long-tail distribution in which a small number of A-list blogs dominate the rest of the blogosphere. Rather, for any one long-term field of interest, for any one short-term topic or issue within this field, for any one group of bloggers discussing this issue and interlinking with one another, there is a long-tail-style distribution across a range of sites, developed ad hoc or accumulated over time. While clearly this does not mean that every news blogger has an equal ability to make their voice heard, at the same time it also does not deny that there is a potential for a news blogger, even a minor one, with an interesting point of view to be recognized, linked to and thereby amplified by their more visible peers. If the blogosphere is democratic, it is so because it offers this potential; but there is no more guarantee that this potential will be realized than there is a guarantee in society-wide democratic processes that all voices, however obscure, will be equally well represented.

BLOGGERS AND COMMENTERS

Perceptions of the blogosphere as an echo chamber full of mutual self-agreement are also undermined by the often vigorous argument which may be found not only between bloggers themselves as they link to one another, but also between the users who respond to blog posts by adding their own comments to them. More so than mainstream news websites, blogs (including news and political blogs) popularized the idea of offering a comment section in which readers could encourage, criticize or otherwise extend the discussion offered by the blog author.

Such functionality has also been added more recently to the pages of many mainstream news sites, where it often continues to suffer from half-hearted implementations. Mainstream news organizations tend to be concerned in the first place about their corporate image and the need to minimize the potential for negative repercussions from inappropriate commentary. They therefore often employ pre-moderation approaches which – unless supported by substantial staff numbers – necessarily slow the flow of comments and discussion. Their highly visible nature also means that even in the face of restrictive moderation practices comment functionality attracts hundreds and thousands of respondents, reducing the coherence of the discussion.

By contrast, due to their significantly more limited public exposure, independent news blogs are usually able to use only light-touch post-moderation, ensuring that new comments appear immediately on the site and that a free-flowing discussion between commenters (including the original blogger) can ensue. Additionally, as most mid-size blogs will only attract some tens or hundreds of commenters, it is more likely that a genuine community of regular participants, familiar with each other's views, can develop; although it may remain spirited and controversial, this tends to improve the quality of debate as it limits the presence of drive-by commenters who are not interested in genuine discussion and seek only to state their own opinions (see, for example, Butler, 2001 and Leskovec et al., 2008 for discussions of the inverse relationship between community size and community quality). Many blogs (and blogging platforms) also allow for the creation of persistent commenter identities, which further help to maintain community discussion across individual blog posts, or even generate persistent URLs for individual comments, enabling bloggers and commenters on the site (or on another blog) to refer specifically to a particular comment. Ultimately, this enables the development of a distributed discussion between bloggers and commenters across a range of sites, conducted through mutual interlinking.

While such commenting practices have largely been restricted to blogs and news sites themselves, the growth of social media platforms and their integration into the wider online media ecology over recent years has seen a gradual transition of news commentary and discussion beyond the blogosphere. Bloggers (as well as mainstream news organizations and their journalists) now regularly share links to their latest articles through Facebook and Twitter; as a result, follow-on commentary on such articles is as likely to take place in these social media spaces as it is on the pages of the blog or news site itself. Especially for flat, open and public social networks such as Twitter (where with few exceptions any user can follow any other, without the need for a follower relationship to be approved by the other party), and less so for the more privately focused Facebook, this has the potential both to broaden the range of active commenters, and to lead to a more real-time, but also more ephemeral discussion of individual articles. The complex nature of interpersonal networks in social media systems (see, for example, Bruns and Moe, 2014, for an outline of different layers of communication on Twitter) further problematizes our understanding of how news stories, and news blog stories, are shared and commented upon in social media spaces, and how this impacts on wider public debate.

Importantly, the transition of news discussion from blogs to spaces such as Twitter also affects the power relations between participants. Where the sites of news organizations as well as news bloggers are clearly controlled by their respective authors and operators, who have the ability to determine both what articles are published and what user comments are made visible, the same is clearly not the case for social media. Twitter and similar platforms provide a third-party space in which news organizations, journalists, bloggers and commenters encounter and engage one another simply as comparatively equal users; only Twitter has the ability to shut down the public discussion of a poorly researched article, for example, and it is unlikely to do so. Coupled with the enhanced publicness of social media debate, this appears to have resulted in a greater level of engagement, especially between professional journalists and their readers; where in the past they may have chosen to ignore a critical blog article about their work rather than leave a comment on the blog, for example, it has now become common to observe interactions between journalists, bloggers, readers and other stakeholders on Twitter (Bruns, 2012).

More generally, social media platforms have also contributed to the further mainstreaming of gatewatching practices: they have enabled a substantially larger number of users – well beyond the population of the blogosphere – to engage in random acts of news sharing, for example by tweeting links (often with some commentary attached) to interesting articles. In the process, gatewatching has become a distributed, communal, crowdsourced practice, and an evaluation of the news-related links tweeted at any one point provides a detailed insight into the issues and topics which currently exercise the general public. Figure 18.1, for example, shows the week-by-week volume of tweeted links to major Australian news websites from mid-2012 to late 2014, pointing to substantial peaks and troughs in activity which can be related to current events in Australian politics and public life.

While such overall patterns provide an indication of general news interests amongst the social media user base, pointing to the everyday routines of gatewatching activities, more focused and heightened patterns of activity can be observed around specific acute events, including political crises, natural disasters and other breaking news. Here, ad hoc publics of users (cf. Bruns and Burgess, 2011) form to collaborate on compiling and curating the information which is relevant to a story, drawing on a diverse range of sources as appropriate to the situation. On Twitter, this is facilitated for example through the mechanism of the hashtag, which enables users to mark their messages as relevant to a specific topic (for example #sandy for the severe Hurricane Sandy which affected the US East Coast in late 2012). Others can then access, follow and track all the messages hashtagged with the same term in one convenient location.

The ad hoc communities which form around hashtagged breaking news events – including journalists, politicians, directly affected participants and other stakeholders, as well as ordinary users – continue to employ standard gatewatching approaches in their activities; in aggregate, their individual actions combine into a collaborative effort to 'work the story' (Bruns and Highfield, 2012), much as the staff in a media newsroom might have done in response to a breaking news story, but on a substantially

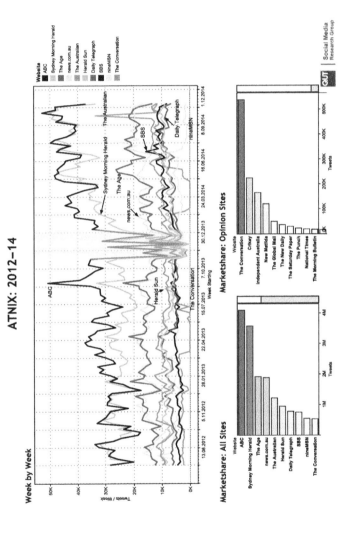

Notes:
The shaded area indicates outage through server maintenance; drop-offs for *The Conversation* are caused by domain name change.
See http://mappingonlinepublics.net/tag/atnix/ for continuing ATNIX updates.

Figure 18.1   Australian Twitter News Index (ATNIX), showing tweets per week linking to key Australian news sites, June 2012 to December 2014

larger basis. News organizations themselves are increasingly reaching out to such self-organizing para-journalistic communities during moments of crisis, in fact essentially inviting them to assist in working a story which, in the confusion which tends to surround such early moments in the development of a story, remains too complex to be made sense of by newsroom staff alone. During riots in the UK in 2011, for instance, journalists used Twitter to both disseminate information and promote their publication's coverage of the unrest, but also to request reliable and accurate accounts and fact-checking from other users (Vis, 2013). The UK *Guardian* newspaper is a notable leader of this collaborative approach to covering breaking news and live events; its extensive use of live blogs to report events as they happen involves the audience in both tone and format, inviting contributions via e-mail, social media and comments on the blog itself (Thurman and Walters, 2013). Although the extent of published reader contributions to the blog varies depending on the type of event, the comments sections for these blogs provide a further space for users to add to the developing story, and interact with other readers and journalists.

FURTHER OUTLOOK

Several key threads for further scholarly research in this area emerge from the preceding discussion. First, the emerging practices of sharing the news through social media provide a fascinating, real-time and large-scale insight into the current preoccupations of the media audience and their active responses to the news. As Figure 18.1 has shown in brief, such patterns may be observed over shorter or longer time frames (similar to Twitter's 'trending topics', but distilled down specifically to an engagement with news content), and can be correlated with the political and other public events of the day. This points both to overall patterns in the distribution of attention across different online news sources, and across different news beats from politics to sport, and to the specific distribution of and response to individual stories, measuring the velocity and tracing the transmission paths by which information travels through the network.

Research in this area may also develop perspectives on the relative importance of specific nodes (users and sites) in the network, revitalizing and updating theories about opinion leadership and information flows (for example, Katz and Lazarsfeld, 1955; Eisenstein, 1994; Bennett and Manheim, 2006) for the contemporary media environment. Crucial to such research will be the development of conceptual models which account for the traditionally recognized stakeholders in public debate – journalists, politicians and other public figures – as well as for the public communication practices of everyday audiences. The Australian Twitter News Index (ATNIX), tracking news sharing in the Australian Twittersphere on a continuing basis since mid-2012, constitutes an early step towards this work (see Bruns et al., 2013), but much more must be done to develop such initiatives.

Second, the transition towards a greater role for social media in sharing and discussing the news raises questions about the future positioning of news blogs, as well as of other news sources. Concerns about 'the death of blogging' are likely to be misplaced: while the blogosphere is no longer growing rapidly, or at all, and while blogs are no longer a particularly 'special' form of online publishing, this must be seen simply as the normalization of a once-novel publishing model rather than as a point of crisis. The fundamental processes of blogging (and of news blogging) have not disappeared per se, but have rather become ordinary to the point of invisibility. Mainstream news sites now regularly refer to their columnists' sections as 'blogs', without using that term to imply any particularly special position; independent commentators express their opinions on websites which are political blogs, but no longer necessarily advertise this fact.

However, what may be lost in this normalization of news and political blogs simply as platforms for commentary is precisely the sense that such blogs constitute their own genre of writing, and their own network of interconnected, interlinked sites – a blogosphere worthy of the term. If social media grow to be a central space for tracking the news and exchanging of political opinion, then it is no more likely that there will continue to be a blogosphere than there is a 'newspaper sphere'; rather, blogs, newspaper sites and news sites may serve simply as sources of information from which individual items are fed, through gatewatching, into the social network to be circulated, discussed and fed back into other forms of media. While the transition to such a model of news flows – with social media as its backbone – has yet to be completed, breaking news events in particular already provide us with early glimpses of where such developments could lead: during times of crisis, as recent natural and human-made disasters have shown, social media already lead the process of news curation.

Finally, considerable further research is necessary on the impact of these changes on the nature of news, and on the news industry. At least since the Indymedia moment, the role of journalism in society has been questioned and critiqued with renewed vigor; the emergence of Independent Media Centers, news and politics blogs, and other alternative online media outlets has challenged mainstream news organizations to a considerable extent, and elicited responses ranging from defensive belligerence to open collaboration. The increase in alternative voices and alternative interpretations of current events, and the speed by which such coverage may be disseminated online, has also led to challenges over what is truth and what is merely 'truthiness', to quote Stephen Colbert (also cf. Zelizer, 2009); such questions are far from being resolved, and may indeed be irresolvable.

What can be examined, however, is how news organizations, as well as individual journalists and other news workers, are positioning themselves in this changing environment. Early, simple divisions between mainstream and alternative media, between news sites and blogs, have long since eroded and given way to the recognition of a complex media ecology in which news organizations and their staff must be present across a range of platforms well beyond their own websites; how such presence is created and operated varies widely and has met a diverse range of responses, however. In addition to the conventional stakeholders in online news (news organizations, journalists, bloggers, commentators, audiences), a new range of third-party participants have also emerged to further complicate the picture: these include the operators of social media spaces themselves (Twitter, Facebook), as well as aggregator and tracker services which utilize the Application Programming Interfaces (APIs) of these platforms to mine news-related activity data. Such data, in turn, trace patterns of news engagement which themselves become new input for the emerging practices of data journalism.

Within this complex media ecology, the collaborative curation of news – and its subsequent analysis – is carried out across multiple platforms by a wide range of user types, from journalists and bloggers to Twitter users casually and infrequently tweeting opinions and links to articles. The roles of, and relationships between, these different contributors have changed over time, in response to the gradual acceptance and normalization of blogging and social media by mainstream news organizations. These changes have also impacted upon how news is presented, particularly breaking news; the importance of social media and blogging, including live blogging on news websites, to covering events as they happen, places continued value on gatewatching and collaboration between the various participants present in this expanded media ecology. Sourcing witness accounts and documentary evidence, verifying information, sharing links and disseminating updates from official sources are roles which, rather than just being the work of the individual journalist, are undertaken by non-professionals in order to further develop the coverage of news in mainstream, alternative and social media. The development of this collaborative rather than confrontational relationship has been a gradual process, and is far from complete; how this relationship evolves, as well as how online news curation continues to change as new practices and new platforms emerge, will remain important subjects for further scholarly research.

FURTHER READING

Allan, Stuart and Einar Thorsen (eds) (2009), *Citizen Journalism: Global Perspectives*, New York: Peter Lang.
Benkler, Yochai (2006), *The Wealth of Networks: How Social Production Transforms Markets and Freedom*, New Haven, CT: Yale University Press.
Bruns, Axel (2005), *Gatewatching: Collaborative Online News Production*, New York: Peter Lang.
Castells, Manuel (2009), *Communication Power*, Oxford: Oxford University Press.
Chadwick, Andrew (2013), *The Hybrid Media System: Politics and Power*, Oxford: Oxford University Press.
Meikle, Graham and Guy Redden (eds) (2010), *News Online: Transformations and Continuities*, London: Palgrave Macmillan.
Zelizer, Barbie (ed.) (2009), *The Changing Faces of Journalism: Tabloidization, Technology and Truthiness*, Milton Park: Routledge.

REFERENCES

Ackland, R. and J. Shorish (2009), 'Network formation in the political blogosphere: an application of agent based simulation and e-research tools', *Computational Economics*, 34 (4), 383–398.
Australian (2007) 'History a better guide than bias'. 12 July. Retrieved 27 January 2014 from http://www.theaustralian.com.au/opinion/editorial-history-a-better-guide-than-bias/story-e6frg6zo-1111113937838.
Benkler, Yochai (2006), *The Wealth of Networks: How Social Production Transforms Markets and Freedom*, New Haven, CT: Yale University Press.
Bennett, W.L. and J.B. Manheim (2006), 'The one-step flow of communication', *Annals of the American Academy of Political and Social Science*, 608 (1), 213–232.
Bruns, Axel (2005), *Gatewatching: Collaborative Online News Production*. New York: Peter Lang.
Bruns, Axel (2006), 'The practice of news blogging', in Axel Bruns and Joanne Jacobs (eds), *Uses of Blogs*, New York: Peter Lang, pp. 11–22.
Bruns, A. (2011), 'Gatekeeping, gatewatching, real-time feedback: new challenges for journalism', *Brazilian Journalism Research*, 7 (2), 117–136. Retrieved 24 May 2013 from http://bjr.libertar.org/index.php/bjr/article/view/355.
Bruns, A. (2012), 'Journalists and Twitter: how Australian news organisations adapt to a new medium', *Media International Australia*, 144, 97–107.
Bruns, Axel and Jean Burgess (2011), 'The use of Twitter hashtags in the formation of *ad hoc* publics', paper presented at the European Consortium for Political Research conference, Reykjavík, 25–27 August. Retrieved 24 May 2013 from http://eprints.qut.edu.au/46515/.
Bruns, Axel and Tim Highfield (2012), 'Blogs, Twitter, and breaking news: the produsage of citizen journalism', in Rebecca A. Lind (ed.), *Produsing Theory in a Digital World: The Intersection of Audiences and Production*, New York: Peter Lang, pp. 15–32.
Bruns, Axel, Tim Highfield and Stephen Harrington (2013), 'Sharing the news: dissemination of links to Australian news sites on Twitter', in Janey Gordon, Paul Rowinski and Gavin Stewart (eds), *Br(e)aking the News*, New York: Peter Lang, pp. 181–209.
Bruns, Axel and Hallvard Moe (2014), 'Structural layers of communication on Twitter', in Katrin Weller, Axel Bruns, Jean Burgess, Merja Mahrt and Cornelius Puschmann (eds), *Twitter and Society*, New York: Peter Lang, pp. 15–28.
Butler, B.S. (2001), 'Membership size, communication activity, and sustainability: a resource-based model of online social structures', *Information Systems Research*, 12 (4), 346–362.
Castells, Manuel (2009), *Communication Power*, Oxford: Oxford University Press.

Deuze, Mark (2009), 'The future of citizen journalism', in Stuart Allan and Einar Thorsen (eds), *Citizen Journalism: Global Perspectives*, New York: Peter Lang, pp. 255–264.

Eisenstein, Cornelia (1994), 'Multi-step flow of communication', in *Meinungsbildung in der Mediengesellschaft: Eine theoretische und empirische Analyse zum M ulti-Step Flow of Communication*, Opladen: Westdeutscher Verlag, pp. 153–175.

Gillmor, Dan (2003), 'Foreword', in Shane Bowman and Chris Willis (eds), *We Media: How Audiences Are Shaping the Future of News and Information*, Reston, VA: Media Center at the American Press Institute, p. vi. Retrieved 24 May 2013 from http://www.hypergene.net/wemedia/download/we_media.pdf.

Highfield, Tim (2011), 'Mapping intermedia news flows: topical discussions in the Australian and French political blogospheres', Doctoral thesis, Queensland University of Technology. Retrieved from http://eprints.qut.edu.au/48115/.

Hindman, Matthew (2009), *The Myth of Digital Democracy*, Princeton, NJ: Princeton University Press.

Hyde, G. (2002), 'Independent Media Centers: cyber-subversion and the alternative press', *First Monday*, 7 (4). Retrieved 24 May 2013 from http://firstmonday.org/ojs/index.php/fm/article/view/944/866.

Katz, Elihu and Paul F. Lazarsfeld (1955), *Personal Influence: The Part Played by People in the Flow of Mass Communications*, New York: Free Press.

Leskovec, J.K., K. Lang, A. Dasgupta and M.W. Mahoney (2008), 'Statistical properties of community structure in large social and information networks', in *Proceedings of the 17th International Conference on World Wide Web*, New York: ACM, pp. 695–704.

Meikle, Graham (2002), *Future Active: Media Activism and the Internet*, New York: Routledge.

Moe, H. (2010), 'Everyone a pamphleteer? Reconsidering comparisons of mediated public participation in the print age and the digital era', *Media, Culture and Society*, 32 (4), 691–700.

Morozov, Evgeny (2011), *The Net Delusion: How Not to Liberate the World*, London: Allen Lane.

Nahon, K., J. Hemsley, S. Walker and M. Hussain (2011), 'Fifteen minutes of fame: the power of blogs in the lifecycle of viral political information', *Policy and Internet*, 3 (1), 1–28.

Nuernbergk, Christian (2013), 'Anschlusskommunikation in der Netzwerköffentlichkeit: Ein inhalts- und netzwerkanalytischer Vergleich der Kommunikation im "Social Web" zum G8-Gipfel von Heiligendamm', Baden-Baden: Nomos.

Pariser, Eli (2011), *The Filter Bubble: What the Internet is Hiding from You*, London: Penguin.

Platon, S. and M. Deuze (2003), 'Indymedia journalism: a radical way of making, selecting and sharing news?', *Journalism*, 4 (3), 336–355.

Rheingold, Howard (1993), *The Virtual Community: Homesteading on the Electronic Frontier*, New York: Basic Books.

Schmidt, Jan-Hinrik (2013), *Social Media*, Wiesbaden: Springer Fachmedien.

Shirky, Clay (2008), *Here Comes Everybody: How Change Happens When People Come Together*, London: Penguin.

Singer, Jane B. (2006), 'Journalists and news bloggers: complements, contradictions, and challenges', in Axel Bruns and Joanne Jacobs (eds), *Uses of Blogs*, New York: Peter Lang, pp. 23–32.

Sunstein, Cass R. (2007), *Republic.com 2.0*, Princeton, NJ: Princeton University.

Thurman, N. and Walters, A. (2013), 'Live blogging – digital journalism's pivotal platform? A case study of the production, consumption, and form of live blogs at Guardian.co.uk', *Digital Journalism*, 1 (1), 82–101.

Vis, F. (2013), 'Twitter as a reporting tool for breaking news: Journalists tweeting the 2011 UK riots', *Digital Journalism*, 1(1), 27–47.

Zelizer, Barbie (ed.) (2009), *The Changing Faces of Journalism: Tabloidization, Technology and Truthiness*, Milton Park: Routledge.

19. Research on the political implications of political entertainment

Michael A. Xenos

INTRODUCTION

Although not always recognized as such, the increasing prominence of political comedy within contemporary political communication is intimately related to the fundamental forces that have given rise to digital politics. Among the first to recognize this were Richard Davis and Diana Owen, who in their 1998 book, *New Media and American Politics*, identified political entertainment programming (alongside the Internet and other factors) as a key part of the 'new media' environment, which they saw as marked by the increasing political relevance of numerous media forms previously not explicitly associated with politics. Thus, during the 2000 US presidential election, we witnessed not only dramatic advancements in the realm of online campaigning, but also the regular appearances of the major candidates on a variety of late-night and daytime talk show programs as well as the special 'Indecision 2000' episodes of *The Daily Show with Jon Stewart*. But the relationship between digital and humorous politics is more than just a historical coincidence. As Bruce Williams and Michael Delli Carpini point out, albeit in conjunction with a number of other social and cultural factors, the same technological advances that have brought about more obviously digital forms of politics and political communication (increasing bandwidth and channel capacities, new logics of communication production and distribution, and so forth) have effectively 'destabilized the media regime of the mid-twentieth century', and along with it strong distinctions between serious political content and mere entertainment (Williams and Carpini, 2011: 21). As a result, contemporary discussions of 'the eroding boundary between news and entertainment' (ibid.) are closely related with the rise of digital politics. Indeed, as early as 2004, scholars were investigating the effects of a growing amount of humorous political content online surrounding political campaigns, and recent research has begun to examine phenomena surrounding user-generated political satire and its increasing circulation through YouTube and social media (Rill and Cardiel, 2013). Political comedy is thus a distinct feature of political communication in the digital era.

In this chapter I provide an overview of research on political entertainment and political comedy, which for my purposes will include research on media effects associated with various kinds of political humor, as well as so-called 'soft news' programming. In doing so I will discuss the principal findings and controversies found within this literature as well as articulate a vision for how I think future research on this topic can best mature and make substantive contributions to the broader fields of communication and political science. I begin by discussing the significance of political comedy and satire in our contemporary communication environment. This will involve consideration of the objective outlines of the increasing place of political humor in political communication over the past couple of decades, as well as further consideration of the relationship between political entertainment and broader features of the world of digital politics just mentioned. From there, I will move on to a summary of the major findings and controversies found in the literature on political comedy, before proceeding to a discussion of what I consider to be the principal challenges and opportunities facing those interested in conducting future research in this area. To be sure, the growth and development of scholarship on political comedy has thus far been impressive, to the point where one can clearly say that such work has 'come of age' as an established area of interest within political communication. At the same time, however, there are a number of obstacles and as yet less developed routes on the path ahead. Thankfully, we can reasonably expect a few good laughs along the way.

THE SIGNIFICANCE OF POLITICAL COMEDY

The use of humor or entertainment as a way to convey political information and arguments is far from new, but notable examples prior to the 1990s are often little more than isolated points of reference. As Matthew Baum explains in his seminal work on the political relevance and impacts of entertainment media, *Soft News Goes to War: Public Opinion and American Foreign Policy in the New Media Age*, until the 1980s, most people learned about politics and foreign policy (a particular emphasis of Baum's work) from newspapers and traditional television news programs (Baum, 2005). For readers with limited first-hand experience of that period, it is worth noting that such a pattern was not simply a matter of viewing preferences. Rather, prior to the 1990s, media professionals and the public simply observed much sharper boundaries between entertainment content and discussion of politics and public affairs. Of course such distinctions were always artificial, and emerged from a historically specific set of cultural and economic factors (see Baym, 2009; Delli Carpini and

Williams, 2001; Williams and Delli Carpini, 2011 for a more detailed discussion of this history), but in empirical terms, they translated into a world in which 'politics' was something primarily encountered only within the context of serious, and distinctly unfunny, media content. In that world, for most people the main source of information about politics and public affairs was the nightly television news broadcast, and though scholars debated whether newspapers were more effective in terms of helping individuals learn about complex policy issues, political entertainment of the kind we regularly encounter today was virtually absent from such discussions (see, for example, Neuman et al., 1992; Robinson and Levy, 1986, 1996).

Over the course of the 1990s, however, the political communication environment began to change. The emergence of cable and satellite television, as well as the Internet, offered more and more bandwidth for all kinds of communication, political and otherwise. This led to the availability of a host of hybrid and niche media products, and the traditional forms of news and entertainment programming became more difficult to identify as purely one or the other. A number of particularly illustrative early examples of this can again be found in Baum's work on soft news (2002, 2003, 2005). These examples – the Persian Gulf War of 1991, the US missile strikes on Afghanistan and Sudan in 1998 (which occurred just days after President Clinton's grand jury testimony regarding allegations of his affair with Monica Lewinsky, spawning many references to the movie *Wag the Dog*), and the events of the 2000 US presidential campaigns – all brought into sharp relief the ways in which coverage of political topics such as elections and foreign policy had, in the 1990s, begun to jump the boundaries between news and entertainment erected in a previous era. Whereas previous wars and earlier discussions of national politics were encountered in only limited and designated 'hard news' spaces, in the 1990s many people in the USA began to learn about similar events through programs like *Entertainment Tonight* and *A Current Affair*. Writing specifically about foreign policy crises but making a point equally applicable to general political content, Baum described the expansion of political coverage at that time as extending 'far beyond traditional network newscasts to include a vast array of soft news programs, including television newsmagazines, "human drama"-oriented local newscasts, tabloid or entertainment news programs, as well as daytime and late-night talk shows' (Baum, 2003: 94). In recent years, this expansion has only continued, to the point where programs such as *The Daily Show*, *The Colbert Report*, and *Saturday Night Live*, as well as similar programs in other countries, have become commonly accepted portions of the political communication landscape, and many traditional news programs have made significant efforts to inject entertainment into their previously more staid offerings in an effort to compete for viewers in an increasingly crowded media system.

Perhaps unsurprisingly, a large number of citizens, particularly those under 30, followed the overflow of political coverage beyond the confines of traditional 'hard news' content that began in the 1990s. Looking to research on the growth of soft news in the 1990s and early 2000s, we see that by 2002 television ratings as well as survey data supported the claim that audiences for soft news programs were 'quite large, both in an absolute sense, and relative to network and cable news programming' (Baum, 2005: 62). Indeed, in a widely cited report from 2000 that enflamed worries about the potentially negative effects of political comedy on democratic processes, the Pew Research Center for the People and the Press (2000) released survey data results showing that 47 percent of US adults aged 18 to 29 sometimes or regularly learned about the presidential campaign from late-night talk shows and political comedy programs. While later research dispelled the myth that such trends were evidence of young people abandoning traditional news sources in favor of exclusive attention to more entertaining political fare (see Young and Tisinger, 2006), these and more recent data confirm that, in contrast to what held in the 'broadcast era', a non-trivial proportion of the public appears to be increasingly consuming a diet of political content that is strongly laced with humor (Kohut, 2012).

Before considering the implications of the rise of 'infotainment' programming further, however, it is important to take note of some important considerations that help to place these developments in context. Though quite popular, particularly among young people, political entertainment programs have not necessarily resulted in a net expansion of the number of people attentive to political information and communication. In fact, scholars such as Markus Prior (2007) would argue that the increasing availability of all kinds of entertainment programming (including, but by no means limited to, political entertainment) creates a world in which those with little interest in politics have an easier time of avoiding public affairs content in any form, creating significant audience fragmentation based on basic levels of individual interest in politics. Thus despite findings suggesting that political entertainment is often complementary with, or even helps to promote, attention to politics and public affairs in other media among the otherwise inattentive (Young and Tisinger, 2006), broader patterns still point to an overall audience for news and politics that is shrinking, rather than growing, relative to the size of audiences interested in media focused purely on entertainment. As Prior concludes in his sweeping analysis of media choice and political engagement from the mid-twentieth century to the early twenty-first century, the 'flight from the news by entertainment' caused by increased levels of media choice 'is a more profound influence than the slowing of this flight through info-tainment' (Prior, 2007: 275). This is of course not to say that the rise of political entertainment and political comedy are not significant. To the contrary, as will be discussed shortly, in some ways these considerations may even make the rise of infotainment all the more significant. It is, however, important not to lose sight of the basic outlines of the existing audiences for political entertainment, especially in terms of their place within broader currents of socio-technical change and other dynamics within contemporary political communication.

Even if one takes great care not to overstate the amount of political entertainment being consumed by various sectors of the population, however, its significance as a research topic is found not so much in the size of the audiences or user-bases for such content (although, again, these are certainly not small in absolute terms), but in the extent to which these underlying dynamics reflect and speak to a variety of research questions, both timeless and contemporary. As will be discussed in more detail in the next section, research on political entertainment has helped to shed light on the complex processes by which citizens gain knowledge about politi-cal figures and issues, form political opinions and attitudes, and become more actively engaged as citizens in a communication environment that has been indelibly affected by the growth of digital technologies. In some cases, such as research exploring the sense in which political comedy can serve as a counter-current to broader patterns of interest-based fragmen-tation just discussed, work in this area helps to identify, and forces us to consider, patterns that are central to understanding democratic processes in a communication environment defined by digital networks. In many other cases, such research simply helps to provide a fresh way of looking at timeless questions concerning topics such as persuasion, political norms, and political participation. Across the board, however, because of the sense in which the rise of political entertainment is intertwined with the rise of the digital age itself, research on the implications of humorous political communication will very likely continue to grow well into the future.

MAJOR FINDINGS AND CONTROVERSIES IN SCHOLARSHIP ON POLITICAL ENTERTAINMENT

Research on political comedy and political entertainment covers a wide array of possible media effects, and has given rise to significant contro-versy concerning the normative implications of these effects. Taking the classic typology of media effects offered by Chaffee (1980) as a guide, one may describe research in this area as focused on content that combines humor with political information or arguments (regardless of medium); as exploring effects dealing with cognitive, attitudinal, as well as behavioral outcomes; and as typically conducted at the individual level of analysis. Such effects are not always direct or entirely straightforward, but they often involve core aspects of citizen engagement (Street et al., 2011). Because of this, research in this area has, since the earliest studies reviewed here, involved a vibrant discussion of whether the rise of political entertainment should be interpreted as a boon or bane for democratic processes. In this section, I will review the major findings in this literature concerning the various kinds of effects (cognitive, attitudinal, behavioral), as well as provide a guide to debates over normative questions, such as whether we should consider someone like *The Daily Show*'s Jon Stewart as a saint of democracy or, as Hart and Hartelius have colorfully declared, a sinner, whose 'sins against the Church of Democracy' are 'so heinous that he should be branded an infidel and made to wear sackcloth and ashes for at least two years, during which time he would not be allowed to emcee the Oscars, throw out the first pitch at the Yankee's [*sic*] game, or eat at the Time-Warner commissary' (Hart and Hartelius, 2007: 263).

Studies of the potential for political comedy exposure to facilitate greater political learning make up one of the largest areas of effects research focused on political entertainment. Most research in this area draws on Baum's notion of political entertainment as a potential 'gateway' to political awareness or information that can help facilitate political learning for individuals who might not otherwise attend to political information in the news (2003, 2005). The idea behind this concept is that by 'piggybacking' political information onto humorous or entertaining messages, political comedy stimulates individuals who might otherwise remain uninterested in politics to pay attention to and acquire factual information about the topics treated in a particular piece of political entertainment or satire (ibid.). Along these lines, researchers have documented this 'gateway effect' with respect to political comedy with evidence from both experimental as well as survey-based studies (for example, Feldman and Young, 2008; Xenos and Becker, 2009). These studies focus on the extent to which political comedy can facilitate learning, especially for those low in intrinsic political interest, 'contributing to an equalizing effect over time' between political sophisticates and those less interested in politics and public affairs (Young and Tisinger, 2006: 116). Other studies have taken a similar tack but focused instead on the ways in which learning is differentially stimulated based on variables such as age (that is, greater learning among younger individuals) or predispositions toward humor as a preferred mode of communication (which can also enhance learning from comedic content), rather than intrinsic interest in politics (for example, Cao, 2008; Matthes, 2013). These studies contribute to an optimistic interpretation of political comedy as a positive force in helping to create an informed citizenry.

Not all research on the effects of exposure to political comedy on political knowledge, however, supports such a straightforwardly optimistic interpretation. Indeed, a number of scholars have taken a more skeptical approach to these effects, typically through careful scrutiny of the kinds of political knowledge and learning that political comedy may facilitate. For example, in response to Baum's early work on soft news, Prior argued that while such programming may be effective in getting the attention of people who might otherwise ignore politics and foreign policy topics, 'this attention does not translate reliably into a learning effect' (Prior, 2003: 162). Pursuing a similar line of argument, Hollander (2005) makes a distinction between mere 'recognition' of information as opposed to 'recall of actual information', in interpreting his findings from a study of political entertainment programming and campaign information. Perhaps the clearest articulation of reservations concerning the effects of viewing political comedy on knowledge, however, comes from work by Baek and Wojcieszak (2009) in which item response theory (IRT) is used to show that such effects are mainly found when investigating 'widely known, thus relatively easy, political facts and issues' (p. 797). Thus despite the intuitive appeal and large body of empirical support for the basic argument surrounding the gateway hypothesis, there are also good reasons to temper overly strong expectations about the power of political comedy to facilitate political learning.

Aside from effects on knowledge, researchers have also examined the power of political humor to affect a variety of attitudes and opinions. Research along these lines has examined the extent to which exposure to political entertainment content has effects on particular opinions about specific political figures or issues, as well as effects on more generalized attitudes such as efficacy and trust. Though a number of studies have explored both kinds of effects simultaneously, given that they involve different theoretical arguments and hold different kinds of implications for democratic processes, I will review each separately.

To understand the power of political comedy to affect specific attitudes or opinions, researchers have turned to a variety of concepts borrowed from scholarship on persuasion and media effects, often from an information-processing perspective. One of the central assumptions of this approach is that 'getting a joke' is fundamentally an exercise in processing information that takes effort. The classic example of this is seen in the reconciliation of punch lines in simple jokes, but in the case of more complicated satire, such as the work of Jon Stewart and Stephen Colbert, it is clear that following the message carefully can require even more mental energy. Based on this assumption, researchers have posited that humor may increase the persuasive power of a political message by reducing one's ability to counter-argue any political claims or opinions contained in the humorous content with which one does not already agree. This is hypothesized to occur either because 'getting the joke' takes up finite cognitive energies, or because simply knowing that something is humorous decreases the motivation for effortful information processing (this is known as a 'discounting effect'), but there is slightly more evidence for the latter (Nabi et al., 2007; Young, 2008). In either case, though the mechanisms may differ somewhat, the ultimate effect is for the attitude or opinion held by the hearer of the joke to move in the direction implied by the comedic message, sometimes much later, in what is referred to as a 'sleeper effect' (Nabi et al., 2007: 49).

Other research on the effects of viewing political entertainment content on specific attitudes have explored less direct routes to persuasion or opinion formation. For example, in a study of the effects of US presidential candidate appearances on the late-night and daytime talk show circuit in 2000, Moy et al. explored the extent to which attitudes toward major political figures may be indirectly affected by political entertainment programming through priming effects (Moy et al., 2006). Specifically, using survey data collected in a rolling cross-section design (the National Annenberg Election Survey), Moy et al. identified a significant increase in the extent to which personality factors (relative to policy or issue concerns) drove public attitudes toward George W. Bush, immediately following his personal appearance on *The Late Show with David Letterman*, resulting in a temporary increase in overall support for Bush among the somewhat left-leaning *Late Show* audience (ibid.). Other studies have suggested that the effects of viewing political humor on attitudes and opinions may be conditioned by the political predispositions of the viewer. In these studies, political opinions held by individuals who are exposed to political comedy are indeed affected by that exposure, but these effects vary based on political partisanship or ideology such that something more complicated than a simple unidirectional persuasive effect explains changes in attitudes (Baumgartner and Morris, 2012; LaMarre et al., 2009; Xenos et al., 2011). Thus the effects of exposure to political entertainment content on attitudes and opinions can sometimes appear to have simply moved opinions in one direction or another, but in reality this is a more subtle result of complex interactions between the content of a comedic message and viewer predispositions, perceptions, and beliefs.

Still more research on the effects of exposure to political comedy on attitudes has focused more on generalized attitudes related to broader democratic processes, as opposed to more distinct opinions about specific individuals or political issues. Most work in this area examines whether political comedy has positive or negative effects on attitudes of political trust and efficacy, and indeed there is certainly a healthy amount of disagreement over this question. On the one hand, scholars such as Baumgartner and Morris (2006), as well as Hart and Hartelius (2007), argue that political comedy and satire, by creating a steady set of negative messages that often impugn the leadership abilities of major government and political figures, may be doing significant harm to attitudes believed to be fundamental to healthy democratic processes. Specifically they argue that programs such as *The Daily Show* serve to erode trust in the media and the electoral process, while fostering a general sense of political alienation among viewers. In contrast, other studies have found positive relationships between regular viewership of political entertainment programming and personal political efficacy, understood as one's sense of confidence in one's ability to understand and engage with the world of politics, as well as general political trust (for example, Becker, 2011; Hoffman and Young, 2011). Those with more optimistic expectations regarding the effects of political entertainment on attitudes like efficacy and trust typically argue that political humor, particularly as practiced on programs such as *The Daily Show*, helps to facilitate good democratic citizenship through modeling critical thinking and cultivating a healthy skepticism among viewers. Though findings are certainly mixed in this area, there is slightly more evidence for positive effects on democratic attitudes, especially when one considers research on participatory behaviors, which will be discussed later.

At the same time, however, more recent work has explored possible effects on other kinds of democratic attitudes, beyond the basics of trust and efficacy, with results that further complicate an optimistic interpretation of these effects. In a study examining the effects of exposure to political satire online, for instance, researchers found that participants assigned to view a satirical version of political content on the web were more likely to engage in partisan selective exposure and registered significantly lower levels of political tolerance, as compared to participants assigned to view comparable non-comic materials or (the control condition) no materials at all (Stroud and Muddiman, 2013). Thus while the overall pattern of findings has been somewhat mixed, research has documented a number of connections between viewing humorous political content and general attitudes that individuals may hold about political participation, major democratic institutions, and their fellow citizens.

Finally, in addition to studying the effects of political comedy on political knowledge and attitudes, researchers have also investigated whether viewing humorous political content has any appreciable effect on concrete political behaviors. In particular, research in this area has focused on various forms of political engagement, including standard forms of electoral participation (such as voting, and involvement with political campaigns), as well as more informal acts of political engagement such as political discussion (Cao and Brewer, 2008; Hoffman and Young, 2011; Moy et al., 2005). As alluded to earlier, research on political comedy and political engagement or participation tends to support the arguments of those who view the rise of political entertainment as beneficial to democratic processes. That is, contrary to concerns such as those expressed by Baumgartner and Morris, who cautioned that political comedy viewership 'may dampen participation among an already cynical audience . . . by contributing to a sense of alienation from the political process' (Baumgartner and Morris, 2006: 362–363), work in this area has generally documented positive relationships between political comedy viewing and political engagement (Cao and Brewer, 2008; Hoffman and Young, 2011; Moy et al., 2005). This work suggests that independent of other kinds of effects, one positive effect of political entertainment content is that (similar to traditional news programming) it may help to increase individuals' feelings of political efficacy, which in turn facilitates greater levels of political participation (Hoffman and Young, 2011). Though arguably the smallest of the three different areas of effects study in the political comedy literature, research on the effects of political entertainment viewership on political engagement is thus in many ways also the most consistent, with a clear set of relatively uniform findings.

As one can easily see from the preceding discussion, debates over whether the rise of political entertainment and political comedy is beneficial or detrimental to healthy democratic processes are a common theme encountered in studies spanning the full range of political humor media effects reviewed here. Much like broader debates over the normative implications of the rise of the internet, conflicting interpretations of what political humor may mean for democracy are often represented as a clash between optimists and pessimists. In the case of the pessimists, whether the roots of critique extend back to the work of Postman (1986), or Patterson (2000), or elsewhere, in many ways the basic concern is that political entertainment may detract from the kind of detached and rational approach to citizenship envisioned by progressive era reformers. Along these lines, pessimists worry that political entertainment may draw citizens away from more serious 'hard news' and information, while promoting a relatively thin engagement with politics and cynical (or distrustful) attitudes toward major democratic institutions. For their part, optimists tend to focus on positive trends found within the empirical literature (such as 'gateway effect' patterns for political knowledge and positive effects on political engagement), while also arguing that in many ways cynicism may be an appropriate or even beneficial ethic in our contemporary political and media systems. As Lance Bennett argued in a 'mock trial' of Jon Stewart held at the 2006 meetings of the US-based National Communication Association (the transcripts of which were later reprinted in *Critical Studies in Media Communication*), 'when the prevailing tone of public life is cynical, the best defense and response may be a probing and illuminating form of cynicism', such as that demonstrated and promoted by figures like Jon Stewart (Bennett, 2007: 282). As is also true in the case of the wider literature on digital media and digital politics, the frame of democratic optimism versus pessimism is a useful heuristic tool, particularly for introducing new students or researchers to the literature, but as will be discussed in the conclusion of this chapter, a careful accounting of available findings often reveals its limitations. Such limitations notwithstanding, however, this frame forms an important point of orientation within the literature on political entertainment media.

TAKING STOCK AND LOOKING AHEAD: COMMENTARY AND SUGGESTIONS FOR FUTURE RESEARCH

Without question, the current wave of literature on political entertainment and political comedy has made significant strides since its beginnings in the 1990s. In particular, this literature has seen progress in terms of both the quantity and quality of empirical studies, as well as the increasing sophistication of the theoretical frameworks used to understand the core dynamics involved in relationships between exposure to political comedy and relevant outcome variables. In a steadily growing body of studies, researchers interested in making sense of the potential political implications of political humor as a broad class of communication forms have drawn on major theoretical frameworks from a variety of social science disciplines, and carefully honed the application of these frameworks to a number of specific questions about the effects of exposure to political entertainment. These developments have transformed what was originally a small group of relatively exploratory investigations into a literature in its own right, with increasing potential in terms of being able to offer original insights back to areas of scholarship to which its early studies were indebted.

This question of what research on political entertainment may ultimately contribute to broader literatures in communication and political science is central to my own particular perspective on this area of research, and strongly informs my assessment of promising paths for future research. To be clear, extant research on political entertainment has certainly already made significant contributions to these literatures simply by virtue of illuminating important recent developments within our communication environments involving the increasing prevalence of humor and satire in political communication. But given the connections between political entertainment and broader tides of socio-technical change discussed earlier, I believe research on political humor has the potential to offer much more than just an understanding of what happens, all else equal, when we inject humor into a particular piece of political communication. Indeed, by virtue of its intimate relationship with underlying tectonic shifts in mediated communication, I think there is a reasonable potential for research on political entertainment to help provide useful insights into the wider world of contemporary political communication from which it has emerged, and that this is a worthy goal for future research in this area. What do patterns of political entertainment media use potentially reveal about how contemporary citizens engage with politics in an age of new media? How do hybrid forms of news and entertainment fit within a wider landscape of changing media structures and patterns? What does the prominence of political comedy and satire, with the documented effects of such media discussed earlier, tell us about the overall health of contemporary democratic processes? To be sure, a number of works in this literature have already explored this register, so to speak (for example, Baym, 2009; van Zoonen, 2004; Williams and Delli Carpini, 2011). With specific investments in a couple of key areas, however, I think that further development of the kind of scholarly discussion about political entertainment initiated in these works could be a promising direction for future research.

One particularly promising avenue for future work in this area that could help to bring about more of a macro-level discussion is comparative research that explores political entertainment and its effects in a variety of countries and media systems. Like much work within political communication, most of the research reviewed above is focused on the USA. But readers familiar with media offerings outside of the USA would certainly agree that there are both similarities and differences in terms of the form and function of political entertainment in such countries. On the one hand, the international popularity of *The Daily Show* appears to have inspired a number of programs in other countries that adopt a very similar format and structure. For example, Germans are able to enjoy humorous and irreverent discussions of public affairs topics through *Heute-Show*, while Pakistanis can turn to *The Real News* for their own version of satirical news. At the same time, there are also a number of distinct ways in which entertainment and politics are hybridized much differently in other countries. Take the singularly unique approach to political entertainment offered by Italy's Beppe Grillo, for example, or to consider a less dramatic but still qualitatively different example, Britain's current events game show, *Have I Got News for You*. Combined with insights from previous research in comparative political communication on features of the media systems in different countries (see, for example, Esser and Pfetsch, 2004; Hallin and Mancini, 2004), as well as contemporary research on the ways in which profound changes in media systems are being experienced around the world (Bennett, 2000), research on these variations and similarities could yield a much more comprehensive understanding of the macro-level implications of political entertainment as a new media form.

A related priority for future research on political entertainment and political comedy would be to increase efforts at what has been called 'systematic normative assessment' in the broader political communication literature (for example, Rinke et al., 2013). Going beyond relatively surface-level connections between particular processes or outcome variables and seemingly generic concepts drawn from 'democratic theory', systematic normative assessment involves a careful and thorough review of how certain empirical findings may be interpreted differently according to a number of distinct (and plural) democratic theories (Althaus, 2012). Such an approach would enable scholars of political comedy not only to establish the significance of particular criterion variables, such as political knowledge or participation, but also to actually arrive at broader, on balance, discussions of how the rise of political entertainment may be affecting political communication as a whole. Indeed, an explicit turn toward more systematic normative assessment in research on political comedy could facilitate, for example, discussions of the extent to which gains in knowledge and participation may or may not be offset by increases in selective exposure and decreases in careful argument scrutiny and political tolerance. In doing so, scholars of political entertainment could find a useful framework for helping existing back-and-forth conversations between 'optimists' and 'pessimists' to move forward in a more productive direction.

In conclusion, the overall assessment of the literature on political entertainment offered here is that while there has been tremendous advancement since the 1990s, there is still a great deal of potential for further growth and development of this literature in the future. Thus far, progress has been primarily concentrated on the development of increasingly sophisticated and effective explanations for particular kinds of effects at the individual level. In other words, research in this area has moved from simply treating humorous political content as 'just another input variable in a media effects equation' (Young, 2008: 133), to developing a mature body of theory-driven studies devoted to exploration of the unique properties of political comedy and the implications of these for political knowledge, persuasion, and attitude change, as well as political engagement. To be sure, work of this kind is important and there are certainly good reasons to further explore and refine scholarship focused on these kinds of effects. On its own, however, progress in this direction still does not provide direct insights into the kind of macro-level questions encountered as one moves beyond the level of the individual media user. In my view, by pursuing such questions, research in this area stands the best chance of further expanding its reach beyond the confines of a particular set of media forms and becoming more integrated with broader conversations in political communication. Though such an approach may seem somewhat daunting, I think that reflection on the connections between political humor and digital politics in general, as well as progress on the specific research priorities articulated above, can both contribute to further advancement in this exciting (as well as entertaining) area of research.

FURTHER READING

Baum, M.A. (2005). *Soft News Goes to War: Public Opinion and American Foreign Policy in the New Media Age*. Princeton, NJ: Princeton University Press.

Baym, G. (2009). *From Cronkite to Colbert: The Evolution of Broadcast News*. New York: Oxford University Press.

Cao, X. and Brewer, P. (2008). 'Political comedy shows and public participation in politics', *International Journal of Public Opinion Research*, 20(1), 90–99.

Davis, R. and Owen, D. (1998). *New Media and American Politics*. New York: Oxford University Press.

Delli Carpini, M.X. and Williams, B.A. (2001). 'Let us infotain you: politics in the new media environment', in Bennett, W. Lance and Entman, Robert (eds), *Mediated Politics: Communication in the Future of Democracy*. New York: Cambridge University Press, pp.160–181.

Feldman, L. and Young, D.G. (2008). 'Late-night comedy as a gateway to traditional news: an analysis of time trends in news attention among late-night comedy viewers during the 2004 presidential primaries'. *Political Communication*, 25(4), 401–422.

Williams, B.A. and Delli Carpini, M.X. (2011). *After Broadcast News: Media Regimes, Democracy, and the New Information Environment*. New York: Cambridge University Press.

Young, Dannagal Goldthwaite (2008). 'The privileged role of the late-night joke: exploring humor's role in disrupting argument scrutiny'. *Media Psychology*, 11(1), 119–142.

REFERENCES

Althaus, Scott (2012). 'What's good and bad in political communication research: normative standards for evaluating media and citizen performance', in Semetko, Holli and Scammell, Margaret (eds), *The Sage Handbook of Political Communication*. London, UK: Sage Publications, pp. 97–112.

Baek, Y.M. and Wojcieszak, M.E. (2009). 'Don't expect too much! Learning from late-night comedy and knowledge item difficulty', *Communication Research*, 36(6), 783–809.

Baum, M.A. (2002). 'Sex, lies, and war: how soft news brings foreign policy to the inattentive public', *American Political Science Review*, 96(1), 91–109.

Baum, M.A. (2003). 'Soft news and political knowledge: evidence of absence or absence of evidence?', *Political Communication*, 20(2), 173–190.

Baum, M.A. (2005). *Soft News Goes to War: Public Opinion and American Foreign Policy in the New Media Age*. Princeton, NJ: Princeton University Press.

Baumgartner, J. and Morris, J. (2006). 'The Daily Show effect: candidate evaluations, efficacy, and American youth', *American Politics Research*, 34(3), 341–367.

Baym, G. (2009). *From Cronkite to Colbert: The Evolution of Broadcast News*. New York: Oxford University Press.

Becker, A.B. (2011). 'Political humor as democratic relief? The effects of exposure to comedy and straight news on trust and efficacy', *Atlantic Journal of Communication*, 19(5), 235–250.

Bennett, W.L. (2000). 'Introduction: communication and civic engagement in comparative perspective', *Political Communication*, 17(4), 307–312.

Bennett, W.L. (2007). 'Relief in hard times: a defense of Jon Stewart's comedy in an age of cynicism', *Critical Studies in Media Communication*, 24(3), 278–283.

Cao, X. (2008). 'Political comedy shows and knowledge about primary campaigns: the moderating effects of age and education', *Mass Communication and Society*, 11(1), 43–61.

Cao, X. and Brewer, P. (2008). 'Political comedy shows and public participation in politics', *International Journal of Public Opinion Research*, 20(1), 90–99.

Chaffee, Steven H. (1980). 'Mass media effects: new research perspectives', in Wilhoit, C.G. and de Bock, H. (eds), *Mass Communication Review Yearbook*, Vol. 1. Beverly Hills, CA: Sage, pp. 77–108.

Davis, R. and Owen, D. (1998). *New Media and American Politics*. New York: Oxford University Press.

Delli Carpini, M.X. and Williams, B.A. (2001). 'Let us infotain you: politics in the new media environment', in Bennett, W. Lance and Entman, Robert (eds), *Mediated Politics: Communication in the Future of Democracy*, New York: Cambridge University Press, pp. 160–181.

Esser, F. and Pfetsch, B. (2004). *Comparing Political Communication: Theories, Cases, and Challenges*. Cambridge: Cambridge University Press.

Feldman, L. and Young, D.G. (2008). 'Late-night comedy as a gateway to traditional news: an analysis of time trends in news attention among late-night comedy viewers during the 2004 presidential primaries', *Political Communication*, 25(4), 401–422.

Hallin, D.C. and Mancini, P. (2004). *Comparing Media Systems: Three Models of Media and Politics*. New York: Cambridge University Press.

Hart, R.P. and Hartelius, E.J. (2007). 'The political sins of Jon Stewart', *Critical Studies in Media Communication*, 24(3), 263–272.

Hoffman, L.H. and Young, D.G. (2011). 'Satire, punch lines, and the nightly news: untangling media effects on political participation', *Communication Research Reports*, 28(2), 159–168.

Hollander, B.A. (2005). 'Late-night learning: do entertainment programs increase political campaign knowledge for young viewers?', *Journal of Broadcasting Electronic Media*, 49(4), 402–415.

Kohut, Andrew (2012). 'Cable leads the pack as a campaign news source', Pew Center for the People and the Press. Available at http://www.people-press.org/files/legacy-pdf/2012%20 Communicating%20Release.pdf (accessed 6 June 2012).

LaMarre, H.L., Landreville, K.D. and Beam, M.A. (2009). 'The irony of satire: political ideology and the motivation to see what you want to see in the Colbert Report', *International Journal of Press/Politics*, 14(2), 212–231.

Matthes, J. (2013). 'Elaboration or distraction? Knowledge acquisition from thematically related and unrelated humor in political speeches', *International Journal of Public Opinion Research*, 25(3), 291–302.

Moy, P., Xenos, M. and Hess, V.K. (2005). 'Communication and citizenship: mapping the political effects of infotainment', *Mass Communication and Society*, 8(2), 111–131.

Moy, P., Xenos, M. and Hess, V. (2006). 'Priming effects of late-night comedy', *International Journal of Public Opinion Research*, 18(2), 198–210.

Nabi, R.L., Moyer-Gusé, E. and Byrne, S. (2007). 'All joking aside: a serious investigation into the persuasive effect of funny social issue messages', *Communication Monographs*, 74(1), 29–54.

Neuman, W.R., Just, M.R. and Crigler, A.N. (1992). *Common Knowledge: News and the Construction of Political Meaning*. Chicago, IL: University of Chicago Press.

Patterson, T.E. (2000). 'Doing well and doing good: how soft news and critical journalism are shrinking the news audience and weakening democracy-and what news outlets can do about it', Research Report, Cambridge, MA: Joan Shorenstein Center on the Press, Politics, and Public Policy. Available at http://www.uky.edu/AS/PoliSci/Peffley/pdf/475PattersonSoftnews(1-11-01).pdf (accessed June 1, 2013).

Pew Research Center for the People and the Press (2000). 'Audiences fragmented and skeptical: the tough job of communicating with voters'. Available at http:/people-press.org/reports/display.php3?ReportID=200 (accessed June 1, 2013).

Postman, N. (1986). *Amusing Ourselves to Death: Public Discourse in the Age of Show Business*. New York: Penguin.

Prior, M. (2003). 'Any good news in soft news? The impact of soft news preference on political knowledge', *Political Communication*, 20(2), 149–171.

Prior, M. (2007). *Post-Broadcast Democracy: How Media Choice Increases Inequality in Political Involvement and Polarizes Elections*. New York: Cambridge University Press.

Rill, L.A. and Cardiel, C.L.B. (2013). 'Funny, ha-ha: the impact of user-generated political satire on political attitudes', *American Behavioral Scientist*, 57(12), 1738–1756.

Rinke, E.M., Wessler, H., Löb, C. and Weinmann, C. (2013). 'Deliberative qualities of generic news frames: assessing the democratic value of strategic game and contestation framing in election campaign coverage', *Political Communication*, 30(3), 474–494.

Robinson, J.P. and Levy, M.R. (1986). *The Main Source: Learning from Television News*. New York: Sage Publications.

Robinson, J.P. and Levy, M.R. (1996). 'News media use and the informed public: a 1990s update', *Journal of Communication*, 46(2), 129–135.

Street, J., Inthorn, S. and Scott, M. (2011). 'Playing at politics? Popular culture as political engagement', *Parliamentary Affairs*, 65(2), 338–358.

Stroud, N.J. and Muddiman, A. (2013). 'Selective exposure, tolerance, and satirical news', *International Journal of Public Opinion Research*, 25(3), 271–290.

Van Zoonen, L. (2004). *Entertaining the Citizen: When Politics and Popular Culture Converge*. Lanham, MD: Rowman & Littlefield.

Williams, B.A. and Delli Carpini, M.X. (2011). *After Broadcast News: Media Regimes, Democracy, and the New Information Environment*. New York: Cambridge University Press.

Xenos, M. and Becker, A. (2009). 'Moments of Zen: effects of the Daily Show on information seeking and political learning', *Political Communication*, 26(3), 317–332.

Xenos, M.A., Moy, P. and Becker, A.B. (2011). Making sense of the Daily Show: understanding the role of partisan heuristics in political comedy effects. In Amarasingham,

A. (ed.), *The Stewart/Colbert Effect: Essays on the Real Impacts of Fake News*. Jefferson, NC: McFarland & Company, pp. 47–62.

Young, Dannagal Goldthwaite (2008). 'The privileged role of the late-night joke: exploring humor's role in disrupting argument scrutiny', *Media Psychology*, 11(1), 119–142.

Young, D. (2012). 'Laughter, learning, or enlightenment? Viewing and avoidance motivations behind the Daily Show and the Colbert Report', *Journal of Broadcasting and Electronic Media*, 57(2), 153–169.

20. Journalism, gatekeeping and interactivity
Neil Thurman

Gatekeeping is one of the most inclusive research traditions in the field of journalism studies. In its investigations into the processes 'by which the vast array of potential news messages are winnowed, shaped, and prodded into these few that are actually transmitted by news media' (Shoemaker et al., 2001: 233) it accommodates political and economic influences, as well as organizational routines and practices; the influence of the audience, outside sources and technology; and journalists' individual characteristics and collective professional values. However, changes in how technology and the audience – individually and collectively – are taking on journalistic gatekeeping functions; how established gatekeeping routines have changed in response to information from the public and about their news consumption behaviour; and some of the political and economic influences on gatekeeping in the online news environment have not, yet, been fully reflected in the academic literature.

In this chapter I will discuss these technological and social influences on journalistic gatekeeping by reflecting on my own research in these areas over the last decade or so. The chapter begins with a review of the literature on gatekeeping as it applies to journalism. I will then use the concepts of 'adaptive' and 'conversational' interactivity to frame the discussions that follow on how technology and the audience are impacting journalistic gatekeeping. The chapter concludes with a discussion of some of the consequences of the full spectrum of forces – political and economic, as well as social and technological – acting on contemporary, mainstream news producers; as well some suggestions for how they may better accommodate to those forces. Finally I give some suggestions for future research in these areas.

Digital, networked media have made it possible for news publishers to give their audiences a relatively high degree of control over the stories they consume and how those stories are delivered and presented. Technology has also enabled publications to invite and distribute reader contributions in modes, quantities and with timeliness unsupportable in the past. These technological affordances all come under the general term 'interactivity'. It is useful, however, to distinguish between:

- 'navigational' interactivity, where the 'user is allowed to navigate in a more or less structured way through the site's content' (Deuze, 2003);
- 'conversational' interactivity (Jensen, 1998), which allows the user to interact with journalists and other users (Deuze, 2003); and
- 'adaptive' (Deuze, 2003) or 'registrational' (Jensen, 1998) interactivity, where a set of technological features adapt media content, its delivery and arrangement to individual users' explicitly registered and/or implicitly determined preferences (Thurman, 2011).

This chapter will focus on the 'conversational' and 'adaptive' forms of interactivity. Although the non-linear navigational structures of news websites and apps do change the way audiences interact with journalism online, especially when compared with the linearity of broadcasting, my own research has focused to a greater extent on how conversational and adaptive interactivity are influencing the gatekeeping processes that determine whether information about an event will become a message, and whether an audience member will pay attention to any resulting news item.

JOURNALISTIC GATEKEEPING: AN OVERVIEW

The psychologist Kurt Lewin (1947) is credited with developing the concept of gatekeeping in his work on food consumption habits. Lewin modelled the metaphorical 'channels' through which food travelled to reach the dining table, the 'gates' through which it had to pass (the 'gate' to the supermarket shelf, for example), and the 'gate-keepers' or 'impartial rules' that controlled those gates (such as supermarket buyers or personal dietary rules). Lewin's theory included the concept of 'forces', which determine whether an item will pass through a gate. A positive force could be a food's health benefits, a negative force its lack of freshness.

Although Lewin (1951: 187) suggested that gatekeeping could be applied to 'the travelling of a news item through certain communication channels', he did not live to apply his concept in the domain of communication. That task fell, initially, to one of his former research assistants, David Manning White, who analysed how a wire editor – whom he called 'Mr Gates' – decided which copy to include in the local US newspaper for which he worked. White found Mr Gates's selection decisions to be 'highly subjective' (White, 1950: 386).

The investigation of possible psychological and cultural influences on such 'subjective' selection decisions has driven a number of subsequent gatekeeping studies. Galtung and Ruge (1965), for example, famously proposed that events were more likely to be newsworthy to the likes of Mr Gates if they had attributes such as timeliness, magnitude, clarity, cultural relevance, consonance with expectations, and novelty. A considerable amount of work has also been undertaken on journalists' personalities (see, for example, Henningham, 1997) and backgrounds and how those characteristics might influence the production of news content. Gans (1979), for example, suggested that journalists practising in the USA were guided by eight values including ethnocentrism, small town pastoralism, individualism and moderatism; with information pertaining to deviations from these values more likely to pass the gates than information that was about behaviour consistent with the status quo.

Such personal characteristics are, of course, not the only ones that influence whether information about an event will become a 'message' or whether an audience member will pay attention to any resulting news item. Cognitive approaches have proposed that communication that is concrete, emotionally interesting and imagery-producing is more likely to pass the gate (Nisbett and Ross, 1980: 45). Narrative structure – for example, whether a news story is conventionally resolved (Bennett, 1988: 24) – may also have an effect.

While White's original study focused on a single 'news processor', later scholarship has broadened the scope of gatekeeping research to include, for example, 'news gatherers' (Bass, 1969: 72), such as writers, reporters and local editors. And the early focus by White and others on editors' individual subjectivity has also been widened to include organizational contexts and routines, including time pressures and production constraints. Such routines have been shown to be a significant influence on gatekeeping decisions, even trumping individual journalists' characteristics (see, for example, Cassidy, 2006). These 'repeated practices' emerge, Shoemaker and Reese (1996) argue, from three sources: journalists' sense of their audience, the newsroom culture and organization in which they work, and their sources of news.

Audience volume and demographics should be important to gatekeepers in media systems where the audience is a market and a product to be sold to advertisers. However, some scholars have dismissed the notion that the audience could directly influence gatekeeping decisions. Gieber (1960), for example, wrote that the selection of news had 'no direct relationship to the wants of readers'. News values have, as Sumpter (2000) and others suggest, instead been influenced by journalists' 'construction' of audience demand. Such views make some sense in light of the incomplete data journalists have traditionally had about audience tastes; data which came from sources such as relatively small samples of TV and radio audiences, surveys based on readers' (potentially fallible) recollections of what they had read, and circulation statistics that relate to media artefacts in their entirety rather than the individual stories they contain. Contemporary work – even that with the traditional media as its focus – has, however, attributed greater influence to the audience (see, for example, Fahmy, 2005). And the era of online news has ushered in new practices where websites are able to adapt to audience demand in much more direct ways.

The second dimension in Shoemaker and Reese's framework is newsroom culture and organization. Factors here include ethical procedures, style guidelines and the pressure of deadlines. In relation to gatekeeping, deadlines, for example, have been shown to influence what gets selected for inclusion (White, 1950), and the stories written by reporters (Dunwoody, 1978). Journalists' notions about their professional role may also have an influence on gatekeeping decisions. Traditional values of independence and objectivity (Arant and Meyer, 1998) may have militated against the use of certain types of sources, such as members of the public. Although the presence of other media in the marketplace can in some circumstances lead to greater diversity, we have evidence stretching back more than 30 years of how media gatekeepers monitor and imitate one another. Crouse wrote about this in 1972 in relation to political reporting, and Boczkowski (2010) has described, in some detail, the culture of imitation in online news. Such imitation may become more common as news organizations, their profit margins falling and their news production moving online, decide to downsize (Paterson, 2001). The aspect of newsroom culture and organization that will be one of the two main foci of this chapter is the technological change that has been taking place in news detection, creation, packaging and distribution. Whilst some, such as Williams and Carpini (2004), believe that such changes have prompted the 'collapse of gatekeeping' in the 'new media environment', others argue that there is a significant continuity between the routines of online news sites and their print or broadcast parents (Arant and Anderson, 2001).

Shoemaker and Reese's third dimension is journalists' sources. Although Bass (1969) refined the gatekeeping concept by extending its scope beyond news processors to news gatherers, his work did not, however, consider how what he termed 'raw news' came into existence. The role of sources – including the public relations departments of corporations and governments – in providing such 'news' in a form that appeals to the news gatherers is now well recognized (see, for example, Gandy, 1982), with the result that gatekeeping theory in the news domain has extended beyond the boundaries of traditional journalistic institutions. There is now considerable journalistic scholarship on news sources, usefully defined by Sigal (1973) as enterprise (such as investigative reporting or direct witnessing), routine (such as public information and staged events), and informal (such as other media or off-the-record briefings). Although studies have consistently shown that journalists rely heavily on official news sources (see, for example, Schiffer, 2006), some have suggested that online journalism may be reducing the media's reliance on such sources (Williams and Carpini, 2004), including in crisis coverage (see, for example, Thurman and Rodgers, 2014).

This very brief summary of more than six decades of research into journalistic gatekeeping can only hint at how, over the last decade or so, members of the public – via processes that this chapter classifies as 'conversational interactivity' – have begun to change journalists' sense of their audience, their sources of news, and how they collectively work. In order to explore these contemporary concerns more deeply, in the section that follows I will reflect on my own research over the last decade.

CONVERSATIONAL INTERACTIVITY

Back in 2004 when I first surveyed online editors' attitudes to conversational interactivity (Thurman, 2008), only one of the British national news sites surveyed hosted real blogs (those with comments enabled) and one national newspaper website had no formats for readers to contribute at all. Where readers could contribute, pre-moderation was the norm, applied in 80 per cent of cases. In this sense, the media was retaining a traditional gatekeeping role, with journalists acting as message filters. Editors' attitudes to conversational interactivity were mixed, with comments like this from the then editor of Telegraph.co.uk: 'This idea with blogs and particularly wikis that you can go in and edit stuff and all join the party. It is a load of fun but it just detracts from what a traditional idea of journalism is. I think we have to be quite careful' (Thurman, 2008).

Editors were concerned about the ways in which non-professionally produced content challenged journalism's professional norms. They expressed particular concern over its news value; standards of spelling, punctuation, accuracy and balance; and the influence of blogs on the mainstream media. There was, however, an understanding of the benefits of users' submissions, although this was framed by editors within existing journalistic norms and practices. Contributions from the public were seen as source material for stories, and as a way of increasing loyalty as well as the depth and diversity of coverage.

Perhaps unexpectedly, my study did not uncover any fundamental prejudice against user media amongst editors, contrary to some of the hyperbole flying around at the time, such as this quote from Danah Boyd (2004): 'Blogging has terrified mainstream media for a while now.

Journalists want to know if blogs are going to degrade their profession, open up new possibilities or otherwise challenge their authority'. Instead, my findings were consistent with Pablo Boczkowski's view (2004: 4) that innovation in newsrooms unfolds in a 'gradual and ongoing fashion' and is 'shaped by combinations of initial conditions and local contingencies'. Specifically, my study found that time and resources, the legal environment, the management and professional preparedness of journalists, and news sites' technical infrastructure (the 'local conditions' referred to by Boczkowski) were the key determining factors in mainstream sites' adoption of conversational interactivity.

About 18 months after my first survey, a follow-up study (Hermida and Thurman, 2008) revealed a significant increase in conversational interactivity. The number of blogs recorded had jumped from 7 to 118 and there had been considerable adoption of 'Comments on stories' and 'Have your says'. At this time the taxonomy developed in my first study was expanded to include a new format, 'Reader blogs', introduced at the website of Britain's biggest-selling newspaper, *The Sun*. This format was a radical departure from the traditional publishing model, as it sought to present news and comment on current events from the point of view of the audience. While news organizations were providing more opportunities for participation, my second study also found evidence that they were retaining a traditional gatekeeping role. Moderation and/or registration remained the norm as editors' concerns over reputation, trust and legal liabilities persisted. This said, editors were relatively open to conversational interactivity. One described user media as a 'phenomenon you can't ignore', another said he 'firmly believed in the great conversation', and one editor explained that he was 'very interested in unlocking' information from his 'very knowledgeable' readers (Hermida and Thurman, 2008). But there was a hidden agenda in news sites' decisions to open up to readers. Self-interest emerged as a strong motivator. Some editors were fearful of being 'left behind', and there was also a worry that, if they did not give their staff a 'piece of property on the Internet', journalists might develop a community of readers by blogging elsewhere (Hermida and Thurman, 2008). This follow-up study confirmed both the desire of publications to get the 'right user-generated content' that fitted their brand's values (Hermida and Thurman, 2008), and the considerable resource implications of moderation. It also questioned the extent to which readers wanted to contribute – and whether that mattered.

My third study, based on a survey conducted in May 2008 (Thurman and Hermida, 2010), showed a continuing adoption of conversational interactivity and, perhaps surprisingly, evidence of a more relaxed attitude to moderation. Despite ongoing concerns, the websites of three national newspapers[1] all published readers' comments without registration or pre-moderation. My study proposed that the shift away from moderation was a result of the increase in opportunities readers had to participate. With more choice, news websites perceived that readers were less likely to participate if barriers to participation (like registration) existed, or if they didn't get the immediate, positive feedback provided by instant publication.

Although there was a continuing increase in opportunities for readers to contribute over the three years of this work, textual contributions were in the main still limited to short comments on subjects or stories determined by professional editors.[2] There was little in the way of longer-form contributions, or opportunities for readers to set the agenda. Therefore my final study suggested that the media was creating an architecture of publication for material from the audience, rather than an architecture of participation. Where opportunities for readers to set the agenda did exist (for example in readers' blogs;[3] or at message boards[4]) they often seemed to be part of what Bowman and Willis (2003) described as 'closed-off annex(es) where readers can talk and discuss, as long as the media companies don't have to be involved'.

Attempts to create genuinely open spaces where readers can set the agenda were few and far between. *The Times*'s 'Your World' travel site was one, but after initial external investment to get it running (it was sponsored by BMW), the site atrophied without ongoing support and management, and it was eventually taken down. A similar feature at *The Guardian* – 'Been there'[5] – was and is a much more successful example of the mainstream media allowing readers to set the agenda. Unlike at *The Times*, there are no restrictions on length, and users can edit and update other submissions. Furthermore, readers can aggregate other readers' tips to create travel guides, hence performing a real editorial role for the first time. Here, conversational interactivity has gone beyond simply publishing material from users and instead emphasizes the sharing and remixing of content. However, we must not forget that this feature was outside what most journalists would consider to be 'news'. In the softer area of lifestyle it was, perhaps, considered more acceptable for publications to cede control.

Over the last decade, then, editors have increasingly been making gatekeeping decisions on information submitted by their readers. Interviews with some of those editors have provided insights into their beliefs and motivations and helped to explain their actions as gatekeepers. My research has shown – in its quantification of the volume of information that readers supply and how much of that 'information' is ultimately published as news items – that the public have, to a significant extent, supplemented traditional sources. It has also shown, however, that such a change is not – as some have suggested – a 'collapse of gate-keeping' (Williams and Carpini, 2004), but rather, in the prominence (or lack of) given to news items that emerge from public sources and the selection decisions that take place in the communication channel along which they travel, there is significant continuity between the routines of online news editors and their print and broadcast counterparts.

ADAPTIVE INTERACTIVITY

My brief review of more than half a decade of journalistic gatekeeping literature shows an almost exclusive concern with human gatekeepers and the psychological, cultural, organizational and technological influences on their behaviour. There is, by contrast, very little research on how – via processes that this chapter defines as 'adaptive interactivity' – the audience, individually and collectively, are taking on journalistic gatekeeping functions as media content is automatically adapted to users' explicitly registered and/or implicitly determined preferences.

I have been researching such adaptive interactivity since 2007, surveying a range of news websites and talking to their editors. My findings show that adaptive interactivity is increasingly common with, at the time of my most recent survey (Thurman and Schifferes, 2012), sites offering, on average, 11 different forms.

There is an important distinction to be made between those forms of adaptive interactivity that are active and those that are passive. With the active form, users register their own content preferences. The passive form infers preferences from data collected, for example, via a registration process or via the use of software that monitors user activity. My research shows that, since 2007, there has been a decline in the category of adaptive interactivity that demands the most input from users, what I call 'MyPages'. By contrast, I found significant growth in some of the passive forms, in particular what I call 'social collaborative filtering'. Indeed, although active forms of personalization were still more common, passive forms grew faster in percentage terms (Thurman and Schifferes, 2012). My taxonomy of adaptive interactivity shows the sheer variety of approaches (18 in all), indicative of the ongoing search among news providers to find the most effective types of adaptive interactivity, balancing the need for precise matching of content to users' interests with the need to make the process of setting up the active forms as easy as possible for the audience.

One of the most notable changes since 2007 has been the sharp increase in adaptive interactivity that uses recommendations from social networking sites, what I call 'social collaborative filtering', for example the Facebook 'Activity Feed' plug-in through which users receive recommendations from their Facebook 'Friends'. A problem with this form of socially powered adaptive interactivity, however, is the infrequency with which content in these plug-ins updates, one reason being that the average Facebook user posts an average of just 2–3 links to stories on news sites a year (Thurman and Schifferes, 2012). It is clear then that the increasing use of social media is prompting news websites to adopt new forms of adaptive interactivity at a rapid rate, but that such developments are still in their early stages.

Over the years news providers have made considerable efforts constructing 'MyPage' functionality to allow users to assemble whole pages of news adapted to their preferences. These 'MyPages' are, however, in decline, dropped by the websites of the *Washington Post*, the *Sun*, the *Telegraph* and being phased out by the *New York Times* and the *Wall Street Journal*. It seems that the uptake of these relatively demanding services has not been sufficient to justify their continuing existence. The editors I interviewed in my research put this down to audience passivity and the difficulty users have in accurately predicting their content preferences in the dynamic news domain.

In contrast to 'MyPages', mobile editions that include some adaptive interactivity have been growing. I found that, on average, news sites provided adaptive apps for at least two devices and more than half had an adaptive mobile version of their site. This is not surprising, given the growing numbers of smartphone users. Indeed there are reasons why smartphones might be particularly good platforms for adaptive information delivery. Firstly, due to their smaller screens and input devices, their browsing capabilities are limited; and secondly, their locative capacity lends itself to content adaptation based on place. It is surprising then that most mobile editions and apps were relatively static in nature, with a minimum of adaptive interactivity (Thurman and Schifferes, 2012). On average, they offered barely one and a half different forms of adaptive interactivity, compared with an average of more than 20 for news sites' full web editions. This thirteenfold difference may be explained by the fact that most of the apps were first generation, but the notion that mobile devices such as the iPad are better suited to passive consumption may also be a factor.

Staying with the idea of the passive user, I will end this section by moving on to rises in passive forms of adaptive interactivity, in particular what I call 'contextual recommendations' and 'aggregated collaborative filtering'. Aggregated collaborative filtering is where selections of news stories or other content (such as readers' comments) are automatically filtered by popularity. Variables include 'most read', 'most e-mailed', and 'most commented'. This form of adaptive interactivity is almost universal, popular with readers and editors alike. The reasons? Firstly, it is passive, requiring no effort from readers; and secondly, editors like it because it increases page views, but in doing so usually reinforces their editorial judgement, as many of the stories recommended have already been selected on the front and section pages. Another form of passive adaptive interactivity has also been growing, what I call 'contextual recommendations'. This is where lists of links or aggregations of content are created algorithmically, based on context. These developments are important because they are indicative of a move away from the traditional concept of the journalist as gatekeeper, deciding which news stories are presented to the public, when and with what priority. As we have seen, this gatekeeping role is increasingly being replaced by algorithms, users and crowds. Algorithmic gatekeeping in particular raises questions about accountability and transparency when some of the companies that offer the enabling technology promise publishers that their recommendations can be skewed for commercial purposes.

DISCUSSION

My own studies of conversational interactivity have shown that in the middle of the first decade of the third millennium, with online news well established, gatekeeping – at least in the mainstream media – was far from collapse. Although the public had joined traditional sources as suppliers of information, their contributions were still subject to many of journalism's long-established individual and organizational routines. Such routines limited, to a degree, the amount of information from the public that eventually appeared as news items, as well as the visibility of those news items. However, in the four years that followed (2004–2008), gatekeeping routines shifted. The channels down which information from the public travelled were enlarged and began to carry increasing volumes of material; and the gates within those channels became easier to penetrate.

I believe that the catalysts for these changes were:

1. Professional and personal influences on journalists, such as a desire to explore new forms of practice, to meet audience expectations, and to integrate quality content whatever the source.
2. Organizational influences, such as internal politics, a desire to retain staff, fear of market marginalization, and other editorial and commercial interests.

3. Wider societal developments in technology and media consumption patterns.

So, although gatekeeping routines have changed within the journalistic field, they have done so relatively slowly as a result of the power of continuity within the institution itself. Furthermore, the changes that have taken place cannot be ascribed to any single political, economic, social, organizational or individual factor but result from the complex interactions of all these influences.

My studies of adaptive interactivity have shown that between 2007 and 2010 there was a significant and consistent growth in mainstream online news publishers' deployment of technologies that adapt content to users' explicitly registered and (in particular) implicitly expressed preferences. This deployment not only allowed audiences to act as gatekeepers in their own right, but also opened a strong feedback channel between audiences and professional gatekeepers and gave computer programs – often developed and controlled externally – gatekeeping responsibilities.

However, as with the deployment of user-generated content initiatives, the adoption of adaptive interactivity did not result in a collapse of gatekeeping in the mainstream media. Rather, these mechanisms were deployed in ways that provided a high degree of continuity with existing editorial practices. Sites still offered, in the main, edited selections of material with multiple opportunities for serendipitous discovery and for journalists to demonstrate the 'value' their core editorial function provided (Thurman, 2011: 412).

The principal conclusion to be drawn from these observations is that gatekeeping models need revision. Shoemaker and Vos's (2009: 125) model of gatekeeping (reproduced as Figure 20.1) is one of the most contemporary and reflects the ability audiences now have to redistribute news items among themselves and to act, collaboratively, to filter news items into lists of the most popular and, as a result, give those items enhanced priority.

Shoemaker and Vos's model does not, however, fully reflect the reality of contemporary online news artefacts and how they are used. To do so the model would need to:

1. Show how the audience – by using Twitter, or subscribing to RSS feeds or e-mail newsletters – can influence which 'observers, participants, interested parties, and experts or commentators' they are exposed to via the source channel.
2. Add a feedback loop between readers and the media channel. This would reflect how readers can influence the news content they receive

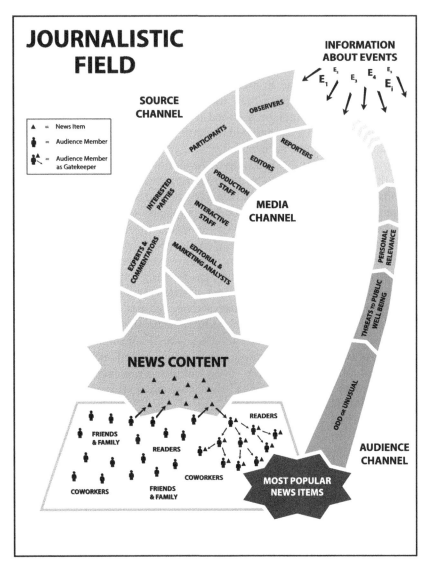

Figure 20.1   Source, media and audience channels in the gatekeeping process

by, for example, customizing 'Home' or 'My' pages, changing their geographical location, interacting with non-linear features, linking news sites with their social network profiles, and interacting with content in particular ways.

3. Show how the gates in the media channel are not – as the model currently has it – mediated exclusively by human operators. The gatekeeping roles of computer programs should be included in the diagram, and in such a way as to show their common existence outside the media institution.

Changes in journalistic gatekeeping routines have, of course, taken place at a time when traditional print and broadcast news providers were attempting to develop new online services (Thurman and Lupton, 2008; Thurman and Myllylahti, 2009), expand into new geographical markets (Thurman, 2007; Thurman et al., 2011) and introduce new ways of charging for their services (Herbert and Thurman, 2007). However, the introduction of these new services, the expansion into new markets, and experiments with new ways of charging for these services, have not resulted in a fundamental transformation of the financial fortunes of news providers.

We could say then that wider social, technological, political and economic forces have had a greater influence on the financial viability of, and audience appetite for, traditional providers' news products than any changes that have taken place in their professional and organizational routines and the resulting news artefacts.

These technological, political, economic and social forces – digitization, changing work patterns, globalization, market liberalization, and so on – are well known. The consequence for traditional news providers has been a slow erosion in the revenues they receive, because of a failure to replace audience attention lost from their traditional print or broadcast platforms with an equivalent amount of attention from their online operations,[6] and because of structural changes in the advertising market that have reduced the value of the space, and the audience, they sell.

I have fewer insights into possible solutions to the problems facing the journalistic field, because many of the innovations to process and product that have been introduced by the mainstream news media have failed to counter the external pressures they face. I do, however, offer some suggestions. Many of the innovations that have taken place have, to a large extent, been within existing organizational norms (Hermida and Thurman, 2008: 353; Thurman and Myllylahti, 2009), tortuous (Thurman and Hermida, 2010: 61), subtle (Thurman, 2011: 412) and restrictive (Thurman et al., 2011), implying an unsustainable level of complacency and inertia. Instead, established news providers should ensure that they:

1. Focus on the specific needs of their audience.
2. Adapt style and structure better to the online medium.
3. Become more selective about how to innovate, basing decisions on evidence.
4. Continually evaluate commercial and technology partnerships to ensure they are in the best interests of the organization.
5. Avoid token gestures and poor execution in favour of well-designed and supported developments.
6. Invest in original content and ongoing research and development.

Why, though, does it matter whether the institution we know as journalism is sustained? Despite the potential of collaborative and open source news (as described, for example, by Bruns, 2009), questions remain about its scope and scalability. The vast majority of original news reporting – some 95 per cent (Pew, 2010) – still emerges from traditional news providers; and that percentage is undoubtedly even higher for investigative, international and other forms of news that are expensive to conduct. Furthermore, although some successful collaborative and open source news channels have emerged – Slashdot, Wikinews, NowPublic, Spot.us, Newsvine, Reddit, and so on – the limited interest I have found that users had in actively interacting with the collaborative publishing and selection tools provided by mainstream news providers prompts questions over collaborative news' scalability.

SOME PRIORITIES FOR FUTURE RESEARCH

Research cited in this chapter has shown how computer algorithms are making decisions on news prioritization and presentation. These mechanisms are difficult to detect and describe because they operate without user involvement and use closely guarded proprietary algorithms, often outside the direct control of the news sites that host the services they provide. The companies involved in providing some of the enabling technology to news websites include Daylife, Evri, Autonomy, Aggregate Knowledge, Blogrunner, Digg, Loomia, Moreover and OneSpot. The outsourcing and automation of gatekeeping processes is worthy of further investigation. Further research should look to reveal the logic behind the computer algorithms that are increasingly determining how news is prioritized and presented. It should identify the sources used for the contextual recommendations provided by many of these companies, and ask what decisions have been made about classification and indexing. Such questions cannot be answered by content analysis alone. In addition, representatives of the companies involved would need to be questioned directly. The results of such an investigation would inform important questions about bias and homogeneity in news output. Do, for example, the mechanisms of adaptive interactivity increase content diversity in online news by taking away some of the control journalists have had over news selection? And what biases are built into the automated systems of news prioritization and the systems of classification and indexing on which they rely?

Gatekeeping literature on how media ownership and market forces influence media content tends to assume the persistence of recent historical media models. The result has been a focus on comparative studies looking at the differences between, for example, market-orientated and public-service-orientated journalism (Beam, 2003) or between chain-owned and independent media (Gaziano, 1989). The external pressures on newspapers are now such that we are likely to see changes in strategy such as switching to online-only publishing. Newspapers and magazines that have already made this decision include *Christian Science Monitor*, Madison's *Capital Times*, *Newsweek*, *Seattle Post-Intelligencer*, *SmartMoney* magazine and *Ann Arbor News*. My own case study of the Finnish newspaper *Taloussanomat* (Thurman and Myllylahti, 2009) showed that when the title went online-only it lost at least 75 per cent of its revenues. Staffing levels dropped, initially by 40 per cent, later even further. The consequences were a shift to popularism, an increase in utilitarian content, and a reduction in journalists' use of enterprise sources to the detriment of news quality and diversity. Gatekeeping scholars could profitably build on my preliminary investigation into the effects of the online-only model on media content, not where the online channel is part of a larger multi-platform news operation, but where it is the only channel to market.

NOTES

1. The independent.co.uk, ft.com and mirror.co.uk.
2. Limits, where they existed, were between 60 and 300 words.
3. Such as those hosted at the websites of the *Sun, Daily Star* and *Telegraph*.
4. 'Message boards' were hosted by the websites of: the *Daily Star, Mirror, Financial Times, Guardian, Telegraph* and *Standard*.
5. http://www.ivebeenthere.co.uk/
6. Between 2004 and 2011 the total annual minutes spent reading by the aggregated UK print and online readerships of 12 UK national newspapers fell by 27 per cent (Thurman, 2014).

372 Handbook of digital politics

FURTHER READING

Anderson, C.W. (2012) 'Towards a sociology of computational and algorithmic journalism', *New Media and Society*. DOI: 10.1177/1461444812465137.
Barzilai-Nahon, K. (2009) 'Gatekeeping revisited: a critical review', *Annual Review of Information Science and Technology*, 43, 433–478.
Lee, Angela M., Lewis, Seth C. and Powers, M. (2012) 'Audience clicks and news placement: a study of time-lagged influence in online journalism', *Communication Research*. doi: 10.1177/0093650212467031.
Pariser, Eli (2011) *The Filter Bubble*, London: Penguin Books.
Singer, Jane B. (2013) 'User-generated visibility: secondary gatekeeping in a shared media space', *New Media and Society* DOI: 10.1177/1461444813477833.

REFERENCES

Arant, M. David and Anderson, Janna Quitney (2001) 'Newspaper online editors support traditional standards', *Newspaper Research Journal*, 22(4), 57–69.
Arant, M. David and Meyer, Philip (1998) 'Public and traditional journalism: a shift in values?', *Journal of Mass Media Ethics: Exploring Questions of Media Morality*, 13(4), 205–218.
Bass, Abraham Z. (1969) 'Refining the "gatekeeper" concept: a UN radio case study', *Journalism Quarterly*, 46, 69–72.
Beam, Randal A. (2003) 'Content differences between daily newspapers with strong and weak market orientations', *Journalism and Mass Communication Quarterly*, 80(2), 368–390.
Bennett, W. Lance (1988) *News: The Politics of Illusion*, 2nd edn, New York: Longman.
Boczkowski, Pablo (2004) *Digitizing the News: Innovation in Online Newspapers*, Cambridge, MA: MIT Press.
Boczkowski, Pablo (2010) *News at Work: Imitation in an Age of Information Abundance*, Chicago, IL: University of Chicago Press.
Bowman, Shayne and Willis, Chris (2003) 'We media: how audiences are shaping the future of news and information', Media Center, 21 September. http://www.hypergene.net/wemedia/weblog.php (accessed November 2006).
Boyd, Dana (2004) 'Demeaning bloggers: the *NY Times* is running scared', *Many 2 Many: a Group Weblog on Social Software*, 26 July, http://www.corante.com/many/archives/2004/07/26/demeaning_bloggers_the_nytimes_is_running_scared.php (accessed April 2005).
Bruns, Axel (2009) *Gatewatching: Collaborative Online News Production*, New York: Peter Lang.
Cassidy, William P. (2006) 'Gatekeeping similar for online, print journalists', *Newspaper Research Journal*, 27(2), 6–23.
Crouse, Timothy (1972) *The Boys on the Bus: Riding with the Campaign Press Corps*, New York: Random House.
Deuze, Mark (2003) 'The web and its journalisms: considering the consequences of different types of newsmedia online', *New Media and Society*, 5(2), 203–230.
Dunwoody, Sharon (1978) 'Science writers at work', Bloomington, IN: School of Journalism, Centre for New Communications, Indiana University.
Fahmy, Shahira (2005) 'Photojournalists' and photoeditors' attitudes and perceptions: the visual coverage of 9/11 and the Afghan War', *Visual Communication Quarterly*, 12 (3–4), 146–163.
Galtung, Johan and Ruge, Mari Holmboe (1965) 'The structure of foreign news', *Journal of Peace Research*, 2(1), 64–90.

Gandy, Oscar H. (1982) *Beyond Agenda Setting: Information Subsidies and Public Policy*, Norwood, NJ: Ablex.

Gans, Herbert J. (1979) *Deciding What's News*, New York: Pantheon.

Gaziano, Cecilie (1989) 'Chain newspaper homogeneity and presidential endorsements, 1972–1988', *Journalism Quarterly*, 66(4), 836–845.

Gieber, Walter (1960) 'How the "gatekeepers" view local civil liberties news', *Journalism Quarterly*, 37(1), 199–205.

Henningham, John (1997) 'The journalist's personality: an exploratory study', *Journalism and Mass Communications Quarterly*, 74(3), 615–624.

Herbert, Jack and Thurman, Neil (2007) 'Paid content strategies for news websites: an empirical study of British newspapers' online business models', *Journalism Practice*, 1(2), 208–226.

Hermida, Alfred and Thurman, Neil (2008) 'A clash of cultures: the integration of user-generated content within professional journalistic frameworks at British newspaper websites', *Journalism Practice*, 2(3), 343–356.

Jenson, Jens F. (1998) '"Interactivity": tracking a new concept in media and communication studies', *Nordicom Review*, 19(1), 185–204.

Lewin, Kurt (1947) 'Frontiers in Group Dynamics II. Channels of group life: social planning and action research', *Human Relations*, 1, 143–153.

Lewin, Kurt (1951) *Field Theory in Social Science: Selected Theoretical Papers*, New York: Harper.

Nisbett, Richard and Ross, Lee (1980) *Human Inference: Strategies and Shortcomings of Social Judgement*, New York: Prentice-Hall.

Paterson, Chris A. (2001) 'The transference of frames in global television', in Reese, Stephen D., Gandy, Oscar H. and Grant, August E. (eds), *Framing Public Life: Perspectives on Media and our Understanding of the Social World*, Mahwah, NJ: Lawrence Erlbaum Associates, pp. 337–353.

Pew (2010) 'How news happens', 11 January. http://www.journalism.org/analysis_report/how_news_happens (accessed 23 May 2013).

Schiffer, Adam J. (2006) 'Blogswarms and press norms: news coverage of the Downing Street Memo controversy', *Journalism and Mass Communication Quarterly*, 83(3), 494–510.

Shoemaker, Pamela J., Eichholz, Martin, Kin, Eunyi and Wrigley, Brenda (2001) 'Individual and routine forces in gatekeeping', *Journalism and Mass Communication Quarterly*, 78(2), 233–246.

Shoemaker, Pamela J. and Reese, Stephen D. (1996) *Mediating the Message: Theories of Influences on Mass Media Content*, 2nd edn, White Plains, NY: Longman.

Shoemaker, Pamela J. and Vos, Timothy P. (2009) *Gatekeeping Theory*, New York: Routledge.

Sigal, Leon V. (1973) *Reporters and Officials: The Organisation and Politics of Newsmaking*, Lexington, MA: D.C. Heath.

Sumpter, Randall S. (2000) 'Daily newspaper editors' audience construction routines: a case study', *Critical Studies in Media Communication*, 17(3), 334–346.

Thurman, Neil (2007) 'The globalization of journalism online: a transatlantic study of news websites and their international readers', *Journalism: Theory, Practice and Criticism*, 8(3), 285–307.

Thurman, Neil (2008) 'Forums for citizen journalists? Adoption of user generated content initiatives by online news media', *New Media and Society*, 10(1), 139–157.

Thurman, Neil (2011) 'Making "The Daily Me": technology, economics and habit in the mainstream assimilation of personalized news', *Journalism: Theory, Practice and Criticism*, 12(4), 395–415.

Thurman, Neil (2014) 'Newspaper consumption in the digital age: measuring multi-channel audience attention and brand popularity', *Digital Journalism*, 2(2), 156–178.

Thurman, Neil and Hermida, Alfred (2010) 'Gotcha: how newsroom norms are shaping participatory journalism online', in Tunney, Sean and Monaghan, Garrett (eds), *Web Journalism: A New Form of Citizenship*, Eastbourne: Sussex Academic Press, pp. 46–62.

Thurman, Neil and Lupton, Ben (2008) 'Convergence calls: multimedia storytelling at British news websites', *Convergence: The International Journal of Research into New Media Technologies*, 14(4), 439–455.

Thurman, Neil and Myllylahti, Merja (2009) 'Taking the paper out of news. a case study of *Taloussanomat*, Europe's first online-only newspaper', *Journalism Studies*, 10(5), 691–708.

Thurman, Neil, Pascal, Jean-Christophe and Bradshaw, Paul (2011) 'Can Big Media do "Big Society"? A critical case study of commercial, convergent hyperlocal news', paper presented to the Future of Journalism Conference, 8–9 September, Cardiff University, UK.

Thurman, Neil and Rodgers, James (2014) 'Citizen journalism in real time? Live blogging and crisis events', in Allan, Stuart and Thorsen, Einar (eds), *Citizen Journalism: Global Perspectives*, Vol. 2, New York: Peter Lang, pp. 81–95.

Thurman, Neil and Schifferes, Steve (2012) 'The future of personalization at news websites: lessons from a longitudinal study', *Journalism Studies*, 13(5–6), 775–790.

White, David Manning (1950) 'The "gatekeeper": a case study in the selection of news', *Journalism Quarterly*, 27, 383–391.

Williams, Bruce A. and Carpini, Michael X. Delli (2004) 'Monica and Bill all the time and everywhere: the collapse of gatekeeping and agenda setting in the new media environment', *American Behavioral Scientist*, 47(9), 1208–1230.

PART VI

INTERNET GOVERNANCE

21. Internet governance, rights and democratic legitimacy
Giles Moss

From the revelations about government internet surveillance disclosed by Edward Snowden to regular controversies about how private companies such as Apple, Facebook and Google regulate their platforms, it seems as though the libertarian ethos of the early internet is long behind us. Early libertarian commentators had viewed the internet, or 'cyberspace' as it was known, as an anarchic and ungovernable space which was beyond the purview of national governments. Yet states across the world have increased their control over the internet in recent years, often working with private intermediaries (internet service providers, search engines and software platform owners) which have become increasingly important sites of regulatory control.

The most important question of internet governance today, it seems, is not whether the internet can be governed, but how it should be governed. Normative debates about the values that should guide internet governance are most often expressed in terms of individual rights, such as freedom of expression, privacy, security, access to knowledge, and so on (Drake and Jørgensen, 2006; Klang and Murray, 2005; see also Association for Progressive Communication, 2013; Internet Rights and Principles Coalition, 2013). But while the language of rights is often shared among the actors involved in internet governance, be they states, private corporations or civil society groups, there is significant disagreement about the meaning of rights, the balance to be struck when rights conflict (as they inevitably do), and the question of how rights are best realized in practice.

Given the uncertain and contested nature of rights, it might be tempting to dismiss them as little more than empty words or ideological smokescreens, the meaning and application of which will ultimately be decided by the most powerful. Following the political theorist Seyla Benhabib (2008, 2009; see also Forst, 2010; Habermas, 1998), I argue instead that rights are valuable 'cosmopolitan norms' that provide a shared normative vocabulary for internet governance, which can empower civil society groups and bind those states and private actors that sign up to them. However, given scope for legitimate variation in the interpretation of rights, their translation in particular contexts is necessarily also a political matter, which to be legitimate must be democratic and supported by processes of deliberation (Benhabib, 2008, 2009). Viewed this way, the crucial question is how democratic, rather than just how effective, existing practices of internet governance are at national, regional and global levels (Freedman, 2012; Sylvain, 2010). In the absence of inclusive and deliberative public participation, internet governance will be beset by legitimation problems.

Before beginning let me make a brief note on terminology and how the terms 'policy', 'regulation' and 'governance' will be used in this chapter (see Freedman, 2008: 13–15). Policy involves making decisions about the goals to pursue in a particular area. Policy-making is never purely a technical matter: it is an inescapably political process, where different values and interests collide and groups compete to shape policy outcomes. If policy involves establishing general goals, regulation refers to the specific actions and tools that are employed in order to achieve policy ends. These tools range from formal legal regulations to less obvious but still important forms of regulation, such as regulating behaviour through the design of technology or code (Lessig, 1999a, 2006). Regulation is not limited to states and related public authorities; it also involves private companies, and most notably those that have a central position in networks as intermediaries and so are well placed to regulate internet content and use, either on behalf of states or in line with their own commercial interests. Self-regulation is where private companies regulate on a voluntary basis. Co-regulation refers to cases where private actors are given regulatory responsibilities, but states supervise the process and reserve the power to intervene (Marsden, 2011). Governance is a less clearly defined concept. The term is used in this chapter in a general sense to refer to how the internet is governed, and more specifically to indicate that governing the internet is a complex process that is not limited to single national governments working in isolation. As will become clear, governing the internet involves interactions and collaboration both among states (in regional, international and global networks) and between states and various non-state actors (private companies, nongovernmental organizations and civil society groups).

THE LIBERTARIAN INTERNET

A libertarian ethos dominated early discussions of the internet. Unlike previous media of communication, the internet was thought to constitute a unique global space (cyberspace) that was beyond the control of nation states. By its very nature as an open and decentralized global network, the internet seemed to be impervious to government regulation. As John Gilmore famously put it, 'the Net interprets censorship as damage and routes around it' (quoted in Elmer-DeWitt and Jackson, 1993). Equally importantly, early 'cyber-libertarians' argued that states would lack the legitimacy to regulate cyberspace, even if they could do so (Murray, 2010: 56–62). From a libertarian perspective, freedom is best achieved where individuals are left to regulate their own behaviour without interference from the state. Rules and regulations, insofar as they need to exist, should emerge from internet users themselves in a ground-up manner.

The libertarian perspective was expressed in its most dramatic terms in 1996 by the internet activist John Perry Barlow in his 'A Declaration of the Independence of Cyberspace'. Responding to the prospect of greater government regulation of the internet (and, specifically, the US Government's Communications Decency Act in 1996), Barlow (1996) wrote:

> Governments of the Industrial World, you weary giants of flesh and steel, I come from Cyberspace, the new home of Mind. On behalf of the future, I ask you of the past to leave us alone. You are not welcome among us. You have no sovereignty where we gather.
>
> We have no elected government, nor are we likely to have one, so I address you with no greater authority than that with which liberty itself always speaks. I declare the global social space we are building to be naturally independent of the tyrannies you seek to impose on us. You have no moral right to rule us nor do you possess any methods of enforcement we have true reason to fear.
>
> Governments derive their just powers from the consent of the governed. You have neither solicited nor received ours.

The legal scholars David Johnson and David Post (1996) echoed some of Barlow's views, albeit in more sober terms. 'Global computer-based communications cut across territorial borders', they argued, 'creating a new realm of human activity and undermining the feasibility – and legitimacy – of laws based on geographic boundaries' (Johnson and Post, 1996: 1367). Given its different geography, the internet would need to develop its own unique laws and rules, which would derive their legitimacy from the community of users themselves, rather than being imposed from the outside by existing nation states: 'a new boundary, made up of the screens and passwords that separate the virtual world from the "real world" of atoms, emerges. This new boundary defines a distinct Cyberspace that needs and can create its own law and legal institutions' (Johnson and Post, 1996: 1367).

The fact that the development of the internet resulted from state intervention and public investment in the first place always sits somewhat uneasily with pure libertarian accounts of cyberspace. As Robert McChesney (2013: 99) notes, 'The entire realm of digital communications was developed through government subsidized and directed research during the post-World War II decades, often by the military and leading research universities. Had the matter been left to the private sector, the Internet may not have come into existence'. In other respects, however, the early years of the internet did appear to support the libertarian view. Compared with previous media of communication, the internet was relatively free of direct government regulation. John Palfrey (2010) characterizes the early period of the internet's development, from the 1960s to the late 1990s, as that of the 'open internet'. He notes that 'up until the late 1990s, most states tended either to ignore online activities or to regulate them very lightly. When states did pay attention to activities online, they tended to think about and treat them very differently from activities in real-space' (Palfrey, 2010: 2). Given the relative lack of direct regulation, the onus was placed on private actors to self-regulate their own behaviour and to develop shared norms to regulate internet content and use.

The organizations that emerged to govern the internet's underlying technical infrastructure were also different from the state-led organizations characteristic of other media and communications (Bygrave and Bing, 2009; Epstein, 2013). Decisions about the internet's structures and standards have been taken by private, non-governmental bodies made up of technical experts and private companies, rather than by political representatives, government officials and policy experts. The allocation of names and addresses for the internet is coordinated by the Internet Corporation for Assigned Names and Numbers (ICANN), a private, non-profit organization established by the US Government in 1998. Technical documentation and standards for the internet are similarly developed by non-state organizations: the Internet Engineering Task Force (IETF) is responsible for developing internet standards while the World Wide Web Consortium (W3C) is responsible for web standards. Some scholars have praised these organizations for their uniquely ground-up approach, where decisions are arrived at through discussion and consensus (Froomkin, 2003). Others are more critical, questioning for example ICANN's independence from the US Government (Freedman, 2012; Mueller, 2010). However, compared with traditional state-based forms of regulation, it is fair to say that the early governance of the internet's infrastructure was characterized by a distinctive governing style that appeared in keeping with the internet's early libertarian ethos. As David Clark (cited in Lessig, 1999b: 1418), a computer scientist who played a central role in the development of the internet, famously described the IETF's decision-making process: 'We reject kings, presidents, and voting. We believe in rough consensus and running code'.

There is some truth in the libertarian account of the early internet as a space of private self-regulation and limited state control. However, the absence of direct regulation by the state can also be seen in part as the result of a conscious policy decision taken by governments, most notably the US Government, which favoured self-regulation by private actors and the market (Freedman, 2008: 103–5). To this extent, the early libertarian internet was as much a product of political choice as of technological design. I return later to the normative question of how the internet should be governed. In the next section, I describe how a number of writers from the late 1990s onwards began to challenge the libertarian claim that the internet could not be governed. As a result, they developed more sophisticated accounts of how the internet is governed and regulated in practice, identifying the growing role not only of state regulation but also of private corporations as influential intermediaries online.

CODE, CAPITAL AND THE CAPACITY TO REGULATE

From the late 1990s onwards, the early libertarian view of cyberspace as an ungovernable space was put in question by writers who argued that in theory there was no reason why the internet could not be regulated and that we should expect it to become more regulated in practice. Where libertarians assumed that freedom from government control was built into the technical structures of the internet, critics suggested that it was a contingent and vulnerable historical fact. The rejection of libertarianism has generated an increasingly nuanced debate today about how the internet is governed and about whose interests and values tend to prevail in internet governance.

The libertarian perspective assumed that the existing laws of nation states did not apply to cyberspace since it was, in effect, a separate realm that transcended territory. Subsequent critics of libertarianism argued that this view was based on a misconception. Individuals are still very much present in the 'real' world when using the internet: 'Unlike the imaginary worlds of childhood fantasy such as Narnia or Alice's Wonderland, Cyberspace is not somewhere to which we are physically transported' (Murray, 2010: 56). Internet users are therefore subject to the laws of the countries in which they live, just as they are in other respects. So while it may not be possible for states to take actions against internet users living outside their borders, they can apply the law effectively to their own citizens. As Jack Goldsmith and Tim Wu (2006) stressed in their book *Who Controls the Internet? Illusions of a Borderless World*, the internet has done nothing to alter the fundamental political fact – originally captured by Max Weber – that states retain a monopoly on the legitimate use of physical force within their territory that allows them to enforce social order effectively. From the imprisonment of online political dissidents in authoritarian countries to legal actions taken against individuals for copyright infringement or computer misuse in the US and elsewhere, the nation state's power over citizens within its territory is clearly evident. Of course, the political legitimacy of particular states to use this power varies, something which depends, as I argue below, on the democratic credentials of the regime in question (Habermas, 1991, 1997).

In challenging the libertarian belief that the internet could not be controlled, the critics also adopted a broader conception of regulation, which included forms of regulation other than the law. The best-known scholar to develop this argument was Lawrence Lessig in his book *Code and Other Laws of Cyberspace* (Lessig, 1999a). Lessig (1999a, 2006) argued that our behaviour online is regulated not only by the law and the threat of legal sanctions, but also by social norms, market forces and by 'code' (by which he meant the technical instructions embedded in hardware and software). All four of these 'modalities of regulation', as Lessig (2006) refers to them, can enable or constrain particular courses of action. States are therefore able to regulate the behaviour of individuals not just directly through the law, but also indirectly by seeking to change social norms through education, by modifying the market, or by requiring changes to code.

As the title of his book (*Code and Other Laws of Cyberspace*) suggests, Lessig (1999a, 2006) was especially concerned with the regulatory role that code might play. Where libertarians had taken the technology of the internet as a given, Lessig (2006) emphasized how the code of the internet was flexible and labile. Code can be designed to serve various purposes and in ways that can have a significant impact on fundamental values and rights. Writing code, therefore, is never just a technical matter: 'Choices among values, choices about regulation, about control, choices about the definition of spaces of freedom – all this is the stuff of politics' (Lessig, 2006: 76). Lessig (2006) argued that the political stakes are raised by some of the distinctive features of code as a means of regulation. Code and its effects may be difficult for users to detect and understand, particularly for those (the majority of us) who lack the necessary technical literacy. At the same time, code operates automatically, regulating behaviour in advance and without the need for legal action (McIntyre and Scott, 2008). By describing how the internet can be regulated through code, Lessig (2006) emphasized not only how regulation online was possible. If we are not careful, he warned, 'cyberspace will be the most regulable space humans have ever known' (Lessig, 2006: 32).

Where the early internet seemed to fit with the libertarian thesis, the development of internet regulation from the late 1990s onwards appeared to support Lessig's account. By the early 2000s, Palfrey (2010) argued that the early phase of the 'open internet' had given way to a second phase, which he calls 'access denied', as code was increasingly used to regulate internet use and content: 'During this second era, states and others came to think of activities and expression on the Internet as things that needed to be blocked or managed in various ways. The thinking was that certain acts of speech and organizing online needed to be regulated like any other' (Palfrey, 2010: 7). States began to encourage or mandate internet service providers to apply filters on their networks which would block access to certain content. Importantly, the rationale for filtering varied. In Europe, for example, internet service providers employed filters to block access to child-abuse images hosted in other countries, whereas filtering was used for political reasons in authoritarian states in order to censor dissidents and dampen opposition (Palfrey, 2010). Code was also adopted early on in an attempt to address online copyright infringement or so-called 'piracy'. For example, 'digital rights management' was applied to music files in order to try and prevent users from copying and sharing them freely (Edwards et al., 2013). The different policy goals that code may serve raise normative and political questions about what values and interests should guide internet governance. I return to these questions below. The present point is how, as Lessig (1999a, 2006) suggested, code can and was being used alongside law as a form of regulation.

While examples of increasing regulation by code appeared to support Lessig's thesis, some critics questioned whether he had nonetheless overstated his case. Internet regulation in practice appeared less certain and more complex and dynamic than Lessig's model allowed. The legal theorist Andrew Murray (2010: 62 70) agrees with Lessig that the internet can be regulated in theory. However, he argues that Lessig underestimates the agency that internet users have to resist and negotiate the forms of regulation to which they are subject. As an example, Murray (2010: 69–70) points to how internet users resisted the use of digital rights management on music files and how copyright infringement has continued online despite the considerable efforts of governments and rights holders in the cultural industries to eradicate it (see also Edwards et al., 2013). To explain such resistance, Murray (2010) develops an alternative 'network communitarian' perspective. He argues that internet users are not isolated individuals but are always already part of a broader community, which is collectively able to play a more active role in the regulatory process. Indeed, Murray argues that three out of four of Lessig's modalities of regulation (law, norms and the market) are directly influenced by the community: 'Laws are passed by lawmakers elected by the community, markets are merely a reflection of value, demand, supply, and scarcity as reflected by the community in monetary terms and norms are merely the codification of community value' (Murray, 2010: 68). Meanwhile, the community is often able to 'engineer around' Lessig's fourth form of regulation: code (Murray, 2011: 24). Regulation is therefore 'socially mediated', as he puts it, and reliant on the agreement of the community which is regulated. As Murray (2010: 68) concludes, 'The socially mediated modalities of law, norms, and markets draw their legitimacy from the community . . . meaning the regulatory process is in nature a dialogue not an externally imposed set of constraints'.

Murray's account of internet regulation is an important contribution to the debate, especially in underscoring the complex and socially mediated nature of regulation. But then, if Lessig underestimates the potential of the internet community to resist and negotiate forms of regulation, is Murray's description of internet regulation as a form of 'dialogue' between regulators and 'the community' too sanguine? Certainly, in terms of the legitimacy of decision-making in normative terms, democratic control over internet policy-making is essential (as I will argue below). However, it is not clear that such control is always secured in practice: democratic control is not only absent in authoritarian countries, but is also limited in today's 'actually-existing democracies' (Fraser, 1990). Critical political economists of the internet point here to the growing power of certain private corporations in internet governance, the interests of which do not always align with those of the public, but which have significant sway over the direction of internet policy (Freedman, 2012; McChesney, 2013).

While the internet is still often depicted as the libertarians imagined it, as an open and decentralized global space, critical scholars emphasize the growing asymmetries and concentrations of power online. In particular, they point to the central regulatory role that certain large private corporations have assumed by virtue of their position as key intermediaries in the networks that make up the internet. As Uta Kohl (2012: 186) puts it, 'our actions and communications are, in the offline world *often* and in the online world *always*, mediated by third parties; and these third parties in so far as they are natural chokepoints, bottlenecks, gatekeepers or border checkpoints are hugely attractive regulatory targets'. Significantly, the regulatory power of intermediaries is something Murray underscores in his recent writings. He notes that:

> As the sole guardians of access to certain spaces, communities or content they are extremely powerful and carry a significant amount of regulatory gravity. In fact they carry a regulatory weight probably not dissimilar to other state based regulators such as statutory regulators due to their unique ability to control the part of the community or network over which they control access. (Murray, 2011: 27)

It is worth distinguishing among different types of intermediary here (see Kohl, 2012, 2013). Firstly, there are internet service providers, which own and control the networks (both fixed and mobile) through which we gain access to the internet in the first place. Secondly, and only slightly less indispensable for the average internet user, there are search engines and aggregation services which help us to find our way around and navigate the web. Thirdly, there are platforms, such as those provided by Apple, Facebook and Google, which host software applications and content that attract huge numbers of users. As Jonathan Zittrain (2009) has argued, companies which control software applications and data in the 'cloud' are particularly important insofar as we use more closed, controllable and 'tethered' devices such as games consoles, e-book readers, smartphones and internet-connected televisions. All the intermediaries above have significant regulatory power to shape internet content and use, by virtue of their central position within the networks that make up the internet. As Des Freedman (2012: 114) argues, 'if we think of regulation in terms of the ability to structure access to and shape content on the internet, then powerful new regulators appear: not simply Comcast, Verizon and AT&T but Facebook, Yahoo! and of course, Google'.

In some countries, such as China, intermediaries are held legally responsible for the content they host and distribute, with the result that they then must vigilantly monitor and regulate internet content and use (Wong and Dempsey, 2011: 17–18). In Europe, by contrast, intermediaries are not held responsible legally, provided they remove illegal material when notified, and only the original producer of the content is liable (the so-called 'notice and take down' approach). The legal immunity of intermediaries can be viewed as important in protecting free speech online and helping to ensure that intermediaries do not police and censor their networks too forcefully (Balkin, 2008). However, in recent years, intermediaries have been increasingly viewed by governments as significant means of regulatory control that can help, as the Organisation for Economic Co-operation and Development (OECD, 2010) puts it, in 'advancing public policy objectives'. Intermediaries can help both to enforce regulation and in monitoring and policing the activities of its users. Among others things, intermediaries have recently been called upon to help tackle the problem of copyright infringement: search engines have been requested to make it more difficult for users to find sites that facilitate copyright infringement, while government legislation, such as the HADOPI law in France (2009) or the Digital Economy Act in the UK (2010), has required internet service providers to monitor their users, and manage and even suspend access to recidivist infringers (Edwards et al., 2013). Seeking to capture the growing role of private intermediaries in internet regulation, Palfrey (2010) argues that a third stage of internet regulation has emerged from 2005 to the present, which he calls 'access controlled'. During this phase, 'States themselves cannot implement the level of control that they seek over activity on the network directly, so their control strategies have expanded to include pressure on private parties' (Palfrey, 2010: 14).

Private intermediaries are not only able to perform regulation on behalf of states, but can also use their position to further their own commercial interests, and in ways that may not always align with the interests of the public (Freedman, 2012; McChesney, 2013). Consider, for example, heated debates about 'net neutrality' (Marsden, 2010). The net neutrality debate involves a complex set of issues, but at its heart is the question of how internet traffic is managed on networks and whether internet service providers should be able to discriminate among the data streams they carry. Internet service providers may have economic incentives to discriminate in order to privilege their own content and services or that of companies they are affiliated with or to limit access to the content and services of competitors (for example, the blocking of internet telephony services on mobile networks). However, such restrictions limit people's freedom of expression online and ability to receive the content and services of their choice (Balkin, 2008). Similar arguments apply to discrimination in the results generated by search engine companies. Privacy is another example of an area where the commercial interests of corporations (in this case, to collect personal data about users so that it can be used for advertising and marketing purposes) may not correspond with those of the public (Bermejo, 2011).

Thinking back to Murray's (2010) 'network communitarian' account of internet regulation, we might argue that the community has the power to hold private companies in check through the market. If consumers are unhappy with the services they receive, they can seek an alternative provider, while companies will need to compete with one another to satisfy consumer demand. The problem is that market competition may be overstated. Individuals rarely act like the rational consumers assumed by simplistic accounts of free market economics, and they lack the information to do so. Furthermore, they may be faced with limited choice in the market. On the internet, as Freedman (2012: 115) notes, 'One thing that has remained constant is the structure of a "winner takes all" market which systematizes the need for huge concentrations of online and offline capital'. Network effects are particularly important in this context, as McChesney (2013: 132) has argued:

> The internet exhibits what economists term network effects, meaning that just about everyone gains by sharing use of a single service or resource ... The

largest firm in an industry increases its attractiveness by an order of magnitude as it gets a greater market share, and makes it impossible for competitors with declining shares to remain attractive or competitive.

States can and to some extent do regulate private companies in order to ensure that important values or rights are upheld. However, arriving at policies that are in the public interest is reliant on all groups and views being represented fairly in the policy-making process. It is clear that the libertarian view that the internet cannot be regulated is too simplistic: the internet is being regulated by states and by private actors too. The important question is the normative one: by what values should the internet be governed and how is internet policy to be decided most legitimately?

RIGHTS, DEMOCRACY AND LEGITIMACY

Normative debates about how the internet should be governed are most often expressed in the language of human rights (Drake and Jørgensen, 2006; Klang and Murray, 2005). Rights refer to valued ways that individuals should be able to act or deserve to be treated (Martin, 1998: 325), and 'human rights' refer to those particularly fundamental rights that individuals are entitled to because of their humanity (Campbell, 2008). Although rights are normative ideals, they also aspire to being realized in practice. Obligations are placed on states not only to 'respect' the rights of individuals by not interfering with them, but also to 'protect' them from infringement by others, and to 'fulfil' them by taking 'appropriate legislative, budgetary, judicial, and other measures' (Drake and Jørgensen, 2006: 15). States across the world are signatories to various international conventions on human rights, most notably the Universal Declaration of Human Rights (UDHR). Private corporations also increasingly adopt the language of rights as part of their corporate social responsibility. Meanwhile, valuable efforts have been made to interpret the meaning of human rights in relation to the internet. For example, the Association for Progressive Communication (2013) and the Internet Rights and Principles Coalition (2013) have both developed internet rights charters, which cover a range of rights around areas such as freedom of expression and association, access and accessibility, privacy, security, non-discrimination, access to culture and knowledge, political participation, and cultural and linguistic diversity.

Given regular violations of rights around the world, it is easy to be sceptical about the invocation of human rights. At least, the discourse of rights can sometimes appear as just a convenient ideological smokescreen, which powerful states and corporations can employ strategically when it suits them. However, it would be wrong to reduce rights entirely to an ideological role. As the political theorist Seyla Benhabib (2008, 2009) has argued, rights are important 'cosmopolitan norms', which enjoy widespread agreement and provide a shared normative vocabulary for governance at national and global levels. Rights can bind those states and private corporations that sign up to them, whilst also empowering rights-based civil society and activist groups. As Benhabib (2008: 100–101) puts it, 'When states subscribe to various international human rights conventions, a dynamic process is set into motion in civil society and the public sphere. These provisions give rise to a public language of *rights articulation* and *claims-making* for all sorts of civil society actors'. She goes on to note that, 'This new language of public claims-articulation circulates in the unofficial public sphere, and can, and often does, impact further institutional reform and legislation' (Benhabib, 2008: 101).

Even where human rights may be accepted in general terms, there is still significant scope for disagreement about the precise content of rights, the balance to be struck when rights conflict, and the question of how rights are best realized in practice. Consider, for example, freedom of expression (Barendt, 2005). A right to freedom of expression is enshrined in Article 19 of the UDHR: 'everyone has the right to freedom of opinion and expression; this right includes freedom to hold opinions without interference and to seek, receive and impart information and ideas through any media and regardless of frontiers'. Freedom of expression is a widely recognized human right, connected with the self-development of individuals and essential to the functioning of democracy. However, it is not an unqualified right. Libertarians may balk at any restrictions being placed on freedom of speech, but others argue that free speech needs to be balanced against other important rights and values, such as national security, a child's right to protection, and so on. In European countries, for example, there are generally stronger legal regulations against racist, homophobic and other forms of hate speech than is the case in the United States (Vick, 2005).

Where libertarians emphasize the danger that the state poses to freedom of speech, social democrats advocate a more positive role for the state. As described in the previous section, private intermediaries are important regulatory forces online that can have a significant impact on internet content and use. Yet, as illustrated by the net neutrality debate, the commercial interests of intermediaries do not always align with the interests of the public. States may therefore be required to intervene and mandate net neutrality in order to protect and promote freedom of expression (Balkin, 2008). Other forms of intervention by the state may also be justified from a free speech perspective. For example, the right to freedom of expression online is of little substantive value to those individuals who lack internet access and skills. The state may therefore be justified in introducing policies to promote access and digital inclusion: Finland, for example, became the first country in 2010 to make broadband internet access a right for all its citizens (Marsden, 2011: 4). As a final example, state intervention can also be justified in relation to the provision of public service media online (Barendt, 2005: 417–451; Wong and Dempsey, 2011: 12). Libertarians of the right may see public service media as an unnecessary state intervention, which is detrimental to market competition. But in the light of concerns about the future of quality journalism (Baker, 2002; Habermas, 2009: 131–138; McChesney, 2013: 172–216), or how new advertising practices may impair the public sphere through increased personalization of media (Couldry and Turow, 2014), public service media may be justified as promoting freedom of speech overall.

Given variation in the interpretations of rights, we should expect – quite justifiably – for rights to be understood and translated in different ways by publics across countries and contexts. Benhabib (2008: 99) uses the concept of 'democratic iterations' to explain this process:

> [D]emocratic iterations signal a space of interpretation and intervention between context-transcendent norms and the will of democratic majorities. On the one hand, the rights claims that frame democratic politics must be viewed as transcending the specific enactment of democratic majorities in specific polities; on the other hand, such democratic majorities *re-iterate* these principles and incorporate them into the democratic will formation process of the people through contestation, revision, and rejection.

In terms of legitimacy, what is crucial is whether or not decisions about rights are in the public or general interest, as clarified through inclusive democratic procedures where all perspectives and options are openly considered and reflected upon (Habermas, 1991, 1997). As Benhabib (2008: 99) goes on to argue:

> Naturally, if the conversations that contribute to democratic iterations were not carried out by the most inclusive and equal participation of all those whose interests are affected, or if these deliberations did not permit the questioning of the conversational agenda, then the 'iterative' process would be unfair, exclusionary, and illegitimate.

The legitimacy of existing practices of internet governance has been much debated. As already noted, the governance of the internet's infrastructure by private, non-governmental organizations has been lauded by some as a consensual, ground-up form of governance, while others question the degree of independence of organizations such as ICANN (Mueller, 2010; Freedman, 2012). The World Summit on the Information Society

(WSIS), organized by the United Nations, sought to address the global governance of the internet in two high-profile meetings held in Geneva in 2003 and Tunis in 2005. Compared with examples of global governance in other areas, WSIS has been viewed by commentators as being significant in procedural terms: it experimented with a new form of 'networked', 'multistakeholder governance', which included civil-society groups campaigning for human rights in addition to the representatives from states and private companies which one might expect at such summits (Raboy et al., 2010). But then these global processes still remain distant from and somewhat opaque to the public at large. In addition, the actual policy outcomes that resulted from the WSIS process were limited, suggesting that the power still tends to reside elsewhere, with the most powerful individual states and corporations. As Mueller (2010: 59–60) concludes: 'WSIS was not a powerful process in most respects . . . The summit did not succeed in reallocating major sums of money. It did not pass binding treaties or conventions backed up by strong new organizations capable of enforcing them in a way that could reshape global communications'. One notable and positive outcome of WSIS was the establishment of an Internet Governance Forum to continue multi-stakeholder discussions about the internet (the Internet Rights and Principles Coalition, mentioned above for its work on human rights, is an offshoot of the forum). But it remains the case, as Freedman (2012: 113) argues, that so far, 'multistakeholder governance has not proved to be a magic solution to the problems posed by entrenched state and corporate power'.

While these processes of global governance are important, the clearest connection between the public and internet policy-making and regulation is still the established political-representative processes offered by democratic states and regional political associations like the European Union (Freedman, 2012; Sylvain, 2010). Such states can therefore play a central role in the democratic interpretation and translation of rights in particular contexts. The problem, as already noted, is that democratic control and inclusive participation in policy-making is often limited here too. The public's distance from decision-making is especially apparent where decisions about rights are 'outsourced' to private companies. As Marsden (2011: 46) argues, 'governments have outsourced constitutionally fundamental regulation to private agents, with little or no regard for the legitimacy claim other than those founded on the avowed deregulatory alchemy of private actors over bureaucrats'. He writes that, 'The growing gulf between states' preferences for co-regulatory and self-regulatory solution, and citizens' preferences for greater control if not ownership of vital regulated industries, has led to a crisis of legitimacy' (Marsden, 2011: 46).

To avoid legitimation problems, it is crucial that internet governance at national, regional and global levels retains a connection with democratic publics, through inclusive democratic procedures where all perspectives and options are openly considered and reflected upon. Whether such democratization is achieved in practice is of course not unconnected to the debate – considered by other chapters in this volume – about how the internet might help to improve the quality of democracy. As has been argued elsewhere (Coleman and Blumler, 2009; Coleman and Moss, 2012; Moss and Coleman, 2014), the internet has the potential to facilitate more deliberative-democratic forms of participation. Realizing this potential will itself take imaginative policy-making, but such a project could play an important role in helping to translate and legitimate rights online. In fact, if the argument in this chapter is correct, rights relating to democratic participation – given their importance in procedural terms in interpreting and legitimating rights more generally – warrant a certain priority in our thinking about how the internet should be governed.

CONCLUSION

The internet often appears in popular discussions in a Janus-faced way, as a source of either enormous benefits (in terms of democracy, individual freedom, creativity, and so on) or harm (facilitating crime, terrorism, a loss of privacy, and so on). At the same time, there is often a sense that little can be done about this one way or the other, that the internet is following a technologically determined path and is beyond control. However, there is nothing predetermined about the internet's future. While early libertarian commentators viewed cyberspace as an ungovernable space, the internet was always in part a product of policy, and is increasingly subject to various forms of public and private regulation today. As I have argued in this chapter, the crucial issue is no longer whether the internet can be regulated, but how it should be governed and with what values and in whose interests. The public's sense of powerless in relation to the internet's future may reflect a deeper political rather than technological truth, which needs to be acknowledged if it is to be challenged: that is, the public's limited democratic control over the decisions currently being taken about the internet's future.

FUTURE RESEARCH DIRECTIONS

Internet governance is a large and complex field and future research is required on various fronts. There is a need for ongoing work on specific policy and regulatory areas, with certain issues (such as privacy and data protection, or securing universal access to media in future networks) appearing to warrant particular attention. At the same time, empirical research is required which analyses the factors that shape internet policy-making and which examines how the internet is governed and regulated in practice. Finally, we should aim to connect empirical studies to normative debates about how the internet should be governed. Drawing on political theorists such as Benhabib (2008, 2009) and Habermas (1997, 1998), this chapter has stressed both the importance of rights in providing a shared normative vocabulary for internet governance, and the need for deliberative-democratic procedures to legitimate and interpret these rights in particular political contexts. Questions related to digital literacy and democratic citizenship become central here, for internet users need both the ability to understand, and meaningful opportunities to influence and control, the forms of internet regulation to which they are subject.

FURTHER READING

Freedman, D. (2012). Outsourcing internet regulation. In Curran, J., Fenton, N. and Freedman, D. (eds), *Misunderstanding the Internet* (pp. 95–121). London: Routledge.
Lessig, L. (2006). *Code 2.0*. New York: Basic Books.
Marsden, C.T. (2011). *Internet Co-Regulation: European Law, Regulatory Governance and Legitimacy in Cyberspace*, 1st edn. Cambridge: Cambridge University Press.
McChesney, R.W. (2013). *Digital Disconnect: How Capitalism is Turning the Internet against Democracy*. New York: New Press.
Mueller, M.L. (2010). *Networks and States: The Global Politics of Internet Governance*. Cambridge, MA: MIT Press.
Murray, A. (2010). *Information Technology Law: The Law and Society*. Oxford: Oxford University Press.
Palfrey, J.G. (2010). Four phases of internet regulation. *Social Research: An International Quarterly*, 77 (3), Berkman Center Research Publication No. 2010-9; Harvard Public Law Working Paper No. 10-42. Retrieved from: http://ssrn.com/abstract=1658191.

REFERENCES

Association for Progressive Communications (2013). APC internet rights charter. Retrieved from: https://www.apc.org/en/node/5677.
Baker, C.E. (2002). *Media, Markets, and Democracy*. Cambridge: Cambridge University Press.
Balkin, J.M. (2008). The future of free expression in a digital age. *Pepperdine Law Review*, 36, 427–444.
Barendt, E. (2005). *Freedom of Speech*, 2nd edn. Oxford: Oxford University Press.
Barlow, J.P. (1996). A declaration of the independence of cyberspace. Retrieved from: https://projects.eff.org/~barlow/Declaration-Final.html.

Benhabib, S. (2008). The legitimacy of human rights. *Daedalus*, 137 (3), 94–104.
Benhabib, S. (2009). Claiming rights across borders: international human rights and democratic sovereignty. *American Political Science Review*, 103 (4), 691–704.
Bermejo, F. (2011). Online advertising – origins, evolution, and impact on privacy. Open Society Media Program. Retrieved from: https://www.soros.org/reports/mapping-digital-media-online-advertising-origins-evolution-and-impact-privacy.
Bygrave, L.A. and Bing, J. (eds) (2009). *Internet Governance: Infrastructure and Institutions*. Oxford: Oxford University Press.
Campbell, T. (2008). Human rights. In McKinnon, C. (ed.), *Issues in Political Theory* (pp. 194–217). Oxford: Oxford University Press.
Coleman, S. and Blumler, J. (2009). *The Internet and Democratic Citizenship: Theory, Practice, and Policy*. Cambridge: Cambridge University Press.
Coleman, S. and Moss, G. (2012). Under construction: the field of online deliberation research. *Journal of Information Technology and Politics*, 9, 1–15.
Couldry, N. and Turow, J. (2014). Big data, big questions. Advertising, big data and the clearance of the public realm: marketers' new approaches to the content subsidy. *International Journal of Communication*, 8. Retrieved from: http://ijoc.org/index.php/ijoc/article/view/2166.
Drake, W. and Jørgensen, F. (2006) Introduction. In Jørgensen, F. (ed.), *Human Rights in the Global Information Society* (pp. 1–51). Cambridge, MA: MIT Press.
Edwards, L., Klein, B., Lee, D., Moss, G. and Philip, F. (2013). Framing the consumer: copyright regulation and the public. *Convergence: The International Journal of Research into New Media Technologies*, 19 (1), 9–24.
Elmer-DeWitt, P. and Jackson, D.S. (1993). First nation in cyberspace. *Time*, 142 (24), 62.
Epstein, D. (2013). The making of institutions of information governance: the case of the Internet Governance Forum. *Journal of Information Technology*, 28 (2), 137–149.
Forst, R. (2010). The justification of human rights and the basic right to justification: a reflexive approach. *Ethics*, 120 (4), 711–740.
Fraser, N. (1990). Rethinking the public sphere: a contribution to the critique of actually existing democracy. *Social Text*, 25–26, 56–80.
Freedman, D. (2008). *The Politics of Media Policy*. Cambridge: Polity.
Freedman, D. (2012). Outsourcing internet regulation. In Curran, J., Fenton, N. and Freedman, D. (eds), *Misunderstanding the Internet* (pp. 95–121). London: Routledge.
Froomkin, A.M. (2003). Habermas@Discourse. net: toward a critical theory of cyberspace. *Harvard Law Review*, 116 (3), 749.
Goldsmith, J. and Wu, T. (2006). *Who Controls the Internet?: Illusions of a Borderless World*. Oxford: Oxford University Press.
Habermas, J. (1991). *Communication and the Evolution of Society*. Oxford: Polity.
Habermas, J. (1997). *Between Facts and Norms: Contributions to a Discourse Theory of Law and Democracy*. London: Polity.
Habermas, J. (1998). Remarks on legitimation through human rights. *Philosophy and Social Criticism*, 24 (2–3), 157–171.
Habermas, J. (2009). *Europe: The Faltering Project*. London: Polity.
Internet Rights and Principles Coalition (2013). The charter of human rights and principles for the internet. Retrieved from: http://internetrightsandprinciples.org/site/.
Johnson, D.R. and Post, D. (1996). Law and borders: the rise of law in cyberspace. *Stanford Law Review*, 48, 1368.
Klang, M. and Murray, A. (2005). *Human Rights in the Digital Age*. London: GlassHouse Press.
Kohl, U. (2012). The rise and rise of online intermediaries in the governance of the Internet and beyond – connectivity intermediaries. *International Review of Law, Computers and Technology*, 26 (2–3), 185–210.
Kohl, U. (2013). Google: the rise and rise of online intermediaries in the governance of the Internet and beyond (Part 2). *International Journal of Law and Information Technology*, 21 (2), 187–234.

Lessig, L. (1999a). *Code and Other Laws of Cyberspace.* New York: Basic Books.
Lessig, L. (1999b). Open code and open societies: values of internet governance. *Chicago-Kent Law Review,* 74 (3), 1405–1422.
Lessig, L. (2006). *Code 2.0.* New York: Basic Books.
Marsden, C.T. (2010). *Net Neutrality Towards a Co-Regulatory Solution.* London: Bloomsbury Academic.
Marsden, C.T. (2011). *Internet Co-Regulation: European Law, Regulatory Governance and Legitimacy in Cyberspace,* 1st edn. Cambridge: Cambridge University Press.
Martin, R. (1998). Rights. In Craig, E. (ed.), *Routledge Encyclopedia of Philosophy* (pp. 325–331). London: Routledge.
McChesney, R.W. (2013). *Digital Disconnect: How Capitalism is Turning the Internet against Democracy.* New York: New Press.
McIntyre, T.J. and Scott, S. (2008). Internet filtering: rhetoric, legitimacy, accountability and responsibility. In Brownsword, R. and Yeung, K. (eds), *Regulating Technologies* (pp. 109–125). Oxford: Hart.
Moss, G. and Coleman, S. (2014). Deliberative manoeuvres in the digital darkness: e-democracy policy in the UK. *British Journal of Politics and International Relations,* 16 (3), 410–427.
Mueller, M.L. (2010). *Networks and States: The Global Politics of Internet Governance.* Cambridge, MA: MIT Press.
Murray, A. (2010). *Information Technology Law: The Law and Society.* Oxford: Oxford University Press.
Murray, A.D. (2011). Nodes and gravity in virtual space. *Legisprudence,* 5 (2), 195–221.
OECD (2010). The role of internet intermediaries in advancing public policy objectives. Retrieved from: http://www.oecd.org/sti/interneteconomy/theroleofinternetintermediariesinadvancingpublicpolicyobjectives.htm.
Palfrey, J.G. (2010). Four phases of internet regulation. *Social Research: An International Quarterly,* 77(3); Berkman Center Research Publication No. 2010-9; Harvard Public Law Working Paper No. 10-42. Retrieved from: http://ssrn.com/abstract=1658191.
Raboy, M., Landry, N. and Shtern, J. (2010). *Digital Solidarities, Communication Policy and Multi-Stakeholder Global Governance: The Legacy of the World Summit on the Information Society.* New York: Peter Lang.
Sylvain, O. (2010) Internet governance and democratic legitimacy. *Federal Communications Law Journal,* 62 (2), 205–274.
Vick, D. (2005). Regulating hatred. In M. Klang and A. Murray (eds), *Human Rights in the Digital Age* (pp. 41–55). London: GlassHouse Press.
Wong, C. and Dempsey, J. (2011). The media and liability for content on the internet. Open Society Media Program. Retrieved from: http://www.opensocietyfoundations.org/reports/media-and-liability-content-internet.
Zittrain, J. (2009). *The Future of the Internet: And How to Stop It,* new edn. London: Penguin.

22. Social media surveillance
Christian Fuchs

INTRODUCTION

Privacy is not a phenomenon specific to digital media like the Internet. Modern thinking about privacy and surveillance has for a long time been bound up with the media: Warren and Brandeis defined privacy as a right to be left alone and situated this understanding in the context of the tabloid press:

> The press is overstepping in every direction the obvious bounds of propriety and of decency. Gossip is no longer the resource of the idle and of the vicious, but has become a trade, which is pursued with industry as well as effrontery. To satisfy a prurient taste the details of sexual relations are spread broadcast in the columns of the daily papers. To occupy the indolent, column upon column is filled with idle gossip, which can only be procured by intrusion upon the domestic circle. (Warren and Brandeis, 1890: 196)

Privacy has been defined either as the right to be left alone or as the right to determine for oneself which areas of life should be accessible to others – or as a combination of the two (Tavani, 2008). In the context of information processing, privacy plays a role because information about the lives of humans can become publicly available and the question arises: which rules shall regulate the becoming public of such information?

Some scholars have defined surveillance as the systematic gathering and processing of personal data for the management of individuals or groups. Others stress that surveillance tries to bring about or to prevent certain behaviours in groups or individuals by gathering, storing, processing, diffusing, assessing and using data (Fuchs, 2011c). Just like privacy, surveillance is also not a phenomenon specific to digital media, which becomes clear in Foucault's (1977) work, which has shown that surveillance is bound up with the history of control, the prison system, the state and the class-structured economies.

There is a debate between scholars studying privacy and surveillance about the relevance of these two concepts. Whereas some argue that privacy is a liberal and individualistic concept and that surveillance is a more critical concept that can focus on the structural implications of data collection in society, others argue that although privacy advocates may not always be successful in preventing the negative effects of state and corporate surveillance, they at least try to bring about political intervention (Bennett, 2011a, 2011b; Boyd, 2011; Gilliom, 2001; Regan, 1995; Stalder, 2011).

Although privacy and surveillance are not new, the rise of the computer in society has brought about special public concern for both phenomena. The first national data protection Act was passed in 1973 in Sweden and subsequent laws followed in other countries. It is no accident that this happened in the early 1970s, a time when large-scale computer-based data processing took broader effect in society. The rise of computing and an information society is the context for the establishment of data protection and explains the connection of data protection with informational privacy and surveillance.

Scholars have been aware of the privacy and surveillance implications of computing for quite some time and have in this context coined notions such as the 'new surveillance' (Marx, 1988, 2002), 'dataveillance' (Clarke, 1988, 1994), the 'electronic (super)panopticon' (Poster, 1990), 'electronic surveillance' (Lyon, 1994), 'digital surveillance' (Graham and Wood, 2007), the 'world-wide web of surveillance' (Lyon, 1998), and the 'digital enclosure' (Andrejevic, 2004, 2007).

The rise of the Internet took a quantum leap in the mid-1990s when the World Wide Web (WWW) became popular and commercialized. The early 1990s until after the new millennium were times of a general and scholarly Internet optimism, spurred by neoliberalism and the new entrepreneurialism of the Internet economy. Issues relating to privacy, surveillance and data protection were often considered as outmoded and old-fashioned. Typical books of neoliberal 1990s Internet gurus such as Nicholas Negroponte's (1996) *Being Digital* or Kevin Kelly's (1999) *New Rules for the New Economy* do not contain terms such as 'surveillance' or 'data protection'. Internet optimism suffered a drawback when the dot. com crisis took effect in 2000 and resulted in the bankruptcy of many Internet companies that had been founded on venture capital investments that could not be translated into actual profits. The rise of what was somewhat mistakenly called social media – blogs, social networking sites, microblogs, content sharing sites and wikis – spurred new hopes (and foundations of another financial bubble) that have been represented by Google and Facebook, among others. At the same time, a new neoliberal and techno-deterministic round of techno-optimism emerged. At the same time, 9/11 sparked new wars, a new surveillance offensive and an intensification of conservative law-and-order politics. In this context of heightened state and commercial surveillance, new discussions about the societal and ethical implications of online media, and especially social media, emerged.

This chapter gives special focus to debates on social media privacy and surveillance. The next section focuses on the discussion of key characteristics of social media surveillance. The chapter then discusses the economic, political and cultural implications of social media surveillance; and the final section draws some conclusions.

WHAT IS SOCIAL MEDIA SURVEILLANCE?

Is the 'social web' a real change in the WWW or a piece of jargon and marketing ideology? Although Tim O'Reilly surely thinks that Web 2.0 denotes actual changes and says that the crucial fact about it is that users as a collective intelligence co-create the value of platforms like Google, Amazon, Wikipedia and Craigslist (O'Reilly and Battelle, 2009: 1), he admits that the term was mainly created for identifying the need of new economic strategies of Internet companies after the dot.com crisis. So he says in a paper published five years after the creation of the term Web 2.0 that this category was 'a statement about the second coming of the Web after the dotcom bust' at a conference that was 'designed to restore confidence in an industry that had lost its way' (O'Reilly and Battelle, 2009: 1).

The question of how social the web is or has become depends on a profoundly social-theoretical question: what does it mean to be social? Are human beings always social, or only if they interact with others? In sociological theory, there are different concepts of the social, such as Émile Durkheim's social facts, Max Weber's social action, Karl Marx's notion of collaborative work (as for example also employed in the concept of computer-supported collaborative work – CSCW), or Ferdinand Tönnies' notion of community (Fuchs, 2010). Depending on which concept of sociality one employs, one gets different answers to the questions of whether the web is social or not, and whether sociality is a new quality of the web or not. Community aspects of the web certainly did not start with Facebook in 2004, but had been used in the 1980s to describe bulletin board systems like the Whole Earth 'Lectronic Link (WELL) (Rheingold, 1993). Collaborative work, for example the cooperative editing of articles performed on Wikipedia, is rather new as a dominant phenomenon on the WWW, but not new in computing (where the concept of CSCW was already the subject of a conference series that started in December 1986 with the 1st ACM Conference on CSCW in Austin, Texas). Neither is the wiki concept new: the WikiWikiWeb was introduced by Ward Cunningham in 1984. All computing systems, and therefore all web applications, can be considered as social because they store and transmit human knowledge that originates in social relations in society. They are objectifications of society and human social relations. Whenever a human uses a computing system or another medium such as a book (even if they do so alone in a room), they interact with an objectification of knowledge, that is, ideas that are stored as objects in media forms. They are the outcomes of social relations. But not all computing systems and web applications support direct communication in which at least two humans mutually exchange symbols that are interpreted as being meaningful. Amazon, for example, mainly provides information about books and other goods one can buy. It is not primarily a tool of communication, but rather a tool of information. In contrast, Facebook has inbuilt communication features that facilitate direct communication between people (mail system, walls for comments, forums, and so on).

The above discussion shows that it is not a simple question to decide whether and how social the WWW actually is. Therefore a theory of Internet and society is needed that identifies multiple dimensions of sociality (such as cognition, communication and cooperation; see Fuchs, 2008, 2010), based on which the continuities and discontinuities of the development of the Internet can be empirically studied. An important theoretical question is: what are the basic characteristics of online and social media surveillance? Fuchs et al. (2012) identify 14 qualities of Internet surveillance based on more general qualities of Internet communication, that are displayed in Table 22.1.

Daniel Trottier and David Lyon (2012) argue that there are five key features of social media surveillance:

- Collaborative identity construction: with the help of image tagging and wall comments, users contribute to the identity construction of others. Users monitor what others say about their friends, contacts and themselves.
- Social media enable the monitoring of individuals' social networks.
- Social media surveillance makes use of social ties that are visible, measurable and searchable.
- Social media surveillance is confronted with continuously changing interfaces and contents.
- Social media surveillance is surveillance of profiles that hold information from many different social contexts, that is, of 'social convergence'. (Trottier and Lyon, 2012: 102)

Daniel Trottier (2012) argues that social media augments surveillance. 'By sharing not only the same body of information, but also the same interface used to access that information, formerly discrete surveillance practices feed off one another' (Trottier, 2012). On social media, there are:

Table 22.1   Qualities of Internet surveillance

Dimension	Quality of Internet communication	Quality of Internet surveillance
1. Space	*Global communication* Global communication at a distance, global information space	*Global surveillance* Surveillance at a distance is possible from all nodes in the network, not just from a single point; combination and collection of many data items about certain individuals from a global information space
2. Time	*Real-time (synchronous) or asynchronous global communication*	*Real-time surveillance* Surveillance of real time communication, surveillance of stored asynchronous communication, surveillance of communication protocols and multiple data traces
3. Speed	*High-speed data transmission*	*High-speed surveillance* Availability of high-speed surveillance systems
4. Size	*Miniaturization* Storage capacity per chip increases rapidly (Moore's law)	*Surveillance data growth* Ever more data on individuals can be stored for surveillance purposes on storage devices that become smaller and cheaper over time
5. Reproduction	*Data multiplicity* Digital data can be copied easily, cheaply and endlessly; copying does not destroy the original data	*Surveillance data multiplicity* Surveillance becomes easy and cheap; if multiple copies of data exist, specific data are easier to find
6. Sensual modality	*Multimedia* Digital combination of text, sound, image, animation and video in one integrated medium	*Multimodal surveillance* Surveillance of multi-sensual data over one medium
7. Communication flow	*Many-to-many communication*	*Social network surveillance* The multiple personal and professional social networks of individuals become visible and can be traced

Table 22.1   (continued)

Dimension	Quality of Internet communication	Quality of Internet surveillance
8. Information structure	*Hypertext* Networked, interlinked and hypertextual information structures	*Linked surveillance* The links between persons can be easier observed
9. Reception	*Online produsage* Recipients become producers of information (produsers, prosumers)	*Economic exploitation of produsage* Economic exploitation of produsage, new capital accumulation strategies based on active, creative users that are sold to advertisers as produsage commodity, targeted advertising based on continuous surveillance of user-generated content
10. Mode of interaction and sociality	*Online cooperation* Cooperative information production at a distance, information sharing at a distance	*Enclosure of digital commons* Laws that enable the surveillance of sharing and cooperation, intellectual property rights
11. Context	*Decontextualization* Decontextualized information and anonymity (for example, authorship, time and place of production might be unclear)	*Intensification of surveillance* Decontextualization advances speculative and pre-emptive surveillance
12. Reality	*Derealization* The boundaries between actuality and fiction can be blurred	*Intensification of surveillance* Fictive reality might be taken for actual reality by surveillers, which puts people at risk and intensifies surveillance
13. Identity and emotions	*Emotive Internet* The Internet is a very expressive medium that allows identity construction and representation online	*Personalized surveillance* Surveillance of very personal characteristics of individuals and their emotions becomes possible

Table 22.1 (continued)

Dimension	Quality of Internet communication	Quality of Internet surveillance
14. Availability	*Ubiquitous Internet* The Internet has become ubiquitous in all spheres of everyday life	*Ubiquitous surveillance* In a heteronomous society, there is constant and profound surveillance of Internet information and communication for economic, political, judicial and other aims

Source: Fuchs et al. (2012: 16–19).

individuals watching over one another, institutions watching over a key population, businesses watching over their market and investigators watching over populations . . . Individual, institutional, market and investigative scrutiny all rely on the same interface. Thus, familiarity with the site as an interpersonal user facilitates other uses. In addition to relying on the same interface, these practices also rely on the same body of information. This means that personal information that has been uploaded for any particular purpose will potentially be used for several kinds of surveillance. (Trottier, 2012)

Fuchs and Trottier (2013) argue that one feature of social media is that they integrate forms of sociality as well as integrated social roles. Based on a dialectical model, we can identify three levels or stages of social life that form the 'triple C' process model of information: cognition, communication and cooperation (Fuchs, 2008, 2010). Cognition refers to the status and processes of human thought that create and reproduce knowledge. Humans are not isolated monads, but social beings: they exist in and through their relations with other humans. Communication is a social relation between at least two human beings in which there is a mutual exchange of symbols that are interpreted so that the interaction partners give meaning to them. Communication is the social dimension of human existence. It is based on cognition because communication changes the states of knowledge of the participating communication partners. Based on communication, humans can collaborate. Many communication processes do not result in cooperation, but some do. Collaboration or cooperation means that humans create new qualities of social systems or new social systems together. Cooperation is based on communication and cognition: every cooperation process is also a communication and cognition process; every communication process involves also cognition processes.

An important characteristic of social media is the convergence of the three spheres of sociality. Social media are simultaneously media of cognition, communication and cooperation. The publication of content or an idea on a social networking site, a wiki or a blog can become the foundation of communication, which in turn can spur collaboration.

In modern society, human beings act in different capacities in different social roles. Consider the modern middle-class office worker, who also has roles as a husband, father, lover, friend, voter, citizen, child, fan, neighbour, to say nothing of the various associations to which he may belong. In these different roles, humans are expected to behave according to specific rules that govern the various social systems of which modern society is composed (such as the company, the schools, the family, the Church, fan clubs, political parties, and so on). Habermas mentions the following social roles that are constitutive for modern society: employee, consumer, client, citizen (Habermas, 1987: 320). Other roles, such as for example wife, husband, houseworker, immigrant, convict, and so on, can certainly be added. What is constitutive for modern society is not just the separation of spheres and roles, but also the creation of power structures, in which roles are constituted by power relations (as for example employer–employee, state bureaucracy–citizen, citizen of a nation state–immigrant, manager–assistant, dominant gender roles–marginalized gender roles).

Based on these theoretical foundations, Fuchs and Trottier (2013) argue that integrated and converging surveillance is a specific feature of social media surveillance: on social media such as Facebook, various social activities (cognition, communication, cooperation) in different social roles that belong to our behaviour in systems (economy, state) and the lifeworld (political public, civic spheres, private life) are mapped to single profiles. In this mapping process, data about social activities within social roles are generated. This means that a Facebook profile holds: (1) personal data; (2) communicative data; and (3) social network and community data, in relation to: (i) personal roles (friend, lover, relative, father, mother, child, and so on); (ii) civic roles (political public: activist, citizen, civic cultures: audience member, fan, association member, neighbour); and (iii) systemic roles (in politics: voter, citizen, client, politician, bureaucrat; in the economy: worker, manager, owner, purchaser or consumer, and so on). The different social roles and activities tend to converge, as for example in situations where the workplace is also a playground, where friendships and intimate relations are formed and where leisure activities are conducted. This means that social media surveillance is an integrated form of surveillance, in which one finds surveillance of different partly converging activities with the help of profiles that hold a complex, networked multitude of data about humans.

This discussion shows that the question of how to understand online and social media's implications for society is a complex one. Basic theoretical questions that arise in this context are: What is social about social media and the Internet? What is privacy? What is surveillance? What are key features and qualities of Internet privacy and Internet surveillance? What are the key features and qualities of privacy and surveillance on newer Internet platforms such as Facebook and Google? Giving answers to such questions requires profound knowledge and application of social theory to the study of online privacy and surveillance. Based on social theory, empirical social research is needed for studying the implications of privacy and surveillance in the online world. The next section gives an overview of some empirical results.

EMPIRICAL STUDIES OF THREE REALMS OF ONLINE PRIVACY AND SURVEILLANCE: THE ECONOMY, POLITICS AND CULTURE

This section presents results from studies of privacy and surveillance in three realms of the online world: the online economy, online politics and everyday culture online.

A First Realm that Concerns Online Privacy and Surveillance is the Economy

Fuchs (2013b) argues that social media constitute spaces where job applicant surveillance, workplace and workforce surveillance, property surveillance, consumer surveillance and surveillance of competitors converge. Dallas Smythe (1977) argued that in commercial media that are funded by advertising (broadcasting, newspapers), the audience is sold as a commodity to advertisers, who pay for access to audiences. He spoke therefore of audience commodification. Fuchs (2013b, 2011) argues that in social media, the audience has turned into 'prosumers', the social media business model is based on Internet prosumer commodification, and that prosumer surveillance that monitors all user data generated on certain platforms like Facebook (and beyond) is built into this business model. As a consequence, advertising becomes targeted and personalized to user activities and interests.

In an analysis based on Smythe (1977), Sut Jhally and Bill Livant (1986) argued that watching is working and that the living room has become a factory for the production of economic value. Andrejevic (2002) argues that in commercial interactive media, surveillance becomes part of the work of watching that as a consequence turns into the work of being watched. He stresses that users of commercial social media create economic value and that their activity is exploited for economic purposes. Users' conscious communication and creation of content creates unintentional information – data about user behaviour captured by the (commercial) platform in surveillance processes (Andrejevic, 2012: 85) – that they do not control and that is turned into profit via targeted advertising. As a result, they are alienated from their activities and products. The users become separated from the 'means of socialization' (Andrejevic, 2012: 88) that are controlled by commercial companies such as Facebook.

Both Fuchs and Andrejevic stress that usage of commercial social media platforms is a form of value-generating labour. Trebor Scholz (2010) therefore argues that Facebook and the commercial Internet are playgrounds and factories, on which users' play becomes digital labour (play labour = 'playbour'). Production and consumption, labour and play, the public and the private, tend to converge on social media. Consumer surveillance on social media tends at the same time to be producer surveillance.

Social media surveillance also relates to traditional wage labour, especially the hiring process and the monitoring of employees' Internet use. A UK survey conducted by Reppler (N = 300) found that 91 per cent of the surveyed companies use social networking sites to screen prospective employees in the hiring process, and 69 per cent say they have rejected a candidate because what they saw about them on a social networking site (SNS) (http://www.thedrum.co.uk/news/2011/10/24/91-employers-use-social-media-screen-applicants). Forty-nine per cent conduct such screening after they have received applications, 27 per cent after an initial conversation, and 15 per cent after a detailed job interview.

A study conducted by the American Management Association and the ePolicy Institute in 2007 found that more than 28 per cent of the surveyed US companies had fired workers for the misuse of e-mail at work, and almost one-third for the misuse of the Internet; 66 per cent said that they monitor employees' Internet use, and more than 40 per cent that they monitor employees' e-mails.

> The 28% of employers who have fired workers for e-mail misuse did so for the following reasons: violation of any company policy (64%); inappropriate or offensive language (62%); excessive personal use (26%); breach of confidentiality rules (22%); other (12%). The 30% of bosses who have fired workers for Internet misuse cite the following reasons: viewing, downloading, or uploading inappropriate/offensive content (84%); violation of any company policy (48%); excessive personal use (34%); other (9%) . . . Computer monitoring takes many forms, with 45% of employers tracking content, keystrokes, and time spent at the keyboard. Another 43% store and review computer files. In addition, 12%

monitor the blogosphere to see what is being written about the company, and another 10% monitor social networking sites. Of the 43% of companies that monitor e-mail, 73% use technology tools to automatically monitor e-mail and 40% assign an individual to manually read and review e-mail.[1]

The use of social media as tools of applicant and workforce surveillance is a relatively new area of research and concern (Sánchez Abril et al., 2012; Clark and Roberts, 2010; Davison et al., 2012; Davison et al., 2011). The published works on this topic tend to agree that this issue is legally relatively unregulated and that more social scientific and legal research is needed in this area. Sánchez Abril et al. (2012: 69) argue that 'employer intrusion into an employee's personal life threatens the employee's freedom, dignity, and privacy – and may lead to discriminatory practices'. They conducted a survey of 2500 undergraduate students and found that 71 per cent agreed that the following scenario could result in physical, economic or reputational injury in the offline world (p. 104f):

> You called in sick to work because you really wanted to go to your friend's all day graduation party. The next day you see several pictures of you having a great time at the party. Because the pictures are dated you start to worry about whether you might be caught in your lie about being sick. You contact the developers of the social network and ask that the pictures be taken down because the tagging goes so far, it would take you too long to find all the pictures. There was no response from the network. You are stunned to be called in by your supervisor a week later to be advised that you were being 'written up' for taking advantage of sick leave and put on notice that if it happened again you would be terminated. (Sánchez Abril et al., 2012: 104)

Clark and Roberts (2010) argue that notwithstanding all legal debates, employers' monitoring of employees' or applicants' social networking sites profiles is a socially irresponsible practice because such practices allow 'employers to be undetectable voyeurs to very personal information and make employment decisions based on that information' (Clark and Roberts, 2010: 518). Due to the persistence of online information, such monitoring can have negative career effects that persist for years. Also, employers can make inappropriate decisions based on very sensitive information ('she is too conservative or too liberal'; Clark and Roberts, 2010: 51).

Protecting employees and job applicants from decisions based on information derived from social media is important because there is an asymmetrical power relationship between employers or managers and employees or applicants. The existence of this asymmetrical power relationship, in which employers have relative power to decide if employees are hired and fired, requires special protection of workers and applicants.

A Second Realm of Online Privacy and Surveillance has to do with the State, especially the Police

Trottier (2011) observes that the police make use of social media in investigations by accessing publicly available profile information, befriending suspects and obtaining personal information from platforms using warrants. One can add to this the targeted surveillance of suspects with the help of communication surveillance technologies in order to try to prevent terrorism. The result is 'an enhanced police presence in – and scrutiny of – everyday life' (Trottier, 2011). The topic of state, police and secret service surveillance of social media has gained special attention since Edward Snowden revealed in 2011 that the US National Security Agency (NSA) and the UK Government Communications Headquarters (GCHQ) have direct surveillance access to the personal data processed by AOL, Apple, Facebook, Google, Microsoft, Paltalk, Skype and Yahoo!. It shows the existence of a surveillance–industrial complex, in which online corporations, state agencies and private security companies collaborate in order to establish and maintain a political-economic control system (Fuchs, 2014).

Crime on social media is a topic that is often presented by the news media in a sensationalistic manner and by presenting single examples:

> Sex-trio abused schoolgirl (16) . . . and wanted to make her walk the streets . . .
> Sex-trap Internet! For a 16-year-old student a flirt on the network 'Facebook' had obviously terrible consequences. The prosecutor is certain: The girl was raped by three men – and was compelled to walk the streets![2] (*Bild Zeitung*, 9 December 2011)
> Paedo groomed Facebook girls . . . A 19-YEAR-OLD man who police said was part of a paedophile gang has admitted having sex with girls as young as 13 . . . The group used Facebook to groom schoolgirls before meeting up with them, plying them with drink and drugs and sexually abusing them. (*Sun Online*, 27 May 2011[3])

The reality is, however, fairly different than such sensationalistic news reports suggest. The European Union (EU) Kids Online II Survey studied the behaviour of children online in the EU27 countries.[4] Eighty-six per cent of surveyed children (9–16 years old) in the EU27 countries say they never sent a photo or video of themselves to somebody they have not met face-to-face, and 85 per cent say they never sent personal information to somebody they have not met face-to-face (p. 43). Twelve per cent say they have been bothered by something online, and 8 per cent of parents say their children have been bothered by something online (p. 46). Eighty-one per cent say that they have never been bullied online or offline (p. 61), while 19 per cent were bullied at least once (p. 61). Six per cent had been bullied online (in the past 12 months), 3 per cent on a social networking site (p. 63), which shows that bullying primarily takes place offline and that online bullying is a relatively rare phenomenon. Two per cent said that they had been asked on the Internet to show photos or videos of themselves nude (within the past 12 months), and 2 per cent had been asked to talk about sexual acts (p. 75). Nine per cent of all survey children said that they met an online contact offline, and 1 per cent reported being bothered by this (p. 92). Seven per cent said that somebody other than themselves used their password to access their account (p. 100). Overall these results show that the crime that children experience online is of a relatively minor extent. Majid Yar (2010) argues that mass media reports of individual incidents of violent online pornography or the raping or killing of children by strangers they first met online often function as 'signal crimes' that result in moral panics and calls for law-and-order policies, Internet policing and surveillance. Statistics show that such panics do not reflect the actual low level of online crime. According to the Special Eurobarometer Study 371 'Internal Security' of 2011,[5] 46 per cent say that the EU is doing enough to fight cybercrime, whereas 36 per cent think it is not doing enough.

A Third Realm of Online Privacy and Surveillance Concerns Everyday Life, Civil Society and the Lifeworld

Based on John B. Thompson's (2005) argument that there is a mediated new visibility, in which those who hold power are made visible to the many, Goldsmith (2010) argues that social media, especially YouTube and Facebook, make police misconduct more visible in the public. So on the one hand the police have powerful surveillance technologies at hand for monitoring citizens, but on the other hand citizens also use less sophisticated technologies with less reach (mobile phone cameras, video live streams, and so on) in aiming to make police power and violence transparent. There is an asymmetry involved in this usage because the police have more resources, capacities, access possibilities and time for conducting surveillance. Goldsmith discusses the example of YouTube videos of the death of Ian Tomlinson in the London G9 protests in 2009, and of the death of Robert Dziekanski at Vancouver Airport in 2009. In both cases, police violence was involved. Other examples that can be mentioned are the YouTube video of the killing of Neda Soltan by police forces in the 2009 Iranian protests, and two 2011 YouTube videos that show how police officers pepper-sprayed unarmed protestors of the Occupy movement (one filmed in New York, the other at the University of California Davis campus). On the one hand one can argue that these are acts of counter-power and counter-surveillance. On the other hand, visibility on the

Internet is not equally distributed; there is 'an Internet attention economy that is dominated by powerful actors' (Fuchs et al., 2012b: 15).

Acquisti and Gross conducted an online survey of SNS users at Carnegie Mellon University in the USA (N = 294; Acquisti and Gross, 2006) and data mining of 7000 social networking site profiles (Gross et al., 2005). They found that although users are highly concerned about privacy, the amount of personal information they include in their SNS profiles is high: for example, 78 per cent revealed their full name and 99.94 per cent of the profiles were publicly accessible. Barnes (2006) called this phenomenon the 'privacy paradox'. Nosko et al. (2010) analysed 400 Facebook profiles. They conclude that there is a high level of information revelation: mini-feed, profile pictures, birth date, friends, college or university, wall postings, gender, used applications, groups, photos, tagged photos and photo albums were disclosed to the public by 70 per cent or more of the studied profiles.

These results to a certain extent imply that social media users deal carelessly with private data and put themselves at risk. They are however to a certain degree questioned by studies that found that Facebook users feel highly confident in managing Facebook privacy settings. According to Boyd and Hargittai (2010), 51 per cent of the respondents in a study of 18–19-year-old SNS users had changed their privacy settings four or more times, 38 per cent two or three times, 9 per cent once, and only 2 per cent never (N2 = 495, survey conducted in 2010). The Special Eurobarometer Study 359 'Attitudes on Data Protection and Electronic Identity in the European Union' (2011)[6] shows that 51 per cent of European social networking site users have changed the privacy settings of Facebook and other sites (p. 164). Eighty-two per cent find it easy to change privacy settings; 18 per cent find it difficult (p. 166). These data show that most users seem to be aware of how to change the privacy settings. Fuchs (2009) and Beer (2008) argue that many of these studies are too focused on individual users' behaviour and neglect macro contexts such as advertising culture, political economy, surveillance or the 'War on Terror'. They stress that revealing information on social media is a means of communication and is not a problem in itself. Rather, the problem is power structures (for example, companies that spy on their employees or applicants) that make use of such data for negatively impacting upon individuals.

Albrechtslund argues that people's practice of watching each other on social networking sites is 'participatory surveillance': it 'can be seen as empowering, as it is a way to voluntarily engage with other people and construct identities, and it can thus be described as participatory ... participatory surveillance is a way of maintaining friendships by checking up on information other people share' (Albrechtslund, 2008). Andrejevic (2005) speaks of lateral surveillance as 'do-it-yourself monitoring' (p. 487) or 'peer-to-peer monitoring', 'the use of surveillance tools by individuals, rather than by agents of institutions public or private, to keep track of one another', for example in relation to romances, family, friends and acquaintances (p. 488). In contrast to Albrechtslund, Andrejevic does not think that lateral surveillance democratizes surveillance, but argues that it reinforces and replicates 'the imperatives of security and productivity' (Andrejevic, 2005: 487) and 'extends monitoring techniques from the cloistered offices of the Pentagon to the everyday spaces of our homes and offices, from law enforcement and espionage to dating, parenting, and social life. In an era in which everyone is to be considered potentially suspect, we are invited to become spies' (p. 494). Thomas Mathiesen argues that everyday life monitoring today also takes on the form of a synopticon, which is 'an extensive system enabling the many to see and contemplate the few', whereas in the panopticon the few 'see and supervise the many' (Mathiesen, 1997: 219). There is a difference between seeing and supervising: in Mathiesen's concept the many do not have the power to supervise the few, but the few have the power to supervise the many.

CONCLUSION

Many questions regarding online privacy and surveillance are largely unanswered and require theory construction, empirical research and ethical reasoning:

- What are the key features and qualities of online privacy and surveillance?
- What are social media and how do privacy and surveillance shape social media?
- How do contemporary societal contexts, such as the new imperialism, capitalism, neoliberalism, global wars and conflicts, the political economy of the surveillance–industrial complex, and so on, shape online privacy and surveillance?
- What is the role of privacy and surveillance in the context of the online economy? That is, what are key features and empirical realities of phenomena such as digital labour, targeted advertising, online marketing, online business models and value creation, class relations and exploitation online, consumer surveillance online, workforce and workplace surveillance online, and so on.

- What are the implications and empirical realities of the online realm for state surveillance, crime, policing and political activism?
- What are the features and empirical realities of online privacy and surveillance in everyday life and relationships?
- How can online surveillance be resisted and what power asymmetries do counter-surveillance projects that make use of the Internet face (for example WikiLeaks, corporate watchdog projects)? Are there ways of overcoming these asymmetries?
- What are philosophical foundations and principles of computer ethics and how do they relate to the study of online privacy and surveillance?
- What is the difference between privacy impact assessments and societal and ethical impact assessments of information and communication technologies (ICTs)? How can societal and ethical impact assessments be best integrated into research and research projects (for example by requiring all research projects that develop or study ICTs and are funded by national research councils, the European Union, and so on to include a work package about societal and ethical impact assessment)?

Scholars studying online privacy and surveillance often situate themselves and their work in either 'Internet studies' (Consalvo and Ess, 2011; Hunsinger et al., 2010) or 'surveillance studies' (Ball et al., 2012), which reflects two sides of the conceptual integration of the concepts of 'online' and of 'privacy' and 'surveillance'. Both of these new fields claim that they are not disciplines, but interdisciplinary or transdisciplinary fields. Nonetheless each displays the habitus, identity and discipline-making behaviour of a discipline. There are a lot of new interdisciplines and transdisciplines today that claim to be new and unique. In making claims that their fields of studies are unique they, however, separate themselves from other academic communities, fields, scholars and institutions and contribute to academic fragmentation. They also imitate the behaviour of established disciplines, so that 'interdisciplines' and 'transdisciplines' may one day simply become the new disciplines.

I am neither arguing for or arguing against established disciplines or new interdisciplines, but instead think that such categorizations are rather meaningless, and pure expressions of academic power struggles. I therefore contend that the study of online privacy and surveillance should neither be situated in the realm of 'Internet studies' nor in the realm of 'surveillance studies'. I rather think that it is today necessary to invoke another distinction in the social sciences and humanities: namely the one between administrative and critical research. Administrative social research describes reality merely as it is by employing empirical social research that follows basic inductive or deductive schemes, and is instrumental in the legitimatization of powerful institutions. In contrast, Horkheimer (2002) stresses that the goal of a critical theory of society is the transformation of society as a whole (p. 219) so that a 'society without injustice' (p. 221) emerges that is shaped by 'reasonableness, and striving for peace, freedom, and happiness' (p. 222). Horkheimer argues that critical theory wants to enhance the realization of all human potentialities (p. 248); it 'never simply aims at an increase of knowledge as such'. Its goal is man's 'emancipation from slavery' (p. 249) and 'the happiness of all individuals' (p. 248).

Online privacy and surveillance happen in societal contexts that are shaped by fundamental socio-economic inequalities, global crises, global wars and conflicts. Therefore it matters not just that we study the Internet, digital politics, online privacy and surveillance, but that we do so in a non-administrative and critical way.

ACKNOWLEDGEMENT

This chapter is an outcome of the EU FP7 project PACT – Public Perception of Security and Privacy: Assessing Knowledge, Collecting Evidence, Translating Research into Action, grant agreement number 285635, http://www.projectpact.eu/.

NOTES

1. http://press.amanet.org/press-releases/177/2007-electronic-monitoring-surveillance-survey.
2. Source: http://www.bild.de/regional/hamburg/vergewaltigung/sex-trio-missbrauchte-schuelerin-21460618.bild.html. Translation from German: 'Sex-Trio missbrauchte Schülerin (16) ... und wollte sie auf den Strich schicken ... Sex-Falle Internet! Für eine 16-jährige Schülerin hatte ein Flirt im Netzwerk "Facebook" offenbar schreckliche Folgen. Der Staatsanwalt ist sicher: Das Mädchen wurde von drei Männern vergewaltigt – und sollte auf dem Straßenstrich landen!'.
3. http://www.thesun.co.uk/sol/homepage/news/3605422/Man-in-paedophile-gang-admits-grooming-girls-on-Facebook.html.
4. 'European Union Kids Online: enhancing knowledge regarding European children's use, risk and safety online, 2010', http://www.esds.ac.uk/doc/6885%5Cmrdoc%5Cpdf%5C6885_reports.pdf.
5. http://ec.europa.eu/public_opinion/archives/ebs/ebs_371_en.pdf.
6. http://ec.europa.eu/public_opinion/archives/ebs/ebs_359_en.pdf.

FURTHER READING

Allmer, Thomas (2012), *Towards a Critical Theory of Surveillance in Informational Capitalism*, Frankfurt am Main: Peter Lang.
Andrejevic, Mark (2007), *iSpy: Surveillance and Power in the Interactive Era*, Lawrence, KS: University Press of Kansas.
Fuchs, Christian (2008), *Internet and Society. Social Theory in the Information Age*, New York: Routledge.
Fuchs, Christian (2011a), 'New media, web 2.0 and surveillance', *Sociology Compass*, 5 (2), 134–147.
Fuchs, Christian (2011b), 'Teaching and learning guide for new media, web 2.0 and surveillance', *Sociology Compass*, 5 (6), 480–487.
Fuchs, Christian (2013a), *Social Media. A Critical Introduction*, London: Sage.
Fuchs, Christian (2014), 'Social media and the public sphere', *tripleC: Communication, Capitalism and Critique*, 12 (1), 57–101.
Fuchs, Christian, Kees Boersma, Anders Albrechtslund and Marisol Sandoval (eds) (2012a), *Internet and Surveillance: The Challenges of Web 2.0 and Social Media*, New York: Routledge.
Gandy, Oscar H. (1993), *The Panoptic Sort. A Political Economy of Personal Information*, Boulder, CO: Westview Press.
Gandy, Oscar H. (2009), *Coming to Terms with Chance: Engaging Rational Discrimination and Cumulative Disadvantage*, Farnham: Ashgate.
Kelly, Kevin (1999), *New Rules for the New Economy*, New York: Penguin.
Negroponte, Nicholas (1996), *Being Digital*, New York: Vintage Books.
Trottier, Daniel (2012), *Social Media as Surveillance*, Farnham: Ashgate.

REFERENCES

Acquisti, Alessandro and Ralph Gross (2006), 'Imagined communities: awareness, information sharing, and privacy on the Facebook', in Phillipe Golle and George Danezis (eds), *Proceedings of 6th Workshop on Privacy Enhancing Technologies*, Cambridge: Robinson College, pp. 36–58.
Albrechtslund, A. (2008), 'Online social networking as participatory surveillance', *First Monday*, 13 (3). http:l/firstmonday.orglarticlelviewl2142/1949.
Andrejevic, Mark (2002), 'The work of being watched: interactive media and the exploitation of self-disclosure', *Critical Studies in Media Communication*, 19 (2), 230–248.
Andrejevic, Mark (2004), *Reality TV: The Work of Being Watched*, Lanham, MD: Rowman & Littlefield.
Andrejevic, M. (2005), 'The work of watching one another: lateral surveillance, risk, and governance', *Surveillance and Society*, 2 (4), 479–497.
Andrejevic, M. (2007), *iSpy: Surveillance and Power in the Interactive Era*, Lawrence, KS: University Press of Kansas.
Andrejevic, Mark (2012), 'Exploitation in the data mine', in Christian Fuchs, Kees Boersma, Anders Albrechtslund and Marisol Sandoval (eds), *Internet and Surveillance: The Challenges of Web 2.0 and Social Media*, New York: Routledge, pp. 71–88.
Ball, Kirstie, Kevin Haggerty and David Lyon (eds) (2012), *Routledge Handbook of Surveillance Studies*, New York: Routledge.
Barnes, S. (2006), 'A privacy paradox: social networking in the United States', *First Monday*, 11 (9). http://www.firstmonday.org/issues/issue11_9/barnes/index.html.
Beer, D. (2008), 'Social network(ing) sites revisiting the story so far: a response to Danah Boyd and Nicole Ellison', *Journal of Computer-Mediated Communication*, 13 (2), 516–529.

Bennett, C. (2011a), 'In defence of privacy: the concept and the regime', *Surveillance and Society*, 8 (4), 485–496.

Bennett, C. (2011b), 'In further defence of privacy', *Surveillance and Society*, 8 (4), 513–516.

Boyd, D. (2011), 'Social network sites as networked publics: affordances, dynamics and implications *A Networked Self: identity, community and culture on social network sites,* ', in Zizi Papacharissi (ed.),. London: Routledge pp. 39–58.

Boyd, D. and Hargittai, E. (2010), 'Facebook privacy settings: Who cares?', *First Monday*, 15 (8).

Clark, Leigh A. and Sherry J. Roberts (2010), 'Employer's use of social networking sites: a socially irresponsible practice', *Journal of Business Ethics*, 95 (4), 507–525.

Clarke, R. (1988), Information technology and dataveillance, *Communications of the ACM*, 31 (5), 498–512.

Clarke, Roger (1994), 'Dataveillance: delivering "1984"', in Lelia Green and Roger Guinery (eds), *Framing Technology: Society, Choice and Change*, Sydney: Allen & Unwin, pp. 117–130.

Consalvo, Mia and Charles Ess (eds) (2011), *The Handbook of Internet Studies*, Chicester: Wiley-Blackwell.

Davison, Kristl H., Catherine Maraist and Mark N. Bing (2011), 'Fiend or foe? The promise and pitfalls of using social networking sites for HR decisions', *Journal of Business and Psychology*, 26 (2), 153–159.

Davison, Kristl H., Catherine Maraist, R.H. Hamilton and Mark N. Bing (2012), 'To screen or not to screen? Using the Internet for selection decisions', *Employee Responsibilities and Rights Journal*, 24 (1), 1–21.

Foucault, Michel (1977), *Discipline and Punish*, New York: Vintage.

Fuchs, Christian (2008), *Internet and Society: Social Theory in the Information Age*, New York: Routledge.

Fuchs, Christian (2009), *Social Networking Sites and the Surveillance Society*, Salzburg/ Vienna: Forschungsgruppe UTI.

Fuchs, Christian (2010), 'Social software and Web 2.0: their sociological foundations and implications', in San Murugesan (ed.), *Handbook of Research on Web 2.0, 3.0, and X.0: Technologies, Business, and Social Applications. Volume II*, Hershey, PA: IGI-Global, pp. 764–789.

Fuchs, C. (2011c), 'How to define surveillance?', *MATRIZes*, 5 (1), 109–133.

Fuchs, Christian (2013b), 'Political economy and surveillance theory', *Critical Sociology* 39 (5), 671–687.

Fuchs, Christian (2014), 'Social media and the public sphere', *tripleC: Communication, Capitalism and Critique*, 12 (1), 57–101.

Fuchs, Christian, Kees Boersma, Anders Albrechtslund and Marisol Sandoval (2012b), 'Introduction: Internet and surveillance', in Christian Fuchs, Kees Boersma, Anders Albrechtslund and Marisol Sandoval (eds), *Internet and Surveillance: The Challenges of Web 2.0 and Social Media*, New York: Routledge, pp. 1–28.

Fuchs, Christian and Daniel Trottier (2013), 'The Internet as surveilled workplayplace and factory', in Serge Gutwirth, Ronald Leenes and Paul de Hert (eds), *European Data Protection. Coming of Age*, Dordrecht: Springer, pp. 33–57.

Gilliom, John (2001), *Overseers of the Poor*, Chicago, IL: University of Chicago Press.

Goldsmith, Andrew John (2010), 'Policing's new visibility', *British Journal of Criminology*, 50 (5), 914–934.

Graham, Stephen and David Wood (2007), 'Digitizing surveillance: categorization, space, inequality', in Sean P. Her and Josh Greenberg (eds), *The Surveillance Studies Reader*, Maidenhead: Open University Press, pp. 218–230.

Gross, Ralph, Alessandro Acquisti and H. John Heinz III (2005), 'Information revelation and privacy in online social networks', in *Proceedings of the 2005 ACM Workshop on Privacy in the Electronic Society*, New York: ACM Press, pp. 71–80.

Habermas, Jürgen (1987), *The Theory of Communicative Action. Volume 2: Lifeworld and System: A Critique of Functionalist Reason*, Boston, MA: Beacon Press.

Horkheimer, Max (2002), 'Traditional and critical theory', *Critical Theory*, New York: Continuum, pp.188–252.

Hunsinger, Jeremy, Klastrup, Listrup and Matthew Allen (eds) (2010), *International Handbook of Internet Research*, Dordrecht: Springer.

Jhally, S. and B. Livant (1986), 'Watching as working: the valorization of audience consciousness', *Journal of Communication*, 36 (3), 124–143.

Kelly, Kevin (1999), *New Rules for the New Economy*, New York: Penguin.

Lyon, David (1994), *The Electronic Eye: The Rise of Surveillance Society*, Cambridge: Polity.

Lyon, D. (1998), 'The world wide web of surveillance: the Internet and off-world power-flows', *Information, Communication and Society*, 1 (1), 91–105.

Marx, Gary T. (1988), *Undercover: Police Surveillance in America*, Berkeley, CA: University of California Press.

Marx, Gary T. (2002), 'What's new about the "new surveillance"? Classifying for change and continuity', *Surveillance and Society*, 1 (1), 9–29.

Mathiesen, Thomas (1997), 'The viewer society: Michel Foucault's "panopticon" revisited', *Theoretical Criminology*, 1 (2), 215–234.

Negroponte, Nicholas (1996), *Being Digital*, New York: Vintage Books.

Nosko, A., E. Wood and S. Molema (2010), 'All about me. Disclosure in online social networking profiles. The case of Facebook', *Computers in Human Behavior*, 26 (3), 406–418.

O'Reilly, T. and N. Battelle, (2009), 'Web squared. Web 2.0 five years on. Special report', available at http://assets.en.oreilly.com/1/event/28/web2009_websquared-whitepaper.pdf (accessed 30 January 2013).

Poster, Mark (1990), *The Mode of Information*, Cambridge: Polity.

Regan, Priscilla (1995), *Legislating Privacy*, Chapel Hill, NC: University of North Carolina Press.

Rheingold, Howard (1993), *The Virtual Community. Homesteading on the Electronic Frontier*, Cambridge, MA: MIT Press.

Sánchez Abril, Patricia, Avner Levin and Alissa Del Riego (2012), 'Blurred boundaries: social media privacy and the twenty-first-century employee', *American Business Law Journal*, 49 (1), 63–124.

Scholz, Trebor (2010), 'Facebook as playground and factory', in Dylan E. Wittkower (ed.), *Facebook and Philosophy: What's on Your Mind?*, Chicago, IL: Open Court, pp.241–252.

Smythe, Dallas W. (1977), 'Communications: blindspot of Western Marxism', *Canadian Journal of Political and Social Theory*, 1 (3), 1–27.

Stalder, F. (2011), 'Autonomy beyond privacy? A rejoinder to Colin Bennett', *Surveillance and Society*, 8 (4), 508–512.

Tavani, Herman T. (2008), 'Informational privacy: concepts, theories, and controversies', in Kenneth E. Himma and Herman T. Tavani (eds), *The Handbook of Information and Computer Ethics*, Hoboken, NJ: Wiley, pp.131–164.

Thompson, John B. (2005), 'The new visibility', *Theory, Culture and Society*, 22 (6), 31–51.

Trottier, Daniel (2011), 'A research agenda for social media surveillance', *Fast Capitalism*, 8 (1).

Trottier, Daniel (2012), *Social Media as Surveillance*, Farnham: Ashgate.

Trottier, Daniel and David Lyon (2012), 'Key features of social media surveillance', in Christian Fuchs, Kees Boersma, Anders Albrechtslund and Marisol Sandoval (eds), *Internet and Surveillance: The Challenges of Web 2.0 and Social Media*, New York: Routledge, pp.89–105.

Warren, S. and L. Brandeis (1890), 'The right to privacy', *Harvard Law Review*, 4 (5), 193–220.

Yar, Majid (2010), 'Public perceptions and public opinion about Internet crime', in Yvonne Jewkes and Majid Yar (eds), *Handbook of Internet Crime*, Cullompton: Willan, pp.104–119.

PART VII

EXPANDING THE FRONTIERS OF DIGITAL POLITICS RESEARCH

23. Visibility and visualities: 'ways of seeing' politics in the digital media environment
Katy Parry

On contemplating how to approach writing a chapter on visual politics online, my initial thoughts turned to the potential slipperiness of each of these terms. How to think about politics online as distinct from its offline manifestations? How narrowly or broadly to define politics and the political? In terms of the recognized political actors and institutions of official politics and policy-making, or more broadly, as a public space in which meanings, identities and values are contested? How productive is it to separate the visual dimension from other qualities or modalities (text, sound) across a range of digital media forms? Indeed 'the visual' is about more than images alone, so how to place notions of the visual within wider concerns of visibility and visuality?

Such questions around indistinct boundaries and instability form the basis for this chapter, then, and in exploring the slipperiness of these terms I address old and new concerns about visibility, vision and visuality in political communication and culture. Drawing on relevant insights from political studies, media and communications and social movement studies, the present chapter places what has come to be known as 'visual culture studies' at the heart of its approach and spirit, for reasons further explained below. It is in the interplay of online and offline political practices, serious and comedic mediations, and their authoritative or subversive purposes, that senses of convergence and collision exist, and through which new opportunities for analysis emerge. But it is also in the productive sharing of resources and tools from across the academic disciplines that we might better scrutinize, understand and appreciate the varied forms of visual politics online.

The chapter is organized into five sections: first, there is a brief outline of how varied disciplinary perspectives have informed the discussion of politics and mediated imagery, tracing the tensions and anxieties associated with emergent media technologies and the concerns over the corrupting influences identified with such media. Second, elaborating on my argument for a central role of 'visual culture studies' in exploring online politics, the subsequent section outlines notions of visibility and visuality. The following two sections are broadly split into politics 'from below' and politics 'from above'. Arguably, the separation of protest from 'official politics' can create another division of research agendas which belies the multiple civic activities and practices each of us chooses to undertake (or not) in our everyday lives. In the hybrid media environment through which many of us experience politics, both established politicians and protesters appear to be adopting more personalized and expressive forms of engagement (Bennett, 2012; Chadwick, 2013). The sections are designed to be illustrative and exploratory, as it is beyond the scope of the current chapter to provide detailed empirical analysis. The final section sets out future research questions and my concluding comments. Before proceeding to the next scene-setting section, I outline two guiding stipulations.

First, as indicated above, studying visual politics online requires openness to a variety of approaches, based on the research questions and methods that most intrigue and provoke us. In his article entitled 'There are no visual media', W.J.T. Mitchell warns against 'visual culture as the "spectacle" wing of cultural studies', noting that its very promise is in its insistence 'on problematizing, theorizing, critiquing and historicizing the visual process as such' (Mitchell, 2005: 264). A broad interdisciplinary interest in varied cultural practices and artefacts does not equate to a flattening out or homogeneity of approach, or a misrecognition of how one medium's affordances are qualitatively different to another's. As Mitchell has argued elsewhere, 'the opening out of a general field of study does not abolish difference, but makes it available for investigation, as opposed to treating it as a barrier that must be policed and never crossed' (Mitchell, 2002: 173).

Second, the fast-evolving role of the Internet in our everyday lives has further disrupted many of the traditional parameters for studying images in mediated contexts. Such disruptions are not entirely new: for example, Elihu Katz (1988) recognized many years ago how the technological theory of 'disintermediation', or cutting out the middleman, could be applicable for media sociology. Social media use and peer-to-peer sharing online merely represent the latest technological and cultural practices embodying a reinvigorated sense of connectivity and direct communication. The promiscuity of digital media images in a 'cut-and-mix' culture (van Zoonen et al., 2010) creates both possibilities and risks for producers, depicted subjects and audiences who view and share such images. The processes of upload and display add to the ephemeral quality, as images circulate and become divorced from original captions or audio, remediated in ever-mutating 'circuits of culture' (du Gay et al., 1997), seemingly unbounded by shared viewing contexts or clearly defined 'imagined communities' (Anderson, 1991). However the often celebratory claims for disintermediation and a sense of less mediation can be misleading and obfuscate the shifts that digital technologies enable; new and old intermediaries may be adapting and altering their role but they remain 'vitally' important (Thumim, 2012). All digital images encountered via the Internet are mediated in some form; the contexts may be increasingly varied, with images constituted as a mix of both amateur and professional in origin, but this multifarious jumble of image-text circulates within a discursive public space of framing practices, semiotic recipes, rhetorical challenges and ironic gestures.

In an inevitably brief review, the next section provides contextual background to the broad fields of study which inform the chapter, setting out the traditional anxieties, valuable perspectives and emergent tensions through which we might incorporate the study of visual culture in understanding politics and digital media.

THE SPECTRE OF THE SPECTACLE: THE HAUNTING ANXIETIES AROUND THE VISUAL IMAGE IN POLITICAL COMMUNICATION

The role of the image in mediated political communication has long provoked suspicion and unease. With its concerns for governance, democracy and citizenship, political studies provides theoretical tools for assessing the health of the polity and public sphere, often bringing a defensive posture to guarding the integrity of politics against debasing forces (Crick, 1968; Flinders, 2012). Although a concern of political philosophers and critical theorists in earlier centuries, fears of a distracted and passive public or citizenry have more recently been associated with the dominant role of television in political campaigning and as the main source for public knowledge. Throughout the twentieth century, as politicians increasingly addressed publics through the mass medium of television, concerns were raised over the diminishing of political life into a spectacle, distorted by a media-driven shift promoting conflict, sensationalism and inauthentic celebrity politicians. In such 'audience democracies' (Manin, 1997) citizen-viewers are characterized as passive, apathetic spectators, monitoring the actions of political leaders but merely reactive to the theatre of political life performed by a set of interchangeable elites, rather than socially engaged actors with a significant decision-making role (see also Edelman, 1988; Meyer and Hinchman, 2002; Postman, 1987; Putnam, 2000). Writers concerned with the role of media in democracy note trends towards a politics evermore shaped by 'media logic' or 'mediatization', accompanied by an overemphasis on stylization, presentation, performance and image (Blumler and Kavanagh, 1999; Mazzoleni and Schulz, 1999; Strömbäck, 2008).

Where some authors see politics tarnished by the blurring lines of information and entertainment that such 'media logic' brings about, with a public service ethos crowded out by consumer-driven 'infotainment' and lifestyle programming (McChesney, 1999; Thussu, 2007), others see an enhancement for democratic life in the popular engagement with politics encouraged by a variety of formats offering a mix of serious and more playful generic recipes (Corner and Pels, 2003; van Zoonen, 2005; Richardson et al., 2012). The shifting interests and concerns reflect a cultural turn across humanities and social sciences, in which other cultural factors and contested sites of meaning and identity are considered alongside political structures, for example, in debating the role of comedy as a catalyst for civic engagement (Baym and Jones, 2012), or the intersections of celebrity culture and political culture (Street, 1997). Alongside this, there is a turn to the emotional or personal aspects of political life, where the private lives and personal-psychological qualities of political leaders are emphasized in discussion of their political performances across the broader media environment (Corner, 2003; Langer, 2011; Stanyer, 2012). In short, many of the discussed trends in political communication and culture – whether on the appeal to the emotions and entertainment, a personalization of politics, or the role of humour and satire – often include an implicit or explicit concern with the visual or symbolic dimension.

The nature of politics as spectacle is often central to these perspectives, despite proponents rarely engaging with the visual or symbolic forms or properties in any detail. The concerns for a healthy and vibrant public sphere are laudable, but such perspectives are in danger of oversimplifying the identified problem. The 'iconophobia' or 'iconoclasm' at the heart of the Western tradition has now attracted critical attention from authors calling for a rethink towards spectatorship and democracy (Finnegan and Kang, 2004; Green, 2010). In summary, I suggest three potential pitfalls. First, there is often a sense of the emotive image in contest with the rational word; that the expressive, symbolic or affective dimensions work to degrade the crucial rationality of the political realm. This negates the interplay of word and image in mediated communications and the constitutive role of both in interpreting our social worlds. Second, an idealized notion of the citizen is contrasted against the deficiencies of the spectator in an unproductive dichotomization which fails to explore how watching and interpreting are also active and creative: 'The spectator also acts, like the pupil or scholar. She observes, selects, compares, interprets' (Rancière, 2011: 13). Third, a traditional emphasis on official politics which narrowly defines political engagement or political actions is in danger of overlooking alternative roles for citizens outside of recognized political structures (see Coleman and Blumler, 2009: 156). This final point brings the discussion back to the distinctions between the top-down politics of institutions and policy-makers against the 'from-below' character of grass-roots politics and dissent.

Where the visual performance of political leaders has been greeted with unease and suspicion (if examined at all), those writing on protest and dissent provide a much more fruitful scholarly engagement with the sense of the visual and the artistic as powerful cultural tools in political action (for a recent collection see McLagan and McKee, 2012). For those struggling to gain attention on the political stage, 'image politics' (DeLuca, 1999) offer a striking way to promote a cause and spark imagination. Operating with a freedom of expression that traditional party politicians are unlikely to risk, the politics of dissent, or contentious politics, can embrace the symbolic and theatrical, and even be cheered for combating the 'pseudo-events' (Boorstin, 1962) of the political elite with heartfelt and humorous image events and political artwork. Choice of imagery and expression can, of course, have repercussions for those hoping to move from 'outsider' to 'insider' status. Fighting the 'image of power with the power of imagery' (Doerr and Teune, 2012) not only identifies you as part of a collective '99%' (to use the Occupy movement's slogan), but, through an adoption of the symbols of contentious politics, can reinforce your status as a heckler rather than an orator; just as notions of 'the political class' can reinforce a sense of an unreachable elite in the distant echelons of high society.

Finally, visual culture studies or 'image studies', with its ancestry in art history, museum studies and media studies, provides the younger sister to the current melange for this chapter. It is through the visually inflected framing of visual culture studies that we can better recognize the rhetorical, aesthetic, expressive and ironic qualities of politically themed imagery. One particular strength of this perspective is that it reflects on its own analytical usefulness, with key theorists sensitive to the limitations of even labelling media texts as visual media (Bal, 2003; Mitchell, 2005). It is in studying the interplay and interdiscursivity of sound, word and image in a variety of media platforms, formats and genres that we can consider visual images as resources for making meaning within a mix of modalities; that is, as resources for producers, viewers or users. Placing an emphasis on the visual here is an attempt to redress the traditional text-based research bias in political communication.

Given the above summary, this chapter aims to 'scope out' the 'ways of seeing' and ways of understanding politics online. The visual metaphors in the preceding sentence are deliberate: conflations of seeing and believing, and seeing and understanding ('Ah, I see') are thought to betray a favouring of vision as the superior sense through which to access the world (see Jay, 1993 and Mirzoeff, 2011 for historical and critical accounts), and yet it is only more recently that social scientists have embraced an iconic turn or 'pictorial turn' within their research agendas (Mitchell, 1994). Visual experience encompasses a great deal more than pictures or imagery and, indeed, includes the written word in graphic form; yet the study of images has been separated from scientific and literary enquiry historically, marked with a rationalist suspicion dating back to Plato's allegory of the cave (see Jay, 1993: 27; Mitchell, 2005: 86). It is not only material cultural artefacts that are of interest here, it is the way such images interact with the 'pictures in our heads' (Lippmann, 1922), the mental images of our mind's eye which also guide how we see the world and how we place ourselves within social spaces and political structures. Our ways of seeing the world are not only about vision (what the eyes observe), but 'visuality' or 'visualities'; 'what is made visible, who sees what, how seeing, knowing and power are interrelated' (Bal, 2003: 19). Notions of control, knowledge and power are crucial here: visuality is also about a political struggle, the 'right to look' and be seen as citizens (Mirzoeff, 2011). It is necessary then to further outline the intersecting notions of vision, visuality and visibility in relation to both mediated political performances and our ways of seeing the political world.

RECONCILING NOTIONS OF VISIBILITY AND VISUALITY IN POLITICAL COMMUNICATIONS

John B. Thompson's work on the visibility of politicians is central to thinking about how visibility and politics link to visuality and vision, and is often cited by those interested in how our politics has become more mediatized, personalized or intimate. Thompson writes of how the development of mass media technologies in the twentieth century, and especially television as the central platform, linked issues of visibility with vision (Thompson, 1995). In the age of mediated visibility, the field of vision is shaped 'by the distinctive properties of communication media, by a range of social and technical considerations (such as camera angles, editing processes and organizational interests and priorities) and by the new types of interaction that these media make possible' (Thompson, 2005: 35–36). Thompson characterizes mediated visibility as a double-edged sword for politicians; a 'source of a new and distinctive kind of *fragility*' (2005: 42) in a more complex and less controllable information environment, and where public–private boundaries are rewritten.

More recently, Daniel Dayan has argued that historically visibility was desired as an enviable right enjoyed by the few; a form of attention-gaining which publics, as spectators, have been denied: 'Being anonymous has become a stigma, and visibility has become a right frequently and sometimes violently claimed; a right that all sorts of people feel entitled to obtain. The exclusive visibility once conferred upon some is perceived by the anonymous as an injustice in need of redress' (Dayan, 2013: 139). Dayan proposes a 'paradigm of visibility', which, as distinct from the media effects paradigm, 'stresses the role of media in coordinating collective attention' (p. 139). This offers a narrative of 'deprivation followed by a conquest' (p. 139) initially acquired by citizens as a form of visibility offered on a conditional basis (as reality TV contestants, for example), or by violent means, such as terrorist acts designed with media in mind. But this is often 'the wrong type of visibility' (p. 141).

Further along in Dayan's narrative, new media technologies and platforms emerge, such as Facebook and YouTube, in which the quest for visibility can be taken a step further: 'Not only do such media allow publics to acquire visibility, and to acquire visibility on their own terms, but they also allow them to define the visibility of others, to become organizers of visibility' (p. 143). Such 'visibility entrepreneurs' can challenge the narrative offered by mainstream media, promoting debate and even scandal, but they can also encourage a battlefield mentality in which professionals assert their legitimacy against 'uninvited intruders' (p. 145). In either case, this is 'an attempt at interfering with a silencing process' (p. 150), in which various forms of expression can be heard or seen, and compared on an equal footing. New representational practices offer innovative ways for users and spectators to find social and political meanings while negotiating issues of truth, trust and credibility. At the same time, the traditional facilitators of the public sphere face increasing economic pressures to perform such a role. In some cases this means the loss of established newspapers, or in a recent example specific to news images, the *Chicago Sun-Times* sacking its entire photography department (BBC, 2013).

Rather than thinking about conditions of visibility from the politicians' perspectives, Dayan's paradigm of visibility helps us to conceptualize online and physical spaces as contested sites of political meaning, values and identity, in dialogue with other more traditional mediated forms. The question then becomes one of investigating the different patterns of involvement and the kinds of visual display produced and circulated across the mediascape.

PUBLICS AND SPECTATORS: BECOMING VISIBLE CITIZENS IN VISUAL DISPLAYS

Dayan's paradigm of visibility, with its emphasis on the media coordination of collective attention, offers a useful way to think about how citizenship is performed and the reactions of publics to different attempts to disrupt the status quo. In her book, *Revolting Subjects*, Imogen Tyler (2013) explores how certain groups in society are figured as 'revolting' and how such stigmatization can be resisted through aesthetic and political strategies. In the chapter on the August 2011 riots in the UK, feelings of abandonment and alienation expressed by the rioters are intricately linked to their sense of invisibility in the social body. Tyler cites the Guardian–London School of Economics (LSE) research project, Reading the Riots, which interviewed 270 people who had participated in the riots: 'This sense of being invisible was widespread' (Lewis et al., 2011: 25, cited in Tyler, 2013: 197). But not all attempts to combat societal invisibility – and especially those of a chaotic and violent nature – lead watching publics to reassess their prior attitudes; indeed in the case of the riots, the outraged media attention, public fear and harsh sentences handed down to participants all suggest that rather than effecting an 'alternative aesthetics', the rioters 'became the abjects they had been told they were' (Tyler, 2013: 204). One way in which the rioters and bystanders attempted to make sense of the disturbances and looting in their own locales was through the sharing of camera images via Twitter; interestingly, these shared images included captured TV screenshots in addition to on-the-street user-generated images, indicating a sense of involvement through 'practices of remote witnessing' alongside the impulse to capture the here and now (Vis et al., 2013: 396). The sharing of such images also records the 'second screen' phenomenon, whereby users participate via websites whilst also watching television; live media events and elections tend to attract this activity but a regular UK example is the high levels of tweeting during the BBC political panel show *Question Time* (using the #bbcqt hashtag) (Anstead and O'Loughlin, 2011).

This struggle for visibility is often most compelling in the 'image politics' (DeLuca, 1999) of political protest and social movements, with those less powerful utilizing the rhetoric of the visual in order to gain the eyes and ears of the public and, ultimately, society's decision-makers. With the diverse groups under the anti-globalization banner dominating the scene in the later twentieth century, the truly transnational character of protest coalesced in 2011 as a mixture of movements with revolutionary aims (for example, Occupy, 15-M and the 'Arab Spring' demonstrations) took to the streets in spectacular style, claiming their right to be visible and vocal in, at times, carnivalesque displays (see Castells, 2012; Gerbaudo, 2012; Khatib, 2012).

In their analysis of YouTube videos uploaded and shared over the 18 days of the Egyptian uprising in 2011, starting with the mobilization of demonstrators on 25 January, Mohamed Nanabhay and Roxane Farmanfarmaian write of an 'amplified public sphere' created through the complex interplay of the 'inter-related spaces of the physical (protests), the analogue (satellite television and other mainstream media) and the digital (internet and social media)' (Nanabhay and Farmanfarmaian, 2011: 573). Warning against simplistic separations between amateur and mainstream image-making, or singling out social media such as Facebook or Twitter, their study points to a symbiotic relationship between journalists and activists, producers and consumers. Their point on the importance of physical place is reinforced in Paulo Gerbaudo's (2012) *Tweets and the Streets*, with the significance of assembly, solidarity and corporeality emphasized in the reappropriation of public space, such as the 15 May 2011 demonstrations in cities across Spain (and in what became known as the 15-M movement, or *los indignados*). Crucially, the *indignados* of Spain were also expressing their collective indignation, and as Gerbaudo argues, while social media was central in mobilizing the demonstrations and protest camps that followed, it was in the symbolic and material concentration in physical spaces, such as in Puerta del Sol in Madrid, that protesters rediscovered 'a sense of physical communion' (Gerbaudo, 2012: 96). In adopting the chant of 'We are not on Facebook, we are on the streets', this was about appearing as a tangible public, and signals how visibility is central to this debate (p. 96).

In the summer of 2013, Taksim Square in Istanbul, and the adjoining Gezi Park, became the latest symbol of an extended protest and clash with authority, with the initial protests against the square's development mutating into widespread demonstrations against the authoritarian nature of the government. Critically, it is the imposition of a commercial redevelopment in an iconic public space that first sparked the protests, while the reaction of Turkish Prime Minister Recep Tayyip Erdogan was to take a dismissive stance towards the protesters and the solidarity expressed via social media, characterizing Twitter as a 'menace' to society (Shafak, 2013).

In setting up camps and demonstrations in the heart of cities around the world, such 'spectacles of dissent' (D'Arcus, 2006) are both about mobilizing local people and embodying revolutionary zeal in the immediate physical place, but also about amplifying their countercultural message through the documentation and circulation of images and videos online. As Tina Askanius argues, we are seeing an 'aesthetization of public protest' in the theatrical displays and vast body of images and videos created: 'This emerging audio-visual repository of interconnected narratives stages popular contestation within a coherent framework and constitutes the basis from which collective identity formation is forged among activists scattered around the world' (Askanius, 2010: 341). The hope of such activists is to challenge the legitimacy of existing power relations through alternative forms of organization and political identity, adopting visually strong practices that range from the raw and antagonistic to the absurdly humorous. Mobile media devices also allow protesters to subvert the surveillance tactics of police, playfully mimicking their recording practices and even capturing violent police behaviour which has led to public inquiries and arrests (Archibald, 2011; Shaw, 2013). The 15-M and Occupy movements are emblematic of the challenge to democratic authority through networked action, but also of the desire to create an experiential and sensorially rich public space for imaginative reworking of what 'the political' might mean.

POLITICIANS ONLINE: SHARING A HUG AND GETTING THE JOKE

When Barack Obama realized he had secured victory in the 2012 US presidential election, his team chose to announce his second term by tweeting an image of the President hugging his wife Michelle with the simple statement: 'Four more years'. The image became the most retweeted in Twitter's history, signalling the desire of more than 816 000 people who shared the image in those first few days (Ries, 2012) to join in some kind of communicative political action or affinity; but also signalling a political leader comfortable with announcing his victory via social media, before his televised speech. The photograph had been taken months earlier in mid-August by Scout Tufankjian, a campaign photographer, and was selected by digital staffer Laura Olin working as part of Obama's social media team. Olin puts Obama's successful social media campaign down to choosing diligent staff who also 'knew their social-media shit', and letting those staff 'talk to people like people' (cited in Ries, 2012). In this way, Obama's campaign harnessed the visual and verbal potential of online mediation, and achieved the much-sought-after 'virality' (the copying and multiplying of a message or phenomenon through social networks) which generally eludes professional political marketing. The particular expressive qualities of the photograph are significant and strategic: its selection speaks to the personalized and emotional appeal of a president who also appears as a loving husband, embedded in a political style that 'weaves together matter and manner, principle and presentation, in an attractively coherent and credible political performance' (Pels, 2003: 57).

Effective use of social media, as with other media genres, is contingent on embracing socio-technical knowhow and competences in mediated visibility. This enables a projected image of authenticity and integrity and is a recipe that few politicians accomplish, at least with any consistency. Images especially thrive in a digital world of mash-up, montage, juxta-position, repetition and manipulation. While Obama's pictured embrace of his wife went viral and became an iconic image of the campaign, Limor Shifman would distinguish this viral image from the 'memetic video' or image which '*lures extensive creative user engagement* in the form of parody, pastiche, mash-ups or other derivative work' (Shifman, 2012: 190). These images are much more indicative of the participatory culture of the Internet, according to Shifman. The most popular Internet memes, whether user-generated or popular culture-related, are humorous and playful but not necessarily political. Those dealing directly with politics can express a range of expressive modes, from light-hearted mockery to oppositional fervour. Visual memes are particularly suited to travelling across national borders, whether mocking in tone (such as the 'Pepper Spray Cop', http://peppersprayingcop.tumblr.com/, PhotoShopped after Lieutenant John Pike casually sprayed peaceful Occupy protesters in California), or symbolic of a struggle against corruption and state vio-lence, (for example, the 'We are all Khaled Said' Facebook page and the later appropriations of Khaled's photograph as a 'visual injustice symbol' by activists during the Egyptian uprising; Olesen, 2013; see also Khatib, 2012).

The success of the viral dance video by South Korean pop-star, PSY, 'Gangnam Style', subsequently inspired its own political parodies, with 'Mitt Romney Style' and 'Kim Jong Style' examples of 'prosumer'-generated material with political intent (Müller and Kappas, 2013). In some cases, politicians refer to their own memes, signalling that they are in on the joke and conversant in the socio-technological practices of online platforms: On 26 August 2012 Barack Obama signed off from his Reddit 'Ask me anything' session with a reference to the 'Not Bad' meme. The 'Not Bad' meme is based on a photograph of the US President pulling a strange expression known as a 'sturgeon face' during a visit to the UK in May 2011, and which is thought to convey general satisfaction. The com-bination of media platform and intertextual playfulness signal a rare mix of approachability and self-assurance. UK Deputy Prime Minister, Nick Clegg, showed he too could get the joke when a video he recorded apolo-gising to university students for breaking his election pledge on tuition fees was subsequently set to music and released on satirical website 'The

Poke' (http://www.thepoke.co.uk): Clegg gave his consent for the video to be released as a charity single, seemingly content to join in the mocking of his own insincerity. Nevertheless, emergent hostility against a strong cult of visual iconography can signal more than satirical or light-hearted ridiculing of political leaders. The destruction of material posters, murals and statues, along with the subsequent circulation of the images and videos depicting such protests online, provides a symbolic rejection of the visual legacy of a regime; as happened in Syria in March 2011, when footage of the ruling Assad family posters being ripped from buildings was shared across YouTube, Twitter and Facebook (Caldwell, 2011).

FUTURE RESEARCH QUESTIONS AND CONCLUDING COMMENTS

This chapter has so far attempted to set the scene for researching visual politics online. In: (1) outlining the traditional tensions between politics, popular culture and images; (2) emphasizing the interrelated notions of visibility and visuality (including how we appear as political actors and the representational forms employed); and (3) providing some recent illustrations of visual politics 'from below' and 'from above', a number of questions emerge for further consideration and examination:

- How might we better understand imaging practices and online activity (posting, viewing, commenting) as meaningful political participation?
- What kinds of political performance are best suited to the hybrid, potentially global, political information environment? And how do we analyse their effectiveness?
- Are the most visible and visual elements of mediated politics online representative of our political cultures, or do they offer a distortive perspective?

Such questions are best approached with keen attention to representational forms, alongside contexts of production, mediation and consumption. Similarly to the old debates on the spectacle of politics, there is a danger that the current fascination with how images or videos circulate online and their role in mobilizing support leads to a lack of attention given to the actual images as expressive forms: 'Rather remarkably, the systematic study of the structure and the expressive means of the image itself ("image studies") is relatively rarely practised . . . there is a persistent misunderstanding that one can go without insight into the structure of images or other visual artifacts' (Pauwels, 2008: 84). As indicated earlier, pulling together approaches and methods from a variety of research fields can offer an enriched perspective from which to question and problematize the nature of visibility and visuality as encountered in the political realm.

Levels of attentiveness and interest are not assured by the expanse of networked political information available, and participation remains unequal across different communities and socio-economic groups, but the affordances of the Internet-based technologies undoubtedly enable access to an abundance of visual display from around the world, and offer cheap and easy ways to generate new material. Paradoxically, the motivations behind the patterns and practices of re-presentation, linking and sharing might work to question the digital images' supposed inherent ambiguity and the 'post-photographic' disruption to truth claims. Our investment in digital images as forms of communicative expression suggests a more complex picture than a simple characterization of shallow naivety or post-photography scepticism. Whether perceived as compelling evidential material, or as profound and transformative expressions of solidarity or affinity, the role of digital images in political discourse has undoubtedly been enhanced through Internet-based presentation and the resulting collective (and at times disorderly) debates they inspire on meanings, values and identities.

FURTHER READING

Dayan, D. (2013). Conquering visibility, conferring visibility: visibility seekers and media performance. *International Journal of Communication*, 7, 137–153.

Finnegan, C.A. and Kang, J. (2004). 'Sighting' the public: iconoclasm and public sphere theory. *Quarterly Journal of Speech*, 90 (4), 377–402.

Gerbaudo, P. (2012). *Tweets and the Streets: Social Media and Contemporary Activism*. New York: Pluto Press.

Khatib, L. (2012). *Image Politics in the Middle East: The Role of the Visual in Political Struggle*. London: I.B. Tauris.

McLagan, M. and McKee, Y. (2012). *Sensible Politics: The Visual Culture of Nongovernmental Activism*. New York: Zone Books.

Mitchell, W.J.T. (2005). There are no visual media. *Journal of Visual Culture*, 4 (2), 257–266.

Pels, D. (2003). Aesthetic representation and political style: re-balancing identity and difference in media democracy. In Corner, J. and Pels, D. (eds), *Media and the Restyling of Politics* (pp. 41–66). London: Sage.

van Zoonen, L., Vis, F. and Mihelj, S. (2010). Performing citizenship on YouTube: activism, satire and online debate around the anti-Islam video Fitna. *Critical Discourse Studies*, 7 (4), 249–262.

REFERENCES

Anderson, B. (1991). *Imagined Communities: Reflections on the Origin and Spread of Nationalism*. London, UK and New York, USA: Verso.

Anstead, N. and O'Loughlin, B. (2011). The emerging Viewertariat and BBC *Question Time*: television debate and real-time commenting online. *International Journal of Press/Politics*, 16 (4), 440–462.

Archibald, D. (2011). Photography, the police and protest: images of the G-20, London 2009. In Cottle, S and Lester, L. (eds), *Transnational Protests and the Media* (pp. 129–139). Oxford: Peter Lang.

Askanius, T. (2010). Video Activism 2.0 – space, place and audiovisual imagery. In Hedling, E., Hedling, O. and Jönsson, M. (eds), *Regional Aesthetics: Locating Swedish Media* (pp. 337–358). Stockholm: Kungliga Biblioteket.

Bal, M. (2003). Visual essentialism and the object of visual culture. *Journal of Visual Culture*, 2 (1), 5–32.

Baym, G. and Jones, J. (2012). News parody in international perspective: politics, power and resistance. *Popular Communication*, 10 (1–2), 2–13.

BBC (2013). *Chicago Sun-Times* sacks entire photo department. *BBC News US and Canada*, 30 May. Retrieved 7 June 2013 from http://www.bbc.co.uk/news/world-us-canada-22723725.

Bennett, W.L. (2012). The personalization of politics: political identity, social media, and changing patterns of participation. *Annals of the American Academy of Political and Social Science*, 644 (1), 20–39.

Blumler, J.G. and Kavanagh, D. (1999). The third age of political communication: Influences and features. *Political Communication*, 16 (3), 209–230.

Boorstin, D.J. (1962). *The Image*. New York: Atheneum.

Caldwell, L. (2011). The new face of President Asad on YouTube. *Arab Media and Society*, Issue 13, retrieved from http://www.arabmediasociety.com/index.php?article=776&p=0.

Castells, M. (2012). *Networks of Outrage and Hope: Social Movements in the Internet Age*. Cambridge, UK and Malden, MA, USA: Polity.

Chadwick, A. (2013). *The Hybrid Media System: Politics and Power*. Oxford: Oxford University Press.

Coleman, S. and Blumler, J.G. (2009). *The Internet and Democratic Citizenship: Theory, Practice and Policy*. Cambridge: Cambridge University Press.

Corner, J. (2003). Mediated persona and political culture. In Corner, J. and Pels, D. (eds), *Media and the Restyling of Politics* (pp. 67–84). London: Sage.

Corner, J. and Pels, D. (2003). *Media and the Restyling of Politics: Consumerism, Celebrity and Cynicism*. London: Sage.

Crick, B. (1968). *In Defence of Politics*. London: Penguin.

D'Arcus, B. (2006). *Boundaries of Dissent: Protest and State Power in the Media Age*. New York: Routledge.

Dayan, D. (2013). Conquering visibility, conferring visibility: visibility seekers and media performance. *International Journal of Communication*, 7, 137–153.

DeLuca, K. (1999). *Image Politics*. New York: Guilford Press.

Doerr, N. and Teune, S. (2012). The imagery of power facing the power of imagery: towards a visual analysis of social movements. In Fahlenbrach, K., Klimke, M., Scharloth, J. and Wong, L. (eds), *The 'Establishment' Responds: Power and Protest During and After the Cold War* (pp. 43–55). Cambridge, UK and New York, USA: Berghahn Books.

Du Gay, P., Hall, S., Janes, L., Mackay, H. and Negus, K. (1997). *Doing Cultural Studies*. London: Sage/The Open University.

Edelman, M. (1988). *Constructing the Political Spectacle*. Chicago, IL: University of Chicago.

Finnegan, C.A. and Kang, J. (2004). 'Sighting' the public: iconoclasm and public sphere theory. *Quarterly Journal of Speech*, 90 (4), 377–402.

Flinders, M. (2012). *Defending Politics: Why Democracy Matters in the Twenty-First Century*. Oxford: Oxford University Press.

Gerbaudo, P. (2012). *Tweets and the Streets: Social Media and Contemporary Activism*. New York: Pluto Press.

Green, J.E. (2010). *The Eyes of the People: Democracy in an Age of Spectatorship*. New York: Oxford University Press.

Jay, M. (1993). *Downcast Eyes: The Denigration of Vision in Twentieth-Century French Thought*. Berkeley, CA: University of California Press.

Katz, E. (1988). Disintermediation: cutting out the middleman. *Inter Media*, 16 (2), 30–31.

Khatib, L. (2012). *Image Politics in the Middle East: The Role of the Visual in Political Struggle*. London: I.B. Tauris.

Langer, A.I. (2011). *The Personalisation of Politics in the UK: Mediated Leadership from Attlee to Cameron*. Manchester: Manchester University Press.

Lippmann, W. (1922). *Public Opinion*. New York: Macmillan.

Manin, B. (1997). *The Principles of Representative Government*. New York: Cambridge University Press.

Mazzoleni, G. and Schulz, W. (1999). 'Mediatization' of politics: a challenge for democracy?. *Political Communication*, 16 (3), 247–261.

McChesney, R. (1999). *Rich Media, Poor Democracy: Communication Politics in Dubious Times*. Urbana, IL: University of Illinois Press.

McLagan, M. and McKee, Y. (2012). *Sensible Politics: The Visual Culture of Nongovernmental Activism*. New York: Zone Books.

Meyer, T. and Hinchman, L. (2002). *Media Democracy: How the Media Colonize Politics*. Cambridge: Polity.

Mirzoeff, N. (2011). *The Right to Look: A Counterhistory of Visuality*. Durham, NC, USA and London, UK: Duke University Press.

Mitchell, W.J.T. (1994). *Picture Theory: Essays on Verbal and Visual Representation*. Chicago, IL, USA and London, UK: University of Chicago Press.

Mitchell, W.J.T. (2002). Showing seeing: a critique of visual culture. *Journal of Visual Culture*, 1 (2), 165–181.

Mitchell, W.J.T. (2005). There are no visual media. *Journal of Visual Culture*, 4 (2), 257–266.

Müller, M.G. and Kappas, A. (2013). Politics Mitt Romney style: Gangnam style as a cross-cultural visual meme – online citizen creativity and the power of digitally facilitated political prosumer participation. Presented at 63rd Annual International Communication Association Conference, 21 June, London.

Nanabhay, M. and Farmanfarmaian, R. (2011). From spectacle to spectacular: how physical space, social media and mainstream broadcast amplified the public sphere in Egypt's 'revolution'. *Journal of North African Studies*, 16 (4), 573–603.

Olesen, T. (2013). 'We are all Khaled Said': visual injustice symbols in the Egyptian revolution, 2010–2011. In Doerr, N., Mattoni, A. and Teune, S. (eds), *Advances in the Visual Analysis of Social Movements*, Research in Social Movements, Conflicts and Change, Vol. 35 (pp. 3–25). Bingley: Emerald Group.

Pauwels, L. (2008). Visual literacy and visual culture: reflections on developing more varied and explicit visual competencies. *Open Communication Journal*, 2, 79–85.

Pels, D. (2003). Aesthetic representation and political style: re-balancing identity and difference in media democracy. In Corner, J. and Pels, D. (eds), *Media and the Restyling of Politics* (pp. 41–66). London: Sage.

Postman, N. (1987). *Amusing Ourselves to Death*. London: Methuen.

Putnam, R.D. (2000). *Bowling Alone: The Collapse and Revival of American Community*. New York: Simon & Schuster.

Rancière, J. (2011). *The Emancipated Spectator*. London, UK and New York, USA: Verso.

Richardson K., Parry, K. and Corner, J. (2012). *Political Culture and Media Genre: Beyond the News*. Basingstoke: Palgrave.

Ries, B. (2012). The story behind the most viral photo ever. *Daily Beast*, 19 November. Retrieved from http://www.thedailybeast.com/articles/2012/11/19/the-story-behind-the-most-viral-photo-ever.html.

Shafak, E. (2013). The view from Taksim Square. *Guardian*, 4 June. G2 supplement, pp. 6–8.

Shaw, F. (2013). 'Walls of seeing': protest surveillance, embodied boundaries, and counter-surveillance at Occupy Sydney. *Transformations*, 23. Retrieved from http://www.transformationsjournal.org/journal/issue_23/article_04.shtml.

Shifman, L. (2012). An anatomy of a YouTube meme. *New Media and Society*, 14 (2), 187–203.

Stanyer, J. (2012). *Intimate Politics: The Rise of the Celebrity Politician and the Decline of Privacy*. Cambridge: Polity.

Street, J. (1997). *Politics and Popular Culture*. Philadelphia, PA: Temple University Press.

Strömbäck, J. (2008). Four phases of mediatization: an analysis of the mediatization of politics. *International Journal of Press/Politics*, 13 (3), 228–246.

Thompson, J.B. (1995). *The Media and Modernity*. Stanford, CA: Stanford University Press.

Thompson, J.B. (2005). The new visibility. *Theory, Culture and Society*, 22 (6), 31–51.

Thumim, N. (2012). *Self-Representation and Digital Culture*. Basingstoke: Palgrave.

Thussu, D.K. (2007). *News as Entertainment: The Rise of Global Infotainment*. London: Sage.

Tyler, I. (2013). *Revolting Subjects*. London, UK and New York, USA: Zed Books.

van Zoonen, L. (2005). *Entertaining the Citizen: When Politics and Popular Culture Converge*. Lanham: Rowman & Littlefield.

van Zoonen, L., Vis, F. and Mihelj, S. (2010). Performing citizenship on YouTube: activism, satire and online debate around the anti-Islam video Fitna. *Critical Discourse Studies*, 7 (4), 249–262.

Vis, F., Faulkner, S., Parry, K., Manyukhina, Y. and Evans, L. (2013). Twitpic-ing the riots: analysing images shared on Twitter during the 2011 UK riots. In Weller, K., Bruns, A., Puschmann, C., Burgess, J. and Mahrt, M. (eds), *Twitter and Society* (pp. 385–398). New York: Peter Lang.

24. Automated content analysis of online political communication

Ross Petchler and Sandra González-Bailón

INTRODUCTION

Content analysis has a long tradition in the social sciences. It is central to the study of policy preferences (Budge, 2001; Laver et al., 2003), propaganda and mass media (Krippendorff, 2013 [1980]; Krippendorff and Bock, 2008), and social movements (Della Porta and Diani, 2006; Johnston and Noakes, 2005). New computational tools and the increasing availability of digitized documents promise to push forward this line of inquiry by reducing the costs of manual annotation and enabling the analysis of large-scale corpora. In particular, the automated analysis of online political communication may yield insights into political sentiment which offline opinion analysis instruments (such as polls) fail to capture. Online communication is constantly pulsating, generating data that can help us uncover the mechanisms of opinion formation – if the appropriate measurement and validity methods are developed.

Several linguistic peculiarities distinguish online political communication from traditional political texts. For a start, it is often far less formal and structured. In addition, automated content analysis techniques are not always as reliable or as valid as manual annotation, which makes measurements potentially noisy or misleading. With these challenges in mind, we provide an overview of techniques suited to two common content analysis tasks: classifying documents into known categories, and discovering unknown categories from documents (Liu, 2012; Blei, 2013). This second task is more exploratory in nature: it helps to identify topic domains when there are no clear preconceptions of the topics that are discussed in a certain communication environment. The first task, on the other hand, can help to label a large volume of text in a more efficient manner than manual annotation; for instance, when the research question requires identifying the emotional tone of political communication (as positive, negative or neutral) or its ideological slant (liberal or conservative). This chapter focuses on the application of these automated techniques to online political communication, and suggests directions for future research in this domain.

METHODS FOR AUTOMATED CONTENT ANALYSIS

The application of automated text analysis techniques requires the prior acquisition and preprocessing of data. This section discusses the logic of preprocessing texts to then provide an overview of techniques to classify documents in known categories or discover topics when no categories are known.

Acquiring and Preprocessing Online Political Texts

Political scientists have applied automated content analysis techniques to many kinds of offline political texts, including newspaper articles (Young and Soroka, 2012), presidential and legislator statements (Grimmer and King, 2011), legislature floor speeches (Quinn et al., 2010), and treaties (Spirling, 2012). Until recently, though, political texts remained relatively understudied because they were difficult to parse and process for analysis.

Acquiring online political texts is becoming simpler as more sites store and transmit them in machine-readable formats such as Extensible Markup Language (XML) and JavaScript Object Notation (JSON) or make them publicly available via application programming interfaces (APIs). When such options are unavailable, researchers familiar with statistical software or scripting languages can use new packages for auto-mated HyperText Markup Language (HTML) scraping (Python and R, for instance, have built-in packages and libraries). Finally, when neither machine-readable nor easy-to-parse HTML data are available, research-ers can crowdsource data acquisition and parsing via sites like Amazon Mechanical Turk (Berinsky et al., 2012). Overall, these new technologies enable communication scholars to access and study previously unavailable indicators of public opinion.

In order to perform automated content analysis researchers must trans-form texts into structured data that can be quantified (Franzosi, 2004). Prior to the advent of new computational tools this was performed by human coders using a pre-determined scheme (Krippendorf, 2013 [1980]; Neuendorf, 2001). Initially, a codebook is written guided by a research question and a theoretical context. It is iteratively improved until coders no longer notice ambiguities, at which point it is applied to the data set. Automated approaches preprocess the text to reduce the complexity of language, often using a bag-of-words model to eliminate the most frequent words and to reduce words to their morphological roots (Jurafsky and Martin, 2009; Hopkins and King, 2010; Porter, 1980). After preprocess-ing, documents are represented as a document-term matrix in which rows correspond to documents, sentences, or expressions (depending on the unit of analysis), and columns correspond to words or tokens. Cells can contain continuous values (representing how frequently each term occurs in each document) or binary values (representing whether each term occurs in the document).

Which of the two approaches is more appropriate (to code documents manually or to apply automated preprocessing) depends on the complexity of the research question at hand, the number of documents collected, and the tolerance for error. Although manually annotated data remain the gold standard for content analysis, the sections that follow focus mostly on cases in which data are automatically preprocessed, since this is more common when dealing with large volumes of text.

Once online political texts are converted to a structured form, several methods for automated content analysis can be applied. We divide these methods into two groups to reflect the two most common content analysis tasks: classifying documents into known categories, and discovering theoretically important categories from the content. The former task encompasses techniques such as lexicon-based classification and supervised learning. The latter task encompasses unsupervised learning and the analysis of text as networks of concepts. The following sections explain the details of these techniques and highlight their relative strengths and weaknesses to help communication researchers choose the approach best suited to their specific data and research question.

Classifying Documents into Known Categories

The goal of supervised content analysis techniques is to classify documents into a number of known categories. For example, news articles may have left-leaning or right-leaning ideological biases (Gentzkow and Shapiro, 2010) or have positive or negative coverage (Eshbaugh-Soha, 2010). This section offers an overview of the techniques that allow that sort of classification. There are two main methods. The first is a lexicon-based approach, which uses relative keyword frequencies to measure the prevalence of each category in a document. The second is supervised learning, which uses a training data set of manually annotated documents to classify new, unlabeled documents.

Lexicon-based classification
The lexicon (or dictionary)-based approach to document classification is the simplest automated content analysis technique (Liu, 2012). It is based on a list of words and phrases and their associated category labels. For example, a lexicon for classifying micro-blog posts according to sentiment

may map the words 'good' and 'beautiful' to the positive category and the words 'bad' and 'ugly' to the negative category. A lexicon for classifying blog posts according to ideological subject may map the words 'health-care' and 'environment' to the left-leaning category and 'foreign policy' and 'taxes' to the right-leaning category.

Off-the-shelf lexicons include the Linguistic Inquiry and Word Count, or LIWC (Pennebaker et al., 2001), and the General Inquirer (Wilson et al., 2005). Not all lexicons are based on binary categories. Some senti-ment lexicons have positive, neutral, and negative terms, measured on a several points scale. The Affective Norms for English Words (ANEW) lexicon, for instance, labels words and phrases according to psychomet-ric categories which rate words on three emotional dimensions: valence, arousal, and dominance (Bradley and Lang, 1999; Osgood et al., 1957). This lexicon helps to analyze documents by counting the relative fre-quency with which words appear and averaging the scores associated to each word in each dimension, from 0 to 9. This approach has been applied effectively to extract sentiment measures from a number of online data sources (Dodds and Danforth, 2009; Dodds et al., 2011).

The success of a lexicon-based content analysis relies on the quality of the lexicon; that is, how appropriate it is in the context of the spe-cific research question and data being analyzed (González-Bailón and Paltoglou, 2015). Using 'off-the-shelf' lexicons compiled with generic research goals may produce poor results when applied to specific types of political communication (Loughran and McDonald, 2011). It is always best to generate lexicons specific to a research question, and there are three main approaches for doing so. The first is to manually annotate the sentiment of all adjectives in a dictionary of all the words in a corpus, in line with the information domain under scrutiny (that is, 'warming' can be labeled differently if used in environmental policy or foreign affairs communication). This is time-consuming but tunes the lexicon to spe-cific communication contexts. Researchers concerned with efficiency as well as accuracy have used online crowdsourcing platforms such as CrowdFlower, Amazon Mechanical Turk, and Taskcn to quickly and accurately label large sentiment lexicons. For example, Dodds et al. (2011) created a lexicon of 10 222 words by merging the 5000 most frequently occurring words in a Tweet corpus, Google Books, music lyrics, and the *New York Times*; they then used Amazon Mechanical Turk to obtain 50 sentiment ratings of each word on a nine-point scale from negative to positive. They found that the sentiment lexicon labeled by crowdsourcing workers was highly correlated with the ANEW lexicon.

The second way to generate a sentiment lexicon is dictionary-based. The general approach is to manually label the sentiment of a small set of seed words, and then search a dictionary (the most frequently used is WordNet; see Miller et al., 1990) for their synonyms and antonyms; these snowballed terms are then labeled with the same or opposite sentiment as the corresponding seed word and then are added to the set of seed words. The process is iterated until no words remain unlabeled. For example, the seed word 'excellent' is labeled positive; synonyms such as 'beautiful', 'fabulous', and 'marvelous' are labeled as positive as well; while antonyms such as 'awful', 'rotten', and 'terrible' are labeled as negative. An example of a lexicon generated using the dictionary-based approach is Sentiment Lexicon, constructed by Hu and Liu (2004). This dictionary-based approach quickly generates a large list of labeled sentiment words, but requires manual cleaning and ignores ambiguity due to context, which is particularly important in the analysis of political communication.

The third way to generate a sentiment lexicon is corpus-based. The general approach is to manually label the sentiment of a small set of seed words and then define linguistic rules to identify similar or dissimilar sentiment words. A seed word may be 'beautiful' and its label 'positive'; linguistic rules based on connective words (such as 'and' or 'but') help to assign labels to subsequent words. For instance, if a document in a corpus contains the phrase 'The car is beautiful and spacious' then the term 'spacious' could be assigned the label 'positive' based on the connective word 'and'. Conversely, if a document in a corpus contains the phrase 'The car is spacious but difficult to drive' then the term 'spacious' could be assigned the label 'negative' based on the connective word 'but'. This methodology requires clearly defined linguistic rules in order to achieve good results; and linguistic rules assume sentiment consistency across documents, which is not necessarily the case for most empirical domains: the same word can express opposite sentiments in different communication contexts (Liu, 2012). Overall, though, the corpus-based methodology to lexicon generation is useful in two cases: to discover other sentiment words and their orientations on the basis of a hand-made seed list; and to adapt a general-purpose lexicon to a specific communication domain. The corpus-based approach is less useful for building a general-purpose sentiment lexicon than the dictionary-based approach because dictionaries encompass more words.

These three techniques are based on different assumptions that affect the results they produce. None of these sentiment lexicons is perfect because they are too general to suit the specific needs of different communication domains. In addition, certain words and phrases in online political communication are too informal, specific, or novel (and therefore infrequent) to be contained in existing lexicons. A corpus-based technique can capture and label these distinct words; for instance, Brody and Diakopoulos (2011) find that lengthened words in microblog posts (for example, 'looove') are strongly associated with subjectivity and sentiment; and Derks et al. (2007) find that emoticons (for example, ':)') strengthen the intensity of online communication. Researchers have already incorporated the peculiarities of online communication into their sentiment models (Paltoglou et al., 2010; Paltoglou and Thelwall, 2012), but often additional manual labeling is needed to add other novel words to the seed list. These limitations make validation a crucial component of automated content analysis (Grimmer and Stewart, 2013). Having the appropriate validation strategies in place is necessary to increase confidence in measurement.

Supervised learning
The second main approach to document classification using pre-existing categories is supervised learning. Supervised algorithms learn from a training data set of manually annotated documents how to classify new, unlabeled documents. The supervised learning approach has three steps. First, it constructs a training data set. Second, it applies an automated algorithm to determine the relationships between features of the training data set and the categories that are used to classify documents. And third, it predicts (or assigns) categories for unlabeled documents and validates that classification. The remainder of this section reviews these three steps in turn.

The first step in supervised learning is to construct a training data set. As described above, this involves transforming unstructured textual data into structured quantitative data. In addition to preprocessing, it is common for researchers to manually code documents for features that the bag-of-words model ignores; for instance, they may add features accounting for the source or the author of a document. The larger the training data set, the more information supervised learning algorithms have with which to make predictions, but scaling up can be computationally costly. The specific research question and data source inform the balance between the need for a large training data set and the costs of compiling training data.

The second step in supervised learning is to apply an algorithm that will associate text features to each category in the classification scheme. There are many different algorithms and the field of machine learning and natural language processing is quickly growing in this area; Hastie et al. (2009) offer a good overview of the techniques available. Each model has specific characteristics and parameters, which makes a general discussion difficult, but popular algorithms include (multinomial) logistic regression, the naive Bayes classifier (Maron and Kuhns, 1960), random forests (Breiman, 2001), support vector machines (Cortes and Vapnik, 1995), and neural networks (Bishop, 1995). Each of these algorithms uses the information gathered from the training data to assign new examples of text into the classification categories.

This assignment takes place in the third and final step, where supervised approaches predict the categories for unlabeled documents and validate the results. A model that performs well will replicate the results of manual coding, which still offers the gold standard; a model that performs poorly will fail to replicate these results. The standard method to validate models is cross-validation. This entails splitting the labeled documents into equally sized groups (usually about ten) and then predicting the categories of the observations in each group using the pooled observations in the other groups. This method avoids overfitting to data because it focuses on out-of-sample prediction. Overall, the supervised approach systematically performs better than unsupervised approaches in the analysis of online communication because it is able to capture more accurately the contextual features of the text and language used (González-Bailón and Paltoglou, 2015).

Discovering Categories and Topics from Documents

Unsupervised learning
In contrast to supervised approaches, unsupervised techniques do not require manually annotated training data; consequently, they are much less costly to implement. They are good exploratory techniques but their results can be difficult to evaluate: concepts such as validity and consistency compared to human labeling do not immediately apply because these techniques are used, in part, to overcome the lack of predefined labels or categories – hence their exploratory nature. This section briefly discusses three categories of unsupervised techniques: cluster analysis, dimensionality reduction, and topic modeling.

The goal of cluster analysis is to partition a corpus of documents into groups of similar documents, where 'similar' is measured in terms of word frequency distributions. The most widely used clustering algorithm is k-means (MacQueen, 1967), which partitions documents into k disjoint groups by minimizing the sum of the squared Euclidean distances within clusters; distance is measured as the number of words that any two documents share. Other clustering algorithms use different distance metrics or objective functions (which are used to optimize or find the best clustering classification out of all possible classifications). Given that few papers provide guidance on which similarity metrics, objective functions, or optimization algorithms to choose, Grimmer and Stewart (2013) caution social scientists from importing clustering methods developed in other, more technical fields like machine learning. The computer-assisted cluster analysis technique suggested by Grimmer and King (2011) offers a more intuitive tool for the task of fully automated cluster analysis.

The goal of dimensionality reduction is to shorten the number of terms in the term-document space while maintaining the structure of the corpus. One dimensionality reduction technique is principal component analysis, which transforms a document-term matrix into linearly uncorrelated variables that correspond to the latent semantic topics in the data set. The technique is not different from more conventional uses in multivariate modeling where a subset of variables are selected to represent a larger data set (Dunteman, 1989). A related dimensionality reduction technique is multidimensional scaling, which projects a corpus of documents into N-dimensional space such that the distances between documents correspond to dissimilarities between them. These methods provide good intuition of the topics that characterize a corpus of text but are best used as exploratory techniques; principal component analysis, in particular, is a typical data reduction step performed prior to subsequent, more substantive analysis.

Finally, the goal of topic modeling is to represent each document as a mixture of topics. Each topic is a probability mass function over words that reflect a distinct information domain. For instance, the topic 'foreign policy' may assign high probabilities to words such as 'war', 'treaty', and 'Iraq'; while the topic 'economy' may assign high probabilities to words such as 'unemployment', 'GDP', and 'labor'. The most widely used topic model is called latent Dirichlet allocation (LDA) (Blei et al., 2003). This technique has recently been applied to the analysis of newspaper content to dissect the framing of policies (DiMaggio et al., 2013). The method provides a new computational lens into the structure of texts and, as the authors state:

> finding the right lens is different than evaluating a statistical model based on a population sample. The point is not to correctly estimate population parameters, but to identify the lens through which one can see the data more clearly. Just as different lenses may be more appropriate for long-distance or middle-range vision, different models may be more appropriate depending on the analyst's substantive focus. (ibid.: 20)

Again, the crucial step in the analysis comes with validation; that is, with the substantive interpretation of the themes identified.

Network Representations of Text

As the sections above have illustrated, content analysis is essentially a relational exercise: words that relate to the same topic are associated by co-appearing frequently in documents and they tend to cluster; likewise, positive words tend to be connected to other positive words, and as shown above, language connectors might change the affective tone of words by setting them in a different linguistic context. Networks offer a mathematical representation of the relational nature of language, and provide yet another tool for the analysis of its structure. Networks have been used to model narratives, and to analyze identity formation (Bearman and Stovel, 2000); to represent mental models (Carley and Palmquist, 1992); and to map semantic associations (Borge-Holthoefer and Arenas, 2010). A network approach has also been used with Twitter data to identify entities by looking at the co-occurrence of words and the clusters that emerge from those connections (Mathiesen et al., 2012). The nodes in these networks are words; what changes depending on the approach is the definition of the links that connect those words: co-occurrence is one of the options, but links can also be used to track the temporal evolution of narratives, as when political movements change their framing or candidates change their positions during an election campaign. These networks can be constructed and visualized using standard network analysis tools.

One of the by-products of generating a dictionary-based lexicon (discussed above) is that the method also creates a network of words that researchers can use to label the strength as well as the sign of the sentiment expressed. For example, Kamps et al. (2004) determined the strength and sentiment of words according to their distances in WordNet from labeled seed words; in this case, two words are linked if they are synonyms, and distance is measured as the number of links that need to be crossed to go from one word to another. Blair-Goldensohn et al. (2008) also used WordNet to construct a network of positive, negative, and neutral sentiment words, and then labeled the strength of the words using a propagation algorithm: starting from a seed word, its sign (positive, negative, or neutral) is propagated to all its neighboring words in the network (its synonyms); following a majority rule, that sign is further propagated to the neighbors of the neighbors, and so on, recursively, until all words have a sign assigned – the valence of which gets weaker the further apart the word is from its seed. These network-based techniques help to extend the dictionary-based approach by suggesting measures of sentiment strength.

APPLICATIONS TO THE ANALYSIS OF ONLINE POLITICAL COMMUNICATION

Sentiment in Online Political Talk

When applying sentiment analysis to political communication, it is important to remember that different methods inherit different assumptions from psychological theories of emotions. The ANEW lexicon, for instance, derives from now classic psychological research suggesting that three dimensions account for variance in the expression of emotion: valence (which ranges from pleasant to unpleasant), arousal (which ranges from calm to excited), and dominance (which ranges from domination to control; Osgood et al., 1957). Neurological research, on the other hand, suggests that five emotional dimensions underlie most brain activity: fear, disgust, anger, happiness, and sadness (Murphy et al., 2003). Reducing the breadth of human emotions to just a few dimensions is arguably a crude simplification, but necessary to make problems tractable; however, it also introduces measurement error that has to be taken into consideration when operationalizing research questions about the affective tone of political communication.

Sentiment analysis of online political communication must take into account not only measurement error but also sampling bias. Internet users, and in particular those present in social media, are typically not representative of the population: they tend to be female, young, and urban (Duggan and Brenner, 2013); in addition, the bias might be more or less important depending on the context and subject of communication. For some dimensions of public opinion, the bias might not matter, but for others it can be crucial. Again, it is only through validity tests that the measures of public opinion extracted from online communication can be relied upon (Grimmer and Stewart, 2013). The increasing number of Internet users who join social media sites and discuss politics means that the volume of online political communication is growing, and the profile of users involved is changing. Analyses of how online sentiment changes over time must therefore account for these non-stationary characteristics, typically by comparing short, adjacent periods of online communication rather than the entire history of communication on a given site.

The assumptions made by automated methods about emotional mechanisms and the nature of the samples analyzed demand a thoughtful research design when studying online communication. In many cases basic methods produce useful results that rival more sophisticated approaches; in particular, simple word frequencies and analysis of how the volume of communication fluctuates over time often yield good insights while preserving efficiency. Carvalho et al. (2011), for instance, found that in some cases these basic descriptive statistics predict sentiment as accurately as more advanced statistical techniques. This suggests that exploratory analysis can be crucial to avoid rushing into the implementation of more complex solutions when a simpler, more intuitive approach would perform as well.

In addition to the lexicons introduced above, a number of alternative approaches have also been developed to facilitate the study of online communication. These include OpinionFinder (OF), which rates expressions as strongly or weakly subjective (Wilson et al., 2005); and the Profile of Mood States (POMS) questionnaire (Lorr et al., 2003), in which respondents rate each of 65 adjectives on a five-point scale. The questionnaire produces emotion scores in six dimensions: Tension–Anxiety, Anger–Hostility, Fatigue–Inertia, Depression–Dejection, Vigor–Activity, and Confusion–Bewilderment. Like ANEW, the POMS lexicon is suited for analyzing more complex emotions in online communication; the OF lexicon, like LIWC, is used for simpler tasks such as the identification of polarity in sentiment analysis. Other prominent lexicons optimized for the analysis of online communication include SentiWordNet (Adrea and Sebastiani, 2006) and SentiStrength (Thelwall et al., 2010). SentiStrength is particularly useful for online political communication because it includes misspellings and emoticons which abound in online talk.

Recent empirical applications of these approaches include Connor et al. (2010), Bollen et al. (2011) and Castillo et al. (2013). Connor et al. (2010) derive sentiment valence from Twitter posts using a subjectivity lexicon based on a two-step polarity classification. They compare Twitter sentiment to consumer confidence and election polling data. They find high correlations (between 0.7 and 0.8) and evidence that smoothed Twitter sentiment predicts consumer confidence (but not election) poll results with relatively high accuracy. However, Bollen et al. (2011) find that the intersection of a tweet corpus and their subjectivity lexicon is not a good leading indicator of the direction of shifts in the Dow Jones Industrial Average. This highlights how sentiment analysis of online communication may not work in all contexts: some lexicons are better suited to particular problem domains, such as consumer confidence, but not financial markets. Finally, Castillo et al. (2013) apply the SentiStrength lexicon to measure sentiment in cable news coverage; although this is traditional media content, the data were accessed through a software company that develops applications for smartphones and tablets that display extra information about TV shows, including captions of content.

Unsupervised Learning Applications

As explained above, many unsupervised learning methods are used as exploratory tools rather than testing techniques, and thus are less common in published literature on online communication. Nevertheless, a few prominent examples exist, although many are still peripheral to the core research questions of political communication.

Turney (2002), for instance, classifies online reviews as positive or negative by estimating the semantic orientation of sentences containing adjectives or adverbs. Specifically, the paper makes use of the pointwise mutual information–information retrieval (PMI-IR) algorithm to measure the number of co-occurrences between words and the seed words 'excellent' and 'poor' on AltaVista search engine results. This co-occurrence frequency determines the semantic orientation of words, and thus can be used to rate online reviews as positive or negative.

Quinn et al. (2010) use a technique similar to LDA in order to analyze the daily legislative attention given to various topics in 118 000 United States Senate floor speeches from 1997 to 2004. They found 42 topics, the most prominent being legislative procedures, armed forces, social welfare, environment, and commercial infrastructure. Yano et al. (2009) use LDA in order to model topics in political blog posts and their corresponding comments sections. They found five topics: religion, (election) primary, Iraq War, energy, and domestic policy. Associated with each topic are a set of words that appeared in blog posts and a set of words that appeared in comments. Additionally, the authors predict which users are likely to comment on which blogs. Finally, another recent example applies the same method to the analysis of issue salience in the Russian blogosphere (Kolstova and Koltcov, 2013). The authors use the method to identify a shift in topics during the political protests that took place during the parliamentary and presidential elections in late 2011 and early 2012.

In sum, unsupervised methods are less frequently used because they are exploratory techniques employed to charter communication domains that lack predefined boundaries. They are good for estimating the structure of a corpus of text when no a priori classifications exist, but they still require a posteriori theoretical and subjective labeling of categories. This stands in contrast to supervised techniques: whereas manual annotation is the starting point for supervised techniques, it is the ending point for the unsupervised approach.

FUTURE LINES OF WORK

This chapter has given an overview of techniques for the automated analysis of large-scale texts, especially as they are generated in online communication. Although this is a massive area of research, and is fast evolving, a few facts have already been established. One is the consistent evidence that the effectiveness of automated classifiers is not independent from the communication domain being analyzed: the meaning of words or their emotional load varies with the context in which they are used. More work is required to build tailored dictionaries that can capture the nuances of political communication as it takes place in different information contexts; for this, supervised-learning approaches offer the most accurate (and promising) solutions. Likewise, more work is needed to consolidate validation strategies, for instance by measuring the strength to which online measures of public opinion are correlated with more traditional measures, such as polls and surveys. A more systematic account of the efficiency and robustness of different algorithms is also needed: some corpora of text are better analyzed by certain techniques than others. Supervised methods, for instance, are more appropriate for content expressed in Twitter messages, whereas for longer communication, such as blog entries, unsupervised methods might be more appropriate. More research is needed to assess the robustness of each method for different data sources, as facilitated by online communication. In any case, the appropriateness of each technique has to be assessed in the light of each particular research question.

Validation is a crucial step in the application of automated content analysis, and this implies finding ways of assessing the accuracy, precision, and reliability of automated classifiers as compared to human coding. For instance, researchers who choose a lexicon-based approach face several design considerations. The first is what type of lexicon to generate or adapt. Some sentiment lexicons have binary categories (positive and negative), some have ternary categories (positive, negative, and neutral), and some have ordinal categories (-5 to $+5$, for example). A second design consideration is what word features to include in a lexicon. Some lexicons simply have word valence (ranging from positive to negative), while others have additional features such as arousal (ranging from calm to excited). The type of sentiment lexicon a researcher chooses should be based on the features a lexicon offers and the specific research question they seek to answer.

Researchers who choose a lexicon-based approach also face several implementation considerations. Most of these have to do with how to detect and resolve the complexities of text. The algorithm that implements a lexicon-based approach should often not just naively match words but also be sensitive to their local context. For instance, it should be aware of negating words (such as 'no', 'not', and 'none') and strengthening punctuation (such as exclamation marks, question marks, and ellipses). Some of the lexicons revised in this chapter, such as SentiStrength, already take these language modifiers and intensifiers into account. Good lexicon-based approaches to sentiment detection do not just rely on word matching: they are also sensitive to how the local context of each word affects the overall sentiment.

The advantage of automated content analysis is that it helps to scale up the amount of text analyzed by lowering the costs of coding and the efficiency of document classification; but it still needs to be reliable. Many sentiment lexicons are based on psychological theories of language use but it is still unclear whether these psychometric instruments work for written communication and large-scale text analysis. In addition, these techniques are still not very good at capturing essential features of political talk, such as sarcasm. A document may contain many strong sentiment words but the author might actually have intended the opposite sentiment to that captured by the automated approach. This means that automated methods might be more appropriate when applied to text in which sarcasm and figurative language are rarely used, for instance news reports; communication through social media, on the other hand, might be more vulnerable to measurement error. As the tools for automated content analysis become more prevalent in communication research, more unified standards for evaluation and assessment will have to be consolidated. The advantages of automated methods are, overall, too great to dismiss.

ACKNOWLEDGEMENTS

We would like to thank the participants of the workshop 'Extracting Public Opinion Indicators from Online Communication', sponsored by the Oxford John Fell Fund under project 113/365 while the authors were based at the University of Oxford. We are especially grateful to Scott Blinder, Javier Borge-Holthoefer, Andreas Kaltenbrunner, Patrick McSharry, Karo Moilanen, and Georgios Paltoglou for insightful discussions.

LEARNING MORE

Methods for automated content analysis are fast evolving, and any list of available resources is likely to be soon outdated. What follows are a few recommendations on where to start to learn more about the methods and applications of automated tools. Rather than an exhaustive list, these references offer entry points to what is a vast and quickly expanding area of research.

FURTHER READING

Krippendorff (2013 [1980]). Now in its third edition, this book is a classic in content analysis, a long-standing reference that precedes the explosion of automated methods for the analysis of large-scale data. Even though the book does not consider emerging methods, the discussion on validity and reliability still applies.

Liu (2012). This monograph is one of the most up-to-date reviews of opinion mining methods. It offers an accessible discussion of state-of-the art tools for automated content analysis, and it defines basic terminology as well as research standards.

Däubler et al. (2012). This research note offers an interesting comparison of the validity of automated versus human coding in identifying basic units of text analysis. The discussion considers how automated methods offer an improvement to human coding schemes without loss of validity.

Grimmer and Stewart (2013). This article offers an interesting overview of methods that analyze text at the document level. In addition to discussing in an accessible way the basic features of different approaches, the article also emphasizes the need to develop new validation methods.

Dodds and Danforth (2009). One of the first examples that used unsupervised methods to extrapolate opinion measures from large-scale communication. It offers a good schematic example of how unsupervised methods work, and how it can be applied to several data sets to track aggregated sentiment dynamics.

Tools for Content Analysis

- R packages:
    - ReadMe: http://gking.harvard.edu/readme
    - TextMining: http://cran.r-project.org/web/packages/tm/vignettes/tm.pdf
    - LDA Topic Modeling: http://www.cs.princeton.edu/~blei/topicmodeling.html
    - TextTools: http://www.rtexttools.com/
- Other software:
    - LexiCoder: http://www.lexicoder.com
    - SentiStrength: http://sentistrength.wlv.ac.uk
    - LIWC: http://www.liwc.net.

REFERENCES

Adrea, E. and Sebastiani, F. (2006). SentiWordNet: a publicly available lexical resource for opinion mining. Paper presented at the 5th Conference on Language Resources and Evaluation, Genoa, Italy.

Bearman, P. and Stovel, K. (2000). Becoming a Nazi: a model for narrative networks. *Poetics*, 27(1), 69–90.

Berinsky, A.J., Huber, G.A. and Lenz, G.S. (2012). Evaluating online labor markets for experimental research: Amazon.com's Mechanical Turk. *Political Analysis*, 20(3), 351–368. doi: 10.1093/pan/mpr057.

Bishop, C.M. (1995). *Neural Networks for Pattern Recognition*. New York: Oxford University Press.

Blair-Goldensohn, S., Hannan, K., McDonald, R., Neylon, T., Reis, G.A. and Reynar, J. (2008). Building a sentiment summarizer for local service reviews. WWW Workshop on NLP in the Information Explosion Era, Beijing.

Blei, D. (2013). Topic modeling and digital humanities. *Journal of Digital Humanities*, 2(1).

Blei, D., Ng, A. and Jordan, M. (2003). Latent dirichlet allocation. *Journal of Machine Learning and Research*, 3, 993–1022.

Bollen, J., Mao, H. and Zeng, X.-J. (2011). Twitter mood predicts the stock market. *Journal of Computational Science*, 2(1), 1–8.

Borge-Holthoefer, J. and Arenas, A. (2010). Semantic networks: structure and dynamics. *Entropy*, 12(5), 1264–1302.

Bradley, M.M. and Lang, P.J. (1999). *Affective Norms for English Words (ANEW): Instruction Manual and Affective Ratings*. Gainesville, FL.

Breiman, L. (2001). Random forests. *Machine Learning*, 45(1), 5–32. doi: 10.1023/A:1010933404324.

Brody, S. and Diakopoulos, N. (2011). Cooooooooooooooollllllllllllllll!!!!!!!!!!!!!!: using word lengthening to detect sentiment in microblogs. Paper presented at the Conference on Empirical Methods in Natural Language Processing, Stroudsburg, PA.

Budge, I. (ed.). (2001). *Mapping Policy Preferences. Estimates for Parties, Electors and Governments 1945–1998*. Oxford: Oxford University Press.

Carley, K. and Palmquist, M. (1992). Extracting, representing, and analyzing mental models. *Social Forces*, 70(3), 601–637. doi: 10.2307/2579746.

Carvalho, P., Sarmento, L., Teixeira, J. and Silva, M.J. (2011). Liars and saviors in a sentiment annotated corpus of comments to political debates. *Proceedings of the 49th Annual Meeting of the Association for Computational Linguistics*. Stroudsburg, PA: Association for Computational Linguistics.

Castillo, C., de Francisci Morales, G., Mendoza, M. and Khan, N. (2013). Says who? Automatic text-based content analysis of television news. Paper presented at the MNLP Workshop, ACM International Conference on Information and Knowledge Management, San Francisco, CA.

Connor, B.O., Balasubramanyan, R., Routledge, B.R. and Smith, N.A. (2010). From Tweets to polls: linking text sentiment to public opinion time series. *Fourth International AAAI Conference on Weblogs and Social Media*. Menlo Park, CA: AAAI.

Cortes, C. and Vapnik, V. (1995). Support-vector networks. *Machine Learning*, 20(3), 273–297. doi: 10.1023/A:1022627411411.

Däubler, T., Benoit, K., Mikhaylov, S. and Laver, M. (2012). Natural sentences as valid units for coded political texts. *British Journal of Political Science*, 42(04), 937–951.

Della Porta, D. and Diani, M. (2006). *Social Movements: An Introduction*, 2nd edn. London: Wiley-Blackwell.

Derks, D., Bos, A.E.R. and von Grumbkow, J. (2007). Emoticons and online message interpretation. *Social Science Computer Review*, 26, 379–388. doi: 10.1177/0894439307311611.

DiMaggio, P., Nag, M. and Blei, D. (2013). Exploiting affinities between topic modeling and the sociological perspective on culture: application to newspaper coverage of government arts funding in the US. *Poetics*, 41(6), 570–606.

Dodds, P.S. and Danforth, C.M. (2009). Measuring the happiness of large-scale written expression: songs, blogs, and presidents. *Journal of Happiness Studies*, DOI: 10.1007/s10902-009-9150-9.

Dodds, P.S., Harris, K.D., Kloumann, I.M., Bliss, C.A. and Danforth, C.M. (2011). Temporal patterns of happiness and information in a global social network: hedonometrics and Twitter. *PloS ONE*, 6(12), e26752. doi: 10.1371/journal.pone.0026752.

Duggan, M. and Brenner, J. (2013). The demographics of social media users. http://www.pewinternet.org/Reports/2013/Social-media-users.aspx.

Dunteman, G.H. (1989). *Principal Components Analysis*. London: Sage.

Eshbaugh-Soha, M. (2010). The tone of local presidential news coverage. *Political Communication*, 27(2), 121–140. doi: 10.1080/10584600903502623.

Franzosi, R. (2004). *From Words to Numbers. Narrative, Data, and Social Science*. Cambridge: Cambridge University Press.

Gentzkow, M. and Shapiro, J.M. (2010). What drives media slant? Evidence from US daily newspapers. *Econometrica*, 78(1), 35–71.

González-Bailón, S., Banchs, R.E. and Kaltenbrunner, A. (2012). Emotions, public opinion and US presidential approval rates: a 5-year analysis of online political discussions. *Human Communication Research*, 38, 121–143.

González-Bailón, S., Paltoglou, G. (2015). Signals of public opinion in online communication: a comparison of methods and data sources, *The Annals of the American Academy of Political and Social Science*, in press.

Grimmer, J. and King G., (2011). General purpose computer-assisted clustering and conceptualization. *Proceedings of the National Academy of Sciences*, 108(7), 2643–2650.

Grimmer, J. and Stewart, B. (2013). Text as data: the promise and pitfalls of automatic content analysis methods for political texts. *Political Analysis*, 21(3), 267–297.

Hastie, T., Tibshirani, R. and Friedman, J. (2009). *The Elements of Statistical Learning: Data Mining, Inference, and Prediction*, 2nd edn. New York: Springer.

Hopkins, Daniel and King, Gary (2010). Extracting systematic social science meaning from text. *American Journal of Political Science*, 54(1), 229–247.

Hu, M. and Liu, B. (2004). Mining and summarizing customer reviews. *Proceedings of ACM SIGKDD International Conference on Knowledge Discovery and Data Mining (KDD-2004)*. New York City: ACM Press.

Johnston, H. and Noakes, J.A. (eds) (2005). *Frames of Protest. Social Movements and the Framing Perspective*. Lanham, MD: Rowman & Littlefield.

Jurafsky, Dan and Martin, James (2009). *Speech and Natural Language Processing: An Introduction to Natural Language Processing, Computational Linguistics, and Speech Recognition*. Upper Saddle River, NJ: Prentice Hall.

Kamps, J., Marx, M., Mokken, R.J. and de Rijke, M. (2004). Using WordNet to measure semantic orientations of adjectives. Paper presented at the LREC. http://dblp.uni-trier.de/db/conf/lrec/lrec2004.html#KampsMMR04.

Kolstova, O. and Koltcov, S. (2013). Mapping the public agenda with topic modelling: the case of the Russian LiveJournal. *Policy and Internet*, 5(2), 207–227.

Krippendorff, K. (2013 [1980]). *Content Analysis. An Introduction to its Methodology*. Los Angeles, CA: Sage.

Krippendorff, K. and Bock, M.A. (eds) (2008). *The Content Analysis Reader*. Thousand Oaks, CA: Sage.

Laver, M., Benoit, K. and Garry, J. (2003). Extracting policy positions from political texts using words as data. *American Political Science Review*, 97(2), 311–331.

Liu, B. (2012). *Sentiment Analysis and Opinion Mining*. Chicago, IL: Morgan & Claypool.

Lorr, M., McNair, D.M., Heuchert, J.W.P. and Droppleman, L.F. (2003). *POMS: Profile of Mood States*. Toronto: MHS.

Loughran, T. and McDonald, B. (2011). When is a liability not a liability? Textual analysis, dictionaries, and 10-Ks. *Journal of Finance*, 66(1), 35–65.

MacQueen, J. 1967. Some methods for classification and analysis of multivariate observations. *Proceedings of the Fifth Berkeley Symposium on Mathematical Statistics and Probability*, 1: 281–297. London: Cambridge University Press.

Maron, M.E. and Kuhns, J.L. (1960). On relevance, probabilistic indexing and information retrieval. *Journal of the ACM*, 7(3), 216–244. doi: 10.1145/321033.321035.

Mathiesen, J., Yde, P. and Jensen, M.H. (2012). Modular networks of word correlations on Twitter. *Scientific Reports*, 2.

Miller, G.A., Beckwith, R., Fellbaum, C., Gross, D. and Miller, K.J. (1990). Introduction to wordnet: an on-line lexical database. *International Journal of Lexicography*, 3(4), 235–244.

Murphy, F.C., Nimmo-Smith, I. and Lawrence, A.D. (2003). Functional neuroanatomy of emotions: a meta-analysis. *Cognitive, Affective and Behavioral Neuroscience*, 3, 207–233.

Neuendorf, K. (2001). *The Content Analysis Guidebook*. London: Sage.

Osgood, C.E., Suci, G.J. and Tannenbaum, P.H. (1957). *The Measurement of Meaning*, Vol. 47. Urbana, IL, University of Illinois Press.

Paltoglou, G., Gobron, S., Skowron, M., Thelwall, M. and Thalmann, D. (2010). Sentiment analysis of informal textual communication in cyberspace. *Proc. ENGAGE*, 13–23.

Paltoglou, G. and Thelwall, M. (2012). Twitter, MySpace, Digg: unsupervised sentiment analysis in social media. *ACM Transactions on Intelligent Systems and Technology (TIST)*, 3(4):66: 1–66: 19.

Pennebaker, J.W., Booth, R.J. and Francis, M.E. (2001). *Linguistic Inquiry and Word Count: LIWC*. Mahwah, NJ: Erlbaum Publishers.

Porter, M. (1980). An algorithm for suffix stripping. *Program*, 14(3), 130–137.

Quinn, K.M., Monroe, B.L., Colaresi, M., Crespin, M.H. and Radev, D.R. (2010). How to analyze political attention with minimal assumptions and costs. *American Journal of Political Science*, 54(1), 209–228.

Spirling, A. (2012). US treaty making with American indians: institutional change and relative power, 1784–1911. *American Journal of Political Science*, 56(1), 84–97. doi: 10.1111/j.1540-5907.2011.00558.x.

Thelwall, M., Buckley, K. and Paltoglou, G. (2010). Sentiment strength detection in short informal text. *Journal of the American Society for Information Science and Technology*, 61(12), 2544–2558.

Turney, P.D. (2002). Thumbs up or thumbs down? Semantic orientation applied to unsupervised classification of reviews. Paper presented at the *Proceedings of the 40th Annual Meeting on Association for Computational Linguistics*, Philadelphia, PA.

Wilson, T., Hoffmann, P., Somasundaran, S., Kessler, J., Wiebe, J., Choi, Y., . . . Patwardhan, S. (2005). OpinionFinder: a system for subjectivity analysis. *Proceedings of HLT/EMNLP on Interactive Demonstrations*, Vancouver, British Columbia, Canada.

Yano, T., Cohen, W.W. and Smith, N.A. (2009). Predicting response to political blog posts with topic models. *Human Language Technologies: The 2009 Annual Conference of the North American Chapter of the ACL* (pp. 477–485). Association for Computational Linguistics.

Young, L. and Soroka, S. (2012). Affective news: the automated coding of sentiment in political texts. *Political Communication*, 29, 205–231. doi: 10.1080/10584609.2012.671234.

25. On the cutting edge of Big Data: digital politics research in the social computing literature
Deen Freelon

Most of this volume's chapters review studies rooted in political science, communication, and closely related disciplines. Indeed, many reference a small clique of foundational authors in agreement and/or disagreement, including Castells, Benkler, Hindman, Jenkins, Morozov, and Shirky. In the current chapter I diverge from this norm to examine a body of literature only rarely acknowledged by mainstream digital politics scholarship. This literature contains politically relevant research by computer scientists and information scientists and is published under a variety of disciplinary labels, but will be referred to here as 'social computing research'. As its name implies, social computing research includes any aspect of human behavior involving both digital technology and more than one person (Parameswaran and Whinston, 2007; Wang et al., 2007). Politics accounts for a small but thriving subset of this literature, which also encompasses health, business, economics, entertainment, artificial intelligence, and disaster response, among other topics.

Social computing research on politics holds relevance for scholars of digital politics and political communication for two related reasons, one methodological and the other theoretical. First, social computing researchers have for many years led the vanguard in computational and 'Big Data' methods (sometimes in combination with other methods), in which the disciplines of political science and communication have both expressed great interest of late.[1] Reviewing how social computing researchers have applied such methods to politically relevant datasets will help digital politics readers to consider how the methods could be applied to their own research. The field's methods and findings also hold a number of theoretical implications, but its researchers devote only sporadic attention to such concerns. For the benefit of those with a more theoretical scholarly orientation, and perhaps also for some social computing researchers with social science leanings, I explore major theoretical trends in the literature. I conclude with suggestions for future research, focusing on how digital politics researchers can best adapt the insights of social computing research to their own ends.

Before proceeding to these sections, however, it is necessary to more thoroughly describe social computing and its goals, which differ in key ways from those of the social science mainstream. The following section is devoted to this task.

SOCIAL COMPUTING: A BRIEF INTRODUCTION

A caveat before I begin: this section is written from the perspective of one who was trained in and still operates within a social science-based research orientation that emphasizes abstract theory as a guide and justification for empirical work (Fink and Gantz, 1996). My participation in social computing research up to this point in my career has been minimal. Accordingly, the description of social computing I offer here is intended as an introduction for those of a similar scholarly orientation to me, which I imagine will include many if not most of this volume's audience.

In a widely cited overview article, Wang et al. (2007) define social computing as 'computational facilitation of social studies and human social dynamics as well as the design and use of ICT [information and communication technology] technologies that consider social context' (p. 79). A similar characterization by Parameswaran and Whinston (2007) establishes social computing as a highly ubiquitous activity of study:

> Social computing shifts computing to the edges of the network, and empower [*sic*] individual users with relatively low technological sophistication in using the Web to manifest their creativity, engage in social interaction, contribute their expertise, share content, collectively build new tools, disseminate information and propaganda, and assimilate collective bargaining power. (p. 763)

Both of the above quotes emphasize the two essential elements of social computing: digital tools ('computing', broadly construed) and social interaction. Of course, researchers in communication, sociology, anthropology, and other social-scientific disciplines have explored topics such as 'computer-mediated communication' and 'cyberculture' for decades. This similarity in subject matter invites the question of how social computing research differs from approaches with which we are more familiar.

The main difference between social computing research and research traditions grounded in social science is as paradigmatic as that between social science and critical theory (Fink and Gantz, 1996; Potter et al., 1993). Whereas social science's goals are to explain empirical outcomes while promulgating theory, and critical theory's is to foment social change, social computing research is devoted to the development of new techniques for organizing, analyzing, and improving the user experience and output of social computing software. As such, social computing studies are usually published in highly technical articles that focus on methods, analysis, and evaluation at the expense of what we would consider 'theory' (Freelon, 2014). The call for papers for the 2014 conference on Computer-Supported Collaborative Work (CSCW), a prominent social computing publication venue, expresses this in its introduction: 'We invite submissions that detail existing practices, inform the design or deployment of systems, or introduce novel systems, interaction techniques, or algorithms' (CSCW, n.d.). Further evidence for this claim can be seen in the strong presence of employees of well-known technology companies such as Google, Microsoft, and Yahoo on major social computing conferences' program committees. Of course, theory is not always entirely absent: some articles include a few theoretical references of relevance to the project at hand, but the discussions tend to be much shorter than in most social science fields. And articles are often accepted without referencing any social science theories at all.

In addition to downplaying theory, social computing research relies heavily on computational methods such as social network analysis, machine learning, computational linguistics, and algorithmic preprocessing of raw web data. These methods are common in computer science and information science, disciplines which many (though by no means all) social computing researchers call home. Programming serves at least two major purposes in social computing: (1) to develop and improve digital platforms for social interaction; and (2) to evaluate their performance efficiently and at scale. Qualitative methods such as ethnography and in-depth interviews are occasionally seen, often as part of a multi-method approach with one or more computational methods. However, such studies are relatively rare, as the next section will demonstrate. The field places the highest value on research techniques and metrics that can be implemented algorithmically. The ability to visualize quantitative results in intuitive and innovative ways is also highly prized.

The final characteristic of social computing research of relevance to the digital politics researcher may seem rather obvious: the field is not principally concerned with politics per se, but rather with social computer use. In other words, social computing researchers typically analyze political cases to make broader points about social computing systems and affordances rather than about politics. Matters of system development and algorithm optimization almost always come first, and broader implications for politics are discussed secondarily if at all. Consequently, the results sections of social computing research papers often leave many implications of theoretical interest unexplored. Later in this chapter I will attempt to reclaim some of these implications in order to clarify their value for students of political science and communication. But first, I will examine in detail the most common methods social computing researchers employ.

METHODS IN SOCIAL COMPUTING RESEARCH ON POLITICS

Social computing research is sometimes published in journals, but many of the most relevant studies for our purposes are published in the archived proceedings of prominent conferences in computer science, information science, and human–computer interaction. Haphazardly selecting papers from these conferences would bias my discussion, so instead I chose them using a systematic and replicable method. First, I focused on conferences and publications sponsored by the Association of Computing Machinery (ACM) and the Institute of Electrical and Electronic Engineers (IEEE), the premier professional organizations in computer science and computer engineering. Using Google Scholar, I searched for the term 'political' exclusively within such outlets. I then ranked the results in descending order by number of citations in order to capture the most widely referenced articles. Being interested only in articles that address politics as a central concern, I qualitatively assessed the most-cited items in each list, flagging articles that empirically analyze political messages, opinions, attitudes, and/or other content as their main focus. (Thus, for example, I excluded articles that analyzed political content as only one of three or more other content categories.) I continued this process until I had flagged 20 articles within each group, for a total of 40 articles (see Table 25.1). These form the basis of the discussions in this section and the next.

After finalizing the sample, I informally classified the articles based on the methods they employed.[2] Three methodological categories were used: traditional quantitative, qualitative, and computational. The traditional quantitative category includes long-established quantitative methods in social science such as surveys, experiments, content analysis, and statistical analysis of secondary data. The qualitative category includes in-depth interviews, field observations, and close readings of texts, among others. The computational category includes any method that entailed the creation of original source code whose purpose was to collect, preprocess, or analyze data. The rest of the chapter will focus mainly on this last category, as the others are much more familiar to scholars of digital politics.

Unsurprisingly, the most prevalent methodological category throughout the sample was by far computational (29/40, 72.5 percent), followed by traditional quantitative (19/40, 47.5 percent) and then qualitative (5/40, 12.5 percent). All but one of the qualitative studies used a mixed

Table 25.1 The methods of 40 highly cited social computing research papers

Authors	Title	Traditional quantitative methods	Qualitative methods	Computational methods
Adamic and Glance, 2005	The political blogosphere and the 2004 US election: divided they blog			x
Awadallah et al., 2010	Language-model-based pro/con classification of political text			x
Baumer et al., 2009	MetaViz: visualizing computationally identified metaphors in political blogs	x		x
Baumer et al., 2010	America is like Metamucil: fostering critical and creative thinking about metaphor in political blogs	x		x
Bélanger and Carter, 2010	The digital divide and internet voting acceptance	x		
Conover et al., 2011	Predicting the political alignment of Twitter users	x		x
DeNardis and Tam, 2007	Interoperability and democracy: a political basis for open document standards		x	
Diakopoulos and Shamma, 2010	Characterizing debate performance via aggregated Twitter sentiment	x		
Diaz-Aviles et al., 2012	Taking the pulse of political emotions in Latin America based on social web streams			x
Fang et al., 2010	Mining contrastive opinions on political texts using cross-perspective topic model			x

Table 25.1 (continued)

Authors	Title	Traditional quantitative methods	Qualitative methods	Computational methods
Fisher et al., 2010	Egovernment services use and impact through public libraries: preliminary findings from a national study of public access computing in public libraries	x	x	
Furuholt and Wahid, 2008	E-government challenges and the role of political leadership in Indonesia: the case of Sragen		x	
Garcia et al., 2010	Political polarization and popularity in online participatory media: an integrated approach			x
Golbeck and Hansen, 2011	Computing political preference among Twitter followers			x
Gulati et al., 2012	Understanding the impact of political structure, governance and public policy on e-government	x		
Hong and Nadler, 2011	Does the early bird move the polls?: The use of the social media tool 'Twitter' by US politicians and its impact on public opinion			x
Jiang and Argamon, 2008	Exploiting subjectivity analysis in blogs to improve political leaning categorization	x		x
Jürgens et al., 2011	Small worlds with a difference: new gatekeepers and the filtering of political information on Twitter			x
Kannabiran and Petersen, 2010	Politics at the interface: a Foucauldian power analysis		x	
Kaschesky and Riedl, 2011	Tracing opinion-formation on political issues on the Internet: a model and methodology for qualitative analysis and results	x	x	

Reference	Title		
Kaschesky et al., 2011	Opinion mining in social media: modeling, simulating, and visualizing political opinion formation in the web	x	
Kim et al., 2007	Toward a model of political participation among young adults: the role of local groups and ICT use		x
Kim et al., 2012	Automatic detection of conflicts in spoken conversations: ratings and analysis of broadcast political debates	x	x
Mascaro et al., 2012	Tweet recall: examining real-time civic discourse on Twitter	x	x
Munson and Resnick, 2010	Presenting diverse political opinions: how and how much	x	x
Nahon and Hemsley, 2011	Democracy. com: a tale of political blogs and content	x	
Park et al., 2011	The politics of comments: predicting political orientation of news stories with commenters' sentiment patterns	x	x
Ratkiewicz et al., 2011	Truthy: mapping the spread of astroturf in microblog streams	x	
Sarmento et al., 2009	Automatic creation of a reference corpus for political opinion mining in user-generated content	x	x
Singh et al., 2010	Mining the blogosphere from a socio-political perspective	x	
Skoric et al., 2012	Tweets and votes: a study of the 2011 Singapore general election	x	
Stieglitz and Dang-Xuan, 2012	Political communication and influence through microblogging: an empirical analysis of sentiment in Twitter messages and retweet behavior	x	

Table 25.1 (continued)

Authors	Title	Traditional quantitative methods	Qualitative methods	Computational methods
Ulicny et al., 2010	Metrics for monitoring a social-political blogosphere: a Malaysian case study			x
Vallina-Rodriguez et al., 2012	Los twindignados: the rise of the indignados movement on Twitter	x		x
Wallsten, 2011	Beyond agenda setting: the role of political blogs as sources in newspaper coverage of government	x		x
Weber et al., 2012	Mining web query logs to analyze political issues			
Wei and Yan, 2010	Knowledge production and political participation: reconsidering the knowledge gap theory in the Web 2. environment	x		x
Younus et al., 2011	What do the average Twitterers say: a Twitter model for public opinion analysis in the face of major political events	x		
Zhang et al., 2009	Gender difference analysis of political web forums: an experiment on an international Islamic women's forum			x
Total		19	5	29

methodology which combined either multiple qualitative methods or one qualitative method with one of the other types. The authors used a variety of traditional quantitative methods, with surveys and content analyses being the two most popular. A large minority of studies employing traditional quantitative methods complemented them with computational methods (8/19, 42.1 percent). Based on this highly cited sample, it would seem that social computing publication venues welcome political research that is methodologically traditional, although computational methods are more common.

I classified an extremely heterogeneous collection of methods as 'computational' in accordance with the operational definition given above. These fall into three general subcategories: data collection, preprocessing, and analysis.

Data Collection

Nearly all major social media services, including Twitter, Facebook, and YouTube, offer application programming interfaces (APIs) through which large amounts of data can be harvested computationally. A variety of options exist for doing so, from simple desktop-based solutions such as NodeXL to Linux-based servers such as 140dev and yourTwapperKeeper to writing a script in the programming language of one's choice. Some articles that analyzed social media content briefly described their data collection process, including such details as the language and specific API used (Mascaro et al., 2012; Ratkiewicz et al., 2011; Skoric et al., 2012), while others did not (Diaz-Aviles et al., 2012; Garcia et al., 2009; Golbeck and Hansen, 2011; Jürgens et al., 2011; Vallina-Rodriguez et al., 2012). Interfacing with APIs to extract data is evidently such a routine activity in social computing research that documenting it in detail is optional. Studies that examined content from sources without APIs – blogs, for example – usually used their own custom web-scraping scripts (Adamic and Glance, 2005; Nahon and Hemsley, 2011; Ulicny et al., 2010).

For researchers in disciplines such as political science and communication that are relatively new to computational methods, this lack of detail on data collection methods is unfortunate. I do not intend to imply that it is the responsibility of social computing researchers to educate outsiders on the elementary aspects of social media data collection, but only observe that beginners will not learn much about how to collect social media data from articles in the field. Textbooks on social media analysis (Leetaru, 2012; Russell, 2013) are more helpful in this regard, but their utility will inevitably decrease with time due to the rapid developmental pace of social media platforms. Some enterprising political science and communication researchers will be able to teach themselves effectively using such resources, but until computational methods become a disciplinary priority, social scientists' ability to collect and analyze social media data will remain limited.

Preprocessing

Preprocessing encompasses a miscellany of techniques to convert raw text and other content collected from the web into research-grade data suitable for quantitative and qualitative analysis. Examples include manipulating social media posts into formats suitable for calculating descriptive statistics (Mascaro et al., 2012), social network analysis (Adamic and Glance, 2005; Conover et al., 2011; Jürgens et al., 2011; Ratkiewicz et al., 2011), simple time-series plots (Vallina-Rodriguez et al., 2012), statistical analysis incorporating non-social media data (Golbeck and Hansen, 2011; Skoric et al., 2012), automated content analysis (Diaz-Aviles et al., 2012; Stieglitz and Dang-Xuan, 2012), and analysis of metadata such as 'likes' or star ratings (Garcia et al., 2012). Like data collection, computational preprocessing requires programming skills by definition, but while the former is a rote task that rarely changes substantially between projects, the latter is completely open-ended. Indeed, creativity in preprocessing determines the kinds of analyses that can be applied to one's data; as such it is more akin to an art than a science.

The articles in the sample exemplify the dizzying range of choices researchers face when preprocessing their data. For example, in using social network methods to analyze relationships between social media users, a preprocessing script may count '@-mentions', retweets relationships, and/or follow relationships as tie indicators, among other features (Conover et al., 2011; Golbeck and Hansen, 2011; Jürgens et al., 2011; Ratkiewicz et al., 2011). The findings of the ensuing social network analysis will obviously differ based on which tie indicators are used. Similarly, most types of automated text analysis require some preprocessing to allow the algorithms to output intelligible results. In a sentiment analysis of political tweets, Stieglitz and Dang-Xuan (2012) imported Twitter-specific jargon and emoticons from their dataset into a dictionary of positively and negatively valenced terms which they used to classify tweets as positive or negative in tone. Using a similar dictionary-based technique, Diaz-Aviles et al. (2012) assembled 'profiles' of tweets and blog posts mentioning 18 Latin American presidents to analyze the online sentiments associated with each. More sophisticated automated techniques such as supervised and unsupervised learning require even more extensive preprocessing. After removing very common words that contain little informational value (called stopwords), raw documents are often disaggregated into clusters of one-, two-, or three-word phrases called 'n-grams' which learning algorithms analyze directly. The choice of which stopwords, types of n-grams, and algorithms to use will all influence the end results. For example, Fang et al. (2012) attempted to quantify the ideological distance between differing political opinions in newspapers and in statements by US senators. To prepare their data for analysis, they used verbs, adjectives, and adverbs as opinion descriptors and retained certain opinion-relevant terms such as 'should' and 'must' that would otherwise be considered stopwords. In a very different research context, Zhang et al. (2009) extracted unigrams and bigrams from an Islamic women's web forum to examine gender differences in content and writing style using supervised learning.

Analysis

Programming usually plays some part in the analysis phase of studies that use computational methods. Complex and creative visualizations produced using specialized code libraries often appear in the results. Most of these tools are applied to communication content – tweets, blog posts, video transcripts, news articles – that do not require direct inter-action with participants. The most common computational analytical methods for texts among the sample are dictionary (or corpus)-based approaches, unsupervised learning, supervised learning, and network analysis.[3]

Dictionary-based approaches use either predefined or custom word collections representing different concepts to classify texts. For example, a dictionary of positive emotions might include terms such as 'love', 'awesome', 'happy', and 'best', and the software might measure positivity as the number of such terms within each text. This technique was used in several articles to analyze social media users' positive and negative feelings toward political issues and politicians (Diaz-Aviles et al., 2012; Garcia et al., 2012; Sarmento et al., 2009; Stieglitz and Dang-Xuan, 2012). Unsupervised learning approaches attempt to detect latent structure in texts inductively and automatically; one of its applications to politics research is the identification of topics mentioned in political texts (Fang et al., 2012). Supervised learning, in contrast, is a deductive method whose goal is to identify pre-established content categories automatically. It often begins with a traditional content analysis, the results of which the algorithm uses as exemplars to classify previously unexamined texts. Several social computing research teams have used supervised learning to predict the political leanings of social media users (Conover et al., 2011; Jiang and Argamon, 2008; Park et al., 2011). Finally, network methods have proven themselves quite versatile, with applications in the study of political spam (Ratkiewicz et al., 2011), communication patterns among political bloggers (Adamic and Glance, 2005; Nahon and Hemsley, 2011; Ulicny et al., 2010), and political gatekeeping in social media (Jürgens et al., 2011).

This very brief survey was intended to highlight some of the ways in which computational methods have been used to study political topics. The kinds of research questions social computing scholars pursue using these methods are limited by their field-specific concerns; thus, there are many opportunities for innovative work by enterprising scholars in other fields with different concerns. The following section substantiates this point more fully.

THEORY IN SOCIAL COMPUTING RESEARCH ON POLITICS

There is a great deal of variation in how social computing research addresses theoretical concerns. Two broad approaches to theory are apparent in the current sample. The first is an *explicit* approach that closely resembles the norm in social science: relevant theoretical contributions from prior research are explored in an in-depth literature review, and then empirical research questions and/or hypotheses are derived from them. The depth of these literature reviews varies widely, as we shall see. The second approach is *implicit* in that theoretical concerns about politics are not discussed at all, but the methods or findings could be integrated into theory-based research by innovative authors. This section will first discuss the theoretical implications of explicitly theoretical papers, and then offer suggestions as to how implicitly theoretical work can inform existing theoretical traditions.

Explicitly Theoretical Work

Social computing research that explicitly incorporates theory does so in a similar fashion to social science. In fact, some such papers are comparable in their theoretical rigor to traditional political science and communication fare (for example, Munson and Resnick, 2010; Nahon and Hemsley, 2011; Wei and Yan, 2010). However, most mention theoretical concerns only in passing: these will typically cite a small number of classic theoretical pieces without exploring much or any of the recent empirical work they have inspired (for example, Adamic and Glance, 2005; Baumer et al., 2009; Kaschesky and Riedl, 2011; Weber et al., 2012). I do not intend to fault the less theoretical pieces here; as explained earlier, social computing and social science have different goals. But observing trends in how the former field uses prior research is important for social scientists who may be interested in building on its studies or in submitting papers to social computing publication venues.

Only one cluster of theories attracted attention from more than one or two papers: online political polarization, homophily, and selective exposure. The research on this topic fell into two categories: studies of online content and evaluations of design interventions. The content-based research analyzed text and metadata from YouTube, the American political blogosphere, Twitter, online newspaper comments, and Yahoo!'s search query logs. Most of these studies found clear evidence of online homophily: for example, that the blogosphere is divided in terms of hyperlinking patterns (Adamic and Glance, 2005); liberal blogs tend to link primarily to liberal election videos and *mutatis mutandis* for conservatives (Nahon and Hemsley, 2011); the Twitter followers of media outlets tend to skew liberal or conservative (Golbeck and Hansen, 2011); and liberals and conservatives tend to use ideologically distinctive queries in search engines (Weber et al., 2012). The design intervention studies evaluated the effects of systems designed to promote exposure to opinion-challenging content (Munson and Resnick, 2010) and critical thinking about politics (Baumer et al., 2009; Baumer et al., 2010). Unsurprisingly, all three of these studies reported some degree of success in their stated goals.

The remaining explicitly theoretical pieces covered a hodgepodge of theoretical concerns. Kaschesky and Riedl (2011) justified their research examining how opinions form and diffuse online partly by reference to the public sphere and deliberation. Along somewhat similar lines, Wei and Yan (2010) grounded their survey-based study of online knowledge production in the knowledge gap and political participation literatures. Bélanger and Carter (2010) invoked the digital divide in a study of US attitudes toward Internet voting, finding that younger and more affluent citizens are more favorably disposed toward it. DeNardis and Tam (2007) offered a legalistic analysis of global ICT standards based on democratic theory, ultimately recommending open document formats for public institutions. In the sole study grounded in critical theory, Kannabiran and Petersen (2010) presented a Foucauldian reading of Facebook's interface.

Implicitly Theoretical Work

Most of the studies reviewed for this chapter did not discuss theory in any substantial way (although some of these cited social science papers to discuss their empirical results). A few lacked literature reviews altogether (Jiang and Argamon, 2008; Jürgens et al., 2011; Ratkiewicz et al., 2011). Those that included them tended to focus on previous studies' methodological efficiency and range of application, and they generally framed their contributions in those terms as well (Diakopoulos and Shamma, 2010; Diaz-Aviles et al., 2012; Fang et al., 2012; Garcia et al., 2012; Kaschesky et al., 2011; Kim et al., 2012; Sarmento et al., 2009; Skoric et al., 2012; Younus et al., 2011; Zhang et al., 2009). In a representative example, Awadallah et al. (2010) presented a new method for classifying political debate arguments as pro or con. Much previous work in the area had at that point been context-independent; for example, judging a statement as inherently positive or negative, whereas pro/con judgments depend upon how the debate position is phrased. Further, previous work had also required manually classified training data, which is time-consuming and expensive. Awadallah's approach was both context-sensitive and fully automatic, which constitute substantive contributions in the social computing research tradition.

Perhaps the best way to demonstrate the value of implicitly theoretical work is to describe its attempted goals, many of which fall into one or more of three categories: classification, forecasting, and description. Classification, the largest category, includes studies that aim to fully or partially automate the process of labeling digital content (mostly but not exclusively text). Some of the classification tasks in this sample include labeling political texts as positive or negative (which is also known as sentiment analysis) (Diakopoulos and Shamma, 2010; Diaz-Aviles et al., 2012; Garcia et al., 2012; Sarmento et al., 2009), pro or con (Awadallah et al., 2010), subjective or objective (Younus et al., 2011), and liberal or conservative (Conover et al., 2011; Fang et al., 2012; Golbeck and Hansen, 2011; Jiang and Argamon, 2008). Forecasting studies seek to predict patterns or outcomes in the digital realm or offline; examples include elections (Skoric et al., 2012), public opinion polls (Diaz-Aviles et al., 2012; Hong and Nadler, 2011), and the diffusion of political opinions online (Kaschesky and Riedl, 2011; Kaschesky et al., 2011). Descriptive studies are similar to their counterparts in social science except that they use very little or no theory (and sometimes no prior research at all) to guide them. As a result, their attempts to discover how platforms such as Twitter were used in particular contexts vary widely in their methodological specifics (Mascaro et al., 2012; Vallina-Rodriguez et al., 2012).

Each of these categories is implicitly theoretical in its own way. Classification studies do not go quite far enough to qualify as social science; their goal is typically to optimize algorithmic performance rather than to contribute to theory. From a social science perspective they resemble extended method sections, full of details on each of the classification task's steps and the results of various evaluation metrics. This simile clarifies the theoretical implications of advanced classification studies to social science: any theory that requires classification could potentially make use of their methodological innovations. For example, the ability to classify political ideology algorithmically could enable theoretically-oriented studies of political polarization and deliberation to analyze sample or population sizes in the millions. Similarly, an automated system for quantifying political sentiment in social media posts could help researchers better theorize how voters react to targeted political messages outside of experimental settings (for more on the uses of sentiment analysis in digital politics research, see Petchler and González-Bailón, Chapter 24 in this volume). Forecasting is more the province of natural scientists and economists than most of social science, which is more concerned with explanation.[4] That said, we should recall that forecasting encompasses within it correlation and time precedence, which are two of Babbie's (2012) three essential components of causation. The remaining component, the elimination of potential alternative causes, then becomes the task of the social scientist. In the rush to build models that can predict elections based on user-generated data (for example), it is the social scientist rather than the social computing researcher who will be interested in why the model works. Finally, most descriptive studies would not pass muster in most social science journals because of their long-standing bias against atheoretical work. Nevertheless, they can still offer the social scientist a sense of the methodological possibilities afforded by new social computing platforms, which could then be incorporated into research questions and/or hypotheses that build theory.

CONCLUSION AND FUTURE WORK

As I have shown, social computing research has produced much of interest to the digital politics researcher. The field has employed computational methods and Big Data since the 1990s, and still conducts much of the cutting-edge research in these areas. In contrast, political science and communication are still very firmly invested in their traditional methods, which are not always optimally suited for analyzing digital data.

Engagement with the best social computing research studies has been and will continue to be essential for all social scientists interested in applying computational methods in their home disciplines. The field's theoretical contributions are not always as obvious, but with a bit of work, students of digital politics will be able to profitably draw upon them for inspiration.

I close this chapter with two general recommendations for social scientists who find this sort of work valuable. The first is simply to learn a programming language suitable for manipulating and analyzing large datasets. While researchers can conduct a few descriptive analyses on large datasets without knowing how to program, most research-grade operations require the ability to work directly with code. Collaborating with social computing researchers may work well for some projects, but as we have seen, they have different standards for what constitutes a contribution (and corresponding publication incentives). Moreover, social scientists can recognize theoretically relevant patterns in data that computer scientists cannot; thus it greatly benefits the former to know how to explore large-scale datasets firsthand. (Imagine having to rely on statisticians for all your statistics!) For the beginning computational researcher I recommend learning the Python programming language, both because it offers a number of libraries and modules specifically for collecting, preprocessing, and analyzing data; and also because its growing popularity in academic circles offers critical support for new learners. The statistical language and programming environment R offers a wider variety of statistical models than Python, but also has a steeper learning curve.

As computational research becomes more accepted in the disciplines in which digital politics research is conducted, graduate faculties should strongly consider how best to teach its methods to their students. Very few communication departments in the USA currently teach computational methods in any systematic fashion, and I suspect the situation is not substantially different in political science. Few US communication departments have any experts in computational methods on faculty, and fewer still have more than one. Some of these experts, such as Benjamin Mako Hill (University of Washington) and Sandra Gonzalez-Bailón (University of Pennsylvania), received their graduate departments in fields other than communication. Others, such as Drew Margolin (now at Cornell) and me, trained in communication departments that do not emphasize computational methods as a core teaching strength. In light of the paramount importance and ubiquity of digital communication data, I submit that computational methods should become one of communication's premier research methods, on par with survey methods, content analysis, experiments, and in-depth interviews. And just as every doctoral student need not learn how to conduct and analyze surveys, not everyone needs to learn computational methods; but it ought to be one of communication's major areas of methodological specialization. A detailed explanation of how to achieve this outcome lies beyond the scope of this chapter, but at a minimum, committed departments will need to thoroughly revise their hiring practices, tenure guidelines, graduate curricula, and departmental resources (including purchasing appropriate hardware, software, and data subscriptions).

My second recommendation pertains to the construct validity of digital traces. Construct validity is the extent to which an operationalized metric actually measures the underlying concept it is intended to measure (Babbie, 2012). As I have documented elsewhere (Freelon, 2014), social computing research studies do not always amply demonstrate the construct validity of the traces they use as metrics. To take an example from the current sample, Ulicny et al. (2010) purport to measure four concepts of academic and practical relevance in the Malaysian blogosphere: relevance, specificity, timeliness, and credibility. Without any reference to prior literature, they define these concepts in terms of manifest digital traces, including use of a real name, network authority, number of comments, and number of unique nouns, among others. Not only are these metrics biased in favor of what can be collected and measured easily, but there is no discussion of whether the metrics are comprehensive, and if not, which aspects of the underlying concepts might be omitted. While a lack of attention to construct validity is by no means universal in social computing research, it is common (Fang et al., 2012; Garcia et al., 2009; Jürgens et al., 2011; Mascaro et al., 2012; Younus et al., 2011).

Social science research on politics is ultimately concerned with abstract concepts such as power, influence, preference, ideology, and homophily, among many others. Traces such as retweets, Facebook 'likes', social media follow relationships, and hyperlink patterns are only interesting inasmuch as they faithfully relate to such concepts. Yet just as we should avoid studying traces for their own sake, so should we also refrain from simply assuming that retweets are always endorsements, hyperlinks always signify authority, and 'likes' always imply approval. Credible arguments for these positions should be articulated and substantiated. In some cases, it will be possible to make logical arguments on the basis of a trace's inherent properties, as in the observation that retweets represent peer-to-peer information propagation. But whenever possible, a trace's imputed meaning should draw on empirical observation. Close qualitative observation of how traces are used can help to fulfill this purpose (Boyd et al., 2010).

The rise of computational techniques in social science has barely begun, and digital politics scholars (myself included) still have much to learn. Social computing researchers offer some of the most methodologically sophisticated work currently available, and many of them are interested in very familiar subject matter. For these reasons, we would do well to learn what we can from them.

NOTES

1. Consider for example the panels held on the topic of Big Data and/or computational methods at the 2013 annual meetings of the International Communication Association (ICA) and the Association of Educators in Journalism and Mass Communication (AEJMC), as well as the conference theme of the 2014 annual meeting of the American Political Science Association (APSA); 'Politics After the Digital Revolution'.
2. I chose not to conduct a formal content analysis here mainly due to the great diversity of methods comprising the 'computational' category, which proved difficult for a non-expert coder to identify consistently.
3. Readers interested in more in-depth discussions of these methods than I offer here are recommended to consult Graesser et al. (2010) and Petchler and González-Bailón (Chapter 24 in this volume).
4. For more on the differences between scientific prediction and explanation, see Shmueli and Koppius (2011).

FURTHER READING

On Social Computing

Parameswaran, M. and Whinston, A.B. (2007). Social computing: an overview. *Communications of the Association for Information Systems*, 19(37), 762–780.
Tian, Y., Srivastava, J., Huang, T. and Contractor, N. (2010). Social multimedia computing. *Computer*, 43(8): 27–36.
Wang, F.-Y., Carley, K.M., Zeng D. and Mao, W. (2007). Social computing: from social informatics to social intelligence. *Intelligent Systems, IEEE*, 22(2), 79–83.

Learning Computational Methods

Cogliati, J., Aikens, M., Morante, K., Cogliati, E., Brown, J.A., Oppengaard, J. and Hell, B. (2013). *Non-Programmer's Tutorial for Python 3*. Wikibooks. Available at http://en.wikibooks.org/wiki/Non-Programmer's_Tutorial_for_Python_3.
Russell, M.A. (2013). *Mining the Social Web*. Sebastopol, CA: O'Reilly Media.
Stanton, J. (2013). *Introduction to Data Science*. Available at http://jsresearch.net/.

REFERENCES

Adamic, L.A. and Glance, N. (2005). The political blogosphere and the 2004 US election: divided they blog. In Proceedings of the 3rd International Workshop on Link Discovery (pp. 36–43).

Awadallah, R., Ramanath, M. and Weikum, G. (2010). Language-model-based pro/con classification of political text. In Proceedings of the 33rd International ACM SIGIR Conference on Research and Development in Information Retrieval (pp. 747–748). New York: ACM. doi:10.1145/1835449.1835596.

Babbie, E. (2012). The Practice of Social Research. Belmont, CA: Cengage Learning.

Baumer, E.P.S., Sinclair, J., Hubin, D. and Tomlinson, B. (2009). metaViz: Visualizing Computationally Identified Metaphors in Political Blogs. In International Conference on Computational Science and Engineering, 2009. CSE '09, Vol. 4 (pp. 389–394). doi:10.1109/CSE.2009.482.

Baumer, E.P.S., Sinclair, J. and Tomlinson, B. (2010). America is like Metamucil: fostering critical and creative thinking about metaphor in political blogs. In Proceedings of the SIGCHI Conference on Human Factors in Computing Systems (pp. 1437–1446). New York: ACM. doi:10.1145/1753326.1753541.

Bélanger, F. and Carter, L. (2010). The digital divide and internet voting acceptance. In Fourth International Conference on Digital Society, 2010. ICDS'10 (pp. 307–310). Retrieved from http://ieeexplore.ieee.org/xpls/abs_all.jsp?arnumber=5432779.

Boyd, D., Golder, S. and Lotan, G. (2010). Tweet, tweet, retweet: conversational aspects of retweeting on twitter. In 43rd Hawaii International Conference on System Sciences (HICSS) (pp. 1–10).

Conover, M.D., Goncalves, B., Ratkiewicz, J., Flammini, A. and Menczer, F. (2011). Predicting the political alignment of Twitter users. In Privacy, Security, Risk and Trust (PASSAT), 2011 IEEE Third International Conference on Social Computing (SocialCom) (pp. 192 –199). doi:10.1109/PASSAT/SocialCom.2011.34.

CSCW (n.d.). Call for participation papers. ACM. Retrieved from http://cscw.acm.org/participation_papers.html.

DeNardis, L. and Tam, E. (2007). Interoperability and democracy: a political basis for open document standards. In 5th International Conference on Standardization and Innovation in Information Technology, 2007. SIIT 2007 (pp. 171–180). doi:10.1109/SIIT.2007.4629327.

Diakopoulos, N.A. and Shamma, D.A. (2010). Characterizing debate performance via aggregated Twitter sentiment. In Proceedings of the SIGCHI Conference on Human Factors in Computing Systems (pp. 1195–1198). New York: ACM. doi:10.1145/1753326.1753504.

Diaz-Aviles, E., Orellana-Rodriguez, C. and Nejdl, W. (2012). Taking the pulse of political emotions in Latin America based on social web streams. In 2012 Eighth Latin American Web Congress (LA-WEB) (pp. 40–47). doi:10.1109/LA-WEB.2012.9.

Fang, Y., Si, L., Somasundaram, N. and Yu, Z. (2012). Mining contrastive opinions on political texts using cross-perspective topic model. In Proceedings of the Fifth ACM International Conference on Web Search and Data Mining (pp. 63–72). Retrieved from http://dl.acm.org/citation.cfm?id=2124306.

Fink, E.J. and Gantz, W. (1996). A content analysis of three mass communication research traditions: social science, interpretive studies, and critical analysis. Journalism and Mass Communication Quarterly, 73(1), 114–134. doi:10.1177/107769909607300111.

Fisher, K.E., Becker, S. and Crandall, M. (2010, January). eGovernment services use and impact through public libraries: preliminary findings from a national study of public access computing in public libraries (pp. 1–10). IEEE.

Freelon, D. (2014). On the interpretation of digital trace data in communication and social computing research. Journal of Broadcasting and Electronic Media, 58(1), 59–75.

Furuholt, B. and Wahid, F. (2008, January). E-government challenges and the role of political leadership in Indonesia: the case of Sragen. In Proceedings of the 41st Annual Hawaii International Conference on System Sciences, IEEE.

Garcia, A., Standlee, A., Beckhoff, J. and Cui, Y. (2009). Ethnographic approaches to the Internet and computer-mediated communication. Journal of Contemporary Ethnography, 38(1), 52–84.

Garcia, D., Mendez, F., Serdült, U. and Schweitzer, F. (2012). Political polarization and popularity in online participatory media: an integrated approach. In Proceedings of the first Edition Workshop on Politics, Elections and Data (pp. 3–10). Retrieved from http://dl.acm.org/citation.cfm?id=2389665.

Golbeck, J. and Hansen, D. (2011). Computing political preference among Twitter followers. In Proceedings of the 2011 Annual Conference on Human Factors in Computing Systems (pp. 1105–1108). New York: ACM. doi:10.1145/1978942.1979106.

Graesser, A.C., McNamara, D.S. and Louwerse, M.M. (2010). Methods of automated text analysis. In Kamil, M.L., Pearson, D., Moje, E.B. and Afflerbach, P. (eds), Handbook of Reading Research, Vol. 4 (pp. 34–53). New York: Routledge.

Gulati, G.J., Yates, D.J. and Williams, C.B. (2012, January). Understanding the impact of political structure, governance and public policy on e-government. In 2012 45th Hawaii International Conference on System Sciences (HICSS) (pp. 2541–2550). IEEE.

Hong, S. and Nadler, D. (2011). Does the early bird move the polls?: The use of the social media tool 'Twitter' by US politicians and its impact on public opinion. In Proceedings of the 12th Annual International Digital Government Research Conference: Digital Government Innovation in Challenging Times (pp. 182–186). New York: ACM. doi:10.1145/2037556.2037583.

Jiang, M. and Argamon, S. (2008). Exploiting subjectivity analysis in blogs to improve political leaning categorization. In Proceedings of the 31st Annual International ACM SIGIR Conference on Research and Development in Information Retrieval (pp. 725–726). New York: ACM. doi:10.1145/1390334.1390472.

Jürgens, P., Jungherr, A. and Schoen, H. (2011). Small worlds with a difference: new gatekeepers and the filtering of political information on Twitter. Proceedings of the ACM WebSci'11 (pp. 14–17).

Kannabiran, G. and Petersen, M.G. (2010). Politics at the interface: a Foucauldian power analysis. In Proceedings of the 6th Nordic Conference on Human–Computer Interaction: Extending Boundaries (pp. 695–698). New York: ACM. doi:10.1145/1868914.1869007.

Kaschesky, M. and Riedl, R. (2011). Tracing opinion-formation on political issues on the Internet: a model and methodology for qualitative analysis and results. In 2011 44th Hawaii International Conference on System Sciences (HICSS) (pp. 1–10). doi:10.1109/HICSS.2011.456.

Kaschesky, M., Sobkowicz, P. and Bouchard, G. (2011). Opinion mining in social media: modeling, simulating, and visualizing political opinion formation in the Web. In Proceedings of the 12th Annual International Digital Government Research Conference: Digital Government Innovation in Challenging Times (pp. 317–326). New York: ACM. doi:10.1145/2037556.2037607.

Kim, B.J., Kavanaugh, A. and Pérez-Quiñones, M. (2007). Toward a model of political participation among young adults: the role of local groups and ICT use. In Proceedings of the 1st International Conference on Theory and Practice of Electronic Governance (pp. 205–212). ACM.

Kim, S., Valente, F. and Vinciarelli, A. (2012). Automatic detection of conflicts in spoken conversations: ratings and analysis of broadcast political debates. In 2012 IEEE International Conference on Acoustics, Speech and Signal Processing (ICASSP) (pp. 5089–5092). doi:10.1109/ICASSP.2012.6289065.

Leetaru, K. (2012). Data Mining Methods for the Content Analyst: An Introduction to the Computational Analysis of Content. London: Routledge.

Mascaro, C., Black, A. and Goggins, S. (2012). Tweet recall: examining real-time civic discourse on Twitter. In Proceedings of the 17th ACM International Conference on Supporting Group Work (pp. 307–308). Retrieved from http://dl.acm.org/citation.cfm?id=2389233.

Munson, S.A. and Resnick, P. (2010). Presenting diverse political opinions: how and how much. In Proceedings of the 28th International Conference on Human Factors in Computing Systems (pp. 1457–1466).

Nahon, K. and Hemsley, J. (2011). Democracy.com: a tale of political blogs and content. In 2011 44th Hawaii International Conference on System Sciences (HICSS) (pp. 1–11). doi:10.1109/HICSS.2011.140.

Parameswaran, M. and Whinston, A.B. (2007). Social computing: an overview. Communications of the Association for Information Systems, 19(37), 762–780.

Park, S., Ko, M., Kim, J., Liu, Y. and Song, J. (2011). The politics of comments: predicting political orientation of news stories with commenters' sentiment patterns. In Proceedings of the ACM 2011 Conference on Computer Supported Cooperative Work (pp. 113–122). New York: ACM. doi:10.1145/1958824.1958842.

Potter, W.J., Cooper, R. and Dupagne, M. (1993). The three paradigms of mass media research in mainstream communication journals. Communication Theory, 3(4), 317–335. doi:10.1111/j.1468-2885.1993.tb00077.x.

Ratkiewicz, J., Conover, M., Meiss, M., Goncalves, B., Patil, S., Flammini, A. and Menczer, F. (2011). Truthy: mapping the spread of astroturf in microblog streams. In Proceedings of the 20th International Conference Companion on World Wide Web (pp. 249–252). New York: ACM. doi:10.1145/1963192.1963301.

Russell, M.A. (2013). Mining the Social Web. Stebastopol, CA: O'Reilly Media.

Sarmento, L., Carvalho, P., Silva, M.J. and de Oliveira, E. (2009). Automatic creation of a reference corpus for political opinion mining in user-generated content. In Proceedings of the 1st International CIKM Workshop on Topic-Sentiment Analysis for Mass Opinion (pp. 29–36). New York: ACM. doi:10.1145/1651461.1651468.

Shmueli, G. and Koppius, O. (2011). Predictive analytics in information systems research. Management Information Systems Quarterly, 35(3), 553–572.

Singh, V.K., Mahata, D. and Adhikari, R. (2010). Mining the blogosphere from a socio-political perspective. In 2010 International Conference on Computer Information Systems and Industrial Management Applications (CISIM) (pp. 365–370). IEEE.

Skoric, M., Poor, N., Achananuparp, P., Lim, E.-P. and Jiang, J. (2012). Tweets and votes: a study of the 2011 Singapore general election. In 2012 45th Hawaii International Conference on System Science (HICSS) (pp. 2583–2591). doi:10.1109/HICSS.2012.607.

Stieglitz, S. and Dang-Xuan, L. (2012). Political communication and influence through microblogging: an empirical analysis of sentiment in Twitter messages and retweet behavior. In Hawaii International Conference on System Sciences (Vol. 0, pp. 3500–3509). Los Alamitos, CA: IEEE Computer Society. doi:http://doi.ieeecomputersociety.org/10.1109/HICSS.2012.476.

Ulicny, B., Kokar, M.M. and Matheus, C.J. (2010). Metrics for monitoring a social-political blogosphere: a Malaysian case study. IEEE Internet Computing, 14(2), 34–44. doi:10.1109/MIC.2010.22.

Vallina-Rodriguez, N., Scellato, S., Haddadi, H., Forsell, C., Crowcroft, J. and Mascolo, C. (2012). Los Twindignados: the rise of the indignados movement on Twitter. In Privacy, Security, Risk and Trust (PASSAT), 2012 International Conference on Social Computing (SocialCom) (pp. 496–501). doi:10.1109/SocialCom-PASSAT.2012.120.

Wallsten, K. (2011, January). Beyond agenda setting: the role of political blogs as sources in newspaper coverage of government. In 2011 44th Hawaii International Conference on System Sciences (HICSS) (pp. 1–10). IEEE.

Wang, F.-Y., Carley, K.M., Zeng, D. and Mao, W. (2007). Social computing: from social informatics to social intelligence. Intelligent Systems, IEEE, 22(2), 79–83.

Weber, I., Garimella, V.R.K. and Borra, E. (2012). Mining web query logs to analyze political issues. In Proceedings of the 3rd Annual ACM Web Science Conference (pp. 330–334). New York: ACM. doi:10.1145/2380718.2380761.

Wei, L. and Yan, Y. (2010). Knowledge production and political participation: reconsidering the knowledge gap theory in the Web 2.0 environment. In 2010 The 2nd IEEE International

Conference on Information Management and Engineering (ICIME) (pp. 239–243). doi:10.1109/ICIME.2010.5477878.

Younus, A., Qureshi, M.A., Asar, F.F., Azam, M., Saeed, M. and Touheed, N. (2011). What do the average Twitterers say: a Twitter model for public opinion analysis in the face of major political events. In 2011 International Conference on Advances in Social Networks Analysis and Mining (ASONAM) (pp. 618–623). doi:10.1109/ASONAM.2011.85.

Zhang, Y., Dang, Y. and Chen, H. (2009). Gender difference analysis of political web forums: an experiment on an international Islamic women's forum. In IEEE International Conference on Intelligence and Security Informatics, 2009. ISI '09 (pp. 61–64). doi:10.1109/ISI.2009.5137272.

Index

absentee voting study, Los Angeles 107–8, 110
abusive posting behavior 255
'access controlled', to internet 386
accountability networks 57
action-at-a-distance, anonymous 41
active minorities
   dictating agenda and style of debate 254
ad hominem attacks, flaming, 'flame-fests' 255
administrative social research 410
adolescents, Swedish, in four groups standby youth, active, unengaged, disillusioned 232
advertising targeted 403
advocacy organizations 23, 169
aesthetics, effects on readers 159
Affective Norms for English Words (ANEW) 436
   lexicons 436
Affordable Care Act ('Obama care') 289
African-Americans
   higher-level digital content creation 209
   prevention from voting 68
   success story 75
African-Americans, poor, less success 75–6
agenda-setting 137–40
algorithm application, in supervised learning 438
algorithmic gatekeeping 366
algorithmic preprocessing of raw web data 453
algorithms
   naive Bayes classifier, random forests, neural networks 438–9
alternative politics 19, 21
Amazon Mechanical Turk 434
   for large sentiment lexicons 436

American Life Project 109
American Management Association surveying employees 404
analysis of e-petitions 146
analysis of studies using computational methods 461–2
ANEW lexicon 442
Anger-Hostility, emotion score 443
anonymity as liberating, on Internet, without prejudices 254
anti-capitalist direct action groups, London 2008 169
   G20 London summit 186–7
anti-capitalist slogans
   four horsemen of the economic apocalypse 177
anti-globalization protesters, activists 70, 325
anti-road-pricing petition, Blair (2007) 141
Apple technology 70, 377
application programme interface (API), 337, 434
   NodeXL 140dev 459
   We The People 146
apps and Big Data, powerful tools 146
Arab countries, changes in government 46
Arab Spring 2011, 35, 70, 186–6
   Facebook, Twitter 72, 310
   sustained protests using digital media 169
arenas of democratic talk, Internet as solution to 266
argument visualization (AV) 273–4
   Decision Structured Deliberation system (DSD) 274
   Deliberatorium from Massachusetts Institute of Technology (MIT) 274
   media coverage 161
argumentation visualization 151, 159, 273–4

Association for Progressive
    Communication (2013)
    internet rights charters 387
Association of Computing Machinery
    (ACM) 454
audience
    commodification 403
    as mass media consumers 5
    for news and politics
        shrinking rather than growing
            343
    popular cultural content 277
Australian GetUp!
    non-governmental organization
        (NGO) sector 140
Australian political blogospheres 330
Australian Twitter News Index
    (ATNIX) 336
    showing tweets per week by domain
        name change 334
authentication, weak or strong 275–6
authoritarian states
    filtering, censoring dissidents 383
    Internet use to reinforce control of
        their citizens 54
automated analysis of large-scale
        texts
    in online communications 444–5
automated content analysis methods
        433–41
availability 401

ballot counting, correctness of 104
ballot transit time problem 107
bankruptcy of Internet companies 396
Barlow, John Perry
    'A Declaration of the Independence
        of Cyberspace' 379
Battle of Seattle, 1999, march in turtle
        costumes 179
Bay Area, New Communalist
        movement 71
BBC Global News Audiences team
    social media analytics 310
BBC *Question Time,* integrated
    audience comments 314
BBC Radio 5 Music, closure stopping,
    e-petition 143
bias and homogeneity in news output
    from news websites 371

Big Brother and *Wife Swap* forums
    diversity of opinions 256
Big Data 283, 286, 451–68
    computational social science 300
    institutionalization by political
        authorities 316
    methods for research 309
    too small 80
'biological networks' 283
blog hosting platforms, Blogger and
        Wordpress 327
blogger arrests 62
blogger killing by Egyptian police
        41
bloggers and commenters 41–2,
        331–5
blogging and news blogging
    ordinary to point of invisibility now
        336
blogging technology, thought sharing
        worldwide 327
blogosphere
    division in terms of hyper-linking
        patterns 463
blogs, influence on mainstream media
        361
blogs, potential degrading of
        journalism 362
blogs number jump, 7 to 118 362
Blue State Digital, Joe Rospars,
        co-founder 122
breaking news, higher patterns of
        activity 333
British Broadcasting Association *see*
        BBC
broadband internet access in Finland
        right for all citizens, 2010 389
Burke, Edmund, philosopher
    on press as Fourth Estate 53
Burma government, closure of Internet
        service 62
Bush, George W. US President
    online field organizing 118
        public attitudes to personal
            appearance on *The Late
            Show* 347
    re-election effort 121
    service records 327
business organizations and work, Fifth
        Estate 60

cable and satellite television and
    Internet 342
California Ideology 70
campaign communications, targeted
    126
Campaign Ideas Forum (CIF),
    Australian GetUp! 140
campaigning changes 129
campaigns, 41
    appealing to youth, Harry Potter
        Alliance 212
Canadian legislation (C30)m reaction
    to bill 61–2
capitalism, condemnation 193
categories and topics from,
    documents discovering 439–40
    unlabeled documents 438
category discovering from content 435
Catholic Relief
    non-governmental organization
        (NGO) 189
censor and control efforts, for Fifth
    Estate 62
censorship, form of, structuring of
    discussion 274
'censorship as damage' 378
challenges for today's youth
    unemployment, underemployment
        200
child-abuse images from other
        countries
    European filters 383
child pornography 22
children in school, civic curricula 214
child's right to protection 388
China
    active blog users, increase in 41
    intermediaries' legal responsibility
        for content
    networked individuals
        information online for
            accountability of government
            59
    regulation of internet content 385
Chinese blogosphere,
    playful commentary, political
        challenge 41
citizens
    disengaged, least confident,
        informed, vocal 277
    dogmatic and inflexible, closed to
        new information 264
    not on line, disenfranchisement 52
    unconfident, no policy opinions 264
citizenship
    class in digital age 79
    class, tension in market-based
        economies 79
    orientation group differences 237
    orientation, results 233–4
    reactive or proactive 18
citizen types, unconfident, and overly
        dogmatic 276
civic and political citizenship, denial
        from Internet 76
'civic cultures' 30–31
civic education 200, 213–15
civic engagement 250
    decline in 18–19
civic gaming, digital game exposure
        212
civic identities of citizens, in
        industrialized democracies 200
civic identity acting out
    becoming a vegetarian, consumer
        choices 203
civic identity, empowering 30
civic participation 21
civic spaces, interaction for citizens 18
civil disobedience 177
civil rights movement in USA 68
civil society and the public sphere
        19–21
civil-society groups, campaign for
        human rights 390
claims-making 388
class, gender, ethnic inequality
    equality of voice between discussers
        269
class divisions in American society 76
classic connective action 194
classic deliberators, search for best
        argument 268
classification and indexing systems
    automated systems of, biases 371
classification, decisions about, from
        news websites 371
classification studies 465
class inequality, analysis of Big Data
        80

clear goals need 236
Clegg, Nick, UK Deputy Prime
    Minister
    video joke 427–8
climate change, addressing 157, 170
Climate Change Conference in
    Copenhagen
    crowd-enabled Twitter network
        190
Clinton, Hillary, competition with
    Obama 122
Clinton-Lewinsky scandal 342
    blog Drudge Report 327
clustering algorithm, *k*-means 439
code regulation of internet use and
    content 382–3
cognition, communication,
    cooperation
    process model of information 401
Cold War, US and former Soviet
    Union 71
collaborative editing, Wikipedia
    156
collaborative network organizations
    (CNOs)
    information products and services,
        Wikipedia 60
collaborative news curation 325–37
collective action 119–24, 179–87, 191
    coordination for 5
    digital activism 77, 169, 172
    frames 178–9
        logic, social network relationships
            182
    networks 186–91
    organizationally-brokered networks
        188
collective identifications 21, 169
    formal organizations 179
'comments on storie', considerable
    adoption of 362
commenting practices 332
commercial control by press 53
commercial interests of corporations
    collecting personal data 386
commercial platforms
    Tumblr, Pinterest, Facebook,
        Twitter 130
commercial redevelopment, in iconic
    public space 425

commercial social media users
    exploitation for economic purposes
        404
commercial surveillance of social
    media sites 7
common good, or particular interests
    251
communication between citizens
    significant role 230, 281, 401
communication channels 358, 422
    concrete, emotionally interesting,
        imagery-producing 359
communication flow 399
Communication Power, by Manuel
    Castells 37
communication surveillance
    technologies National Security
        Agency (NSA) US
    surveillance access to personal data
        406
Communications Decency Act, 1996
    US Government 379
communicative power shifts 53, 57
communicative power with
    government
    of Fifth Estate 51
communicative rationality, Habermas,
    Jürgen 251
communicative relationships
    between government and governed
        265
communicative spaces
    no limitations of time, space, access
        247
communitarians, strengthening
    collective ties 268
community legitimacy 384
community notion, Ferdinande
    Tönnies 397
comparative research on voting advice
    applications (VAAs) 98
Compendium project
    open-source software tool 154,
        274
computational approaches 286–8
computational linguistics 453
computational methods 459
    how best to teach students 466–7
computational preprocessing,
    programming skills need 460

computational techniques in social
science 283, 467
computer ethics 410
computer misuse, nation state's power
382
computer supported argument
visualization (CSAV)
deliberative technology 161–2
hypertext, solving 'wicked problems'
153
information design, sense of detail,
sense of whole, in arguments
158–9
mass media 160–61
mass online deliberation 156–8
technology, emphasis on visual 9,
159
work, 'argument network'
visualization 156
Computer-Supported Collaborative
Work (CSCW) 153, 397–8, 453
as deliberative technology 151–6
computing and social interaction 453
computing research papers, social
455–8
conflict of opinion in political debate
151
Confusion-Bewilderment, emotion
score 443
connection across nation-state
boundaries 37
connective action 9–10, 179–87, 191,
194
to collective action 175
crowd-enabled networks 188
logic of 169–94
networks 181, 186–91, 188
connective and collective action
networks 188
ConsiderIt tool 161–2
consumer behaviour, societally
discriminatory 224
contemporary youth in Internet era
222–3
content analysis, in social sciences 433
content, non-professionally produced
361
absence of journalistic control 371
contraceptive mandate 289
control of Internet access 62

controversies, understanding 288–91
conversational interactivity 358, 361–4
increase 2008 362
co-occurrence of hashtags
as indicator of pubic sentiment
288
co-occurrences between words 444
cooperation, new social systems 401
copyright 61
nation state's power 382
infringement, 'piracy' 385
code adoption 383
corruption incidences 57
'counter-democracy' 19
'counter-politics' 44
counter-surveillance 407, 410
creative Internet use, higher level of
political engagement 231
crime on social media 406
critical mass of citizens online 52–7
cross-sectional design of studies 239
crowd-enabled connective action 187
crowdsourcing platforms,
CrowdFlower 436
CSAV *see* computer supported
argument visualization
cultural engagement 211–13
cultural studies, 'convergence culture'
36
cultures, participatory, properties of
211
current events game show, Britain
Have I Got News for You 352
cybercrime 407
cyberspace, regulation of 378–9

Daily Mail journalist Jan Moir
homophobic commentary on death
of Irish pop star Stephen
Gateley 315
dance video, viral, South Korean pop-
star, PSY 427
Darth Vader-suited riot police,
Pittsburgh 193
data collection 459–60
data multiplicity 399
'data visualization' 282–3
Davis, Richard
New Media and American Politics
1998 340

deadline pressures, influences on
    selection 360
Dean, Howard
    online grassroots mobilization 118
    primary campaign 120–21
Death Star proposal, humorous
    e-petition 140
debate and deliberation by citizens
    154 160
Debatepedia 157
Defense Advanced Research Projects
    Agency (DARPA)
    military origin of Internet 71
deference disinclination, to voices of
    authority 7
deliberation, formal 264–5
    everyday political talk in the
        Internet-based public sphere
        249–50
deliberation in democracies 125,
    151–62, 248
deliberative design, technical
    considerations 271–6
deliberative online environment
    well-designed for feeling safe 268
deliberative outcomes, online or offline
    267
deliberative quality principles 266–71,
    276
deliberative tools, majority
    asynchronous
    challenges to 272–3
Deliberatorium from Massachusetts
    Institute of Technology (MIT)
    161, 275
Deme interface, fostering informed
    debate 272
Democracia real YA! 187
    Spanish city nodes 171
democracy 387–91
    participation of citizens 17
democracy definition, three macro-
    conditions 39
democratic citizenship 153
democratic control 390
    absence in authoritarian countries
        384
democratic participation in Internet
    70, 391
democratic processes, pluralist 53

democratic theory, researchers 151
democratic values 247, 249
democratization, and digital media
    potential 45–6
demographics 359
    priorities for research 47
Depression-Dejection, emotion score
    443
design interventions, evaluation of 463
designers of online deliberate spaces,
    key factors 271, 278
determining factors in mainstream sites
    adopting conversation interactivity
    362
developing world
    few computers, Inter, more cell
        telephones 109
dictionary-based generation of
    sentiment lexicon 436–7
dictionary-based lexicon, WordNet
    441
digital campaigning 118–32
digital citizenship, political, civil, social
    68–9
digital communication technology
    transcendence of geography 211
digital content creation 75, 201–2
digital divides 103, 108–10
Digital Economy Act, UK, 2010 385
digital environment, lawlessness of 7
digital games
    Second Life, Ultima Online, World
        of Warcraft 212
digital inequality (digital divide) 72–4
    and citizenship 79
    race, ethnicity, gender, age 68
digital media conversations,
    facilitating 284
digital media, effects on politics,
    diffusion, differential, conditional
    effects 124–5
digital media environment
    'ways of seeing' politics 417–29
digital media interaction, link to
    political 200
digital media, method of use 45
digital media strategy, Obama
    campaign, 2008 206
digital politics gap 76–8, 80
    exclusionary segregation 69

digital politics, motives for 44–6
digital politics research 12
  in social computing literature 451
digital rights management, to music
  files 383
digital social interaction, 'solo sphere'
  28
digital technologies 51, 54, 58, 205
digitally networked action (DNA)
  173–4
dimensionality reduction 439–40
direct recording electronic (DRE)
  voting system 108
direct witnessing, news sources 360
Direct.gov, e-petition 136–7
disabled children, funding for,
  e-petition 143
discourse ethics, Jürgen Habermas 251
discursive equality, all participants are
  equal 253–5
discussion of subject matter, affect on
  deliberative quality 268–9
discussion platforms, need for
  complexity 273–4
diversity of opinion 255–6
DNA, nature of 184, 193
document classification into known
  categories
  left or right leaning ideological basis
    435
  positive or negative coverage 435
Dow Jones Industrial Average 443
Downing Street e-petitions 136–7, 139,
  143
Durkheim, Emile, sociology as a
  distinctive science 302
  study of society 282
Dutch 'Stemwijzer' 91
duty and obligation, decreased
  experience 203

echo chamber debate, news
  blogosphere 329–31
economic justice and environmental
  networks
  Germany, Sweden, UK 189
economy, online privacy and
  surveillance 403
e-democratic innovations, and digital
  divide 144

editors' concerns over reputation,
  trust, legal liabilities 362
education and research, e-learning
  networks 61
effects of technology on citizenship,
  'new' media 160
egalitarian participation, ideology 67,
  70
Egyptian revolution, 2011 protests
  199, 425
Election Assistance Commission
  (EAC) US
  voting review in US 111
elections 6, 18, 465
electronic ballot scanner 108
electronic petitioning, 'e-petitioning'
  136
E-Liberate system, Roberts's Rules of
  Order 275
elite democracy 18
elites and non-elites, digital production
  gap 75
emotive image versus rational word
  420
emotion, expression of, variance
  valence, arousal, dominance 442
emotions, fear, disgust, anger,
  happiness, sadness 442–3
employers as undetectable voyeurs 405
employers' e-mail monitoring 404
entertainment Internet, relation to
  politicization 236
entertainment media, 'fandoms' 211
  political relevance 341
enthusiasts and sceptics 25–6
entrepreneurs, whizz-kid, social media
  age 201
environmental direct activists 169
e-petitions 136–47
  impacts on public policy but
    controversial 145
  motivation of participants 147
  priorities for future research 146
  support for policy changes 9
equal distribution of voice indicator,
  discursive equality 254
equality of access for e-petitioners
  144
ethical questions, processing and
  harvesting of data 314

EU Profiler
  academic project, European 158
  independence from any party, claims
    party 158
  Puzzled By Policy
Europe, intermediaries, no legal
    responsibility 385
European Commission-funded project,
    WeGov
  popular social media 160
European Union
  crisis 23
  data, youth Internet similarity 222
  survey of Internet use, 2012
    northern countries higher than
      southern 109
  member states, VAAs running 90
European young, daily use of Internet
    109
Eurovision Song Contest, audiences
    responses 310
event coordination, protest calendars
    184–5
everyday political talk in the
    Internet-based
  public sphere 247–58
everyday political talk, online 257
example of argument visualization
    (AV)
  Compendium 274
expatriate voters 107
explicit or *implicit* approach, to
    theoretical concerns 462
expression intensity, Sandra Fluke case
    294–7
Extensible Markup Language (XML)
    434

Facebook
  challenges to 60
  data on individuals 27
  data theft 27
  information on individual users,
    huge amounts 314
  low interest in politics 124
  personal data, communicative data,
    social network and community
    data, civic roles 402
  users 365
face-to-face contexts 20

fairness of voting online 108
Fatigue-Inertia, emotion score 443
features of political talk, sarcasm 445
feedback mechanisms to elites 285
Fifth Estate 52, 59
  balance of power change 63
  critical mass of users 63
  empowerment of individuals 63
  holding others to account 55
  impact on policy 62–3
  interaction with 61–2
  Internet as significant political
    resource 54
  Internet-enabled networked
    individuals 53
  rising force 51–65
  potentially effective examples 55
  technology for power limitation of
    59
financial rescue policies, world
    economy crisis 169
Finland, newspaper, *Taloussanomat*
    371
Firefox web browser 60
firing for misuse of e-mail and Internet
    at work 404–5
Florida, 'stand your ground' state
  legal protection for use of deadly
    force 290
Fluke, Sandra, law student in US
  contraception mandate on health
    care plans 287–288
  hashtags, support and call to action
    288, 292–3, 295, 298
  keywords on Twitter 291
  Rush Limbaugh 289–90
  Washington DC conflict 289–91
  women's rights activist 287–8
Follower counts 288
'forces' concept in gatekeeping 358
forecasting studies, digital prediction
  elections, public opinion polls
    464–5
formal organizations with resources
    180–81, 191
Four Horsemen of financial
    apocalypse
  radical, violent, negative press
    coverage 177
  'storm the banks' 178

Fourth and Fifth Estate, murky
    relationship 60
Fourth Estate, the press, in eighteenth
    century 53
freedom of expression, right to
    needs to be balanced 377, 388
free market fundamentalism 67
Free Tibet matching cymbal band 193
French political blogospheres 330
French political theorist, Pierre
    Rosanvallon (2008) 44
Friends of the Earth, activism 169
future research, on emergence of Fifth
    Estate 64–5

G20 Meltdown, London, financial
    crisis, anarcho-socialist
    demonstrations 175, 177, 190
G20 policy statement, 2010 170
G20 protests, on world financial crisis
    193
gadgets owned, high levels of control
    75
'Gangman Style', political parodies
    427
gas explosion in Nanjing, China, 2010
    little or no coverage in traditional
        media 63
gatekeeping, research tradition in
    journalism studies 357–71
gatewatching 325–37, 326
    curating news as published by other
        sites 328–9, 333
'gateway effect' patterns 350
gender inequality 46
gender, religion, nationhood, attitudes
    to 310
General Election manifesto of Labour,
    2005
    road pricing 141
General Enquirer, off-the-shelf
    lexicons 436
genocide in Africa
    Harry Potter Alliance campaign
        against 212
genocide in Darfur 193
German Bundestag, parliament 136
German political blogosphere's
    coverage
    2007 G8 summit 330

German system of e-petitions 138
German VAA 'Wahl-O-Mat', most
    successful 90
Germans, political humor, *Heute-Show*
    351–2
global communication 399
globalization 38, 193, 369
global North and South, digital divide
    74
Google effect on internet 27, 60
government and democracy on the
    Line 58–9
government control, libertarian
    freedom from, cyberspace 381
government interest in data, risks
    316
government internet surveillance 377
government-led e-petitions 142
government regulation, HADOPI law
    in France 2009
    on user monitoring 385
government regulations on the internet
    379
government research
    in development of digital
        communications 379–80
grass roots politics 421
'Great Firewall of China', control of
    Internet content 62
Greece, austerity measures 176
graphical representations computer-
    generated 159
Guardian newspaper, UK
    notable leader covering breaking
        news, live events 335

Habermas, Jürgen 311
    deliberative democracy 247, 311
    *The Structural Transformation of the
        Public Sphere* 248–9
Hackathon, at the White House,
    powerful tools 146
Haitian earthquake relief, Harry
    Potter Alliance 212
Harry Potter Alliance
    'fandom' raising money for
        charitable causes 211–12
hashtag clusters for Sandra Fluke,
    Trayvon Martin 292–4
    proportional volume 295–6

'hashtags'
  call for activism against Limbaugh
    292
  components analysis 293
  everyday political expression 286
    use in Sandra Fluke, Trayvon
      Martin cases
    user-generated keywords around
      # symbol 286
    #sandrafluke, #slutgate 294
'hate map'
  US tweets using racially offensive
    language 313
hate speech and homophobia 388
health service patient, information
  sourcing on Internet 56
healthy democratic processes 349
  significant harm to 348
hemp and marijuana slogans 193
Her Majesty's (HM) Government's
  e-petitions
  creating and signing online 51
high speed data transmission 399
Hillsborough tragedy, e-petition
  143
horizontal communication in society,
  the internet 22
House Oversight and Government
  Reform Committee 289
'human drama'-oriented newscasts
  342
human gatekeepers 364
human memory limitation, on
  argumentative material 273
human rights 387–91
human thought and knowledge 401
humor, increase the persuasive power
  of political message 347, 351
humorous petitions 139–40
humorous political content online
  340
hybrid spaces of deliberation, mass-
  media audiences online 277
hypertext, graphical 154
HyperText Markup Language
  (HTML) 434
hypertext research 156
hypertext, solving 'wicked problems'
  153
hypertext tool, gIBIS system 154

IBIS and non-IBIS 156
'iconophobia' 420
identity construction, collaborative
  398
'image politics' 421
  political protest and social
    movements 424
imitation in online news 360
impact for e-petitions on public policy
  142
incentives and disincentives, shift in
  balance 44
income inequality 209
'Indecision 2000'
  The Daily Show with Jon Stewart
    340
Independent Media Center (IMC),
  Seattle 325, 336
indexing, decision about, from news
  websites 371
individual Internet use, political
  engagement in youth 225–7
individual rights 78, 377
individuality demands 203
individuals living overseas 107
individuals with disabilities, Internet
  voting 113–14
Indymedia 325–7, 336
influence of e-petitions, unequal
  participation 141
informal political actors 41, 255
informal political talk, public sphere
  ideals 247
information acquisition, fast and easy
  226
informational Internet, relation to
  politicization 236
informational use of media
  newspaper and online news 283
information and communication
  technologies (ICTs) 53, 128, 151
  impact assessment of 410
information derived from social media
  protection of employees, job
    applicants 405
'information design', cognitive
  science, interface design, visual
  communication 159
information empowerment of citizens
  25

information leaks online 43
information overload 159
information searching and sourcing,
    Fifth Estate 55–6
information sharing 408–9
    YouTube and Facebook 184–5
information sources. not fully trusted
    26, 55
information technology, researchers
    151
'Infotainment' 343–4, 420
innovation diffusion research 282
innovation in newsrooms 362
Institute of Electrical and Electronic
    Engineers (IEEE)
    computer science and computer
        engineering 454
institutions, networked 57–8
institutions of power, national policy
    discussion 269–70
interactional Internet, relation to
    politicization 236
interactive moderation 274–5
intermediaries, private
    further their commercial interests 386
    in internet regulation 385–6
    regulation on behalf of states 386
international online networks 40–41
Internet
    access, absence of 109
        variations across age, race, ethnic
            divides, countries 104
    adoption 72–3
    architecture, horizontal, non-
        hierarchical 70
    censorship 59
    civic space 17–32
    corporate ownership, capitalist
        features 70
    democratic accountability 52
    emergence 1
    enabling, to hold government and
        press accountable 51
    environment, fast pace of change
        239
    future research 391–2
    internet governance 11–12
        capacity and legitimacy to govern
            377–92
        value guidance 377

interrelated technologies 128–9
marginalization of high-quality
    journalism 60
medium promoting activity,
    mobilization, debate 311
multi-channel, infinite-capacity
    203
optimism, dot com crisis 2000 396
privacy 403
private self-regulation
    limited state control 380
public space 6
regulation
    by code 383
    by states and private actors 387
    from late 1990s 382–3
rights charters 387
service providers 385–6
    regulatory control 377
surveillance 403
    qualities of 399–401
tool for authoritarian control as in
    China, Belarus 26
type of activity, methods of access
    228
use
    pessimists and optimists 239–40
    positive use 240
utopianism 67
voting 103–15
    access to ballot box 103–4
    blacks and whites 113
    Estonia and Switzerland 103
    Ireland and Netherlands,
        discontinued 103
    security risks 9, 112
    technologies to prevent mistakes
        108
    young voters 105–7, 223
weapon of elite 78
Internet-based activities. fundamental
    threats 110–11
Internet Corporation for Assigned
    Names and Numbers (ICANN)
    non-profit organization 380
    psychological empowerment in
        adolescents 231
Internet Engineering Task Force
    (IETF)
    developing internet standards 380

Stuart Brand, co-founder of WELL 71
Internet Rights and Principles
    Coalition (2013) 390
Internet surveillance, qualities of
    399–401
internet users, subject to laws of
    countries 381
Internet use types
    relating to political engagement
        228–30
Internet voting trials, UK, Norway,
    Estonia 106–7
interpersonal communication, effect
    on voting 99
    for community activism, political
        engagement 283
interpersonal influence 281
intra-family dynamics 213
'inventive' message expression, Twitter
    tracking 286
investigative reporting, news sources
    360
Iran against mainstream politics
    online campaigning 41
Iraq War online protesters against
    confidence in influencing fellow
        citizens, not government 270
Iraq War protests, international
    growth of, 2003 43
Issue-Based Information System (IBIS)
    software tool and approach 154
Issue Map and Argument Network
    styles 155
Istanbul, Taksim Square, 2013
    demonstrations against government
        425
Italian comedian-activist, Beppo Grillo
    352
    anti-government network 176
Italy, crisis-ridden. media practices of
    activist workers 30
item response theory (IRT) 346

JavaScript Object Notation (JSON)
    434
journalism 11
    gatekeeping and interactivity
        357–71
    quality 389

journalist, role in society, questioning
    336
journalists as message filters 361
journalists' personalities 359
journalists' sources 360

key political issues of day 151
keyword use, proportional volume
    for Sandra Fluke and Trayvon
        Martin 291, 293
keywords, frequency of use 288
King, Martin Luther
    assassination in Memphis,
        Tennessee, 1968 68
    support for striking sanitation
        workers 68

language connectors 441
language of rights 377
language-processing 301
latecomers to a discussion online,
    OpenDCN 273
late-night talk shows 342–3
'leapfrogging', African-Americans 76
left-leaning or right-leaning category
    436
legal approval of Barlow, David
    Johnson, David Post 1996 379
legal immunity of intermediaries
    free speech protection online 385
    no forceful policing and censoring
        385
legislative attention to topics 444
legitimacy 387–91
legitimation problems, democratic
    procedures 391
Lessig, Lawrence
    Code and Other Laws of Cyberspace,
        2006 382
Lewin, Kurt, psychologist (1947)
    food consumption habits and
        gatekeeping 358
lexicon-based classification 435–8,
    445
liberal blogs, links to liberal election
    videos 463
liberal-individualists
    self-expression, self-actualization
        268
liberation of the individual 71

libertarian ethos of early internet
378–81
'cyberspace' 377
light-touch post-moderation 331
Limbaugh, Rush, radio talk show host
289, 291
apology for insult 290
conservative talk show, 2012, insult
to Sandra Fluke 289–90
Linguistic Enquiry and Word County
(LIWC)
off-the-shelf lexicons 436
linguistic rules 437
listening as well as speaking 251
literacy limited 73
literature reviews 464
logic behind computer algorithms
370–71
logic of collective action 180–82
logic of connective action, recognition
of digital media 183
London 2012 Olympic Games
data in Arabic, Russian, Persian,
English 310
London Riots 2011, convicts
petition for loss of welfare benefits
144
London School of Economics (LSE)
research project, Reading the Riots
424
low-income individuals, hardware
costs 73

machine learning 439, 453
Madrid train bombings, March
2004
texting for anti-government rallies
59
Magna Carta 1406
formalization of petitioning the
monarch 147
mainstream news organizations 331
Malaysian blogosphere
relevance, specificity, timeliness,
credibility 467
manifestos of competing parties 88
Mapping Controversies on Science for
politics (MACOSPOL) 157
mapping networks of message
retweeting 289

'maps' of argument and evidence 274
'market fundamentalism' 79
market liberalization 369
marketing and advertising departments
of commercial organizations 160
marketing on social media 28
marriage, same-sex, rights, Harry
Potter Alliance 212
Martin, Trayvon, unarmed
African-American
hashtags on 294, 296–7, 299–300
Keywords on Twitter 291, 297,
killed by neighbourhood watch
volunteer, George Zimmerman
287–8, 290–92
mass communication, and
interpersonal conversation 281
mass media 4–5, 38, 60
meaning of rights 377–8
measurement error 442
media
chain-owned and independent 371
communication encouragement
230
consumption patterns 367
headlines, sensational 265
role of, on political opportunity
44–5
sharing 211
technologies, Facebook, YouTube
423
media devices, mobile
against surveillance tactics of police
426
media ownership and market forces
influence on media content 371
mental and social forces, interplay of
282
message platforms 284
metadata, increased use
for measuring social media reactions
314
methodology and analysis, future of
313–14
Middle East, North Africa, pre-
democracy protests, 2011 59
Military and Overseas Voter
Empowerment (MOVE) Act
113–14
military personnel 107

Millennial generation born after 1980
201
   cable television, video games 202–3
   different world for 203
miniaturization 399
mixed-media ecosystem, role played
   by 160
mobile phone users in USA
   while watching TV, percentages 317
'modalities of regulation' 382
moderation and registration 362
moderation criteria for e-petitions 138
moderation practices, governmental
   platforms 274
monitoring of Internet use 404
multidimensional scaling 440
multi-issue organizations 190
multimedia 24, 399
'My Page' functionality, in decline
   by *Washington Post, Sun, Telegraph,
   New York Times,Wall Street
   Journal* 365

name-calling, aggression, irrelevancy,
   misogyny, homophobia, racism
   on internet 311
Narnia, Alice in Wonderland 381
National Annenberg Election Survey
   347
national economies, interpenetration
   202
national policy proposal discussion,
   deliberation 269–70
national political actors, Mumsnet,
   local UK mothers 41
national security 388
National Security Agency (NSA) US
   security revelations 110
   surveillance of citizens 26
natural disasters 310, 333
natural language processing
   technologies 309
'navigational' interactivity 358
negative and positive aspects of
   Internet use
   on political engagement 227
negative messages, erosion of trust in
   electoral process 348
negative repercussions from
   inappropriate commentary 331

negative sentiment on political
   engagement
   Internet, 'dystopian' perspective 226
negative sentiment on social media
   Public relations (PR) people,
   intervention by 160
neoliberal capitalism, class inequalities
   69
neoliberalism 19, 67, 396
   virtual poll tax and citizenship
   78–80
net neutrality debate 386, 388
Netherlands VAA 'KiesKompas' 92
Netiquette 255
'network communitarian' perspective
   383–4
networked audience, demands of 315
networked individualism 203
networked institutions and individuals
   58
networked politics
   2008 Obama presidential campaign
   38
network filters, access blocking 383
network mapping 297–300
network representation of text
   440–41
network, retweet based on tweets
   Fluke and Martin cases 297
'network society' 37–8
networking as Fifth Estate action 56
networking hubs for collective action
   190
new blogs to news on Twitter 325–37
new content , social delivery, political
   learning 205
new markets, expansion into 369
New Media Division. 2008, Obama's
   campaign 122
news
   and comment, from audience
   viewpoint 362
   bloggers 330, 332
   engagement in gatewatching
   327–9
   consumption behavior 357
   gatherers, writers, reporters. local
   editors 359
   newsgroups, partisan, attack-
   orientated 255

newspapers, loss of established ones
  Chicago Sun-Times 423
  hidden agenda 362
  prioritization 371
  organizations 332, 335
  providers, traditional,
      consequences 369–70
  sensationalistic, on paedophilia
      406–7
  sources, public information 360
  staged events 360
  through digital media, aims to
      reach young 200
newsroom culture and organization
    360
newsworthy events
  timeliness, magnitude, clarity,
      cultural relevance, novelty 359
new wars spiking, 9/11 396
'noise' in media environment 45
non-governmental e-petitions 147
non-governmental organization
    (NGO) sector 40–41
  actions 169
  VAA development 90
non-professionally produced content
  standard of spelling, news, value,
      punctuation, accuracy, balance
      361
non-state power centres 45–6
'Not Bad' meme
  photo of US President in UK 427

Obama, Barack, US President
  digital campaigning 121–3
  early youth following 199, 208
  election campaigns 118, 199
  election victory 2008 23
  Going Inside the Cave 122
  president and loving husband
      426–7
  tweeted image of wife hugging
      426–7
  victory in 2012 US presidential
      election 426
  youth outreach campaign 206
Occupy movement in USA, 2011 23,
    41, 43, 173, 187
  social networking platforms 172

Occupy Rome protests, 2011 190
Occupy Wall Street movement in
    USA, 2011, 199
  model of Arab Spring 185–6
  sustained protests using digital
      media 169
old media, radio, TV, effects of 160
online activity 126, 224, 428
online and offline activism
  youth participation, one or other
      210
online banking threats 111
online behaviour
  regulation by social norms, market
      forces and 'code' 382
online bullying 406–7
online deliberation, case for 267
online campaigning 23
online communicative forms, variety
    for youth 210
online communty for physicians in
    USA, Sermo 56
online content production, creation of
    blogs 74
online content research 463
online cooperation 400
online crime 407
online debates 251–2
online deliberation 151, 247
  comparisons with face-to-face 250,
      256–7
online discussion forums
  on European Union (EU) policy
      269
online electoral politics, less for poor
    working class 77
online environments, political
    expression, conversation 282
online fundraising, small-dollar,
  John McCain, Bill Bradley 120
online interface system, Vote Builder
  Democratic Party 122
online journalism and official news
    sources 361
online marketing 409
online news consumption
  political behaviors, connection 231
online-only publishers
  *Christian Science Monitor, and
    others in USA* 371

online-only publishing, switching to
371
online political communication,
analysis of 441–4
automated content analysis 433–47
online political dissidents,
imprisonment 382
online political equality 74
online political expression,
computational approaches 281–302
online political talk 10–11, 71, 283–4,
300
analysis and assessment 250–57
online political texts, convert to
structural form 434–5
online politics research 300
online polling organizations 51
online privacy and surveillance
economy, politics, culture, key
features 403–9
online services, new 369
online social networks
interest-based communities for
youth 202
online space, for deliberative political
practice 266
online surveillance and privacy 12
online voting advice applications,
emerging research field 87–99
Open DCN project, 'informed
discussion' tool 272
open government initiatives 52, 59
Open Net Initiative 62–3
open source software, powering
protest networks 184
open spaces for readers to set agenda
The Times, 'Your World' travel site
363
Opinion Finder (OF) 443
opinions
fellow citizens 153
formation 286
leadership 281
mining, sentiment analysis 160
new media influence 161
polarization 255
organizational influences in journalism
366
organized politics, individuals relating
differently 191

Organizing for Action
mobilization-based election
campaigns 123
Owen, Diana, New Media and
American Politics 340

Pakistanis, political humor, *The Real
News* 352
Palestinian peace advocates 193
panel survey methods, on political talk
283
Parmenides platform
policy proposal and justification
157
participant authentication 275–6
passive and active forms, of adaptive
interactivity 364–6
peer discussion influence 213
peer-to-peer monitoring 409
pepper-spraying of unarmed protestors
New York, California University
407
personal action frames 175, 178–9
personal communication technologies
for sharing 175
personal identification number (PIN)
for voting online 112
personal information, selling of
commercial use by Google 27
personalization of media, increased
389
pessimists' and optimists' views
on political entertainment or
comedy 349–50
petition site '38 Degrees' 51
petition to provide cancer drugs on
National Health Service (NHS)
to Scottish Parliament 142–3
petitions, individual, to government
ministers 142
Pew Internet 109
Pew Research Center for the People
and the Press 343
report, USA, 'connected viewer'
306
photo identification for voting 103
physical space, importance of 425
pilot trials of internet voting, UK, US,
Norway
security concerns, voter privacy 103

Pittsburgh, Free Tibet matching cymbal band 193
plastic bags, mandatory charge, e-petition 143
platforms, Apple, Facebook, Google hosting software applications 385
pledges, pre-election, monitoring of by VAAs 90
pointwise mutual-information information-retrieval PMI-IR algorithm 444
police activism 410
police misconduct, visibility from social media 407–8
police use of social media 406
Policy Commons Argument Visualization Tool 155
policy impact of e-petitions 146–7
policy-makers, VAA development 90
policy preferences 87, 89
policy profile of users of VAA
of users and parties, salience and variability 88
policy-related public deliberation
relating background information 273
policy related to copyright in knowledge economy 154
policy think tanks 190
policy to sell off national forests, e-petition against 143
political activity online, translation into voting 41, 105, 192
political actors, network actors 40
political affiliation, interest-based 45
political and media organizations
horizontal and vertical axis 3–4
political and media theorists
two-screen viewing questions
political behaviour and action 213
political blog posts
corresponding comments section 444
political bloggers (Guido Fawkes) 41–2, 284, 327
political campaigning 118–24
political comedy, beneficial or detrimental 349
digital politics 340

political comments on line, from young people 106
political communication
changing environment 3–7, 342
homophilic, attraction is challenge to democracy 267
researchers 151
scholars on expansion of public sphere 247
visual image in 419
political content, as PPF , London 2009, or Occupy 175
humorous effect on concrete political behaviors 349
social media, quality of 310–11
political conversations, informal knowledge increasing 250
political crises, higher patterns of activity 333
political debate articles as pro or con 464
political decision-makers 153
'average' citizen removal from 312
political democracy, contemporary challenges to 264
political engagement 44, 210
among youth 221–40, 224
decrease of, crisis of democracy 223–4
redefinition of, by internet 29
political entertainment
beneficial or detrimental 349
comparative research with other countries 351–2
help to facilitate political learning 345–6
implications of 340–53
new media form 352
programming 340
political everyday talk about politics 266
political exchanges, partisan, prejudiced, uncivil 266
political information, mainstream media 4–5, 160
during campaign 127
political institutions and politicians 4
Internet usage 221

political interest and political
    knowledge 127–8
digital skills 93
no learning effect from humour 346
political life diminishment by television
    419
political mobilization 39, 286
political participation, from digital
    media use 67, 128
political parties
    monitoring reputation of leaders,
        policies 310
political performance, how effective
    428
political practices
    drive for self-promotion, self-
        revelation 28
    revolutionized by digital media
        35
political protest since 2010, age of
    social media 46
political role of Internet in society
    54
political satire 340
political science and communication
    465
political scientists
    on significance of television with
        two-screen viewing 306, 311
political socialization 35, 200, 213–14
political talk
    everyday 248–52
    online 291–300
    situating 282–3
    via digital media 284–5
political viewpoints, exposure to,
    online 213
politicians, televised appeals to voters
    audience was a sitting target 5–6,
        426–8
politics
    communication about digital
        campaigning 130
    of dissent 421
    interest in, barrier for young 209
    mediated, representative or distorted
        428
    technology 2
    'ways of seeing' in digital media
        environment 417–29

poll taxes and literacy tests
    in southern USA to discourage
        voting 68
poll taxes, to prevent people voting 79
pollution levels in cities, collective
    intelligence on 57
popular support courting, by
    politicians 6
populist discourses 26
positive and negative words 358,
    436–7
positive coalitions 44
postal voting 105
'postmaterialist' world for youth 203
post-privacy era 26
power and inequality 80
power between media actors and
    public 312
power of interpretation, audience
    possession 312
power of political comedy
    to facilitate political learning 346
power relations between participants
    332
power strategies, communicative 55–7
presidential campaigns, of US 2000
    342
press censorship 59
Press Complaints Commission 315
press independence from government
    53
press, radio, television, mass media
    53
press, role of creating a public 281
press, tabloid, bounds of propriety,
    overstepping 395
primary school student blog, on school
    lunch picture 56
principal components analysis 440
printing press in Europe, official
    languages 1
priority areas for future research
    158–62
privacy and surveillance, computers,
    concern 396, 403
privacy settings, changes to 408
privacy violations 22
private and public morality 7
private corporations
    corporate social responsibility 387

private corporations in internet
governance 384
production constraints 359
professional and personal influences
on journalists
for audience expectations 366
professionalization of movement
organizations 181
Profile of Mood States (POMS),
questionnaire 443
pro-segregation remarks 327
protest networks, Wall Street to
Madrid to Cairo 182–3
protest themes, personalized, across
national boundaries 176
protest, transnational character of,
Arab Spring, Occupy 424
protests in Heiligendamm 330
psychological empowerment 230–31
public action as personal expression
184
public contributions, source material
for stories 361
public service media, provision online
389
public space reappropriation
Spain, May 2011 (15-M movement)
425
public sphere 20–21
active citizens, autonomous
communicative spaces 249
deliberation 248–9
levels and fields, Jürgen Habermas
249
publics and spectators
visual citizens in visual displays
424–6
Put People First (PPF) campaign
172–3
financial crisis, London march
169–70
Oxfam, Tearfund, Catholic Relief,
World Wildlife Fund 189
social technologies 175–6
Putin, Vladimir, affection for Siberian
tigers 210
Puzzled By Policy
academic project, European 158
focus on issue of immigration in EU
158

Python programming language
466

quantitative and qualitative analysis
460
Queensland Parliament, all local
governments in UK 136
Question Time BBC TV
high levels of tweeting 424
patterns of interaction 309
Twitter as part of programme's
format 309

race gaps 75–6
racial dimension of digital divide 109
racial disparities in digital media,
closing of 215
racism, legal regulations against 388
rational-critical debates 251, 253, 311
rationality, decline of, in public sphere
26
'raw news' 360
raw text into research grade data 460
'reader blogs' *The Sun* 362
reader comments, pre-moderation
media with gatekeeping role 251,
361
readers' comments publication
on websites of three national
newspapers 362–3
real time or asynchronicity
people participation on own terms,
reflective outcomes 272–3
real-time communications 399
reciprocal exchange on *Guardian*
newspaper 253
reciprocity level 251–3
of political conversations 252
recommendations for social scientists
466
learn a programming language 466
recording violent police behaviour
Occupy, 15-M 426
reflexivity 253
regulation to achieve policy ends 378
regulatory control of intermediaries of
internet 385
regulatory power of intermediaries 384
'reinforcement politics' 54
'relational power' 37

Reporters Sans Frontiers 62–3
representative democracy 136
republican models of democracy 18
Republican Party, online platforms,
    voter databases 121
research
    new, priorities for 47, 236–9
    on digital campaigning 129–32
    on e-petitions impact 140–44
    on Internet use 229–30
    on political entertainment 351
Research, Twittersphere and
    mainstream news media 257
resignation of Trent Lott, US House
    Senate Leader 327
resistance
    to digital rights management on
        music files 383
resource mobilization theory (RMT)
    181
respect for all arguments 265
retweet network
    for Sandra Fluke, major hashtags
        298
    for Trayvon Martin, major hashtags
        299
revision for gatekeeping models 367–9
rights and 'human rights' 387–8
'rights-bearing' citizenship 203
rioters, sharing of camera images on
    Twitter 424
riots in the UK, 2011, journalists using
    Twitter 335
riots in UK, August 2011
    outraged media attention, public
        fear, harsh sentences 424
    rioters sense of invisibility in social
        body 424
rising demands for expression 203
risk assessment, on by-mail absentee
    voting 111–12
risks to public health, economic
    stability, social order 316
role of comedy in civic engagement 420
Russian blogosphere, issue salience 444

sampling bias 442
Sandra Fluke, Washington DC
    conflict
    contraceptive mandate 289–90

San Francisco hippy counterculture
    movement 1960s, northern
    California 71
Save the Children, activism 169
scepticism, in online public discussion
    270
schoolgirl grooming cases 406
schools, teaching role for online
    interpretation 214
'science of the social' 281–302
Scottish Parliament 136
    e-petitions 138
search engines 53
    for navigation of web 385
    regulatory control 377
'second screen' phenomenon, websites
    and television 424
security 377
    concerns of Internet voting 104–5,
        110–13
    confidentiality, anonymity,
        eligibility 112–13
seed words, synonyms and antonyms
    437
self-defense claim from George
    Zimmerman
    Florida's 'stand your ground' law
        287–8
self-expression engagement 213
self-motivating participation 183
self-organizing roots 187
self-regulation 378
sentiment analysis 160
    cable new coverage 443
    online political communication
        441–3
    positive or negative words 464
Sentiment Lexicon 437, 445
sentiment valence from Twitter posts
    443
significance of political comedy 341–4
Silicon Valley, fruit farming area,
    Northern California 69
Silicon Valley Ideology 79
    active perpetuation of inequalities
        80
    class inequality 67–81
    neoliberal system 69–72
    political content posting 74
skills and online practices 208–10

Sleep Train Mattress Centers, on Rush Limbaugh 289–90
slogan 'eat the rich', Jean-Jacques Rousseau 177
small group discussion
  on affirmative actions, nuclear power Arab-Israeli conflict, US industry 283
smartphone, laptop or tablet, creating online content 306
smartphone users
  good platforms for adaptive information delivery 365
social accountability across many sectors 52–3, 57
social action, Max Weber 397
social class divisions, with political activity 68, 77
'social collaborative filtering'
  Facebook 'Activity Feed' plug-in 364–5
social computer use 12, 313, 452–4
social computing research 455, 458
  human behavior and digital technology 451
  methods of 456–7
  politics 454–2
  theoretical concerns 462
social contract between developers, administrators, contributors 271
social coordination 24
social developments in technology, media consumption patterns 367
social dysfunction 181
social facts, Émile Durkheim 397
social foundations. of future digital politics 35–48
social grouping of people deliberating 271
social media 12, 396
  attention, Trayvon case 290
  candidates' digital exposure increase 285
  development and use, twenty-first century 199
  different forms of civic participation 30
  effective use 427
  mass diffusion, 2006

Twitter, YouTube, Facebook, *Time Magazine* 70, 119
monitoring consumer sentiment 160
monitoring, understanding of opinion polls 316
platforms 306
  barrier lowering for direct action 315
  democratic deliberation 277
surveillance 395–411
  key features 398
technology, democratic impact 153
use by African-Americans 75
users, risk for 408
social media on internet 19
  agricultural diseases, animal food shortage,
    public protests against badger culls 316
social movement, anti-war, intra-party 'netroots' 120–21
social movement organizations (SMOs) 23, 181
social networking sites (SNS) 54, 61, 205, 408, 453
  Facebook, Twitter 156, 160
  linked to civic engagement, political action 285–6
social networks digitally mediated 183
social norms of deliberation 251–2
social policy as argumentative process 153
social science research on politics, abstract concepts 467
social scientists 439, 465
  VAA development 90
social technologies, loose public networks 187
socio-democratic groups, traditional empowerment 92
socio-economic status 208–10
  impact on youth engagement 200, 209
soft news 343
  Persian Gulf War 1991 342
software platform owners, regulatory control 377
source, media, audience channels in gatekeeping process 368
spaces for online deliberation 264–78

Spain, 15M mobilizations 171, 173
Spain, *los indignados* 170–72, 185–6,
    199, 425
  sustained protests using digital
    media 169
spectator as pupil or scholar 420
state governance and digital media 7
state surveillance and crime 410
stopwords 461
Streaming Twitter API 287
structured data that can be quantified
    434–5
students' challenges to teachers 61
style guidelines 360
subjects, engaging, intellectually
    challenging 269
substantial equality, respecting equal
    voice 254–5
substantive debate engendering online
    271–2
Sun Microsystems 71
'Super Bowl Sunday', Nestlé's
    marketing team 160
'Super Tuesday', Obama campaign,
    2008 122
supervised learning 435, 438–9
  deductive method 461–2
supporters' use of websites 120
surveillance enhancement
  crime-fighting, political suppression,
    routine monitoring 24
surveillance of citizens 395
surveillance of undergraduate students
    405
survey research on two-screen viewing,
    Ofcom
  use of Internet while watching TV
    308
Swedish youth, evidence of civic
    participation 205
Syria, Assad family, poster ripping,
    2011 428

tablets and smartphones, reaction to
    media events 318
talking into acts, votes 282
Talking Politics, William Gamson,
    1992 283
Tarde, Gabriel, nineteenth century
    French sociologist 301

conversation at centre of
    sociological enquiry 281
  *The Laws of Imitation (Les Lois de
    l'imitation)* 282
targeted advertising 409
  commercial company, Facebook
    404
'targeted sharing program' 128
targeting in campaigns 126
Taskcn, for large sentiment lexicons
    436
Tea Party movement, US 43, 193
Tearfund, non-governmental
    organization (NGO) 189
technical documentation for internet
    non-state organizations 380
technological advances, youth interest
    222
technological changes, 'networked
    society' 130
technological convergence of
    computers
  digital media 23–4
technology-enabled networking 180
technology, influence for youth 208
technology, lightweight, for online
    publishing 327
Telegraph editor, comment on blogs,
    wikis and journalism 361
television, central focus 306
television in political campaigning,
    dominant role 419
television newsmagazines 342
template policy content 126
tension between politics, popular
    culture, images 428
Tension-Anxiety, emotion score 443
terrorism, efforts to prevent 406
text messaging, Obama campaign,
    2008 206
texting and tweeting 54
The Colbert Report 342
The Daily Show with Jon Stewart 342
  erosion of trust in media 348
  saint of democracy or sinner 345
The Guardian
  'Been there', readers' setting agenda
    363
the Press, Fourth Estate. holding to
    account 59–60

theoretical work, explicit 462
third party mediation 384
time pressures 359
time spent on Internet, by adolescents
    228, 232
tools and techniques in political science
    data production, processing
        techniques 313
topic model, latent Dirichlet allocation
    (LDA) 440
topic modelling goal 439–40
traditional providers' news product
    369
traditional values of media
    independence and objectivity 360
training data set construction
    in supervised learning 438
transcripts for latecomers, Unchat 272
transdisciplines 410
transmedia 24
transnational corporations, rising
    power 202
transparency and accountability 152
Trenchard, John (1731), definition of
    politics 2
trust 39–40, 152
Turing, Alan, scientist and
    code-breaker
    e-petition for pardon 143
Turkish Prime Minister
    Recep Tayyip Erdogan, reaction
        425
turnout and internet voting 104–5
TV Licensing Authority, UK,
    'chatterboxing' 306
tweets, geotagged 313–14
Tweets on the Streets, Paulo
    Gerbaudo, 2012 425
Twitter
    celebrities campaign, complaints to
        BBC 315
    conversations 10
    hashtags 60
    information dissemination, triggered
        by media happenings, news
        coverage 285
    low interest in politics 124
    'menace' to society 425
    sites blocking 62
    third-party space 332–3

two-screen politics
    evidence, theory and challenges
        306–19
    'group-viewing' 312
    response of traditional broadcasters
        314–15
two-screen viewing
    challenges 313–19
    change in audience experience 312
    data and existing literature 307–10
    digital divide 316–18
types of Internet use
    communicative, social 229
    creative, expressive, finance-
        managerial 229
    informational, interactional 229
typography and graphic design 159

UK Children With Diabetes Advocacy
    Group 56
UK Dept for Environment, Food and
    Rural Affairs (Defra)
    horizon scanning software for social
        media analysis 316
UK Government Communications
    Headquarters (GCHQ)
    surveillance access to personal data
        406
Ukraine, 2014, government challenges
    59
Unchat, Noveck 2003 275
    experimental real-time discussion
        tool for small-group discussion
        272
Uniform and Overseas Civilian
    Absentee Voting Act (UOCAVA)
    USA 105
    Internet voting 113–14
United Nations Foundation, Climate
    Change portal 157
United States attitudes toward Internet
    voting 463
United States Department of State,
    Opinion Space 158
United States journalists, values
    of, ethnocentrism, small town
    pastoralism, individualism,
    moderation 359
United States missile strikes on
    Afghanistan and Sudan, 1998 342

United States presidential campaigns,
    use of Internet and digital media 9
United States presidential candidate
    appearances
  on late-night and daytime talk
        shows, 2000 347
  online campaigning 340
United States television programs
  A Current Affair 342
  Entertainment Tonight 342
Universal Declaration of Human
    Rights (UDHR) 387
universities
  campus grids, digital library
        collections online courses 61
university governance 61
unsupervised learning 439–40, 443–4,
    461
user-generated content (UGC) 24
user media 361
'utopian' perspective, potential of
    Internet 226

VAAs (voting advice applications)
  Dutch 'Stemwijzer', very popular
        90–92
  effect on users, increased voter
        turnout 94
  in-depth interviews on usage 99
  influence on electoral outcomes
        94–5, 98
  inspiration for further information
        95
  qualitative research on 99
  trustworthiness of advice 97
  usage, 'digital divide' problem
        92–4
  voting advice calculation 96
validation in automated content
    analysis 445
Vancouver Airport
  death of Robert Dziekanski, 2009
        407
videocassette recorder (VCR)
  programming in 1980s 199
video games, The Sims 212
viewing experience, for broader social
    deliberation 277
views, accounting for 265
Vigor-Activity, emotion score 443

Vikileaks 62
viral communication students,
    memes 176
virtual city of Alphaville, virtual social
    problems 212
virtual poll taxes 79
visibility and visuality
  'visibility entrepreneurs', 423
  in political communications 422–4
visual culture studies
  ancestry in art history, museum
        studies, media studies 421
visual iconography 428
visual image in political
    communication 419–22
visual language and graphics
  for depiction of argumentation 159
visual language, designing
  color textures, shapes 159
visual political communication 12
voices not online, under-representation
    316–17
voluntary work of youth 224
volunteer recruitment and mobilization
    120
vote, as political participation, young
    adult turnout low 206
vote on issue preferences 89–90
Voter Activation Network (VAN) 122
voter database, Democratic Party 122
voter turnout in American presidential
    elections
  1972–2012 (by age) 207
  voter turnout rates 87
voting advice applications (VAAs) 9
  accessibility of, online 89
  effect on electoral process 88
  informed voting, health of
        democracy 90
  internet platforms run during
        elections 87–99
  methods used 88
  policy preferences of voters 88
  short questionnaires for preferences
        87–99
  socio-demographic or political
        background of users 88
voting and involvement with political
    campaigns 349
voting by e-mail in USA 103

voting, by-mail, modest gains in turnout 105
voting, for people with disabilities 105
voting in general election 44
voting on iPads 114
voting technology, privacy protection 107

wage labour, traditional, hiring, monitoring of prospective employees 404
Washington DC-centric affair, Sandra Fluke 295
We The People e-petitions 139
Web 2.0 technologies 325
web-based tool, policy Commons 154
Weber, Max
    state monopoly on legitimate use of force 381–2
web links 184–5
website surfing in 1990s 199
Welsh Assembly e-petitions 138
Whole Earth 'Lectronic Link (WELL)
    online engagement with news and current events 325
Wikipedia 60
    dangers of 26
WikiWikiWeb, Ward Cunningham 1984 397
'winner takes all' markets 386–7
Wired magazine, California 69–70
WordNet 437
Wordpress, YouTube, technology platforms 156
working class, faltering interpersonal community interactions, traditional
    less exposure for young 202
World Summit on the Information Society (WSIS), United Nations Internet Governance Forum 390
    meetings, Geneva, 2003, Tunis, 2005 389–90
World Trade Organization (WTO)
    controversial 1999 meeting in Seattle
    activists using Web2.0 technology 325
World Vision, activism 169

World Wide Web (WWW) 118, 325
    1990s 396
    e-democracy 266
World Wide Web Consortium (W3C), web standards 380
World Wildlife Fund
    non-governmental organization (NGO) 189
writing code, as regulation 382
WUNC and PPF
    London, Toronto and Pittsburgh 189, 193
    worthiness, unit, numbers, commitment 170–72

young people
    civic identity 201–4
    digital communication in political life 199
    'digital natives' 199
    engagement in politics, potential 200
    'Net Generation', life style difference 223
    prominence in protest politics 199
young people of color, digital technology role 208
YourView, Australia
    online platform for public political debate 157
youth
    civic engagement 9–10, 199–215
    contemporary culture
    fruitfulness for democratic governance 238
    digital media 214–15
    divergence from parents in political activism 213
    engagement, on social networking sites 213
    involvement, in US Presidential election, 2008 206
    news 204–5
    participation effect of Internet voting 104
    politically engaged, informational and creative Internet 223–5, 236
    socialization into civic and political life 214

turnout, increase since 2000, non-
    white youth 206
YouTube, low interest in politics 124
YouTube videos and Tweets 74
    death of Ian Tomlinson in London
        G9 protests, 2009 407

zero-order correlations
    political engagement and Internet
        use 235
    relation to politicization 236
Zimmerman, George, killer of Trayvon
    Martin 290–91

Printed and bound by CPI Group (UK) Ltd, Croydon, CR0 4YY

16/04/2025

14658378-0003